ISNM
International Series of Numerical Mathematics
Vol. 129

Hyperbolic Problems: Theory, Numerics, Applications

Seventh International Conference in Zürich, February 1998
Volume I

Edited by

Michael Fey
Rolf Jeltsch

Springer Basel AG

Editors:

Rolf Jeltsch
Seminar for Applied Mathematics
ETH Zentrum
8092 Zürich

Michael Fey
Seminar for Applied Mathematics
ETH Zentrum
8092 Zürich

1991 Mathematics Subject Classification 35LXX; 35L65; 65N12

A CIP catalogue record for this book is available from the Library of Congress, Washington D.C., USA

Die Deutsche Bibliothek – CIP-Einheitsaufnahme

Hyperbolic problems: theory, numerics, applications : seventh international
conference in Zürich, February 1998 / ed. by Michael Fey ; Rolf Jeltsch. –
Basel ; Boston ; Berlin : Birkhäuser.

Vol. 1. – (1999)
   (International series of numerical mathematics ; Vol. 129)

   ISBN 978-3-0348-9742-6   ISBN 978-3-0348-8720-5(eBook)
   DOI 10.1007/978-3-0348-8720-5

© *1999* Springer Basel AG

Originally published by *Birkhäuser* Verlag, Basel, Switzerland *1999*

Printed on acid-free paper produced of chlorine-free pulp. TCF ∞
Cover design: Heinz Hiltbrunner, Basel

9 8 7 6 5 4 3 2 1

Contents

Volume 1

Contents

Volume 2

International Series of Numerical Mathematics
Vol. 129, © 1999 Birkhäuser Verlag Basel/Switzerland

Discrete Kinetic Schemes for Systems of Conservation Laws

D. Aregba-Driollet and R. Natalini

Abstract. We present here a class of numerical schemes for general multidimensional systems of conservation laws. These schemes are based on a discrete kinetic approximation which includes the relaxation schemes by S. Jin and Z. Xin. We give, by a Chapman–Enskog type expansion, a suitable multidimensional generalization of the Whitham's stability subcharacteristic condition. In the scalar multidimensional case we establish the rigorous convergence of the approximated solutions to the unique entropy solution of the equilibrium Cauchy problem.

1. Introduction

In this paper we present a new class of numerical schemes for multidimensional hyperbolic systems of conservation laws. These schemes are based on a discrete kinetic approximation. Consider a weak solution $u : \mathbb{R}^D \times [0, T] \to \mathbb{R}^K$ to the Cauchy problem

$$\partial_t u + \sum_{d=1}^{D} \partial_{x_d} A_d(u) = 0, \tag{1}$$

$$u(x, 0) = u_0(x), \tag{2}$$

where the system is hyperbolic (for instance symmetrizable) and the flux functions A_d are locally Lipschitz continuous on $\mathbb{R}^K$ with values in $\mathbb{R}^K$. We approximate problem (1), (2) by a sequence of semilinear systems

$$\partial_t f^\epsilon + \sum_{d=1}^{D} \Lambda_d \partial_{x_d} f^\epsilon = \frac{1}{\epsilon} \left(M(Pf^\epsilon) - f^\epsilon \right), \tag{3}$$

with Cauchy data

$$f^\epsilon(x, 0) = f_0^\epsilon(x). \tag{4}$$

Here ϵ is a positive number, Λ_d are real diagonal $L \times L$ matrices, P is a $K \times L$ constant real matrix and M is a Lipschitz continuous function defined on $\mathbb{R}^K$ with values in $\mathbb{R}^L$. Moreover we suppose the following relations are satisfied for all $u \in I$, for some fixed rectangle I in $\mathbb{R}^K$:

$$\begin{cases} PM(u) = u, \\ P\Lambda_d M(u) = A_d(u), \quad d = 1, \ldots, D. \end{cases} \tag{5}$$

It is easy to see that, if f^ϵ converges in some strong topology to a limit f and if Pf_0^ϵ converges to u_0, then Pf is a solution of problem (1), (2). In fact system (3) is just a BGK approximation for equation (1). Our purpose here is to construct numerical schemes for system (3), in order to obtain a numerical approximation of (1) in the relaxed limit $\epsilon = 0$.

As well-known for general relaxation problems, approximation (3) needs for suitable stability conditions to produce the correct limits. In the framework of general quasilinear hyperbolic relaxation problems this condition is known as the sub-characteristic condition. In Section 2, we shall argue in the spirit of the Chapman–Enskog analysis [15, 10, 4] to find a stability condition for (3).

Actually, in [13] and for the scalar case $K = 1$, convergence of Pf^ϵ towards the Kruzkov entropy solution of (1), (2) has been obtained under a slightly stronger condition where monotonicity properties of the Maxwellian Function M are involved. The main tool in that case is the fact that under this condition the right hand side in system (3) is quasimonotone in the sense of [7] and this implies special comparison and stability properties on the corresponding system. Here we construct numerical schemes which preserve these properties, so that we are able to establish a rigorous convergence theory.

Our approximation framework generalizes to systems the construction presented in [13] for the scalar case, and shares most of the advantages of the relaxation approximation as proposed in [8] (see also [12, 1]): simple formulation even for general multidimensional systems of conservation laws and easy numerical implementation, hyperbolicity, regular approximating solutions. Actually the main advantage, specially in the multidimensional case, of both the approximations seems to be the possibility of avoiding the resolution of local Riemann problems in the design of numerical schemes. Moreover our framework presents some special properties:

- the scalar and the system cases are treated in the same way at the numerical level;
- all the approximating problems are in diagonal form, which is very likely for numerical and theoretical purposes;
- we can easily increase the number of velocities involved in our construction to obtain a clear and simple improvement of the accuracy of the method.

In this sense our work shares most of the spirit of the papers [14, 11], where very flexible and simple schemes, which do not need Riemanns solvers in their construction, were proposed to approximate general multidimensional systems of conservation laws. Let us also observe that the presented algorithms are surely not optimal, but they just illustrate how to construct an efficient and simple approximation even for very complicate systems.

The plan of the article is the following: in Section 2 we establish the stability condition and define the monotone Maxwellian functions. In Section 3 we present some examples of approximation in the class (3) and investigate the meaning of both stability and monotonicity conditions. In Section 4 we present the numerical scheme and in the scalar multidimensional case we state convergence results. Finally we present some numerical experiments.

2. Chapman-Enskog analysis and Monotone Maxwellian functions

In this section we discuss the stability conditions for the discrete kinetic approximation (3).

Let f^ε be a sequence of solutions to (3)–(4) parameterized by ε, for a fixed initial data f_0, which for simplicity we can choose as a local equilibrium, i.e.: $f_0(x) = M(u_0)$ for some $u_0 \in L^\infty(\mathbb{R}^D, \mathbb{R}^K)$. Set

$$u^\varepsilon := Pf^\varepsilon, \quad v_j^\varepsilon := P\Lambda_j f^\varepsilon, \quad j = 1, \ldots, D.$$

Then, from (3) and the compatibility assumptions (5), we have

$$\begin{cases} \partial_t u^\varepsilon + \sum_{j=1}^{D} \partial_{x_j} v_j^\varepsilon = 0, \\ \partial_t v_j^\varepsilon + \sum_{d=1}^{D} \partial_{x_d}(P\Lambda_j\Lambda_d f^\varepsilon) = \dfrac{1}{\varepsilon}(A_j(u^\varepsilon) - v_j^\varepsilon), \quad j = 1, \ldots, D. \end{cases} \quad (6)$$

Consider a formal expansion of f^ε in the form

$$f^\varepsilon = M(u^\varepsilon) + \varepsilon g^\varepsilon + O(\varepsilon^2). \quad (7)$$

Then

$$\begin{aligned} v_d^\varepsilon &= A_d(u^\varepsilon) - \varepsilon\Big(\partial_t v_d^\varepsilon + \sum_{j=1}^{D} \partial_{x_j}(P\Lambda_d\Lambda_j f^\varepsilon)\Big) + O(\varepsilon^2) \\ &= A_d(u^\varepsilon) - \varepsilon\Big(\partial_t v_d^\varepsilon + \sum_{j=1}^{D} \partial_{x_j}(P\Lambda_d\Lambda_j M(u^\varepsilon))\Big) + O(\varepsilon^2). \end{aligned} \quad (8)$$

Reporting in (6) yields

$$\begin{aligned} \partial_t u^\varepsilon &+ \sum_{d=1}^{D} \partial_{x_d} A_d(u^\varepsilon) \\ &= \varepsilon \sum_{d=1}^{D} \partial_{x_d}\Big(\partial_t v_d^\varepsilon + \sum_{j=1}^{D} \partial_{x_j}(P\Lambda_d\Lambda_j M(u^\varepsilon))\Big) + O(\varepsilon^2). \end{aligned} \quad (9)$$

Now, dropping the higher order terms in (8), we have

$$\partial_t v_d^\varepsilon = A_d'(u^\varepsilon)\partial_t u^\varepsilon + O(\varepsilon) = -\sum_{j=1}^{D} A_d'(u^\varepsilon)A_j'(u^\varepsilon)\partial_{x_j} u^\varepsilon + O(\varepsilon). \qquad (10)$$

Then, up to the higher order terms in (9), we obtain

$$\partial_t u^\varepsilon + \sum_{d=1}^{D} \partial_{x_d} A_d(u^\varepsilon) = \varepsilon \sum_{d=1}^{D} \partial_{x_d} \left(\sum_{j=1}^{D} B_{dj}(u^\varepsilon)\partial_{x_j} u^\varepsilon \right) \qquad (11)$$

where

$$B_{dj}(u) := P\Lambda_d\Lambda_j M'(u) - A_d'(u)A_j'(u). \qquad (12)$$

is a $K \times K$ matrix. Therefore we have proved our stability condition.

Proposition 2.1. *The first–order approximation to system (3) takes the form (11) and it is dissipative provided that the following condition is verified:*

$$\sum_{j,d=1}^{D} (B_{dj}(u)\xi_j, \xi_d) \geq 0 , \qquad (13)$$

for all $\xi_1 \in \mathbb{R}^K, \ldots, \xi_D \in \mathbb{R}^K$ and every u belonging to some fixed rectangle $I \subseteq \mathbb{R}^K$.

Let us remind that the expansion (11) cannot be considered in any way as a rigorous asymptotic description of system (3). Actually to prove our rigorous convergence results we need a slightly stronger version of condition (13).

Definition 2.2. *Take $K = 1$ and let $I \subseteq \mathbb{R}$ be a fixed interval. A Lipschitz continuous function $M = M(u) : I \to \mathbb{R}^L$ is a Monotone Maxwellian Function (MMF) for equation (1) and with respect to the interval I if conditions (5) are verified and if moreover*

$$\begin{array}{l} M_i \text{ is a monotone (nondecreasing)} \\ \text{function on } I, \text{ for every } i \in \{1, \ldots, L\}. \end{array} \qquad (14)$$

This condition was used in [13] to show the convergence of approximation (3),(4) at the continuous level and in the multidimensional scalar case. In the following section we present some examples of different approximations according to the choices of the matrices of velocities Λ_j and the local Maxwellian function M, and we investigate both stability and (for $K = 1$) monotonicity conditions. Of course one could also define the monotonicity condition for $K > 1$ but this is not of interest since in practice only scalar conservation laws are concerned with monotonicity.

3. Examples of discrete kinetic approximations

In order to construct the system (3) one has to find P, M and Λ such that relations (5) are satisfied. The examples presented here own a block structure: keeping notations of the introduction we take $L = N \times K$, $P = (I, \ldots, I)$ with N blocks I, the identity matrix in $\mathbb{R}^K$. Each matrix Λ_d is constituted of N diagonal blocks of size $K \times K$:

$$\Lambda_d = diag\left(C_1^{(d)}, \ldots, C_N^{(d)}\right) \quad \text{and} \quad C_n^{(d)} = \lambda_n^{(d)} I, \quad \lambda_n^{(d)} \in \mathbb{R}.$$

In the scalar case one has the following proposition [13]:

Proposition 3.1. *Let $K = 1$ and suppose that M is a MMF. Then (13) is satisfied.*

In the general case, denote B the K.D×K.D matrix defined by the blocks B_{dj}, $d, j = 1, \ldots, D$. The stability condition means that ${}^tB + B$ is symmetric positive for all $u \in I$, some fixed rectangle of $\mathbb{R}^K$.

Example 1.
We first consider the minimal case $N = D + 1$. The system (5) is then a squared linear one. We take

$$C_j^{(d)} = \begin{cases} -\lambda I & \text{if } j = d, \\ \lambda I & \text{if } j = D+1, \\ 0 & \text{otherwise}. \end{cases}$$

The Maxwellian function is then the unique solution of (5). For a one-dimensional system of conservation laws this formulation coincides with the relaxation approximation of [8]. We recall that in this case and for $K = 1$, convergence to the unique entropy solution was proved in [12], and convergence of the associated numerical relaxed schemes was done in [1]. However, in several space dimensions there is no diagonal form for the formulation of [8].

For a one-dimensional system the stability condition (13) is here:

$$\forall \xi \in \mathbb{R}^K, \quad \left(\left(\lambda^2 I - A'^2\right)\xi, \xi\right) \geq 0.$$

If $K = 1$ this coincides with the monotonicity condition. For a two-dimensional system we have

$$B = \frac{1}{3}\begin{pmatrix} 2\lambda^2 + \lambda(-A_1' + 2A_2') - 3A_1'^2 & \lambda^2 + \lambda(A_1' + A_2') - 3A_1'A_2' \\ \lambda^2 + \lambda(A_1' + A_2') - 3A_2'A_1' & 2\lambda^2 + \lambda(2A_1' - A_2') - 3A_2'^2 \end{pmatrix} \tag{15}$$

and we have proved the following result [2]:

Proposition 3.2. *Let us suppose that $K = 1$, $D = 2$ and that the choice of P, Λ_1, Λ_2 and M is as here above. Then the stability condition (13) and the monotonicity condition (14) coincide and can be written as:*

$$\lambda \geq \max(-A_1' - A_2', 2A_1' - A_2', -A_1' + 2A_2') \tag{16}$$

Example 2.

In this example we take a greater number of equations, $N = 2D + 1$, and in view of a more accurate approximation we decompose the Jacobians of the fluxes in positive and negative part. Denoting by B_d the diagonal matrix of eigenvalues of A'_d and by Q_d the associated matrix of the right eigenvectors we set

$$(A'_d)_\pm = Q_d (B_d)_\pm Q_d^{-1} .$$

$$A'_d = (A'_d)_+ - (A'_d)_-, \quad |A'_d| = (A'_d)_+ + (A'_d)_-$$

Then we can define

$$(A_d)_\pm (u) = \int_0^1 (A'_d)_\pm (tu + (1-t)u_0).(u - u_0)dt .$$

with $A_d(u_0) = 0$. Set now

$$C_j^{(d)} = \begin{cases} \lambda_d I \delta_{jd} & \text{if } j = 1, \ldots, D , \\ 0 & \text{if } j = D+1 , \\ -\lambda_{j-(D+1)}\delta_{d,j-(D+1)}I & \text{if } j = D+2, \ldots, 2D+1 . \end{cases}$$

for some $\lambda_d > 0$ and

$$\begin{cases} M_d(u) = (A_d)_+ (u)/\lambda_d & \text{if } d = 1, \ldots, D , \\ M_{D+1}(u) = u - \left(\sum_{d=1}^D \lambda_d^{-1} \left[(A_d)_+ (u) + (A_d)_- (u) \right] \right) , \\ M_d(u) = \left(A_{d-(D+1)} \right)_- (u)/\lambda_{d-(D+1)} & \text{if } d = D+2, \ldots, 2D+1 . \end{cases}$$

In the scalar case and for appropriate discretization (see below), this choice corresponds, in the relaxation limit $\epsilon \to 0$, to the Engquist-Osher numerical scheme [3]. For a one-dimensional system the stability condition (13) is here:

$$\forall \xi \in \mathbb{R}^K, \quad (|A'| (\lambda I - |A'|) \xi, \xi) \geq 0 \tag{17}$$

so that in the 1-D scalar case it coincides with the monotonicity condition.

In the 2-D case we have

$$B = \begin{pmatrix} \lambda_1 |A'_1| - A'^2_1 & -A'_1 A'_2 \\ -A'_2 A'_1 & \lambda_2 |A'_2| - A'^2_2 \end{pmatrix} .$$

For K=1 this matrix is positive if and only if

$$1 - \frac{|A'_1|}{\lambda_1} - \frac{|A'_2|}{\lambda_2} \geq 0, \tag{18}$$

and this is exactly the monotonicity condition. Hence we have the

Proposition 3.3. *Let us suppose that K=1 and that the choice of P, Λ_1, Λ_2 and M is as here above. Then the stability condition (13) and the monotonicity condition (14) coincide and can be written as (18).*

Others examples owning an arbitrary number of equations can be found in [2].

4. Numerical schemes and convergence theory

The flexibility and the efficiency of the method rely on the choice of Λ, M, P, as well as on the choice of the numerical scheme for the kinetic system (3). Of course the discretizations proposed in [8] are available, but somewhat more complicated in the implementation. Here system (3) is splitted into a linear diagonal hyperbolic part and an ordinary differential system. Following usual notations $f_\Delta^{\epsilon,n}$ is the approximate solution at time t_n

$$f_\Delta^{\epsilon,n}(x) = \sum_{\alpha \in \mathbb{Z}^D} f_\alpha^{\epsilon,n} \chi_{I_\alpha}(x) ,$$

where $(I_\alpha)_\alpha$ is a mesh of the space domain. If

$$u_\alpha^0 = (vol(I_\alpha))^{-1} \int_{I_\alpha} u_0(x) dx , \tag{19}$$

then f_0^ϵ is approximated by

$$f_\alpha^{\epsilon,0} = M(u_\alpha^0). \tag{20}$$

For a given $f_\Delta^{\epsilon,n}$, $f_\Delta^{\epsilon,n+1/2}$ is an approximated solution at time t_{n+1} of the problem

$$\partial_t f_l + \sum_{d=1}^{D} \lambda_{ld} \partial_{x_d} f_l = 0, \quad l = 1, \ldots, L , \tag{21}$$

$$f(t_n) = f_\Delta^{\epsilon,n} , \tag{22}$$

where $\Lambda_d = diag(\lambda_{ld})$. As the system is diagonal, we may consider each equation separately. At this stage a great amount of methods is available including finite volumes on unstructured meshes. We present here a finite differences method. The space time domain $\mathbb{R}^D \times [0,T]$ is discretized by a rectangular grid. We suppose that the scheme can be put in conservation form. In the following, the scheme on the linear part will be referred to as (HS) (*homogeneous scheme*).
Then, for all $\alpha \in \mathbb{Z}^D$, we solve on $[t_n, t_{n+1}]$ the ODE

$$F' = \frac{1}{\epsilon} (M(PF) - F) , \tag{23}$$

with initial data

$$F(t_n) = f_\alpha^{\epsilon,n+1/2} . \tag{24}$$

Using (5) we obtain

$$PF' = 0 , \tag{25}$$

so that the solution of (23) can be obtained explicitly. Hence

$$f_\alpha^{\epsilon,n+1} = M(u_\alpha^{\epsilon,n+1/2}) + \exp\left(-\frac{\Delta t_n}{\epsilon}\right) \left[f_\alpha^{\epsilon,n+1/2} - M(u_\alpha^{\epsilon,n+1/2}) \right] , \tag{26}$$

where

$$u^\epsilon = P f^\epsilon .$$

We have then constructed for the semilinear system (3) a wide family of numerical schemes which differ by the choice of the (HS). In the scalar case and when the monotonicity condition is satisfied, we have used the results of [7] to prove that in fact the properties of each scheme are roughly speaking the same as those of the (HS) and the estimates are uniform in ϵ. Following the method of [5] we have then a convergence theorem for fixed ϵ. For the relaxed $\epsilon = 0$ limit of the scheme we prove (see [2]):

Theorem 4.1. *Consider $u_0 \in L^1(\mathbb{R}^D) \cap L^\infty(\mathbb{R}^D) \cap BV(\mathbb{R}^D)$. Suppose M is a Monotone Maxwellian Function on $[-\|u_0\|_\infty, \|u_0\|_\infty]$, and that (HS) is monotone. Set $u = Pf$. The sequence f_Δ^ϵ converges as $\epsilon \to 0$ in $L^\infty((0,T); L^1_{loc}(\mathbb{R})^L)$ to a limit f_Δ which satisfies*

$$\|u_\Delta(t)\|_\infty \le \|u_0\|_\infty \,, \tag{27}$$

$$\|f_\Delta(t) - f_\Delta(t')\|_1 \le C(|t - t'| + \Delta t) \,, \tag{28}$$

$$M(u_\Delta^n) = f_\Delta^n \,. \tag{29}$$

Moreover the resulting numerical scheme is monotone and u_Δ converges to the entropy solution of (1), (2).

5. Numerical experiments

As system (21) is diagonal and linear, all known methods of type upwind and MUSCL apply without difficulty and one obtains *direct explicit* formulations for the schemes. For example if one takes the approximation of Example 2 and solves exactly system (21), then the associated relaxed scheme is the Engquist-Osher scheme, see [3]. Notice that solving exactly system (21) imposes that the time step is in a fixed proportion with the space meshing, so that it does seem convenient to prefer MUSCL discretizations for HS. That is what we have done here, with a minmod limiter.

For the same space discretization one can observe an important improvement of the solution when approximation changes. Figure 1 represent the density in a 1-D Sod shock tube for both approximations exposed in Section 3. In figure 2 we present the isolines of the density at T=0.16 for the two-dimensional Euler equations. The initial data are chosen in order to represent a 'double Sod tube':

$$\begin{cases} \rho_0(x,y) = 0.1 \quad \text{if} \quad xy < 0, \quad 1 \quad \text{otherwise} \,, \\ v_0(x,y) = 0 \,, \\ p_0(x,y) = 0.1 \quad \text{if} \quad xy < 0, \quad 1 \quad \text{otherwise} \,. \end{cases} \tag{30}$$

We have chosen here the second order type approximation (5 K equations) of Example 2, Section 3. The calculation has been performed with a rectangular meshing where $\Delta x = \Delta y = 0.01$ and the CFL condition has been fixed to 0.4. $\lambda = \max(\rho(A_1), \rho(A_2))$. As in the one-dimensional case, shocks are well approached

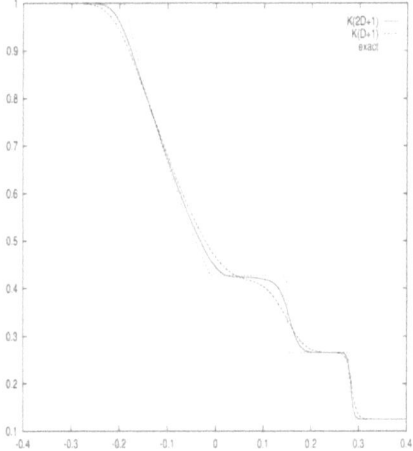

FIGURE 1. Density for a Sod shock tube

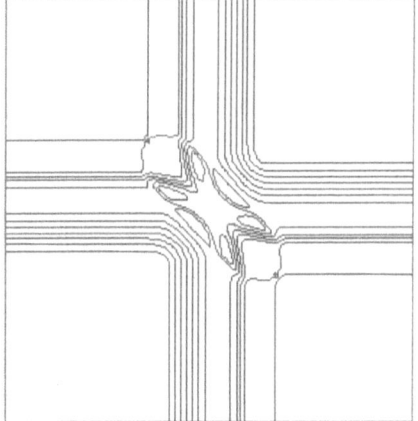

FIGURE 2. 2-D Euler equations: density

while dissipation is present in contact discontinuities. This phenomena has already been observed for kinetic schemes [6].

References

[1] D. Aregba–Driollet and R. Natalini, *Convergence of relaxation schemes for conservation laws,* Appl. Anal., **61** (1996), 163–193.

[2] D. Aregba–Driollet and R. Natalini, *Discrete kinetic schemes for multidimensional conservation laws,* Technical Report **22**, Quaderno IAC, 1997.

[3] Y. Brenier, *Systèmes hyperboliques de lois de conservation,* Cours de DEA 1992/93; Publ. Laboratoire d'analyse numérique, Univ. Pierre et Marie Curie, Paris, 1992.

[4] G.-Q. Chen and C. Levermore and T.P. Liu, *Hyperbolic conservation laws with stiff relaxation terms and entropy,* Comm. Pure Appl. Math. **47** (1994), 787–830.

[5] M. Crandall and A. Majda, *Monotone difference approximations for scalar conservation laws,* Math. of Comp **34** (1980), 1–21.

[6] E. Godlewski and P.A. Raviart, *Numerical approximation of hyperbolic systems of conservation laws,* Springer-Verlag, New York, **1996**.

[7] B. Hanouzet and R. Natalini, *Weakly coupled systems of quasilinear hyperbolic equations,* Diff. Integral Eq., **9** (1996), 1279–1292.

[8] Jin and Z. Xin, *The relaxation schemes for systems of conservation laws in arbitrary space dimensions,* Comm. Pure Appl. Math., **48** (1995), 235–277.

[9] P.L. Lions, B. Perthame and E. Tadmor, *A kinetic formulation of multidimensional scalar conservation laws and related equations,* Journal A.M.S. **7** (1994), 169–191.

[10] T.P. Liu, *Hyperbolic conservation laws with relaxation,* Comm. Math. Phys. **108** (1987), 153–175.

[11] X.D. Liu and P.D. Lax, *Positive schemes for solving multidimensional hyperbolic systems of conservation laws,* Rend. Circ. Mat. Palermo, Proceedings of the VIII International Conference on Waves and Stability in Continuous Media, Part I (Palermo, 1995) **45** (1996), 367–375.

[12] R. Natalini, *Convergence to equilibrium for the relaxation approximations of conservation laws,*Comm. Pure Appl. Math., **49** (1996), 795–823.

[13] R. Natalini, *A discrete kinetic approximation of entropy solutions to multidimensional scalar conservation laws,* J. Differential Equations (1998), to appear.

[14] H. Nessyahu and E. Tadmor, *Nonoscillatory central differencing for hyperbolic conservation laws,* J. Comput. Phys. **87** (1990), 408–463.

[15] J. Whitham, *Linear and nonlinear waves,* Wiley, New York, **1974**.

D. Aregba-Driollet,
Mathématiques Appliquées de Bordeaux,
Université Bordeaux 1,
351 cours de la Libération,
F-33405 Talence cedex, France

R. Natalini,
Istituto per le Applicazioni del Calcolo "M. Picone" ,
Consiglio Nazionale delle Ricerche,
Viale del Policlinico 137,
I-00161 Roma, Italia
E-mail address: `aregba@math.u-bordeaux.fr`, `natalini@iac.rm.cnr.it`

International Series of Numerical Mathematics
Vol. 129, © 1999 Birkhäuser Verlag Basel/Switzerland

A Mixed Finite Volume/Finite Element Method for 2-dimensional Compressible Navier-Stokes Equations on Unstructured Grids

Paul Arminjon and Aziz Madrane

Abstract. To solve flow problems associated with the Navier-Stokes equations, we construct a mixed finite volume/finite element method for the spatial approximation of the convective and diffusive parts of the flux, respectively. The finite volume component of the method is adapted from the authors' construction ([1], [2], [3]), for hyperbolic conservation laws and unstructured triangular or rectangular grids, of 2-dimensional finite volume extensions of the Lax-Friedrichs and Nessyahu-Tadmor central difference schemes, in which the resolution of Riemann problems at cell interfaces is by-passed thanks to the use of the Lax-Friedrichs scheme on two specific staggered grids. Piecewise linear cell interpolants, slope limiters and a 2-step time discretization lead to an oscillation-free second order resolution.

For the viscous terms we use a centred finite element approximation inspired by [9], [11].

Numerical experiments on classical test problems including comparison with other methods lead to fairly competitive results with favourable computing times and sharper shock capture.

1. Introduction. Mathematical modelling

1.1. Introduction

In recent papers ([1], [2]), we have presented, for hyperbolic conservation laws

$$U_t + f(U)_x + g(U)_y = 0$$

a 2-step, 2-dimensional finite volume method inspired by the principle of using the Lax-Friedrichs scheme on two staggered grids at alternate time steps to by-pass the detailed computation of the Riemann problems generated at the cell interfaces. In one dimension, this approach had been applied by several authors e.g. [7] and particularly [10], which presented an elegant non-oscillatory central difference scheme based on this principle and on van Leer's MUSCL [13] piecewise linear cell interpolants with limiters.

In [3], we have presented a first finite volume extension of the Nessyahu-Tadmor one-dimensional difference scheme to 2-dimensional rectangular grids, with several numerical applications, while in [1], [2], [5], and at the 5th Int. Conf. on

Hyperbolic Problems (Stony Brooks, N.Y., June 1994), we had presented finite volume methods of Lax-Friedrichs type for unstructured two-dimensional triangular grids. Comparisons with other methods [2] including a systematic comparison [4], [5] with a well established discontinuous finite element method [12] showed the competitivity of our new method.

   In this paper, we show how our method can be built into an efficient solver for the Navier-Stokes equations. We choose the approach of a mixed Galerkin-type finite volume/finite element method on unstructured triangular grids inspired by Rostand and Stoufflet [9], in which the convective and viscous terms are treated with different techniques: starting from a general Galerkin formulation, we compute the convective terms with our finite volume method, while the viscous terms are handled with a centred finite element procedure proposed in [9].

1.2. Mathematical Modelling

In the sequel, we consider domains of computation related to external flows around bodies; the body is an obstacle which limits the domain of computation by its wall Γ_B. Let $\Omega \subset \mathbb{R}^2$ be the flow domain of interest and Γ be its boundary, we write $\Gamma = \Gamma_B \cup \Gamma_\infty \cup \Gamma_E$, where Γ_B is the body boundary, and Γ_∞ is the (upwind) farfield boundary, and $\Gamma_E = \Gamma_E^1 \cup \Gamma_E^2$ is the exit part of the boundary.

The equations describing two-dimensional compressible viscous flows are given by

$$\frac{\partial U}{\partial t} + \vec{\nabla} \cdot \vec{\mathcal{F}}(U) = \frac{1}{R_e} \vec{\nabla} \cdot \vec{\mathcal{R}}(U) \tag{1}$$

where (see [8] for details)

$$U = (\rho, \rho u, \rho v, E)^T, \quad \vec{\nabla} = \left(\frac{\partial}{\partial x}, \frac{\partial}{\partial y} \right)^T,$$

$$\vec{\mathcal{F}}(U) = \left(\begin{array}{c} F(U) \\ G(U) \end{array} \right), \quad \vec{\mathcal{R}}(U) = \left(\begin{array}{c} R(U) \\ S(U) \end{array} \right). \tag{2}$$

ρ is the density, $\vec{V} = (u, v)$ is the velocity vector , $E = \rho e = \rho \epsilon + \frac{1}{2}\rho(u^2 + v^2)$ is the total energy per unit volume; p is the pressure; with the equation of state written, for a perfect gas, as :

$$p = (\gamma - 1)\rho\epsilon = (\gamma - 1)\rho(e - \frac{1}{2}\|\vec{V}\|^2) \tag{3}$$

ϵ denotes the specific internal energy related to the temperature by :

$$\epsilon = C_v T = e - \frac{1}{2}(\|\vec{V}\|^2) \tag{4}$$

2. Discretization with respect to space and time

2.1. Definitions

Assuming that the computational domain Ω_h is polygonal and bounded in $\mathbb{R}^2$, we start from an arbitrary FEM triangular grid $\mathcal{T}_h$. The discretization of the

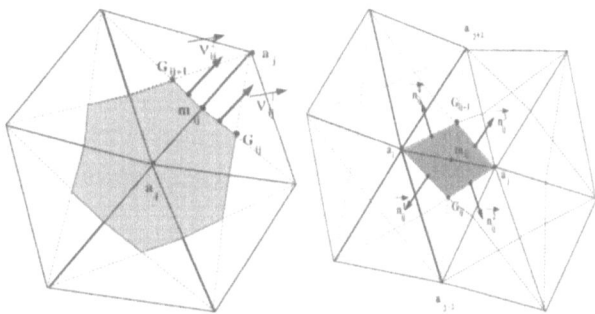

FIGURE 1. Barycentric cell C_i around node a_i and quadrilateral cell $a_i G_{ij} a_j G_{i,j+1}$, with normal vectors, $T_{ij}^1 = a_i a_{j-1} a_j$, $T_{ij}^2 = a_i a_j a_{j+1}$

convective terms is performed with the help of our finite volume method of Lax-Friedrichs type ([1], [2], [5]).

The method is a two-step scheme defined with the help of two alternate, staggered grids; for the first grid, the nodes are the vertices a_i $1 \le i \le nv$ of the triangles $K \in \mathcal{T}_h$, and the finite volume cells are the barycentric cells C_i obtained by joining the midpoints m_{ij} of the sides originating at node a_i to the centroids G_{ij} of the triangles of $\mathcal{T}_h$ which meet at a_i (Fig. 1). For the second grid the nodes are the midpoints m_{ij} of the sides, while the cells are the quadrilaterals $L_{ij} = a_i G_{ij} a_j G_{i,j+1}$ ("diamond cells") obtained by joining two nodes a_i , a_j to the centroids of the two triangles of $\mathcal{T}_h$ of which $a_i a_j$ is a side. Let $U_i^n \cong U(a_i, t^n)$ and $U_{ij}^{n+1} \cong U(m_{ij}, t^{n+1})$ denote the nodal (cell average) values in the first and second grid at time $t = t^n$ and $t = t^{n+1}$, respectively (n even). We have $\Omega_h = \bigcup_{i=1}^{nv} C_i$, $\Omega_h = \bigcup_{k=1}^{ne} L_k$, where nv and ne are the number of vertices and number of edges, respectively.

2.2. Approximation spaces

Let ψ_i be the characteristic function of cell C_i and, for every edge l_k ($1 \le k \le ne$), χ_k is the characteristic function of cell L_k. We define the discrete spaces

$$\mathcal{V}_h = \{v_h | v_h \in C^0(\Omega), v_h|_T \in P_1, \forall T \in \mathcal{T}_h\}$$

$$\mathcal{W}_h = \{v_h | v_h \in L^2(\Omega), v_h|_{C_i} = v_i = const; \ i = 1, \cdots, nv\}$$

$$\mathcal{Z}_h = \{v_h | v_h \text{ is continuous at the edge midpoints; } v_h|_T \in P_1, \forall T \in \mathcal{T}_h\}$$

$$\mathcal{U}_h = \{v_h | v_h \in L^2(\Omega), v_h|_{L_k} = v_k = const; \ k = 1, \cdots, ne\}$$

Any function $f \in \mathcal{V}_h$ is uniquely determined by its values $f(a_i)$ at the vertices, and if $(N_i)_{i=1}^{nv}$ is the basis set of $\mathcal{V}_h$ we have :

$$f(\vec{X}) = \sum_{i=1}^{nv} f(a_i) N_i(\vec{X}) \qquad \vec{X} \in \Omega_h$$

There exists a natural bijection S between the spaces $\mathcal{V}_h$ and $\mathcal{W}_h$ defined by :

$$\forall f \in \mathcal{V}_h \quad, \quad S(f(\vec{X})) = \sum_{i=1}^{nv} f(a_i)\psi_i(\vec{X})$$

with a similar bijection operator B between $\mathcal{Z}_h$ and $\mathcal{U}_h$.

2.3. A mixed finite volume/finite element method for the Navier-Stokes equations

Mixed finite volume/finite element methods are constructed on the principle of a typical Galerkin formulation, but where the convective terms are treated with a finite volume method, while the viscous terms are handled with a finite element approach. This allows for a reduction of the order of the viscous terms from two to one thanks to an integration by parts.

The finite volume method used here, borrowed from ([1]–[5]), is a two-step algorithm using two dual grids, so that our mixed method also involves two steps.

Using the above discrete spaces, we first write a variational formulation of (1):

Find $U_h \in (\mathcal{V}_h)^4$ (or $(\mathcal{Z}_h)^4$ at alternate steps) such that

$$\int_{\Omega_h} \frac{\partial U_h}{\partial t}\varphi_h dxdy + \int_{\Omega_h} \vec{\nabla} \cdot \vec{\mathcal{F}}(U_h)\varphi_h dxdy = \tag{5}$$

$$\frac{1}{R_e}\int_{\Omega_h} \vec{\nabla} \cdot \vec{\mathcal{R}}(U_h)\varphi_h dxdy \qquad \text{for all } \varphi_h \in \mathcal{V}_h(or\,\mathcal{Z}_h)$$

Then a **mixed** finite volume/finite element approximation ([9]) is **defined** by using different techniques to compute the left- and right-hand side integrals in (5).

First step

Starting from known barycentric cell values U_i^n ($1 \le i \le nv$), we compute new values for the diamond cells U_{ij}^{n+1}. Starting from $U_h \in (\mathcal{Z}_h)^4$ in (5), choosing for $\varphi_h \in \mathcal{Z}_h$ the shape function $M_{ij} \in \mathcal{Z}_h$ associated with the midpoint m_{ij} of edge $a_i a_j$, and applying the operator B to the left-hand side, we obtain

$$\int_{Supp\chi_i=L_{ij}} \frac{\partial U_h}{\partial t}dxdy + \int_{Supp\chi_i=L_{ij}} \vec{\nabla} \cdot \vec{\mathcal{F}}(U_h)dxdy =$$

$$\frac{1}{R_e}\int_{SuppM_{ij}=T_{ij}^1 \cup T_{ij}^2} M_{ij}\vec{\nabla} \cdot \vec{\mathcal{R}}(U_h)dxdy \tag{6}$$

Second step

Starting from known cell values U_{ij}^{n+1}, we compute cell values $U_i^{n+2}(1 \le i \le nv)$; given $U_h \in (\mathcal{V}_h)^4$ in (5), choosing for φ_h the shape function $N_i \in \mathcal{V}_h$ associated with node a_i, and applying S to the left-hand side, we obtain

$$\int_{Supp\psi_i=C_i} [\frac{\partial U_h}{\partial t} + \vec{\nabla} \cdot \vec{\mathcal{F}}(U_h)]dxdy = \frac{1}{R_e}\int_{SuppN_i} N_i\vec{\nabla} \cdot \vec{\mathcal{R}}(U_h)dxdy \tag{7}$$

Applying Green's formula to the convective terms of (6),(7), and integration by parts to the diffusive terms, neglecting the boundary integrals obtained in the R.H.S. and observing that

$$\partial C_i = \bigcup_{j \in K(i)} \{\partial C_i \cap \partial C_j\} \cup \{\partial C_i \cap \Gamma_B\} \cup \{\partial C_i \cap \Gamma_\infty\}$$

we obtain

First time step :

$$\int_{L_{ij}} \frac{\partial U_h}{\partial t} dx dy + \int_{\partial L_{ij}} \vec{\mathcal{F}}(U_h) \cdot \vec{n_{ij}} d\sigma = -\frac{1}{Re} \sum_{T \in [T^1_{ij}, T^2_{ij}]} \int_T \vec{\mathcal{R}}(U_h) \cdot \vec{\nabla} M^T_{ij} dx dy$$

Second time step :   (8)

$$\int_{C_i} \frac{\partial U_h}{\partial t} dx dy + \sum_{j \in K(i)} \int_{\partial C_{ij} = \partial C_i \cap \partial C_j} \vec{\mathcal{F}}(U_h) \cdot \vec{\nu_{ij}} d\sigma +$$ (9)

$$\int_{\partial C_i \cap \Gamma_B} \vec{\mathcal{F}}(U_h) \cdot \vec{\nu_i} d\sigma + \int_{\partial C_i \cap \Gamma_\infty} \vec{\mathcal{F}}(U_h) \cdot \vec{\nu_i} d\sigma = -\frac{1}{Re} \sum_{T, a_i \in T} \int_T \vec{\mathcal{R}}(U_h) \cdot \vec{\nabla} N^T_i$$

where $K(i) = \{j : 1 \leq j \leq nv \quad a_j \text{ is a neighbour of } a_i\}$. The convective terms are computed according to our finite volume method [2], [5].

The velocity vector $\vec{V}_T$ on a triangle T is computed as the nodal average, and the viscous fluxes are approximated as follows: for the first step,

$$\sum_{T \in [T^1_{ij}, T^2_{ij}]} \int_T \vec{\mathcal{R}}(U_h) \cdot \vec{\nabla} M^T_{ij} dx dy \equiv \sum_{k=1}^2 Area(T^k_{ij}) \left[R(T^k_{ij}) \frac{\partial M^{T^k_{ij}}_{ij}}{\partial x} + S(T^k_{ij}) \frac{\partial M^{T^k_{ij}}_{ij}}{\partial y} \right]$$

                                                                                       (10)

and for the second step

$$\sum_{T, a_i \in T} \int_T \vec{\mathcal{R}}(U_h) \cdot \vec{\nabla} N^T_i dx dy \equiv \sum_{T, a_i \in T} Area(T)(R(T) \frac{\partial N^T_i}{\partial x} + S(T) \frac{\partial N^T_i}{\partial y}) \quad (11)$$

where the values of $R(T)$ and $S(T)$ are constant on the triangle T, computed from the average velocity vector $\vec{V}_T$.

For second-order time and space accuracy, each step is written as predictor-corrector using MUSCL piecewise linear cell interpolants for the convective terms:

Predictor (First step) :

On each side of the diamond cell L_{ij}, using the Euler equations (neglecting the viscous terms) we define a predicted vector

$$U^{n+1/2}_{a_i G_{ij}} = U^n_{a_i, G_{ij}} - \frac{\Delta t}{2} \{F'(U^n_{a_i, G_{ij}}) P^n_i + G'(U^n_{a_i, G_{ij}}) Q^n_i\}$$ (12)

where

$$U^n_{a_i, G_{ij}} = U^n_i + \frac{1}{2}(x_{G_{ij}} - x_i) P^n_i + \frac{1}{2}(y_{G_{ij}} - y_i) Q^n_i$$ (13)

The corrector can now be written as
Corrector (First step)

$$
Area(L_{ij})U_{ij}^{n+1} - \left\{ \int_{L_{ij}\cap C_i} U_h(x,y,t^n)\,dxdy + \int_{L_{ij}\cap C_j} U_h(x,y,t^n)\,dxdy \right\} +
$$

$$
\Delta t \int_{\partial L_{ij}} \vec{\mathcal{F}}(U_h^{n+\frac{1}{2}})\cdot\overrightarrow{n_{ij}}\,d\sigma = -\frac{\Delta t}{Re}\sum_{T\in[T_{ij}^1,T_{ij}^2]}\int_T \mathcal{R}(U_h)\cdot\vec{\nabla}M_{ij}^T\,dxdy \tag{14}
$$

where the right-hand side is computed according to (10).

The **Second step** of our method is performed in a similar way.

Predictor (second step)

$$
U_{G_{ij}m_{ij}}^{n+\frac{3}{2}} = U_{G_{ij}m_{ij}}^{n+1} - \frac{\Delta t}{2}\left\{ F'(U_{G_{ij}m_{ij}}^{n+1})P_{ij}^{n+1} + G'(U_{G_{ij}m_{ij}}^{n+1})Q_{ij}^{n+1} \right\}. \tag{15}
$$

where

$$
U_{G_{ij}m_{ij}}^{n+1} = U_{ij}^{n+1} + \frac{1}{2}(x_{G_{ij}} - x_{m_{ij}})P_{ij}^{n+1} + \frac{1}{2}(y_{G_{ij}} - y_{m_{ij}})Q_{ij}^{n+1} \tag{16}
$$

Corrector (second step)

$$
Area(C_i)U_i^{n+2} - \sum_{j\in K(i)}\int_{C_i\cap L_{ij}} U(x,y,t^{n+1})\,dxdy \tag{17}
$$

$$
+ \Delta t \sum_{j\in K(i)}\int_{\partial C_i\cap\partial C_j=\Gamma_{ij}} \vec{\mathcal{F}}(U(x,y,t^{n+3/2}))\cdot\overrightarrow{v_{ij}}\,d\sigma + \Delta t \int_{\partial C_i\cap\Gamma_B} \vec{\mathcal{F}}(U_h^{n+1})\cdot\overrightarrow{v_i}\,d\sigma
$$

$$
+ \Delta t \int_{\partial C_i\cap\Gamma_\infty} \vec{\mathcal{F}}(U_h^{n+1})\cdot\overrightarrow{v_i}\,d\sigma = -\frac{\Delta t}{Re}\sum_{T,a_i\in T}\int_T \mathcal{R}(U_h^{n+1})\cdot\vec{\nabla}N_i^T\,dxdy
$$

2.4. Approximation of the slopes

The gradients of the MUSCL piecewise linear cell interpolants are computed with a least-squares technique described in ([2], [14]).

When considering flows for which the viscous terms are dominant (small values of the Reynolds number) the limitation procedure used for the gradients, described in [2], is generally not necessary.

2.5. Treatment of the boundary conditions

Wall boundary: the no-slip condition is taken into account using a strong formulation to compute pressure and energy; to compute the density we equal to zero the first component of the boundary flux in the fourth term of (17).

Inflow and outflow boundaries: at these boundaries, a precise set of compatible exterior data which depend on the flow regime and the velocity direction, is to be specified. For this purpose a plus-minus flux splitting is applied between exterior

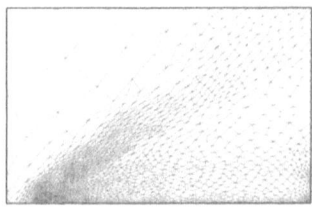

FIGURE 2. Initial grid for the plate problem (1637 vertices) and solution (Mach contours)

data and interior values. The boundary integral (third term of (9)) is evaluated using a non-reflective version of the flux-splitting of Steger and Warming [15]:

$$\int_{\partial C_i \cap \Gamma_\infty} \overrightarrow{\mathcal{F}(\mathcal{U})}_h \cdot \overrightarrow{n}_i d\sigma = \mathcal{A}^+(U_i, \overrightarrow{\mathcal{V}}_{i\infty}) \cdot U_i + \mathcal{A}^-(U_i, \overrightarrow{\mathcal{V}}_{i\infty}) \cdot U_\infty \qquad (18)$$

where $\mathcal{A}$ is the flux Jacobian matrix $\frac{\partial \overrightarrow{\mathcal{F}}(U)}{\partial U} \cdot \overrightarrow{\mathcal{V}} = \frac{\partial F}{\partial U}\mathcal{V}_x + \frac{\partial G}{\partial U}\mathcal{V}_y$, and $\mathcal{A}^+, \mathcal{A}^-$ are the positive and negative parts of $\mathcal{A}$, respectively.

3. Numerical experiments

3.1. Supersonic Navier-Stokes flow past a flat plate

In order to assess the basic qualities of our method, we treated several typical test cases of viscous flows, of which we will present two in this paper.

We first consider the classical test problem of the flow past a flat plate [11], [17] with a Reynolds number $R_e = 1000$, and Mach number M_∞ at the farfield equal to 3. The initial grid (1637 vertices) (Fig. 2) is adapted twice with the help of a grid adaptator developed by Castro Diaz and Hecht at INRIA [16], resulting in a 1726-vertex final grid (Fig. 3).

The results obtained with the initial grid (Fig. 2) are already valuable, as can be seen from a comparison with those obtained in [11], [17], with the same conditions.

In fact our initial results (1637 vertices) can sustain comparison with the final results in [17] (6750 vertices); but the method used in [17] is based on mesh refinement only, while ours also resorts to mesh adaptation.

With the adapted final grid (1726 vertices), we observe (Fig. 3) a very clean capture of the shock and the boundary layer.

The final grid also leads to a substantial improvement in the resolution of the pressure coefficient $C_p = \frac{p - p_\infty}{1/2\rho_\infty ||\overrightarrow{V}_\infty||^2}$ (Fig. 4).

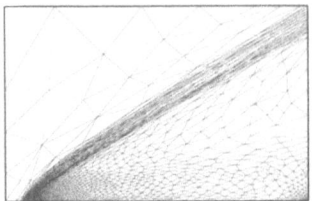 

FIGURE 3. Enriched grid for the plate problem (1726 vertices) and solution (Mach contours)

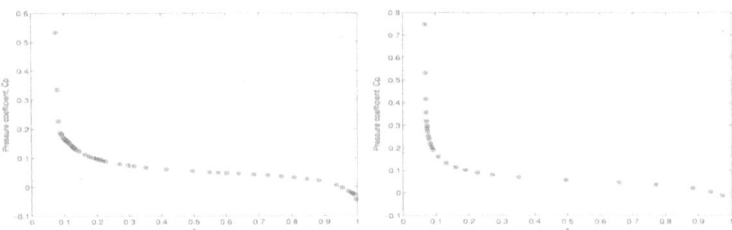

FIGURE 4. Flat plate problem. Pressure coefficient C_p with initial grid and enriched grid

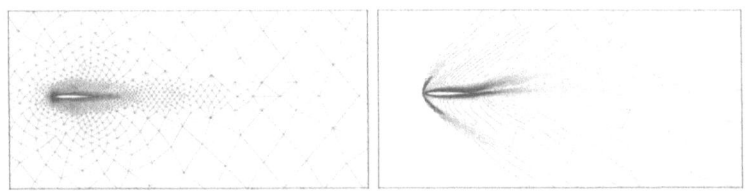

FIGURE 5. NACA0012 : Initial grid (2274 vertices) and Mach contours

3.2. Supersonic flow around a NACA0012 airfoil

We consider a more complex flow around a NACA0012 airfoil, with a farfield Mach number of 2, angle of attack 10 degrees and a Reynolds number of 1000. This case was also studied in [11].

A very thin detached shock is present. The initial grid (2274 vertices, Fig. 5) is adapted 3 times. The final grid contains 7114 nodes (Fig. 6). We present the corresponding Mach contours (Fig. 6); the extra refinement has led to a thin capture of the shock. Here again, the results obtained for this test case can be favourably compared with those of [11]. In particular, we have obtained a sensibly sharper shock capture and smoother Mach isolines, and a very good resolution of the boundary layer, with the corresponding detachment at the trailing edge.

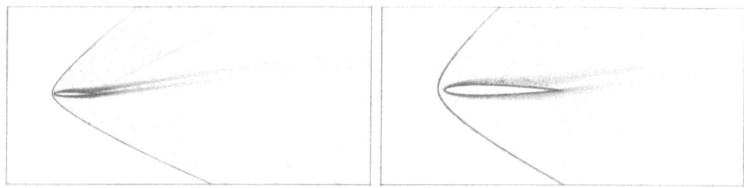

FIGURE 6. NACA0012 : Final grid (7114 vertices) Mach contours with enlarged details.

4. Concluding remarks

Starting from the principle of mixed Finite Volume/Finite Element methods, we have constructed a new method for the two-dimensional compressible Navier-Stokes equations, whose main component lies in our recent staggered grid central finite volume method inspired from the one-dimensional Lax-Friedrichs and Nessyahu-Tadmor difference schemes.

From a theoretical viewpoint, L^∞-weak convergence of our Finite Volume method has been proved first for a linear hyperbolic equation [6], using L^∞-and weighted total variation estimates, and more recently, for a nonlinear scalar hyperbolic equation, with the help of Young measures.

In this paper, we have concentrated on the presentation of the method, and its validation through a comparative study of two classical test problems, where our new method has shown excellent capabilities both for an accurate and oscillation-free, sharp capture of shocks and for the high quality of the boundary layer resolution with an accurate simulation of the boundary layer detachment.

References

[1] P. Arminjon and M.C. Viallon, *Généralisation du schéma de Nessyahu-Tadmor pour une équation hyperbolique à deux dimensions d'espace*, Comptes Rendus de l'Acad. des Sciences, Paris, t. 320, série I , 85–88, January 1995.

[2] P. Arminjon, M.C. Viallon and A. Madrane, *A Finite Volume Extension of the Lax-Friedrichs and Nessyahu-Tadmor Schemes for Conservation Laws on Unstructured Grids*, Int.J. Comp.Fluid Dynamics, **9** (1997), 1–22 .

[3] P.Arminjon, D.Stanescu and M.C. Viallon, *A two-dimensional finite volume extension of the Lax-Friedrichs and Nessyahu-Tadmor schemes for compressible flows*, Proc. of the 6 th. Int. Symp. on Comp. Fluid Dynamics, Lake Tahoe (Nevada), September 4–8, 1995, M. Hafez and K.Oshima, editors, **4**, 7–14.

[4] P. Arminjon, A. Madrane and M.C. Viallon, *Comparison of a finite volume version of the Lax-Friedrichs and Nessyahu-Tadmor schemes and discontinuous finite element methods for compressible flows on unstructured grids*, Proceedings of a Symposium Honoring S.K.Godunov, May 1–2, 1997, The University of Michigan, Ann Arbor (Mi.), B. van Leer, editor, to appear as a special volume in J. Comp. Physics.

[5] P. Arminjon, M.C. Viallon, A. Madrane and H. Kaddouri, *Discontinuous finite elements and a 2-dimensional finite volume generalization of the Lax-Friedrichs and Nessyahu-Tadmor schemes for compressible flows on unstructured grids,* to appear in C.F.D. Review, M. Hafez and K. Oshima, editors, (1997).

[6] M.C. Viallon and P. Arminjon, *Convergence of a finite volume extension of the Nessyahu-Tadmor scheme on unstructured grids for a 2-dimensional linear hyperbolic equation,* Res. Rep. No. 2239, CRM, Univ- de Montréal, January 1995, to appear in SIAM J. Num. Anal. For the nonlinear case: to appear in Num. Mathematik, (1995).

[7] R. Sanders and A. Weiser, *A high order staggered mesh approach for nonlinear hyperbolic system of conservation laws,* JCP 1010, (1992), 314–329. 739CP, The 12th AIAA CFD Conf., 1995.

[8] R. Peyret and T.D. Taylor, *Computational Methods for Fluid Flow,* Springer-Verlag, New-York, Heidelberg, Berlin, (1983).

[9] P. Rostand and B. Stoufflet, *Finite Volume Galerkin Methods for Viscous Gas Dynamics,* INRIA Res. Rep. No. 863, Rocquencourt, 78153 Le Chesnay, France, (1988).

[10] H. Nessyahu and E. Tadmor, *Non-oscillatory Central Differencing for Hyperbolic Conservation Laws,* J. Comp. Physics 87, **2** (1990), 408–463.

[11] B. Palmerio, L.Fezoui, C. Olivier and A. Dervieux, *On TVD criteria for mesh adaption for Euler and Navier-Stokes calculations,* INRIA Res. Rep. No. 1175, Rocquencourt, 78153 Le Chesnay, France, (1990)

[12] J. Jaffré and L. Kaddouri, *Discontinuous finite elements for the Euler equations,* Proc. 3^d. Int. Conf. on Hyperbolic Problems, June 11–15, 1990, Uppsala (Sweden), B. Engquist and B. Gustafsson, editors, Studentlitteratur, Chartwell-Bratt, **2** (1991), 602–610.

[13] B. van Leer, *Towards the Ultimate Conservative Difference Scheme. II. Monotonicity and conservation combined in a second order scheme,* J. Comp. Phys., **14** (1974) 361–370.

[14] S. Champier, *Convergence de schémas numériques type Volumes Finis pour la résolution d'équations hyperboliques,* Thèse, Univ. de St-Étienne, **1992**.

[15] J. Steger, R.F. Warming, *Flux vector splitting for the inviscid gas dynamic with applications to finite-difference methods,* J. of Comp. Phys., **40 (2)** (1981), 263–293.

[16] M.J.Castro Diaz and F.Hecht, *Anisotropic Surface Mesh Generation,* INRIA Res. Rep. No. 2672, October 1995, INRIA, Rocquencourt, 78153 Le Chesnay, France.

[17] B.V.K. Satya Sai, O.C.Zienkiewicz, M.T.Manzari, P.R.M. Lyra and K.Morgan, *General purpose versus special algorithms for high-speed flows with shocks,* Int. J. Num. Meth. in Fluids, **27** (1998), 57–80.

Centre de Recherches Mathématiques,
Université de Montréal,
C.P.6128, Succ. Centre-ville, Montréal, Québec, CANADA, H3C 3J7
E-mail address: `arminjon@crm.umontreal.ca`

International Series of Numerical Mathematics
Vol. 129, © 1999 Birkhäuser Verlag Basel/Switzerland

Large Time Stability of Propagating Phase Boundaries

Fumioki Asakura

Abstract. We study the Cauchy problem for a 2 × 2-system of conservation laws: $v_t - u_x = 0$, $u_t - \sigma(v)_x = 0$ which describe the phase transition. Two constant states satisfying the Maxwell equal-area principle constitute an admissible stationary solution; a small perturbation of these *Maxwell states* will be our initial data. We shall show that: there exists a global in time propagating phase boundary which is admissible in the sense that it satisfies the Abeyaratne-Knowles kinetic condition; the states outside the phase boundary tend to the Maxwell states as time goes to infinity.

1. Phase boundaries

The van der Waals equation of state:

$$p(v) = \frac{RT}{v-b} - \frac{a}{v^2} \quad (v > b)$$

represents homogeneous isothermal liquid-vapor states, but does not agree with the requirements of stable inhomogeneous state composed of a liquid and a vapor part. We know that this inhomogeneous state is in thermal equilibrium at the saturation pressure if and only if the pressure satisfies the Maxwell equal-area principle; the liquid and the vapor state determined by this principle are called the *Maxwell states*. Now a simple question arises: if these Maxwell states are compactly perturbed, then, as time goes on, the states come back or not to the Maxwell states. In this paper, we shall find one way to say "Yes".

We study a 2 × 2-system of conservation laws:

$$v_t - u_x = 0, \quad u_t - \sigma(v)_x = 0. \tag{1}$$

Here $\sigma(v)$ is a C^2- function and there exist α, β ($\alpha < \beta$) such that

$$\sigma'(v) = \begin{cases} > 0 & \text{for} & v < \alpha, \\ < 0 & \text{for} & \alpha < v < \beta, \\ > 0 & \text{for} & v > \beta, \end{cases} \tag{2}$$

and we further assume that

$$\sigma''(v) \neq 0, \quad \text{for} \quad v < \alpha, \ v > \beta. \tag{3}$$

The system of equations is *hyperbolic* for $v < \alpha$, $v > \beta$ and *elliptic* for $\alpha < v < \beta$; the region $\Omega_\alpha = \{(v, u); \ v < \alpha\}$ is called the α-*phase* and $\Omega_\beta = \{(v, u); \ v > \beta\}$ the β-*phase*. We consider the initial data of the form:

$$(v(x, 0), u(x, 0)) = \begin{cases} (v_L(x), u_L(x)) \in \Omega_\alpha & \text{for} \quad x < 0, \\ (v_R(x), u_R(x)) \in \Omega_\beta & \text{for} \quad x > 0. \end{cases} \tag{4}$$

If the initial data $U_R(x) = (v_R(x), u_R(x))$, $U_L(x) = (v_L(x), u_L(x))$ are constant states U_R, U_L, the initial value problem is called the *Riemann problem*. The jump discontinuity of the form:

$$(v(x, t), u(x, t)) = \begin{cases} (v_-, u_-) & \text{for} \quad x < st, \\ (v_+, u_+) & \text{for} \quad x > st, \end{cases} \tag{5}$$

($v_\pm, u_\pm$: constants) is said to be a *phase boundary* provided the states belong to the different phases: $U_- = (v_-, u_-) \in \Omega_\alpha$, $U_+ = (v_+, u_+) \in \Omega_\beta$ and satisfy the Rankine-Hugoniot condition. Sometimes it will be convenient to use the term *phase boundary* to refer the phase front $x = st$. If the propagation speed s is non-zero, then it is called a *propagating phase boundary*. There exists a unique pair of states $U_- \in \Omega_\alpha$, $U_+ \in \Omega_\beta$ satisfying

$$\sigma(v_+) = \sigma(v_-), \quad u_+ = u_-, \quad \int_{v_-}^{v_+} \sigma(v)dv = \sigma(v_\pm)(v_+ - v_-). \tag{6}$$

These states, denoted in particular by $U_- = U_m = (v_m, u_m)$, $U_+ = U_m^* = (v_m^*, u_m^*)$, are called the *Maxwell states* and constitute an important stationary phase boundary. The Maxwell states are admissible in the sense that they minimize the entropy rate for v close to v_m and stationary phase boundaries are not admissible unless $v_- = v_m$ (Hattori's theorem [9], see also Dafermos [6])). This admissibility condition is generalized by Abeyaratne-Knowles [1] in ingenious way which we review briefly in the following.

The system of conservation laws (1) is endowed with the canonical entropy pair (total mechanical energy [11]):

$$\eta(U) = \frac{1}{2}u^2 + \int_0^v \sigma(w)dw, \quad q(U) = -u\sigma(v). \tag{7}$$

Let the interval $[x_1, x_2]$ contain a finite number of jump discontinuities (5). By direct computation, we find that the local entropy is decreasing in time if and only if

$$sf(v_+, v_-) \geq 0 \tag{8}$$

holds at every discontinuity. Here f is defined by

$$f(v_+, v_-) = \int_{v_-}^{v_+} \sigma(v)dv - \frac{1}{2}\{\sigma(v_+) + \sigma(v_-)\}(v_+ - v_-) \tag{9}$$

which is called the *driving traction* in [1]. At a shock wave, the Lax entropy condition implies the above inequality. At a phase boundary, we adopt Abeyaratne-Knowles' *kinetic condition*: there exists is a nondecreasing function $\Phi(f)$ satisfying

$\Phi(0) = 0$ such that the speed of the discontinuity is function of the driving traction:

$$s = \Phi(f(v_+, v_-)) \tag{10}$$

Hence the condition (8) is obviously satisfied at the phase boundary. They also assumed that no new phase occurs from any point in the interior of the α or β-phase, which is called the *nucleation condition*. A solution is said to be *admissible* provided it satisfies both the kinetic condition and the nucleation condition.

The aim of this article is to give a brief account of [3] in which the author studies the existence and large time stability of an admissible global in time solution involving a single propagating phase boundary which is obtained by an initial perturbation of the Maxwell states. We shall show that this perturbation problem is stable if we assume that Φ is a C^2-function and

$$\Phi'(0) > 0. \tag{11}$$

In the following argument, we shall always assume that Φ satisfies the above condition (11) which is not repeatedly stated. Global solutions are obtained by using wave-front tracking method (Bressan [4], Chern [5], Risebro [14]).

Theorem 1.1. *Suppose that the initial perturbation is sufficiently small in total variation i.e. the quantity:*

$$T.V.(U_L(x) - U_m)|_{x<0} + T.V.(U_R(x) - U_m^*)|_{x>0} \tag{12}$$

is sufficiently small. Then there exists a weak global solution with a single phase boundary which is Lipschitz continuous curve in (x, t)-space.

Our phase boundary is admissible in the following sense.

Theorem 1.2. *The phase boundary $x = \chi(t)$ is subsonic, and $U(x, t) \in \Omega_\alpha$ for $x < \chi(t)$ and $U(x, t) \in \Omega_\beta$ for $x > \chi(t)$. The limit:*

$$\lim_{\delta \to \pm 0} U(\chi(t) + \delta, t) = U_\pm(t) \tag{13}$$

exists except countable t at the phase boundary. Moreover, the Rankine-Hugoniot conditions and the kinetic condition

$$\dot{\chi}(t) = \Phi(f(v_+(t), v_-(t))) \tag{14}$$

holds at these points.

Similar results are obtained by LeFloch [12] for the tri-linear material. Finally, we can show that the single phase boundary is asymptotically stable

Theorem 1.3. *Suppose that the initial data satisfy*

$$U(x, 0) = \begin{cases} U_m & \text{for} \quad x < -M \\ U_m^* & \text{for} \quad x > M \quad (M > 0) \end{cases} \tag{15}$$

and the total variation of the initial perturbation is sufficiently small. Then the speed and the states both side of the phase boundary approach those of the Maxwell phase boundary at the rate $t^{-\frac{3}{2}}$ and the total variation of the solution outside of the phase boundary approaches zero at the rate $t^{-\frac{1}{2}}$.

2. The riemann problem

The Riemann problem is the Cauchy problem with the discontinuous initial data of the form (4) with $U_R(x) = (v_R, u_R)$, $U_L(x) = (v_L, u_L)$ where v_L, u_L, v_R, u_R are constants. First suppose that U_L and U_R are contained in the same phase. The Lax existence theorem ([10]) says that: if $|U_L - U_R|$ is sufficiently small, then there exists a self-similar solution which consists of 3 constant regions U_L, U_M, U_R connected by rarefaction waves and shock waves. Moreover the solution of this form is unique and U_M is differentiable with respect to U_L, U_R.

Next suppose that $U_L = (v_L, u_L)$ and $U_R = (v_R, u_R)$ are contained in small neighborhoods of the Maxwell states $U_m = (v_m, u_m)$ and $U_m^* = (v_m, u_m^*)$, respectively. If $|U_m - U_L|$ and $|U_m^* - U_R|$ are sufficiently small, then we can construct a self-similar solutions which consists of 4 constant regions U_L, $U_- \in \Omega_\alpha$ and U_+, $U_R \in \Omega_\beta$, where U_L, U_- are connected by a 1-rarefaction wave or shock wave in Ω_α, U_-, U_+ by a single phase boundary and U_+, U_R by a 2-rarefaction wave or shock wave in Ω_β. However the solution of this form is not unique and these solutions constitute a one parameter family.

Now we consider the kinetic condition (10) with (11). This condition, together with the Rankine-Hugoniot condition, determines v_+ as a function of v_-.

Lemma 2.1. *If $\Phi(0) = 0$ and $\Phi'(0) > 0$, then v_+ and s are functions of v_- in a neighborhood of v_m and*

$$\frac{dv_+}{dv_-}\bigg|_{v_-=v_m} = \frac{\sigma(v_m)}{\sigma(v_m^*)}, \quad \frac{ds}{dv_-}\bigg|_{v_-=v_m} = -\Phi'(0)\sigma(v_m)(v_m^* - v_m). \quad (16)$$

We note that the derivative of v_+ does not depend on the particular value of $\Phi'(0)$ and coincides with the expression coming from the maximum entropy dissipation rate admissibility condition in Hattori [9]. By using these relations, the following existence theorem is obtained.

Theorem 2.2. *If $|U_L - U_m|, |U_R - U_m^*|$ are sufficiently small, then there exists a unique admissible solution which consists of 4 constant regions connected by rarefaction waves, shock waves and a phase boundary. Moreover these constant states are differentiable with respect to the initial data U_L, U_R.*

3. Existence of global solutions

Global solutions are constructed by using wave-front tracking method (Bressan [4], Chern [5], Resebro [14]). For any positive h, we choose a sequence $\{x_j\}_{j=1}^{N_0}$ in R such that the approximation of the initial data by step functions $U_0^h(x)$ satisfies $\sup_{x \neq x_n} |U_0^h(x) - U_0(x)| \leq h$. At each point of discontinuity $x = x_n$, we solve the

Riemann problem with the initial data $U_L = U_0^h(x_n - 0)$, $U_R = U_0^h(x_n + 0)$. If both U_L and U_R belong to the same phase, we adopt the Lax solution; if $U_L \in \Omega_\alpha$ and $U_R \in \Omega_\beta$, we construct the admissible solution following the way in Section 2. We define the approximate solution by combining an approximation for each solution to the Riemann problem; the approximation consists of constant states separated by a phase boundary and discontinuities propagating in the j-th characteristic direction (see Resebro [14] for the details). This approximate solution, denoted by $U^h(x,t)$, is defined as long as the neighboring waves of the approximations collide at $t_1 > 0$. At $t = t_1$, since $U^h(x, t_1 - 0)$ is also a step function, we can construct an approximation in the same way. Then we can extend the approximation to the next collision time $t = t_2 > t_1$. Repeating this construction, we obtain the approximate solution. In order to prove that the approximate solutions thus constructed are global in time and a subsequence converges to a weak solution, we have to show interaction estimates; the proof is carried out in the same manner as Glimm [7], Liu [13] and Chern [5].

Let P be the point of collision in the hyperbolic region, where p waves collide. We call these waves the *incoming waves* which are denoted by $\alpha = (\alpha_2^{(1)}, \ldots, \alpha_2^{(p_2)}, \alpha_1^{(1)} \ldots, \alpha_1^{(p_1)})$ from left to right ($p = p_1 + p_2$). Here the index $i = 1, 2$ says that $\alpha_i^{(k)}$ is an i-wave. The elementary waves issuing from P, denoted by $\epsilon = (\epsilon_1, \epsilon_2)$, are called the *outgoing waves*. We define the amount of interaction at the point of collision P as the following (see also Liu [13] Theorem 2.2) :

$$Q(P) = Q_d(P) + Q_s(P) \tag{17}$$

where

$$Q_d(P) = \sum_{\substack{1 \leq k \leq p_1 \\ 1 \leq l \leq p_2}} |\alpha_1^{(k)} \alpha_2^{(l)}|, \quad Q_s(P) = \sum_{\substack{i=1,2 \\ k \neq l}} Q_s(\alpha_i^{(k)} \alpha_i^{(l)}), \tag{18}$$

$$Q_s(\alpha, \beta) = \begin{cases} 0 & \alpha \geq 0, \beta \geq 0 \\ |\alpha|^3 & |\alpha| \leq |\beta|, \beta \geq 0, \alpha < 0 \\ |\alpha||\beta|^2 & |\alpha| \leq |\beta|, \beta < 0 \\ |\beta||\alpha|^2 & |\beta| \leq |\alpha|, \alpha < 0 \\ |\beta|^3 & |\beta| \leq |\alpha|, \alpha \geq 0, \beta < 0. \end{cases} \tag{19}$$

Lemma 3.1 (Local interaction estimate). *If the amplitude of incoming waves is sufficiently small, then it follows that*

$$\epsilon_i = \sum_k \epsilon_i^{(k)} = \sum_{k=1}^{p_i} \alpha_i^{(k)} + O(1)Q(P), \quad i = 1, 2. \tag{20}$$

Here the first sum denotes the approximation of the outgoing i-wave; $O(1)$ depends only on the system and the two constant states connected by the incoming waves.

Suppose that a phase boundary π and $\beta = (\beta_2^{(1)}, \ldots, \beta_2^{(p_2)}, \beta_1^{(1)} \ldots, \beta_1^{(p_1)})$ enter P, and generate π' and $\beta' = (\beta_1', \beta_2')$. Here we set $Q(P) = 0$. Instead, we denote by $A(\mathrm{P})$ the total amount of incoming waves:

$$A(\mathrm{P}) = \sum_{\substack{i=1,2 \\ 1 \le k_i \le p_i}} |\beta_i^{(k_i)}|. \tag{21}$$

Lemma 3.2 (Local interaction estimate). *If the amplitude of incoming waves β is sufficiently small, then it follows that*

$$\pi' = \pi + O(1)A(\mathrm{P}), \quad \beta_i' = O(1)A(\mathrm{P}), \ i = 1, 2 \tag{22}$$

Here $O(1)$ depends only the two constant states connected by the incoming waves and the system.

Let J be approximate space-like curve and $W(J)$ the collection of (the approximation of) rarefaction waves and shock waves crossing J. We denote by $W^-(J)$ the set of shock waves crossing J. Also $W_A(J)$ is the set of these waves crossing J and approaching the phase boundary. The single phase boundary crossing J is denoted by $\pi(J)$. We set

$$L(J) = \sum_{\alpha \in W(J)} |\alpha|, \quad L_A(J) = \sum_{\alpha \in W_A(J)} |\alpha|.$$

We define the interaction potentials for the waves of the same family as the following.

$$
\begin{aligned}
Q_s(J) &= \sum \{|\alpha|^3; \ \alpha \in W^-(J)\} \\
&\quad + 4\sum \{|\beta\gamma||\beta + \gamma|; \beta, \gamma \in W^-(J), \ \beta \ne \gamma \text{ and of the same family}\} \\
&\quad + 8\sum \{|\epsilon\eta\zeta|; \epsilon, \eta, \zeta \in W^-(J), \text{ distinct and of the same family}\}, \\
Q_a(J) &= \sum \{|\alpha\beta|; \ \alpha \in W_A(J), \beta \in W(J) \text{ and } \alpha \ne \beta\}, \\
Q_{aa}(J) &= \sum \{|\alpha\beta|; \ \alpha, \beta \in W_A(J)\}.
\end{aligned}
$$

In order to estimate the amount of interaction produced by the waves of different family, we define another interaction potentials:

$$
\begin{aligned}
Q_d(J) &= \sum \{|\alpha\beta|; \ \alpha, \beta \in W(J) \text{ and } \alpha \text{ is any } i\text{-wave} \\
&\qquad\qquad \text{lying toward the left of a } j\text{-wave } \beta \text{ with } i > j\}, \\
Q_{da}(J) &= \sum \{|\alpha\beta|; \ \alpha, \beta \in W_A(J) \text{ and } \alpha \ne \beta\}.
\end{aligned}
$$

We set finally

$$Q(J) = Q_s(J) + KQ_a(J) + K^2 Q_{aa}(J) + Q_d(J) + KQ_{da}(J)$$

Lemma 3.3 (Global interaction estimates). *Let $Q(\mathrm{P})$ be the amount of interaction at P defined by (17) and $A(\mathrm{P})$ the amount of waves incoming to P defined by (21).*

Then it follows that

$$Q(\mathrm{P}) \;\leq\; Q(J_1) - Q(J_2) + M(1 + K + K^2)L(J_1)Q(\mathrm{P}), \tag{23}$$
$$A(\mathrm{P}) \;\leq\; L_A(J_1) - L_A(J_2) + MQ(\mathrm{P}). \tag{24}$$

Here M stands for O(1) in Proposition 3.1.

If $L(O) \leq \frac{1}{2}(1 + K + K^2)$, then it follows by using this lemma that the total amount of interaction and that of waves approaching to the phase boundary are uniformly bounded i.e. quantities $\sum_P Q(\mathrm{P})$ and $\sum_P A(\mathrm{P})$ are uniformly bounded. Moreover, $L(J)$ also has a uniform bound.

4. Admissibility via Glimm-Lax Theory

In order to prove that the solution obtained in the previous section is *admissible*, we study the behavior of the solution at the phase boundary. By the same argument in Glimm-Lax [5], an *approximate i-characteristic curve* issuing from (x_h, t_h) are defined in the approximate solutions just constructed by the front-tracking alternative.

Let Λ^h be a closed region in $R \times R_+$ surrounded by an approximate i-characteristic curve χ^h and segments joining points of interaction. We use the following notation; $E(\chi^h)$ the total amount of shock waves entering χ^h, $X^{\pm}(\chi^h)$ the total amount of rarefactions and shock waves crossing χ^h, $Q(\Lambda^h) = \sum_{P \in \Lambda^h} Q(\mathrm{P})$, $C(\Lambda^h) = \sum_{P \in \Lambda^h} C(\mathrm{P})$ where $C(\mathrm{P})$ is the *cancellation* of incoming waves defined by $C_i(\mathrm{P}) = \frac{1}{2}(\sum_{k:i_k=i} |\alpha_{i_k}^{(k)}| - |\sum_{k:i_k=i} \alpha_{i_k}^{(k)}|)$. It follows from the global interaction estimates that, as $h \to 0$, (x_h, t_h) tends to (x_0, t_0) and (a subsequence of) these approximate characteristic curves converges uniformly to a Lipschitz function $\chi_i(t)$ on any bounded interval of time; $\chi_i(t)$ is called a *generalized i-characteristic curve* issuing from (x_0, t_0). Moreover, the derivatives converge pointwisely

$$\dot{\chi}_i(t) = \lim_{h \to 0} \dot{\chi}_i^h(t). \tag{25}$$

with exception of a certain countable set of values of t. We can define a *global* phase boundary issuing from $(0,0)$ in the same manner as those generalized characteristic curves. We first show that

Proposition 4.1. *If the total variation of the initial perturbation (12) is sufficiently small, then the single phase boundary is subsonic.*

Let $(\chi(t_0), t_0)$ be a point on the phase boundary such that there exists a neighborhood Δ of $(\chi(t_0), t_0)$ and a sequence $(\chi^h(t_h), t_h) \in \Delta$ converging to $(\chi(t_0), t_0)$ and $Q(\Delta)$, $C(\Delta)$, $E(\chi^h \cap \Delta)$, $X^{\pm}(\chi^h \cap \Delta)$ are arbitrarily small; we can see that such neighborhood exists except countable t. Since the phase boundary is subsonic and almost stationary, all the waves in Δ are crossing χ^h and hence the amount of waves are arbitrarily small. Then we find that the limits (13) exist and the relation (14) holds.

5. Large time stability

Let $\chi_L = \{(x_L(t), t); \ t \geq 0\}$ and $\chi_R = \{(x_R(t), t); \ t \geq 0\}$ be generalized 1 and 2-characteristic curves issuing from $(-M, 0)$ and $(M, 0)$ respectively such that $U(x, t) = U_m$ for $x < x_L(t)$ and $U(x, t) = U_m^*$ for $x > x_R(t)$. Let $X_j^+(t)$ and $X_j^-(t)$ denote respectively the amount of j-rarefaction and j-shock at time t. The generalized j-characteristic curves through $\{(x_L(t), t)\}$ and $\{(x_R(t), t)\}$, respectively, are denoted by $\chi_j^1(t)$ and $\chi_j^2(t)$ where χ_j^1 lies to the left of χ_j^2. Since the phase boundary is subsonic, there exists $t_1 > t$ such that $\chi_2^1(t)$ and $\chi_1^2(t)$ enter the phase boundary χ before the time t_1 and $t_1 = O(1)t$. For $t \geq t_1$, we define $\chi_1^2(t)$ to be the 1-generalized characteristic curve issuing from $(\chi(t_1), t_1)$ and $\chi_2^1(t)$ the 2-generalized characteristic curve issuing from $(\chi(t_1), t_1)$; if these curves start in rarefaction waves, $\chi_1^2(t)$ ($\chi_2^1(t)$ respectively) is so defined as to run along the right (left, respectively) edge of the rarefaction wave. We set $t_1 = t_1(t)$ and $t^* = t_1(t_1(t))$. Clearly there exists a constant C ($C > 1$) depending only on the system and s_0 such that t^* satisfies

$$t^* = Ct. \tag{26}$$

For each $t > t_1(0)$, $s \geq t_1(t)$ and $j = 1, 2$, we set

$$
\begin{aligned}
\Lambda_j(t) &= \text{region between } \chi_j^1(t) \text{ and } \chi_j^2(t), \\
\widetilde{X}_j(s; t) &= \text{amount of } j\text{-waves outside of } \Lambda_j(t) \text{ at time } s, \\
Q(t, s) &= \text{amount of interaction between } t < t' < s, \\
A(t, s) &= \text{amount of approaching waves between } t < t' < s, \\
H(t, s) &= \text{amount of } i\text{-wave } (i \neq j) \text{ crossing } \chi_j^1(t) \text{ and } \chi_j^2(t) \\
&\quad \text{ between } t < t' < s.
\end{aligned}
$$

We further set

$$X(t) = \sum_{j=1,2} (X_j^+(t) + |X_j^-(t)|), \quad Q(t) = Q(t, \infty).$$

Lemma 5.1. *There exist bounds $O(1)$ such that for any $s \geq t_1(t)$,*

$$
\begin{aligned}
\widetilde{X}_j(s; t), \ H(t, s) &= O(1)Q(t) \quad \text{for} \quad j = 1, 2 & (27) \\
\pi(s') - \pi(s), \ A(s, s') &= O(1)Q(t) \quad \text{for} \quad s' \geq s & (28) \\
X(t_1) &\leq O(1)X(t) + O(1)Q(t) & (29) \\
Q(t_1) &= O(1)X(t_1)^3 + O(1)Q(t)X(t) & (30)
\end{aligned}
$$

Lemma 5.2. *There exist bounds $O(1)$ such that for any $s \geq t^*$ and for any (x_-, t_-), (x_+, t_+) in the wake region such that $x_- < \chi(t_-)$ and $x_+ > \chi(t_+)$*

$$
\begin{aligned}
Q(s) &= O(1)X(t)^3 & (31) \\
|U(x_-, t_-) - U_m|, \ |U(x_+, t_+) - U_m^*| &= O(1)X(t)^3 & (32)
\end{aligned}
$$

With the aid of the preceding lemmas, we can prove the large time stability theorem following the argument in Liu [13] and Asakura [2]. We show that

$$X(t) = O(1)t^{-\frac{1}{2}} \quad \text{and} \quad Q(t) = O(1)t^{-\frac{3}{2}}, \tag{33}$$

which imply Theorem 1.3.

References

[1] R. Abeyaratne and J. K. Knowles, Kinetic relations and the propagation of phase boundaries in solids, Arch. Rational Mech. Anal., **114** (1991), 119–154.

[2] F. Asakura, Asymptotic stability of solutions with a single strong shock wave for hyperbolic systems of conservation laws, Japan J. Industrial and Applied Math., **11** (2), (1994) 225–244.

[3] F. Asakura, Large Time Stability of Phase Boundaries, Preprint.

[4] A. Bressan, Global solutions of systems of conservation laws by wave-front tracking, J. Math. Analysis and App., **170** (1992), 414–432.

[5] I.-L. Chern, Stability theorem and truncation error analysis for the Glimm scheme and for a front tracking method for flows with strong discontinuities, Comm. Pure Appl. Math., **42** (1989), 815–844.

[6] C. M. Dafermos, The entropy rate admissibility criterion for solutions of hyperbolic conservation laws, J. Differential Equations, **14** (1973), 202–212.

[7] J. Glimm, Solutions in the large for nonlinear hyperbolic systems of equations, Comm. Pure Appl. Math., **18** (1965), 697–715.

[8] J. Glimm – P. D. Lax, *Decay of solutions of systems of nonlinear hyperbolic conservation laws*, Amer. Math. Soc. Memoir, No. 101. A.M.S. Providence, **1970**.

[9] H. Hattori, The Riemann problem for a van der Waals fluid with entropy rate admissible criterion – Isothermal Case, Arch. Rational Mech. Anal., **92** (1986), 246–263.

[10] P. D. Lax, Hyperbolic systems of conservation laws II, Comm. Pure Appl. Math., **10** (1957), 537–566.

[11] P. D. Lax, Shock waves and entropy, *E. Zarantonello (ed.), Contributions to nonlinear Functional Analysis*, Academic Press, New York, **1971**, 603–634.

[12] P. LeFloch, Propagating Phase boundaries: formulation of the problem and existence via the Glimm method, Arch. Rational Mech. Anal., **123** (1993), 153–197.

[13] T. P. Liu, Decay to N-waves of solutions of general systems of nonlinear hyperbolic conservation laws. Comm. Pure Appl. Math., **30** (1977), 585–610.

[14] N. H. Risebro, A front-tracking alternative to the random choice method, Proc. Amer. Math. Soc., **117** (1993), 1125–1139.

Faculty of Engineering,
Osaka Electro-Communication University,
Neyagawa, Osaka 572, Japan
E-mail address: asakura@isc.osakac.ac.jp

International Series of Numerical Mathematics
Vol. 129, © 1999 Birkhäuser Verlag Basel/Switzerland

Convergence of Meshless Methods for Conservation Laws Applications to Euler equations

B. Ben Moussa, N. Lanson and J.P. Vila

Abstract. This paper is devoted to analyse new meshless methods. They generalize classical weighted particle methods for conservation laws. We prove that they can be both conservative and consistent. We obtain convergence of the methods in scalar case with the only requirement that the ratio of the smoothing length (or size of the cut-off) to the characteristic size of the mesh be bounded. Applications for Euler equations are proposed.

1. Introduction

Meshless methods have been recently developed for approximation of transient problems for conservation laws in various field of applications such as:

- compressible fluid dynamic (Smooth Particle Hydrodynamics, see [7])
- impact problems and crash simulations (SPH, Kernel Reproducing Methods, Diffuse Approximation)

The recent review in [6] shows the ability of these methods to handle with complex situations. However convergence properties and their links with conservativity are not yet well understood. Even if we restrict ourselves to particle weighted like methods we have to choose among different versions of smoothing length ([8]), kernels ... Renormalization is also a recent tool introduced in ([12]) and ([9]), and seems to give accurate results. According to the literature (see for instance [5]) these tools, although very performing and attractive for applications seem to make the global conservativity property of the original method disappear.

The third author has introduced in [2] and [1] a new class of meshless approximation for conservation laws: the renormalized Meshless Derivative (RMD) which turns out to give accurate approximation of derivative under less restrictive conditions than standard particle weighted methods, in particular the ratio of the size of the mesh to the smoothing length just needs to be $\mathcal{O}(1)$. The convergence is analysed in [1] in the case of linear symmetric hyperbolic systems, and continues in time.

In this paper we investigate the application of these tools to nonlinear conservation laws: we extend the convergence results of [4] to the case of (RMD), and we also apply the method to Euler equations of compressible gas dynamic.

2. Weak formulation and Renormalized Meshless derivative

Let v a regular vector field in $\mathbb{R}^d$. We consider the following model PDE in conservation form:

$$L_v(\Phi) + \text{div } F(x, t, \Phi) = S \tag{1}$$

where F is the flux vector ($\in \mathbb{R}^d$) of the conservation law and L_v is the transport operator given by: L_v :

$$\Phi \longrightarrow L_v(\Phi) = \frac{\partial \Phi}{\partial t} + \sum_{l=1,d} \frac{\partial}{\partial x^l}(v^l \Phi)$$

To get a particle approximation of the equation (1), let us take a set of moving particles $(x_i(t), w_i(t))_{i \in P}$, indexed by $i \in P$, where $x_i(t)$ is the position of the particle and $w_i(t)$ its weight. We classically move the particles along the characteristic curves of the field v and also modify the weights in order to take account of deformations due to the field v:

$$(i) \quad \frac{d}{dt} x_i = v(x_i, t) \qquad (ii) \quad \frac{d}{dt} w_i = div(v(x_i, t)) \, w_i \tag{2}$$

The accuracy of the approximation is connected with the quadrature formula over $\mathbb{R}^d$, given by the particles $(x_i(t), w_i(t))_{i \in P}$:

$$\int_{\mathbb{R}^d} g(x) dx \approx \sum_{j \in P} w_j(t) g(x_j(t))$$

This formula is accurate for any $t > 0$ as soon as it is accurate initially and the particles and weights move according to (2) (see [13]). We provide the space with the discrete scalar product

$$(\varphi, \Psi)_h := \sum_{i \in P} w_i \varphi_i . \Psi_i$$

which is clearly an approximation of the scalar product in $L^2(\mathbb{R}^d)^m$.

We also introduce the linear operator $D_{h,S}$ – we do not need at this stage to specify it as the RMD derivative – which is supposed to approximate strongly the derivative, i.e. for any φ regular enough

$$\sup_{i \in P} \|D_{h,S}\varphi_i - D\varphi_i\| \to 0 \text{ as } h \to 0$$

and let us define $-D_{h,S}^*$ as the adjoint operator of $D_{h,S}$. We thus have

$$(D_{h,S}\varphi, \Psi)_h = -(\varphi, D_{h,S}^*\Psi)_h \tag{3}$$

The weak formulation of (1) is defined by: $\forall \varphi \in \mathbf{C}_0^2(I\!\!R^d \times I\!\!R^{+,*})$

$$\int_{I\!\!R^d \times I\!\!R^+} (\Phi.L_v^*(\varphi) + F(x,t,\Phi).\nabla(\varphi) + S.\varphi)\, dx dt = 0$$

where $-L_v^*$ is the adjoint operator of L_v defined as:

$$L_v^*(\varphi) = \frac{\partial \varphi}{\partial t} + \sum_{l=1,d} v^l \frac{\partial \varphi}{\partial x^l}$$

A discrete version of (1) is provided by just replacing the integration over $I\!\!R^d$ by the discrete scalar product $(.,.)_h$ and the derivative $\nabla(\varphi)$ by its approximation $D_{h,S}\varphi$: $\forall \varphi \in \left[\mathbf{C}_0^2(I\!\!R^d \times I\!\!R^{+,*})\right]^m$

$$\int_{I\!\!R^+} \left[(\Phi^h, L_v^*(\varphi))_h + \sum_{\alpha=1,\dots,d} (F^\alpha(\Phi^h), D_{h,S}^\alpha \varphi)_h + (S + R_h(\Phi^h), \varphi)_h \right] dt = 0 \quad (4)$$

$R_h(\Phi^h)$ is an additional term which represents for example the artificial viscosity. Let us remark that, for any φ sufficiently regular

$$\frac{d}{dt}(\varphi_i) = \frac{d}{dt}(\varphi(x_i(t),t) = L_v^*(\varphi)_i \qquad \frac{d}{dt}(w_i\varphi_i) = w_i L_v(\varphi)_i$$

Making an integration by part with respect to t, we get easily that (4) is true if and only if:

$$\begin{cases} (i) & \dfrac{d}{dt}x_i = v(x_i,t), \qquad\qquad x_i(0) = \xi_i \\[2mm] (ii) & \dfrac{d}{dt}w_i = \mathrm{div}\,(v(x_i,t))\,w_i \qquad w_i(0) = \omega_i \\[2mm] (iii) & \dfrac{d}{dt}(w_i\Phi_i^h) + w_i \displaystyle\sum_{\alpha=1,\dots,d} D_{h,S}^{\alpha,*}(F^\alpha)_i = w_i\left(S_i + R_h(\Phi^h)_i\right) \end{cases} \quad (5)$$

where F_i stands for $F(x_i,t,\Phi_i)$, and $\Phi_j^h(0) = \Phi^0(\xi_j)$.

In a general framework the RMD of a function is given by

$$D_{h,S}^\alpha f(x) = \sum_{j \in P} w_j(t)\,(f(x_j(t)) - f(x))\,A^\alpha(x,x_j(t))$$

We only discuss here the case of RMD based upon particle weighted approximation associated to a cut off ζ with compact support of characteristic size ε_0. The discrete operators are given by:

$$D_{h,S}f_i = \sum_{j \in P} w_j(f_j - f_i)A_{ij} \qquad D_{h,S}^* f_i = \sum_{j \in P} w_j(f_i + f_j)A_{ij} \quad (6)$$

where $A_{ij} := B_{ij}.\nabla \zeta_{ij}$ and the renormalization matrix $B_{ij} := \frac{1}{2}(B_i + B_j)$ is computed according to

$$E_i^{\alpha\beta} = \sum_{j \in P} w_j(x_j^\beta - x_i^\beta)\,\partial^\alpha \zeta_{ij} \qquad\qquad B_i = (E_i)^{-1} \quad (7)$$

It reduces to the standard particle weighted approximation of derivatives if we take $B(x) = 1$. The kernel is taken spherical in order to have $\nabla \zeta_{ij} = -\nabla \zeta_{ji}$, and consequently $A_{ij} = -A_{ji}$. It can be proved (see [2] and [1]) the

Proposition 2.1. *Let us suppose that the ratio $\frac{h}{\varepsilon}$ is $\mathcal{O}(1)$, then the matrix $E(x)$ is invertible, $\|B(x)\|$ is uniformly bounded, and there exists a positive constant $C(T)$ such that for any $\varphi \in W^{2,\infty}(\mathbb{R}^d)$*

$$\|D_{h,S}\varphi(x) - D\varphi(x)\| \leq C\varepsilon_0\|B(x)\|\|D^2\varphi\|_\infty$$

Let us know detail how we can extend the WP schemes which mimic SPH schemes, by using the RMD. We mainly need to precise 2 points:

: the way to introduce artificial viscosity
: the time discretization

Let us first write the method of lines for (14) with no artificial viscosity. We use the notation $n_{ij} = \frac{A_{ij}}{\|A_{ij}\|}$, equation (14) (iii) reads

$$\frac{d}{dt}(w_i \Phi_i) + w_i \sum_{\alpha=1,\ldots,d} (F_i + F_j).n_{ij}\|A_{ij}\| = w_i S_i$$

which introduces naturally the conservation law in the direction n_{ij}:

$$\frac{\partial}{\partial t}(\Phi) + \frac{\partial}{\partial x}(F(x_{ij}, t, \Phi).n_{ij}) = 0 \tag{8}$$

Therefore it is natural to introduce a 1-dimensional finite difference scheme in conservation form associated to (8), which brings a sufficient numerical viscosity. Such a scheme consists in replacing the centered approximation $(F_i + F_j).n_{ij}$ by the numerical flux of a Finite Difference scheme $2g(n_{ij}, \Phi_i, \Phi_j)$, which is required to satisfy:

$$\begin{aligned}(i) \quad & g(n(x), u, u) = F(x, t, u).n(x) \\ (ii) \quad & g(n, u, v) = -g(-n, v, u)\end{aligned}$$

The numerical viscosity $Q(n, u, v)$ is classically defined in the scalar case (i.e. $\Phi \in \mathbb{R}$) as:

$$Q(n(x), u, v) = \frac{F(x, t, u).n(x) - 2g(n(x), u, v) + F(x, t, v).n(x)}{v - u}$$

In the case of an explicit particle scheme the connection with finite 1D difference schemes is rather precise. As in [3], we can establish in a way similar to the one used for finite volume schemes that the particle scheme is a convex combination of 1D finite difference schemes at which we add a controlled error term.

There is a lot of numerical flux well suited for such upwinding, among them we can quote the Lax Friedrichs and the Godunov schemes which belong to the widest class of E scheme – see [11] and [3] for detailed dissipation properties. By introducing the numerical viscosity our numerical scheme, which consists in finding functions $t \in \mathbb{R}^+ \longrightarrow \Phi_i(t) \in \mathbb{R}$, $i \in P$ solutions of the differential system:

$$\frac{d}{dt}(w_i\Phi_i) + w_i \sum_{j \in P} w_l 2g(n_{ij}, \Phi_i, \Phi_j)\|A_{ij}\| = w_i S_i \qquad (9)$$

also reads as:

$$\frac{d}{dt}(w_i\Phi_i) + w_i \sum_{j \in P} w_j \left[(F(x_{ij}, t, \Phi_i) + F(x_{ij}, t, \Phi_j)).n_{ij} + \Pi_{ij}\right] \|A_{ij}\| = w_i S_i$$

where $\Pi_{ij} = Q(n_{ij}, \Phi_i, \Phi_j)(\Phi_i - \Phi_j)$. This form is very closed to the one classically used in SPH literature (see e.g. [7]).

The time discretization is performed via Euler forward scheme:

$$w_i^{n+1}\Phi_i^{n+1} = w_i^n \left(\Phi_i^n - \Delta t \sum_{j \in P} w_l^n 2g(n_{ij}, \Phi_i^n, \Phi_j^n)\|A_{ij}^n\| + \Delta t S_i^n\right) \qquad (10)$$

We can prove the following Theorem:

Theorem 2.2. *Let* Φ *the unique entropic solution of (1), we suppose that the initial condition belongs to* $BV \cap L^\infty$ *and that the numerical flux* g *belong to the class of E-flux, then the numerical solution* Φ^h *given by the scheme (10) converges towards* Φ *in* $L^\infty(0, T; L^1_{loc}(\mathbb{R}^d))$.

2.0.1. SKETCH OF THE PROOF For simplicity we take $S = 0$ and we introduce the following notations

$$F_{i,ij}^n := F(x_{ij}, t^n, \Phi_i^n)$$

$$g_{ij}^n := g(n_{ij}, \Phi_i^n, \Phi_j^n) = \tfrac{1}{2}(F_{i,ij}^n + F_{j,ij}^n).n_{ij} + Q_{ij}^n(\Phi_i^n - \Phi_j^n)$$

We also note the following properties of the RMD

$$\|A_{ij}^n\| \leq \frac{C}{\varepsilon_0^{d+1}} \qquad \sum_{j \in P} w_j^n \|A_{ij}^n\| \leq \frac{C}{\varepsilon_0} \qquad \left\|\sum_{j \in P} w_j^n A_{ij}^n\right\| \leq C \qquad (11)$$

Taking a test function φ we can easily prove that

$$\sum_n \Delta t \left[\left(\Phi^{h,n}, \frac{\varphi^n - \varphi^{n-1}}{\Delta t}\right)_h + (F(\Phi^{h,n}), D_{h,S}\varphi)_h + (R_h(\Phi^{h,n}), \varphi)_h\right] dt = 0$$

$$(12)$$

where

$$R_h(\Phi^{h,n}) := R_{h,DF}(\Phi^{h,n}) + R_{h,Q}(\Phi^{h,n})$$

$$R_{h,DF}(\Phi^{h,n})_i := \sum_{j \in P} w_j^n (F_{i,ij}^n + F_{j,ij}^n - F_i^n - F_j^n).A_{ij}^n$$

$$R_{h,Q}(\Phi^{h,n})_i := \sum_{j \in P} w_j^n Q_{ij}^n (\Phi_i^n - \Phi_j^n)\|A_{ij}^n\|$$

The numerical flux is taken from a E-scheme (e.g. the Godunov scheme), then with the help of (11), it can be proved as in [4] L^∞ and weak estimates on the derivative, for $n\Delta t \le T$:

$$\left\|\overline{\Phi}^n\right\|_\infty \le C(T)$$

$$\sum_{\substack{i,j\in P \\ n\Delta t\le T}} \Delta t w_i^n w_j^n \left(Q_{ij}^n\right)^2 (\Phi_i^n - \Phi_j^n)^2 \|A_{ij}^n\| \le C(T)$$

Making a discrete integration by part we get

$$\left(R_{h,DF}(\Phi^{h,n}), \varphi\right)_h = \sum_{i,j\in P} w_i^n w_j^n (\varphi_j^n - \varphi_i^n)(F_{j,ij}^n - F_j^n).A_{ij}^n$$

which leads to the bound

$$\left|(R_h(\Phi^{h,n}), \varphi)_h\right| \le \varepsilon_0^2 \|F_x\|_\infty \|\varphi_x\|_\infty \sum_{i,j\in P} w_i^n w_j^n \|A_{ij}^n\| \le C\varepsilon_0 \|F_x\|_\infty \|\varphi_x\|_\infty$$

We establish similarly that

$$\left(R_{h,Q}(\Phi^{h,n}), \varphi\right)_h = \frac{1}{2} \sum_{i,j\in P} w_i w_j (\varphi_i - \varphi_j) Q_{ij}^n (\Phi_i^n - \Phi_j^n) \|A_{ij}^n\|$$

We then get

$$\left|\sum_n \Delta t \left(R_{h,Q}(\Phi^{h,n}), \varphi\right)_h\right| \le$$

$$\left(\sum_{\substack{i,j\in P \\ n\Delta t\le T}} \Delta t w_i^n w_j^n \left(Q_{ij}^n\right)^2 (\Phi_i^n - \Phi_j^n)^2 \|A_{ij}^n\|\right)^{\frac{1}{2}} \times$$

$$\left(\sum_{\substack{i,j\in P \\ n\Delta t\le T}} \Delta t w_i^n w_j^n (\varphi_i^n - \varphi_j^n)^2 \|A_{ij}^n\|\right)^{\frac{1}{2}}$$

The second term in the right-hand side is bounded by $meas(spt(\varphi))\|\varphi_x\|_\infty \sqrt{\varepsilon_0}$. We thus conclude that $R_h(\Phi^{h,n}) \to 0$.

That gives the basis for use of Diperna uniqueness result for m-v solution of the conservation law or establish directly an error estimate between the exact solution and the approximate one by use of Kruskov entropy (see e.g. the paper of Perthame in this book).

In practise some important new difficulties arise. We first need to prove the analogous of (12) for the Kruzkov entropy. At the continuous level it reads:

$$L_v(|\Phi - c|) + \text{div } [\text{sgn } (\Phi - c)(F(x,t,\Phi) - F(x,t,c)]$$

$$+\text{sgn}(\Phi - c) [\text{div}_x(F(x,t,c) - c\text{div}_x v] \le 0$$

At the discrete level we only prove

$$L_v(|\Phi^h - c|) + \text{div} \ [\text{sgn}(\Phi^h - c)(F(x, t, \Phi^h) - F(x, t, c)]$$
$$+ \text{sgn}(\Phi^h - c) [\text{div}_{h,x}(F(x, t, c)) - c\text{div}_x v] \leq \nu_h$$

where ν_h is a measure converging towards 0, in a suitable sense and $\text{div}_{h,x}(F(x,t,c))$ an approximation of $\text{div}_x(F(x, t, c))$, uniformly (in h) Lipchitz with respect to c, but only weakly converging towards $\text{div}_x(F(x, t, c))$. That makes difficult a direct application of comparison tools based upon Kruskov-Kutznetsov technics. To overcome the difficulty, we note that if Φ is BV, we also have – by studying carefully the weak convergence of $\text{div}_{h,x}(F(x, t, c))$ and regularizing the sgn function:

$$L_v(|\Phi - c|) + \text{div} \ [\text{sgn} \ (\Phi - c)(F(x, t, \Phi) - F(x, t, c)]$$
$$+ \text{sgn}(\Phi - c) [\text{div}_{h,x}(F(x, t, c) - c\text{div}_x v] \leq \nu_{0,h}$$

where $\nu_{0,h}$ is a measure converging towards 0. With this in mind, comparison technics with Kruskov entropy lead to an error estimate in L^1_{loc}.

3. Application to Euler equations

Thanks to the set of approximation rules described in the previous section we are now able to design the SPH approximation of Euler equations of a compressible fluid. We start by the simplest case of a single compressible gas. Such a fluid satisfies the following Euler system of equations:

$$L_v(\Phi) + \sum_{l=1,d} \frac{\partial}{\partial x^l}(F^l(\Phi)) = 0 \qquad (13)$$

where Φ, the vector of conservative variables and the fluxes F^l are given by:

$$(i) \quad \Phi = \begin{pmatrix} \rho \\ \rho v^1 \\ \rho v^2 \\ E \end{pmatrix} \quad (ii) \quad F^1(\Phi) = \begin{pmatrix} 0 \\ p \\ 0 \\ v^1 p \end{pmatrix} \quad (iii) \quad F^2(\Phi) = \begin{pmatrix} 0 \\ 0 \\ p \\ v^2 p \end{pmatrix}$$

We have supposed for simplicity that the dimension of space is $d = 2$. The equation of state of the fluid gives the pressure as a function $p(\rho, u)$ of the density and the internal energy. The total energy E is defined by:

$$E = \rho(u + \frac{1}{2}\|v\|^2)$$

This system has the same form than the model PDE (1). It follows from (5), the system of ordinary differential equations defined for $i \in P$ by:

$$
\left\{
\begin{array}{lll}
(i) & \dfrac{d}{dt}\, x_i = v(x_i, t), & x_i(0) = \xi_i \\[2ex]
(ii) & \dfrac{d}{dt}\, w_i = \mathrm{div}(v(x_i, t))\, w_i & w_i(0) = w_i \\[2ex]
(iii)' & \dfrac{d}{dt}\, w_i \rho_i = \dfrac{d}{dt}\, m_i = 0 \\[2ex]
(iii)'' & \dfrac{d}{dt}\, v_i = -\displaystyle\sum_{j \in P} \dfrac{m_j}{\rho_i \rho_j}\,(p_i + p_j)A_{ij} - \sum_{j \in P} m_j \Pi_{ij} A_{ij} \\[3ex]
(iii)''' & \dfrac{d}{dt}\, u_i = -\displaystyle\sum_{j \in P} \dfrac{m_j}{\rho_i \rho_j}\, p_j(v_j - v_i)A_{ij} - \dfrac{1}{2}\sum_{j \in P} \Pi_{ij}(v_j - v_i)A_{ij} \\[3ex]
(iv) & \Phi_j(0) = \Phi^0(\xi_j)
\end{array}
\right.
\tag{14}
$$

Equation $(iii)'$ gives $m_i =$ cste and we use it in $(iii)''$ and $(iii)'''$.
In $(iii)''$ and $(iii)'''$ equations Π_{ij} is the artificial viscosity term. It is defined as:

$$
\Pi_{ij} = \left\{
\begin{array}{ll}
\dfrac{\mu_{ij}(\beta \mu_{ij} - \alpha \bar{c})}{1/2(\rho_i + \rho_j)} & \text{if } (v_i - v_j).(x_i - x_j) < 0, \\[2ex]
0 & \text{elsewhere}
\end{array}
\right.
\tag{15}
$$

where $\bar{c}$ is the mean sound velocity and λ_{ij} an approximation of $l\,\mathrm{div}v$ given by:

$$
\mu_{ij} = \frac{\varepsilon(v_i - v_j).(x_i - x_j)}{|x_i - x_j|^2 + \kappa \varepsilon^2}
$$

The equations concerning the density ρ and the size of the cut-off ε are established by using the approximation of $\mathrm{div}v$: $(\mathrm{div}v)_i = \sum_{j \in P} w_j(v_j - v_i)A_{ij}$ and $\varepsilon(x, y) = \frac{\varepsilon(x) + \varepsilon(y)}{2}$. We get:

$$
\frac{d\rho_i}{dt} = -\rho_i \sum_{j \in P} w_j(v_j - v_i)A_{ij} \quad \text{and} \quad \frac{d\varepsilon_i}{dt} = \frac{1}{d}\,\varepsilon_i \sum_{j \in P} w_j(v_j - v_i)A_{ij}
$$

where d is the dimension of space.

4. Numerical results

We first present numerical results on a standard SOD 1D shock tube. The Figure 4 shows the behavior of the velocity and density of two ideal gases with different initial condition for the density. This shock tube has been computed with the ratio $\frac{h}{\varepsilon} = 0.6$ and with renormalized and non-renormalized ($B_i = 1$) methods. Note that the results of the standard scheme do not cope with Rankine Hugoniot condition. The results of the standard scheme are very sensitive to the value of $\frac{h}{\varepsilon}$. Looking more carefully to this problem reveals that practitioners often use values of $\frac{h}{\varepsilon}$ such that on a regular grid $D_{h,s}x = 1$, and $B = Id$. This insures that the effective

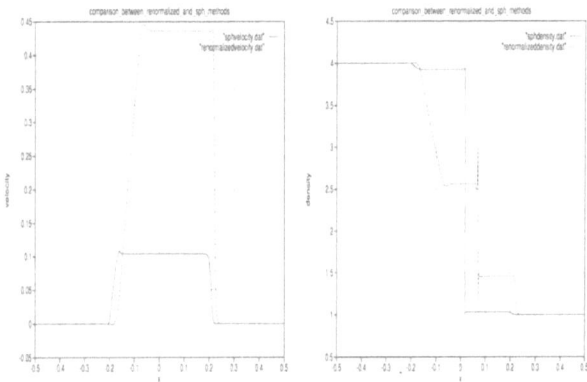

FIGURE 1. Density and velocity profiles

computations are not so far from the desired values. Nevertheless a precise study shows that when we make $h \to 0$ standard method does not converge towards a solution satisfying exactly consistency (even on 1D SOD problem). The new method turns out to be more robust than standard methods, and also less expensive since we can use small values of the ratio $\frac{h}{\varepsilon}$. This makes decrease the number of neighbors (a factor 2 or 3 is possible) and the cost of the method also decrease with a similar ratio. This is particularly true in situation with complex physics since the additional cost due to the computation of Renormalization matrices is no more than the computation of an additional physical unknown.

To show the efficiency of the WRMS in more complex situation, for example in the field of high velocity impacts, let us give an other example where computation of free surfaces is important. We can see in Figure 2 the impact of a copper ball against an aluminum wall initially at rest. The diameter of the ball is 0.5 cm, the width of the wall is 1 cm and the initial distance between the ball and the wall is 1 cm. The initial velocity of the ball is $2.10^5 \times \begin{pmatrix} 1 \\ -1 \end{pmatrix} cm.s^{-1}$. In the l.h.s. of Figure 2 the computation is performed with standard method ($B_i = 1$), in the r.h.s. of Figure 2 the computation is performed with WRMS. On the one hand, the renormalized computation gives smoother shape of the free surface, note also that it also removes the little gap at the interface between copper and aluminum. On the other hand it gives also a lower cost due to the decrease of the number of neighbors and provides efficient results.

References

[1] J.P. Vila, *Meshless Methods for Conservation Laws,* submitted Math. of Comp.

[2] J.P. Vila, *Weighted Particle methods and Smooth Particle Hydrodynamics,* to appear M3AS.

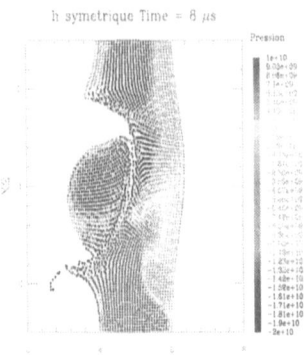

FIGURE 2. Standard Method (left) – Renormalized(right)

[3] S. Benharbit, A. Chalabi and J.P. Vila, *Numerical viscosity, and convergence of finite volume Methods for conservation laws with Boundary conditions*, SIAM J. Numer. Anal., **32 (3)** (1995), 775–796.

[4] B. Ben Moussa and J.P. Vila, *Convergence of SPH method for scalar nonlinear conservation laws*, to appear in SIAM Numerical Analysis.

[5] G.V. Bicknell, *The equations of motion of particles in smoothed particle hydrodynamics*, SIAM J. Sci. Stat. Comput., **12 (5)** (1991), 1198–1206.

[6] *Special issue Gridless methods*, Comput. methods Appl. Mech. Engrg., **139** (1996).

[7] R.A. Gingold and J.J. Monaghan, *Shock simulation by the particle method S.P.H.*, J.C.P.**52** (1983), 374–389.

[8] L. Hernquist and N. Katz, *TREESPH: a unification of SPH with the hierarchical tree method*, The Astr. J. S.S., **70** (1989), 419–446.

[9] G.R. Johnson and S.R. Beissel, *Normalized Smoothing functions for impact computations*, Int. Jour. Num. Methods Eng., (1996).

[10] P.D. Lax and B. Wendroff, *Systems of Conservation Laws*, CPAM, **13** (1960), 217–237.

[11] S. Osher, *Riemann solvers, the entropy condition and difference approximations*, Siam Jour.Num.Anal, **21 (2)** (1984), 217–235.

[12] R.W. Randles and L.D. Libertsky, *Smoothed Particle Hydrodynamics, Some recent improvements and Applications*, Comp. Meth. Appli. Mech. Eng;, **139** (1996), 375–408.

[13] P.A. Raviart, *An analysis of particle methods, Num. method in fluid dynamics*, F. Brezzi ed. Lecture Notes in Math., 1127, Berlin, Springer, **1985**.

Institut National des Sciences Appliquées de Toulouse
Département Génie Mathématique et Modélisation
UMR CNRS 5640
Mathématiques pour l'Industrie et la Physique
F-31077 TOULOUSE CEDEX
E-mail address: `benmou or lanson or vila@gmm.insa-tlse.fr`

International Series of Numerical Mathematics
Vol. 129, © 1999 Birkhäuser Verlag Basel/Switzerland

Multi-dimensional Stability of Propagating Phase Boundaries

Sylvie Benzoni

Abstract. The paper addresses the multi-dimensional stability of propagating phase boundaries in real fluids, from a viewpoint similar to Majda's concerning classical discontinuities. For planar boundaries, a Lopatinski condition is derived and shown to be satisfied in the inviscid case. Some surface waves prevent the uniform Lopatinski condition to be satisfied though. Then it is shown that, for weakly dissipative phase boundaries, the surface waves are stabilized and the uniform Lopatinsky condition is satisfied.

1. Introduction

We are interested in the Euler equations in several dimensions,

$$\partial_t \rho + \nabla \cdot (\rho \, \vec{u}) = 0,$$
$$\partial_t (\rho \, \vec{u}) + \nabla \cdot (\rho \, \vec{u} \otimes \vec{u}) + \nabla p(v) = 0,$$

where ρ denotes the density, $v = 1/\rho$ the specific volume and $\vec{u} \in \mathbb{R}^d$ the velocity of the fluid, endowed with a nonmonotone pressure law $v \mapsto p(v)$. For instance, we may consider a van der Waals isothermal,

$$p(v) = \frac{RT}{v - b} - \frac{a}{v^2},$$

which is, below the critical temperature, increasing in an interval $(v_\star, v^\star)$ and decreasing in $(b, v_\star) \cup (v^\star, +\infty)$. The corresponding system of conservation laws is of mixed type, since the hyperbolic region, $(b, v_\star) \cup (v^\star, +\infty)$, consists of two connected parts, or phases, separated by the so-called spinodal region, $(v_\star, v^\star)$, where the system is elliptic. Very few results are known about the solutions of such systems. However, planar propagating discontinuities of the form:

$$\rho(\vec{x}, t) = \rho_l \text{ and } \vec{u}(\vec{x}, t) = \vec{u_l}, \quad \text{if } \vec{x} \cdot \vec{n} - \sigma t < 0,$$
$$\rho(\vec{x}, t) = \rho_r \text{ and } \vec{u}(\vec{x}, t) = \vec{u_r}, \quad \text{if } \vec{x} \cdot \vec{n} - \sigma t > 0,$$

where $\vec{n}$ is a unit vector, provide simple explicit weak solutions. The "classical" discontinuities are those involving states in the same phase. They consist either of pressure waves, across which the jump of the velocity, $[\vec{u}] = \vec{u_r} - \vec{u_l}$, is parallel to $\vec{n}$, or of vortex sheets, across which we have $[\vec{u}] \cdot \vec{n} = 0$ and $[p] = 0$. The former are associated to the genuinely nonlinear eigenvalues $\lambda_{1,3} = \vec{u} \cdot \vec{n} \pm c$, where

$c := \sqrt{d_\rho p}$ denotes the sound speed, and they satisfy the Lax shock inequalities, whereas the latter are contact discontinuities associated to the eigenvalue $\lambda_2 = \vec{u} \cdot \vec{n}$, of multiplicity $d - 1$. As far as we are concerned, there is another kind of propagating discontinuities, which involve states in two different phases. Although noncharacteristic, if we assume $\vec{u} \cdot \vec{n} \neq \sigma$, they violate the Lax shock inequalities, the number of outgoing modes being equal to the number of incoming modes, i.e.,

$$(\lambda_i^l - \sigma)(\lambda_i^r - \sigma) > 0, \, i = 1, 2, 3.$$

These propagating phase boundaries are thus analogous to undercompressive shocks [3].

The structural stability of all these propagating discontinuities is of great importance for their physical interest. Such a stability amounts to the existence of discontinuous solutions across boundaries close to $\{\vec{x} \cdot \vec{n} = \sigma t\}$, with states that remain close to the original ones on either side. This is a difficult question that was investigated in the 80's by Majda [5] for the classical discontinuities. He transformed the involved Free Boundary problems into – nonstandard – Initial Boundary Value Problems and then based his analysis on a kind of Kreiss-Lopatinsky condition [4]. Roughly speaking, he proved that shock waves were stable whereas contact discontinuities were unstable.

The aim of this work is to provide a similar analysis in the case of propagating phase boundaries. The first difficulty lies in the fact that there are fewer degrees of freedom than for classical discontinuities. As a matter of fact, undercompressive discontinuities have to satisfy the Rankine-Hugoniot conditions as well as an additional jump condition – otherwise they would not even be 1-D stable [3]. Here we shall consider the condition imposed by the macroscopic effects of capillarity and viscosity [6]. The second difficulty is due to the form of this additional condition, which may be written as:

$$E(v_r) - E(v_l) + (v_r - v_l)(p(v_l) + p(v_r))/2 = -I\nu \rho_{l,r}(\vec{u}_{l,r} \cdot \vec{n} - \sigma), \quad (1)$$

where E denotes the free specific energy, defined by $d_v E = -p$, $\nu > 0$ is a nondimensional parameter related to the viscosity and the capillarity coefficients of the fluid, and I is related to the internal structure of the phase boundary. More precisely, we have:

$$I = \int_{-\infty}^{+\infty} v'(\xi)^2 \, d\xi,$$

where $\xi \mapsto v(\xi)$ is the volume component of a *viscous capillary profile* of the phase boundary, that is to say a plane wave of the form

$$\rho(\vec{x}, t) = \rho\left(\frac{\vec{x} \cdot \vec{n} - \sigma t}{\varepsilon}\right), \, \vec{u}(\vec{x}, t) = \vec{u}\left(\frac{\vec{x} \cdot \vec{n} - \sigma t}{\varepsilon}\right),$$

$$\rho(-\infty) = \rho_l, \vec{u}(-\infty) = \vec{u}_l,$$
$$\rho(+\infty) = \rho_r, \vec{u}(+\infty) = \vec{u}_r,$$

solution to the Euler equations supplemented with a capillarity coefficient $\varepsilon^2 > 0$ and a viscosity coefficient $\nu\varepsilon$,

$$\partial_t \rho + \nabla \cdot (\rho\,\vec{u}) = 0,$$
$$\partial_t(\rho\,\vec{u}) + \nabla \cdot (\rho\,\vec{u} \otimes \vec{u}) + \nabla p(v) = \nu\varepsilon\triangle\vec{u} - \varepsilon^2\nabla(\triangle v).$$

Relation (1) cannot be expressed explicitly in terms of the left and right states, except when the viscosity is neglected, i.e., when the right hand side in (1) vanishes.

2. The inviscid case

In this section, we assume $\nu = 0$, which makes the additional jump condition (1) much simpler. Following Majda [5], we reformulate the problem into an Initial Boundary Value Problem, where the boundary conditions come from the whole set of $(d+2)$ jump conditions. After linearization, we derive a generalized Lopatinsky condition. The algebraic version of this condition asks that some $(d+2) \times (d+2)$ determinant, $D(\tau, \eta)$, where $\tau \in \mathbb{C}$ is a frequency and $\eta \in \mathbb{R}^{d-1}$ is a wave vector, does not vanish for Re $\tau > 0$. Actually, we can reduce the problem to a 4×4 determinant depending only on $||\eta|| \in \mathbb{R}$, $\Delta(\tau, ||\eta||)$, which means that the d-dimensional stability of a phase boundary is equivalent to its 2-dimensional stability. Now, $\Delta(\tau, ||\eta||)$ happens to be nicely factorized. More precisely, we first show that, for all $\tau \in \mathbb{C}$ such that Re $\tau > 0$,

$$\Delta(\tau, 0) \neq 0.$$

This implies in particular the 1-D stability. Moreover, there exists an holomorphic function, ϕ_1, homogeneous of degree 1, and an holomorphic function, ϕ_2, homogeneous of degree 2, such that, for all $\eta \neq 0$,

$$\Delta(\tau, ||\eta||) = \phi_1(V)\,\phi_2(V), \tag{2}$$

where:

$$V := \frac{\tau}{i\,||\eta||}.$$

It is then rather easy to show that neither ϕ_1 nor ϕ_2 have any zero in the half-plane

$$\{V \in \mathbb{C}\,;\; \text{Im}\,V < 0\}.$$

This shows the weak linearized stability of nondissipative phase transitions. A stronger stability would hold if the *uniform* Lopatinsky condition were satisfied. The latter asks that, for all $\eta \in \mathbb{R}^{d-1}$, the function $\Delta(\cdot, ||\eta||)$, when extended by continuity to the closed half-plane:

$$\{\tau \in \mathbb{C}\,;\; \text{Re}\,\tau \geq 0\},$$

does not vanish – except of course at point 0 if $\eta = 0$. Unfortunately, this is only true for $\eta = 0$. As a matter of fact, $\Delta(\tau, 0) \neq 0$ for Re $\tau \geq 0$, but there exists $V_0 \in \mathbb{R}^{+*}$, which we can compute explicitly, such that

$$\phi_2(\pm V_0) = 0.$$

The speed V_0 corresponds to *surface waves*, which oscillate in the direction of the boundary and decay exponentially in the transverse direction. Finally, we have the following result in the case $\nu = 0$ – we refer to [1] for a detailed proof.

Theorem 2.1. *Nondissipative planar propagating boundaries,*

$$\begin{aligned}
\rho(\vec{x},t) = \rho_l = 1/v_l \ , \ \vec{u}(\vec{x},t) = \vec{u}_l \,, & \quad \text{if } \vec{x} \cdot \vec{n} - \sigma t < 0 \,, \\
\rho(\vec{x},t) = \rho_r = 1/v_r \ , \ \vec{u}(\vec{x},t) = \vec{u}_r \,, & \quad \text{if } \vec{x} \cdot \vec{n} - \sigma t > 0 \,,
\end{aligned} \tag{3}$$

where v_l and v_r belong to different phases, satisfying the jump conditions:

$$[\rho \vec{u}] = \sigma [\rho] \,, \tag{4}$$

$$[\rho (\vec{u} \cdot \vec{n}) \vec{u} + p \vec{n}] = \sigma [\, \rho \vec{u} \,] \,, \tag{5}$$

$$E(v_r) - E(v_l) + (v_r - v_l)(p(v_l) + p(v_r))/2 = 0 \,, \tag{6}$$

are weakly linearly stable in the sense of Majda. There exist some surface waves, of which the relative speed, V_0, depends on the normal relative velocities $\vec{u}_{l,r} \cdot \vec{n} - \sigma$ as well as the sound speeds $c_{l,r}$. Moreover, we have:

$$V_0 < \sqrt{(\vec{u}_l \cdot \vec{n} - \sigma)(\vec{u}_r \cdot \vec{n} - \sigma)} \,,$$

which means that the surface wave are rather slow.

Remark 2.2. The functions ϕ_1 and ϕ_2 in (2) have other nice properties that are not stated here but will be much useful in the following. The important point is that $\pm V_0$ are the only zeroes of the product $\phi_1 \phi_2$ in the closed half-plane $\{V \in \mathbb{C} \,; \ \text{Im } V \leq 0\}$.

3. Perturbation by means of a small viscosity

We shall show that a small positive viscosity eliminates the surface waves and changes the weak stability into uniform stability.

In the case $\nu > 0$, we have to deal with the integral term I in the additional jump condition (1). This can be done through a bifurcation analysis concerning the ODE governing viscous capillary profiles. We thus find a perturbed Lopatinsky determinant, $\Delta_{\tilde{\nu}}(\tau, \|\eta\|)$, depending on a parameter $\tilde{\nu}$ that is zero if $\nu = 0$ and is positive if $\nu > 0$ is small enough. Then we can show that:

1. if $\tilde{\nu} \geq 0$, for all $\tau \in \mathbb{C} \backslash \{0\}$ such that Re $\tau \geq 0$,

$$\Delta_{\tilde{\nu}}(\tau, 0) \neq 0 \,,$$

2. there exists a holomorphic function, ψ, such that, for all $\tau \in \mathbb{C}$ such that Re $\tau > 0$, for all $\eta \neq 0$,

$$\Delta_{\tilde{\nu}}(\tau, \|\eta\|) = \phi_1(V) \phi_2(V) + \tilde{\nu} \psi(V) \,,$$

with

$$V = \frac{\tau}{i \, \|\eta\|} \,.$$

Point 1 shows in particular that the weakly dissipative phase transitions are uniformly 1-D stable, like the nondissipative ones. Point 2 shows the weak multi-D stability of weakly dissipative phase transitions. Moreover, we can show, under some technical assumptions, that:

3. if $\tilde{\nu} > 0$ is small enough, the function $\phi_1 \phi_2 + \tilde{\nu} \psi$, when extended by continuity to the closed half-plane

$$\{V \in \mathbb{C}\,;\; \text{Im}\, V \leq 0\},$$

does not vanish anywhere in this half-plane.

In other words, the surface waves are stabilized and there are no exploding modes. This shows the uniform multi-D stability of weakly dissipative phase transitions.

Theorem 3.1. *If ν is small enough, planar propagating phase boundaries (3) satisfying the jump conditions (1), (4), (5) are uniformly linearly stable.*

We refer to [2] for more details, in particular concerning the technical assumptions mentioned above, which are satisfied for the van der Waals law of "real" fluids.

As a conclusion, we can say that, in view of Majda's work, weakly dissipative propagating phase boundaries are likely to be structurally – nonlinearly – stable. However, the proof requires careful estimates on the corresponding linearized Initial Boundary Value problems, in order to work out an iteration process. This remains to be done.

References

[1] S. Benzoni-Gavage, *Stability of multi-dimensional phase transitions in a van der Waals fluid,* Nonlinear Anal., **31 (1–2)** (1998), 243–263.

[2] S. Benzoni-Gavage, *Stability of subsonic planar phase transitions in a van der Waals fluid,,* Preprint ENS Lyon **221** (1998).

[3] H. Freistühler, *The Persistence of Ideal Shock Waves,* Appl. Math. Lett., **7 (6)** (1994), 7–11.

[4] H.-O. Kreiss, *Initial boundary value problems for hyperbolic systems,* Comm. Pure and Applied Maths, **23** (1970), 277–298.

[5] A. Majda, *Compressible fluid flow and systems of conservation laws in several space variables,* Springer-Verlag, Berlin, **1984**.

[6] M. Slemrod, *Admissibility criteria for propagating phase boundaries in a van der Waals fluid,* Arch. Rational Mech. Anal., **81** (1983), 301–315.

UMPA, ENS Lyon,
46, allée d'Italie,
F-69364 Lyon Cedex 07, France
E-mail address: `benzoni@umpa.ens-lyon.fr`

International Series of Numerical Mathematics
Vol. 129, © 1999 Birkhäuser Verlag Basel/Switzerland

Travelling Wave Solutions of a Convective Diffusive System with First and Second Order Terms in Nonconservation Form

Christophe Berthon and Frédéric Coquel

Abstract. The present work is devoted to the study of travelling wave solutions of a convective dissipative system, the so-called (k, ϵ) system modelling turbulent compressible flows. This system exhibits non conservative products for both the first order and second order operators. A particular attention is payed to the laminar and turbulent viscosities which ratio dictates the shape of the second order operator. In that aim, we introduce an equivalent system involving nonconservative products that are weighted by this ratio with the additional benefit of a second order operator in conservation form. Two limit cases are distinguished since they yield convective-dissipative systems in full conservation form.

1. Introduction

In the present work, we study for existence travelling wave solutions of a convective-diffusive system governing turbulent compressible flows at high Reynolds numbers. The so-called (k, ϵ) model is derived under that assumption and we consider here the associated set of PDE's, we write for short as:

$$\partial_t \mathbf{u} + \mathbf{A}(\mathbf{u})\partial_x \mathbf{u} = \partial_x \left(\mathbf{D}_0(\mathbf{u})\partial_x \mathbf{u}\right) + (\partial_x \mathbf{u}, \mathbf{D}_1(\mathbf{u})\partial_x \mathbf{u}) + \mathbf{S}(\mathbf{u}), \quad t > 0, x \in \mathbb{R}. \quad (1)$$

The precise shape of the system is given below. Since the Reynolds numbers under consideration are large, the solutions we are interested in are mainly driven by the first order extracted system. This first order system turns out to be hyperbolic but writes in non conservation form (see below and the references therein) so that its shock solutions are not defined in the classical distributional sense. This fact has been already underlined by Forestier-Hérard-Louis [1], Louis [5] and Vatel [4], who have proposed to use the Volpert product (see for instance [2]) to define the generalized jump conditions.

However, following LeFloch [3], the shape of the second order operator dictates the definition of shock solutions. These one are defined when studying travelling wave solutions for the convective-diffusive system (1) (see [3], [6], [7]).

Here, we prove the existence in the large of such travelling wave solutions. The main difficulty comes from the non conservation form met by both the first order and second order operator. We refer the reader to the companion paper [9] for the definition of shock solutions on the basis of the present study.

The shape of the second order operator, we consider below, depends on the ratio of laminar viscosity μ and the turbulent viscosity μ_t. We thus pay a particular attention to the role played by this ratio of these two viscosities. In that aim, we introduce an equivalent system that admits a second order operator in conservation form but at the expense of nonconservative products with the time partial derivative:

$$\mathbf{A}_0(\mathbf{w})\partial_t\mathbf{w} + \mathbf{A}_1(\mathbf{w})\partial_x\mathbf{w} = \partial_x\left(\mathbf{C}(\mathbf{w})\partial_x\mathbf{w}\right) + \mathbf{S}(\mathbf{w}). \tag{2}$$

Such non conservative products involve the ratio of the viscosities as a weight. Two limit cases naturally appear: (i) evanescent turbulence $(\mu_t/\mu \to 0)$ and (ii) fully expanded turbulence $(\mu/\mu_t \to 0)$. In both cases, travelling wave solutions of (1) are shown to converge to travelling wave solutions of fully conservative systems, i.e. systems under the form:

$$\partial_t\mathbf{v} + \partial_x\mathbf{H}(\mathbf{v}) = \partial_x\left(\mathbf{D}(\mathbf{v})\partial_x\mathbf{v}\right) + \mathbf{S}(\mathbf{v}). \tag{3}$$

2. Convective and dissipative operators

In this section, the system of PDE's associated with the (k, ϵ) model is introduced. For the sake of simplicity in the notations, we restrict ourselves to the system with one space variable. Let us underline that the existence result for travelling wave solutions we state below, holds true in higher space dimensions (see the companion paper [9]). The set of governing equations under consideration writes:

$$\partial_t\rho + \partial_x\rho u = 0, \tag{4a}$$

$$\partial_t\rho u + \partial_x(\rho u^2 + p + \frac{2}{3}\rho k) = \partial_x\left((\mu + \mu_t)\partial_x u\right), \tag{4b}$$

$$\partial_t E + \partial_x(E + p + \frac{2}{3}\rho k)u = \partial_x\left((\mu + \mu_t)u\partial_x u\right) + \partial_x(\kappa\partial_x T) + \partial_x(\kappa_k\partial_x k), \tag{4c}$$

$$\partial_t\rho k + \partial_x\rho k u + \frac{2}{3}\rho k\partial_x u = \mu_t\left(\partial_x u\right)^2 + \partial_x(\kappa_k\partial_x k) - \rho\epsilon, \tag{4d}$$

$$\partial_t\rho\epsilon + \partial_x\rho\epsilon u + \frac{2}{3}C_1\rho\epsilon\partial_x u = \mu_t C_1\frac{\epsilon}{k}\left(\partial_x u\right)^2 + \partial_x(\kappa_\epsilon\partial_x\epsilon) - C_2\rho\frac{\epsilon^2}{k}. \tag{4e}$$

Here k denotes the kinetic turbulent energy and ϵ its dissipation rate. We refer the reader for instance to Mohammadi-Pironneau [8] for a derivation of the above equations and the main underlying assumptions. A polytropic pressure law is assumed and the pressure p, involved in (4), reads

$$p = (\gamma - 1)(E - \frac{(\rho u)^2}{2\rho} - \rho k), \qquad \gamma > 1.$$

In (4e), C_1 and C_2 denote two constants of the model, to be chosen in the interval $]1, 2[$ (see [8] for pairs of practical interest). The turbulent and laminar viscosity laws, respectively denoted by μ_t and μ, write

$$\mu_t = C_{\mu_t}\rho\frac{k^2}{\epsilon}, \qquad \mu = C_\mu\tilde{\mu}(\frac{p}{\rho}) \tag{5}$$

where C_{μ_t} and C_μ denote two positive constants and $\tilde{\mu} : \mathbb{R} \rightarrow \mathbb{R}^+$ is a smooth bounded positive function. Furthermore κ, κ_k and κ_ϵ stand for heat conductivity coefficients (see [8] for detailed forms). As far the modelling is concerned, the main assumption in the present work stays in the omission of these diffusive effects, adopting $\kappa = \kappa_k = \kappa_\epsilon = 0$. We refer the reader to [9] for comments about the present simplifying assumption and its practical consequences.

Setting $\mathbf{u} = {}^t(\rho, \rho u, E, \rho k, \rho\epsilon)$, we consider the following state space:

$$\Omega = \left\{\mathbf{u} \in \mathbb{R}^5 \; ; \; \rho > 0, \; \rho u \in \mathbb{R}, \; E - \frac{(\rho u)^2}{2\rho} - \rho k > 0, \; \rho k > 0, \; \rho\epsilon > 0\right\}.$$

To shorten the notations, we rewrite the system (4) in the form

$$\partial_t\mathbf{u} + \mathbf{A}(\mathbf{u})\partial_x\mathbf{u} = \partial_x\left(\mathbf{D}_0(\mathbf{u})\partial_x\mathbf{u}\right) + (\partial_x\mathbf{u}, \mathbf{D}_1(\mathbf{u})\partial_x\mathbf{u}) + \mathbf{S}(\mathbf{u}), \tag{6}$$

where the mappings $\mathbf{A}, \mathbf{D}_0, \mathbf{D}_1 : \Omega \rightarrow \text{Mat}(\mathbb{R}^5, \mathbb{R}^5)$ and $\mathbf{S} : \Omega \rightarrow \mathbb{R}^5$ are easily defined from the developed form (4). The mapping $(.,.) : \mathbb{R}^5 \times \mathbb{R}^5 \rightarrow \mathbb{R}^5$ denotes the contracted product. Let us stress that the source term $\mathbf{S}(\mathbf{u})$, in (6), is a vector valued function involving no partial differential operator:

$$\mathbf{S}(\mathbf{u}) = {}^t\left(0, \; 0, \; 0, \; -\rho(\frac{\epsilon}{k})k, \; -C_2\rho(\frac{\epsilon}{k})\epsilon\right). \tag{7}$$

In (7) the ratio k/ϵ has the dimension of a time and can be thus understood as a relaxation time.

As a consequence of (6) and when noticing that the terms $\{\partial_x\rho ku + 2/3\rho k\partial_x u\}$ and $\{\partial_x\rho\epsilon u + 2/3C_1\rho\epsilon\partial_x u\}$ cannot be put in divergence form, the extracted first order system is not in conservation form. In the same way, the second order operator involving the terms $\mu_t \left(\partial_x u\right)^2$ and $\mu_t C_1 \frac{\epsilon}{k} \left(\partial_x u\right)^2$ is also in nonconservation form. In the next section, the extracted first order system is seen to be hyperbolic over the phase space Ω. The main purpose of the present work is to study for existence travelling wave solutions $\mathbf{u}(\xi)$ with $\xi = x - \sigma t$ of (6) focusing ourselves on the role played by both the laminar and turbulent viscosities. Because of the particular shape of the relaxation source term $\mathbf{S}$ given in (7), the only possible travelling wave solutions necessarily satisfy $\lim_{\xi\rightarrow\pm\infty} k(\xi) = \lim_{\xi\rightarrow\pm\infty} \epsilon(\xi) = 0$ and to allow for more general solutions, we set in the present study $\mathbf{S}(\mathbf{u}) = 0$. Existence and uniqueness of travelling wave solutions with (7) is the matter of a forthcoming work.

3. Main results

In this section, we state several properties of the convective-dissipative system (6). The extracted first order system is first analyzed. Its hyperbolicity properties and its Lax entropy pairs are characterized. Travelling wave solutions are then studied for existence and an equivalent system is exhibited. This equivalent system is in non conservation form in both time and space but with the benefit to admit a second order operator in conservation form. Furthermore, the role played respectively by the laminar and turbulent viscosity is clearly put forward. Two limit cases, namely evanescent turbulence ($\mu_t/\mu \to 0$) and fully expanded turbulence ($\mu/\mu_t \to 0$), naturally appear and respectively yield limit system in full conservation form.

Lemma 3.1.

i) The first order extracted system from (6) is hyperbolic over the phase space Ω. The eigenvalues are respectively

$$u - c, \ u, \ u + c \qquad c^2 = \frac{\gamma p}{\rho} + \frac{10}{9} k, \tag{8}$$

the second eigenvalue u is with third order of multiplicity while the two others are simple. The 1-wave and the 3-wave are genuinely nonlinear and the other fields are linearly degenerate.

ii) u is continuous through a contact discontinuity. The products $\rho k \partial_x u$ and $\rho \epsilon \partial_x u$ are thus usually defined. However these products are not defined in the distributional sense for 1-shock and 3-shock solutions.

iii) Let $\mathbf{u}$ be a smooth solution of the first order extracted system. Then, $\mathbf{u}$ satisfies the following additional conservation laws:

$$\partial_t U(\mathbf{u}) + \partial_x F(\mathbf{u}) = 0, \ U(\mathbf{u}) = \rho \mathcal{F}\left(\frac{p}{\rho^\gamma}, \frac{\rho k}{\rho^{5/3}}, \frac{\rho \epsilon}{\rho^{2C_1/3+1}}\right), \ F(\mathbf{u}) = u U(\mathbf{u}), \tag{9}$$

where $\mathcal{F} : \Omega \to \mathbb{R}$ denotes any smooth mapping : $\mathbb{R}^3 \to \mathbb{R}$ (see [9] for convexity conditions about $U : \Omega \to \mathbb{R}$). Moreover, taking into account dissipative terms, these laws become

$$\partial_t U(\mathbf{u}) + \partial_x F(\mathbf{u}) =$$
$$\frac{\gamma - 1}{\rho^{(\gamma-1)}} \mu(\partial_x u)^2 \ \partial_s \mathcal{F} + \frac{1}{\rho^{2/3}} \mu_t (\partial_x u)^2 \ \partial_{s_t} \mathcal{F} + C_1 \frac{\epsilon}{k} \frac{1}{\rho^{2C_1/3}} \mu_t (\partial_x u)^2 \ \partial_{s_\epsilon} \mathcal{F}, \tag{10}$$

where we have set $s = \dfrac{p}{\rho^\gamma}, \ s_t = \dfrac{\rho k}{\rho^{5/3}}, \ s_\epsilon = \dfrac{\rho \epsilon}{\rho^{2C_1/3+1}}.$

The properties *i)* and *ii)* can be found in [1], [4] and [5]. The property *iii)* completes the previous results concerning Lax entropy pairs and their compatibility with dissipative phenomena. Three particular cases are of interest:

$$\partial_t \rho \frac{p}{\rho^\gamma} + \partial_x \rho \frac{p}{\rho^\gamma} u = \frac{(\gamma-1)}{\rho^{(\gamma-1)}} \mu (\partial_x u)^2 > 0, \tag{11}$$

$$\partial_t \rho \frac{\rho k}{\rho^{5/3}} + \partial_x \rho \frac{\rho k}{\rho^{5/3}} u = \frac{1}{\rho^{2/3}} \mu_t (\partial_x u)^2 > 0 \tag{12}$$

and

$$\partial_t \rho \frac{\epsilon}{k^{C_1}} + \partial_x \rho \frac{\epsilon}{k^{C_1}} u = 0. \tag{13}$$

The analogy between the equations (11) and (12) is clear: $\rho k / \rho^{5/3}$ behaves like a specific turbulent entropy with the adiabatic constant $\gamma_t = 5/3$. From equation (13), ϵ / k^{C_1} turns out to be an advected quantity in the presence of the dissipative phenomena.

The main results of the present work are stated below.

Theorem 3.2.
i) There exists a C^1-diffeomorphism $\Phi : \Omega \to \Omega$ such that, for smooth solutions, the system (4) writes:

$$\mathbf{A}_0(\mathbf{w})\partial_t \mathbf{w} + \mathbf{A}_1(\mathbf{w})\partial_x \mathbf{w} = \partial_x \left(\mathbf{C}(\mathbf{w})\partial_x \mathbf{w} \right) + \mathbf{S}(\mathbf{w}), \tag{14}$$

with $\mathbf{w} = {}^t(\rho, \rho u, E, \frac{\rho k}{\rho^{5/3}}, \rho \frac{k^{C_1}}{\epsilon})$. The matrix $\mathbf{A}_0(\mathbf{w})$ is invertible for all $\mathbf{u} \in \Omega$. The developed form of the system (14) can express:

$$\partial_t \rho + \partial_x \rho u = 0, \tag{15a}$$

$$\partial_t \rho u + \partial_x (\rho u^2 + p + \frac{2}{3}\rho k) = \partial_x \left((\mu + \mu_t) \partial_x u \right), \tag{15b}$$

$$\partial_t E + \partial_x (E + p + \frac{2}{3}\rho k) u = \partial_x \left((\mu + \mu_t) u \partial_x u \right), \tag{15c}$$

$$\rho^{2/3} \mu \left\{ \partial_t \rho \frac{\rho k}{\rho^{5/3}} + \partial_x \rho \frac{\rho k}{\rho^{5/3}} u \right\} - \frac{\rho^{\gamma-1}}{\gamma-1} \mu_t \left\{ \partial_t \rho \frac{p}{\rho^\gamma} + \partial_x \rho \frac{p}{\rho^\gamma} u \right\} = -\rho \epsilon (\mu_t + \mu), \tag{15d}$$

$$\partial_t \rho \frac{\epsilon}{k^{C_1}} + \partial_x \rho \frac{\epsilon}{k^{C_1}} u = -(C_2 - C_1)\rho (\frac{\epsilon}{k}) \frac{\epsilon}{k^{C_1}}. \tag{15e}$$

ii) Let $\mathbf{u}_L \in \Omega$ and $\sigma \in \mathbb{R}$ such that $(u_L - \sigma)/c(\mathbf{u}_L) > 1$ where c is the sound speed defined in (8). Then, there exists a unique state $\mathbf{u}_R \in \Omega$ and a unique travelling wave (up to a translation) denoted by $\hat{\mathbf{u}}(\xi)$ with $\xi = x - \sigma t$, solution of (6) and satisfying $\lim\limits_{\xi \to -\infty} \hat{\mathbf{u}}(\xi) = \mathbf{u}_L$ and $\lim\limits_{\xi \to +\infty} \hat{\mathbf{u}}(\xi) = \mathbf{u}_R$.
Symmetrically, there are global travelling wave solutions of (6) with $\mathbf{u}_R$ and σ such that $(u_R - \sigma)/c(\mathbf{u}_R) < -1$.

The result *i)* of the above theorem allows to write dissipative terms in conservation form but at the expense of non conservative products with the time partial derivative operator. Through this property, the (k, ϵ) model differs from other

nonconservative hyperbolic system recently studied ([2] for example) which are in
the form

$$\partial_t \mathbf{w} + \mathbf{B}(\mathbf{w})\partial_x \mathbf{w} = \partial_x \left(\mathbf{D}(\mathbf{w})\partial_x \mathbf{w}\right). \tag{16}$$

In other words, the dissipative operator is in conservation form, *i.e.* $\mathbf{D}_1 \equiv 0$.

One can notice that there is no second order term in equations (15d) and
(15e). Then (15d) is the only equation in non conservation form in the system (15).
Next equation (15d) writes as a balance equation between the two entropy rates
where the laminar and turbulence viscosities act like weights. The main interest
of the equivalent system (15) lies in the existence of two limit cases associated
respectively with evanescent turbulence and fully expanded turbulence.

Theorem 3.3.
i) **Evanescent turbulence.** *Let* $\eta = C_{\mu_t}$ *(see equation (5)) and denote* $\hat{u}_\eta$ *the
travelling wave solutions of* $(6)_\eta$. *Then* $\{\mu_t/\mu\}(\hat{u}_\eta(\xi))$ *tends to 0 as* η *goes to 0
and this uniformly in* $\xi \in \mathbb{R}$. *Furthermore the travelling wave solutions* $\hat{u}_\eta$ *of
*$(6)_\eta$ *converge uniformly as* η *goes to 0 to travelling wave solutions of the following
system in conservation form:*

$$\partial_t \mathbf{v} + \partial_x \mathbf{H}(\mathbf{v}) = \partial_x \left(\tilde{\mathbf{D}}(\mathbf{v})\partial_x \mathbf{v}\right). \tag{17}$$

Its developed form is given by

$$\partial_t \rho + \partial_x \rho u = 0, \tag{18a}$$

$$\partial_t \rho u + \partial_x \left(\rho u^2 + p + \frac{2}{3}\rho k\right) = \partial_x \left(\mu \partial_x u\right), \tag{18b}$$

$$\partial_t E + \partial_x \left(E + p + \frac{2}{3}\rho k\right)u = \partial_x \left(\mu u \partial_x u\right), \tag{18c}$$

$$\partial_t \rho h \left(\frac{k}{\rho^{2/3}}\right) + \partial_x \rho h(\frac{k}{\rho^{2/3}})u = 0, \tag{18d}$$

$$\partial_t \rho \frac{\epsilon}{k^{C_1}} + \partial_x \rho \frac{\epsilon}{k^{C_1}}u = 0, \tag{18e}$$

where $h : \mathbb{R} \to \mathbb{R}$ *is a strictly monotone function. Therefore, if* $\eta \to 0$, *the systems
(15) and (4) become equivalent to the limit system*

$$\partial_t \mathbf{v} + \partial_x \mathbf{H}(\mathbf{v}) = \partial_x \left(\tilde{\mathbf{D}}(\mathbf{v})\partial_x \mathbf{v}\right) + \tilde{\mathbf{S}}(\mathbf{v}), \tag{19}$$

with the following source term:

$$\tilde{\mathbf{S}}(\mathbf{v}) = {}^t\left(0,\ 0,\ 0,\ \rho^{1/3}h'(\frac{k}{\rho^{2/3}})\epsilon,\ -(C_2 - C_1)\rho\frac{\epsilon^2}{k^{C_1+1}}\right). \tag{20}$$

ii) **Fully expanded turbulence.** *Let* $\eta = C_\mu$ *(see equation (5)) and denote* $\hat{u}_\eta$ *the
travelling wave solutions of* $(6)_\eta$. *Then* $\{\mu/\mu_t\}(\hat{u}_\eta(\xi))$ *tends to 0 as* η *goes to 0
and this uniformly in* $\xi \in \mathbb{R}$. *Furthermore the travelling wave solutions* $\hat{u}_\eta$ *of*

$(6)_\eta$ *converge uniformly as η goes to 0 to travelling wave solutions of the following system in conservation form:*

$$\partial_t \mathbf{w} + \partial_x \mathbf{G}(\mathbf{w}) = \partial_x \left(\hat{\mathbf{D}}(\mathbf{w}) \partial_x \mathbf{w} \right). \tag{21}$$

Its developed form is given by

$$\partial_t \rho + \partial_x \rho u = 0, \tag{22a}$$

$$\partial_t \rho u + \partial_x \left(\rho u^2 + p + \frac{2}{3} \rho k \right) = \partial_x \left(\mu_t \partial_x u \right), \tag{22b}$$

$$\partial_t E + \partial_x \left(E + p + \frac{2}{3} \rho k \right) u = \partial_x \left(\mu_t u \partial_x u \right), \tag{22c}$$

$$\partial_t \rho f \left(\frac{p}{\rho^\gamma} \right) + \partial_x \rho f (\frac{p}{\rho^\gamma}) u = 0, \tag{22d}$$

$$\partial_t \rho \frac{\epsilon}{k^{C_1}} + \partial_x \rho \frac{\epsilon}{k^{C_1}} u = 0, \tag{22e}$$

where $f : \mathbb{R} \to \mathbb{R}$ is a strictly monotone function. Therefore, if $\eta \to 0$, the systems (15) and (4) become equivalent to the limit system

$$\partial_t \mathbf{w} + \partial_x \mathbf{G}(\mathbf{w}) = \partial_x \left(\hat{\mathbf{D}}(\mathbf{w}) \partial_x \mathbf{w} \right) + \hat{\mathbf{S}}(\mathbf{w}), \tag{23}$$

with the following source term:

$$\hat{\mathbf{S}}(\mathbf{w}) = {}^t \left(0,\ 0,\ 0,\ (\gamma - 1)\rho^{2-\gamma} f'(\frac{p}{\rho^\gamma})\epsilon,\ -(C_2 - C_1)\frac{\epsilon^2}{\rho k^{C_1 + 1}} \right). \tag{24}$$

These last results motivate asymptotic expansions respectively in $\mathcal{O}((\frac{\mu_t}{\mu})^2)$ and $\mathcal{O}((\frac{\mu_t}{\mu})^2)$. Such expansions allow to approximate jump relations associated with the limit cases (cf. [9] for more details). This closure relation of the system (14) can be used to develop an approximate Riemann solver.

Acknowledgement. We wish to thank Lionel SAINSAULIEU and Olivier PIRON-NEAU for valuable discussions. The first author wishes to thank Aérospatiale for its partial financial support.

References

[1] A. Forestier, J.M. Herard and X. Louis, *A Godunov type solver to compute turbulent compressible flows*, C.R.A.S. Paris, Série I, **324** (1997), 919–926.

[2] G. Dal Maso, P.G. LeFloch and F. Murat, *Definition and weak stability of noncon-servative products*, J. Math. Pures Appli., **74** (1995), 483–548.

[3] P.G. LeFLoch, *Shock waves for non linear hyperbolic systems in non conservative form*, IMA preprint series No 593 (1989), University of Minnesota.

[4] V. Vatel, Ph.D. thesis (1993), Université Bordeaux I.

[5] X. Louis, Ph.D. thesis (1995), Université Paris VI.

[6] P.A. Raviart and L. Sainsaulieu, *A nonconservative hyperbolic system modelling spray dynamics. Part 2: existence of travelling waves solutions*, Mathematical Methods in Models in Applied Sciences (1195).

[7] L. Sainsaulieu, *Travelling waves solutions of convection-diffusion systems whose convection terms are weakly nonconservative*, SIAM Journal of Applied Math., **55** (1995), 1552–1576.

[8] B. Mohammadi and O. Pironneau, *Analysis of the K-Epsilon Turbulence Model*, Research in Applied Mathematics, **31** (1994), Masson, Paris.

[9] C. Berthon and F. Coquel, in preparation.

ONERA, MFE/DSNA,
B.P.72,
92322 Châtillon Cedex, France
E-mail address: berthon@onera.fr
E-mail address: coquel@onera.fr

LAN-CNRS URA 189, Tour 55-65,
Université Paris VI,
4 Place Jussieu, 75252 Paris Cedex 05, France
E-mail address: berthon@ann.jussieu.fr
E-mail address: coquel@ann.jussieu.fr

International Series of Numerical Mathematics
Vol. 129, © 1999 Birkhäuser Verlag Basel/Switzerland

High Order Central Schemes for Hyperbolic Systems of Conservation Laws

Franca Bianco, Gabriella Puppo and Giovanni Russo

Abstract. In this talk we present third and fourth order central schemes for the approximate solution of quasilinear systems of conservation laws. The schemes are an extension of the second order Nessyahu-Tadmor scheme, and are based on a ENO reconstruction from cell averages, and a numerical computation of the flux on cell boundaries, efficiently obtained by Runge-Kutta schemes with Natural Continuous Extension.

Here we focus on the linear stability analysis of the scheme. The exact CFL condition for linear third and fourth order schemes are derived.

1. Introduction

Central schemes for conservation laws have been a subject of active research in recent years [8, 10, 8, 7, 5, 6, 2]. The main advantage of central schemes over upwind schemes is that they do not require the solution of Riemann problems, or the computation of characteristic velocities of the system.

These features make the central scheme approach very attractive for those systems for which the solution to the Riemann problem is complicated, or when there is no simple analytical expression for the eigenvalues of the Jacobian matrix. This is the case of systems arising, for example, in semiconductor modeling [9, 1]. In that case the NT scheme, suitably modified to incorporate source terms, has been successfully used.

The schemes that we present can be viewed as an extension of the second order Nessyahu-Tadmor scheme [8]. A third order scheme has been presented by Liu and Tadmor [7]. They show that the scheme is *Number of Extrema Decreasing* (NED), and it gives good numerical results both on the scalar equation and on the Euler equations.

The main focus of our work is the development of third and fourth order schemes, which are robust and efficient, so that they can be easily implemented for several systems of conservation laws. The user has to provide only a subroutine for the computation of the flux vector and an estimate of the eigenvalues (necessary to satisfy the stability condition).

This goal is obtained by the combination of two main ingredients: high order ENO reconstruction from cell averages (which provides high order space accuracy

and shock capturing capability), and Runge-Kutta schemes with Natural Continuous Extension (NCE) for the integration of the flux (which provides stability and
high order accuracy in time, without requiring the computation of the Jacobian
or the Hessian of the system).

The paper is divided in two parts. In the first part we describe the general
features of the method, and we summarize the main results presented in [3]. The
second part of the paper is original, and deals with the study of the linear stability
analysis of the linear schemes. The analysis is relevant for the computation of
the exact CFL condition, and helps developing modification to the original ENO
stencil selection mechanism.

2. Description of the method

We describe the method in the case of scalar equation. The extension to systems
is recalled later.

Let us consider the scalar conservation law:

$$u_t + f_x(u) = 0, \tag{1}$$

on an interval I, with suitable boundary conditions. We consider for simplicity
a uniform grid on I of points $\{x_j\}$, $j = 0, \ldots, N$, with $x_{j+1} - x_j = h$. Let k
be the time step, with $u_j^n = u(x_j, t^n)$, $t^n = nk$. Finally, w will denote the computed solution of (1). At time t^n, we start from a piecewise constant function $\overline{w}_j^n$,
representing the cell averages of the computed solution w at time t^n, namely:

$$\overline{w}_j^n = \frac{1}{h} \int_{-h/2}^{h/2} w(x_j + y, t^n) \, dy. \tag{2}$$

From the values $\{\overline{w}_j^n\}_{j=0}^N$ we reconstruct the point values of the function $w(x, t^n)$,
via a suitable non linear piecewise polynomial interpolation. The reconstruction
we use is due to Harten et al. [4].

Let $\mathcal{R}(\overline{w}^n; x)$ be the reconstruction operator, where $\overline{w}^n$ is the vector with
components $\overline{w}_j^n$, $j = 0, \ldots, N$. Then:

$$w(x, t^n) := \mathcal{R}(\overline{w}^n; x) \tag{3}$$

is the function defined on I which will be used as initial data for the n-th time
step. The reconstruction $\mathcal{R}(\overline{w}^n; x)$ is piecewise polynomial in the sense that:

$$\mathcal{R}(\overline{w}^n; x) \in P_j^m(x) \qquad \text{for } x \in \left[x_j - \frac{h}{2}, x_j + \frac{h}{2}\right],$$

where P_j^m is the space of polynomials of degree m defined on the interval $[x_j -
h/2, x_j + h/2]$. Note that in general R will have jump discontinuities at the points
$x_j \pm h/2$.

The solution is updated on a staggered grid. By integrating the conservation law (1) on the cell $[x_j, x_{j+1}] \times [t^n, t^n + k]$, we obtain

$$
\begin{aligned}
\overline{w}_{j+1/2}^{n+1} = \; & \frac{1}{h} \left\{ \int_0^{h/2} w(x_j + y, t^n) \, dy + \int_{-h/2}^0 w(x_{j+1} + y, t^n) \, dy \right\} \\
& - \frac{1}{h} \int_0^k [f(w(x_{j+1}, t^n + \tau)) - f(w(x_j, t^n + \tau))] \, d\tau,
\end{aligned}
\tag{4}
$$

If the time step is subjected to the CFL condition $k \leq h/(2 \max |f'(u)|)$, we can assume that $w(x, t^n + \tau)$ is smooth at x_{j+1} and x_j, since the discontinuities starting at t^n from the staggered grid points $x_{j+1/2}$ have not had the time to reach the cell boundaries.

Then the time integrals can be approximated by a quadrature formula, say:

$$
\frac{1}{k} \int_0^k f(w(x_j, t^n + \tau)) \, d\tau \simeq \sum_{l=0}^L f(w(x_j, t^n + k\tau_l)) \, \omega_l,
\tag{5}
$$

where τ_l and $\omega_l \in [0, 1]$ are the knots and weights of the quadrature formula. Simpson's rule is enough for third and fourth order schemes.

Since w is smooth at x_j, we can evaluate w at the intermediate times $t^n + k\tau_l$ through Taylor expansion or with a Runge Kutta method.

In the following time step we repeat a similar process and go back to the original grid.

2.1. Reconstruction

Reconstruction is a key step in high resolution schemes. The algorithm we consider here was introduced in [4] and it has been widely implemented, see [12] and references therein.

The basic idea of ENO reconstruction is that by a suitable choice of the stencil, the interpolation polynomial will have only small oscillations. The original technique for the selection of the stencil is very sensitive to the data, and this may cause a deterioration of the accuracy. In order to overcome this problem, two modified stencil have been used, one developed by Shu, and an original one (MC) proposed by the authors [3].

2.2. Evaluation of the fluxes

To compute the time integrals of the fluxes in (5), we need to evaluate the function $f(w(x_j, t^n + k\tau_l))$ at the different instants $k\tau_l \in [0, k]$, $l = 0, \ldots, L$. This can be obtained by Taylor expansion [7], or by Runge-Kutta schemes. Here we describe the latter approach.

The evaluation of the field at the j-th grid point can be written as

$$
\begin{cases}
y'(\tau) & = \; F(\tau, y(\tau)) = -f_x(y(x_j, t^n + \tau)) \\
y(\tau = 0) & = \; w(x_j, t^n).
\end{cases}
\tag{6}
$$

Thus the computation of the i-th Runge-Kutta flux requires the evaluation of the x-derivative of f at the intermediate time $t = t^n + c_i k$, where c_i are the coefficients of the RK scheme.

Therefore we must compute all grid values of f at the intermediate time, and perform a piecewise polynomial interpolation of these data to maintain high accuracy and control over oscillations in the evaluation of f_x.

For a method of order $m = 4$, we need a third order accurate three stage Runge-Kutta method. Thus we need to compute three polynomial interpolations for each node τ_l in the quadrature formula appearing in (5).

Fortunately a great saving in computational time can be obtained with the use of Natural Continuous Extensions (NCE) of a Runge-Kutta scheme.

The properties of NCE's which are essential for their application to our scheme are described and proved in [13].

At each time step, we apply the Runge-Kutta scheme only once, and we obtain all intermediate values $w(x_j, t^n + k\tau_l)$ through the evaluation of the appropriate NCE.

2.3. Systems

We consider the system of conservation laws:

$$\mathbf{u}_t + \mathbf{f}_x(\mathbf{u}) = 0, \tag{7}$$

where $\mathbf{u}$ and $\mathbf{f}$ are vectors with M components.

We apply our scheme componentwise. At each time step, we start from the array of cell averages, $\{\overline{w}_{j,i}^n\}, j = 1, \ldots, N; \ i = 1, \ldots, M$. We apply the reconstruction step at each component. The stencil chosen in general will be different for each component. The evaluation of the space contribution appearing in (4) is now straightforward.

For the time evolution, we must compute all components of the Runge-Kutta fluxes $g_i^{(j)}, i = 1, \ldots, M$, before computing the successive fluxes $g_i^{(j+1)}$. No differentiation of the flux function $\mathbf{f}$ is required. We only need an estimate of the maximum characteristic velocity to satisfy the CFL condition.

2.4. Stability analysis

Linear stability analysis of the schemes is performed, in order to identify the linearly stable central schemes and compute the critical Courant number of the scheme. The former information will be used as a guideline for the choice of the stencil in the non linear schemes.

Let us consider a generic central scheme of third and fourth order, applied to the linear equation:

$$u_t + u_x = 0,$$

Such scheme will take the form

$$\overline{w}_{j+\frac{1}{2}}^{n+1} = \sum_{l=0}^{m-1} (\frac{1}{2})^{l+1} \frac{1}{(l+1)!} [\tilde{D}_j^l(\overline{w}^n) + (-1)^i \tilde{D}_{j+1}^l(\overline{w}^n)]$$

$$- \frac{\lambda}{6} \{[w_{j+1}^n + 4w_{j+1}^{n+\frac{1}{2}} + w_{j+1}^{n+1}] - [w_j^n + 4w_j^{n+\frac{1}{2}} + w_j^{n+1}]\}, \qquad (8)$$

where the discrete space derivatives of the field, $\tilde{D}_j^l$, are obtained from cell averages $\{\overline{w}_j^n\}$ by deconvolution [3, 4], and $\lambda = k/h$ denotes the mesh ratio.

A detailed truncation analysis of the schemes shows that with our approach it is necessary to use a m degree interpolation polynomial in the reconstruction step to obtain a method of order m [3].

We shall consider separately third and fourth order schemes.

2.4.1. THIRD ORDER SCHEMES The different stencils will be labeled by the value of $il(j) - j$, where $il(j)$ is the leftmost point of the stencil. The stencil of third order schemes is formed by four points, and must include points j and $j + 1$, therefore the possible stencils are (-2), (-1), and (0).

From deconvolution we obtain:

$$\tilde{D}_j^0(\overline{u}) = D_j^0(\overline{u}) - \frac{1}{24}D_j^2(\overline{u}), \quad \tilde{D}_j^1(\overline{u}) = D_j^1(\overline{u}), \quad \tilde{D}_j^2(\overline{u}) = D_j^2(\overline{u})$$

where $D_j^0(u) = u_j$, and $D_j^k(u)$, $k \geq 1$, denotes the numerical approximation of k-th derivative of u (times h^k). They are obtained by taking the derivatives of the interpolating polynomial of the particular stencil chosen. We list them here:

Stencil (-2)

$$D_j^1(u) = \frac{1}{6}(u_{j-2} - 6u_{j-1} + 3u_j + 2u_{j+1})$$

$$D_j^2(u) = u_{j-1} - 2u_j + u_{j+1}.$$

Stencil (-1)

$$D_j^1(u) = \frac{1}{6}(-2u_{j-1} - 3u_j + 6u_{j+1} - 2u_{j+2})$$

$$D_j^2(u) = u_{j-1} - 2u_j + u_{j+1}.$$

Stencil (0)

$$D_j^1(u) = \frac{1}{6}(-11u_j + 18u_{j+1} - 9u_{j+2} + 2u_{j+3})$$

$$D_j^2(u) = \frac{1}{3}(6u_j - 15u_{j+1} + 12u_{j+2} - 3u_{j+3}).$$

From deconvolution we obtain:

$$\tilde{D}_j^0(\overline{u}) = D_j^0(\overline{u}) - \frac{1}{24}D_j^2(\overline{u})$$

$$\tilde{D}_j^1(\overline{u}) = D_j^1(\overline{u})$$

$$\tilde{D}_j^2(\overline{u}) = D_j^2(\overline{u}).$$

Here we consider three different schemes, namely *Timeux*, RK2-2, and NCERK2. The schemes differ in the computation of the predictor values $w_j^{n+1/2}$ and w_j^{n+1}. The first scheme uses Taylor expansion (as in [7]), the second uses two steps of Runge-Kutta 2, one for $w_j^{n+1/2}$ and one for w_j^{n+1}, and the third scheme uses RK2 for the computation of w_j^{n+1}, and NCE to evaluate $w_j^{n+1/2}$.

Scheme NCERK2 is the one that gives the best results in terms of robustness and efficiency (in the case of systems), therefore we describe in detail only the analysis for this one.

In this case, w^{n+1} is computed by RK2 scheme, and $w^{n+1/2}$ is computed using the corresponding NCE of degree 2. Only two evaluations of the function are needed. The predictor values are given by

$$
\begin{aligned}
w_j^{n+\beta} &= w_j^n + k[b_1(\beta)F(w^n; j) + b_2(\beta)F(K_1; j)], \quad \beta = 1/2, 1 \\
K_{1,j} &= w_j^n + \beta k F(w^n; j),
\end{aligned}
\tag{9}
$$

where $b_i(\beta)$ are the coefficients of the Natural Continuous Extension of Runge-Kutta schemes [13].

Here and in the following, K_1 denotes the vector with components $K_{1,j}$. Similarly, w^n and $\overline{w}^n$ are the vectors with components w_j^n and $\overline{w}_j^n$. Furthermore we recall that $w_j^n = D_j^0 = D_j^0(w^n)$, and

$$
\begin{aligned}
kF(w^n; j) &= -\lambda \tilde{D}_j^1(\overline{w}^n), \\
kF(K_1; j) &= -\lambda D_j^1(K_1).
\end{aligned}
$$

Substituting in (9) we obtain

$$
w_j^{n+\beta} = \tilde{D}_j^0(\overline{w}^n) - \lambda(b_1(\beta)\tilde{D}_j^1(\overline{w}^n) + b_2(\beta)D_j^1(K_1)), \quad \beta = 1/2, 1.
$$

2.4.2. FOURTH ORDER SCHEMES In this case the stencil contains five points. The possible stencils are: (-3), (-2), (-1), and (0).

Stencil (-3)

$$
\begin{aligned}
D_j^1(u) &= \frac{1}{12}u_{j-3} - u_{j-1} + \frac{2}{3}u_j + \frac{1}{4}u_{j+1} \\
D_j^2(u) &= -\frac{3}{4}u_{j-3} - \frac{2}{3}u_{j-2} + \frac{3}{2}u_{j-1} - 2u_j + \frac{11}{12}u_{j+1} \\
D_j^3(u) &= -\frac{3}{4}u_{j-3} + 2u_{j-2} - \frac{3}{2}u_{j-1} + \frac{1}{4}u_{j+1}
\end{aligned}
$$

Stencil (-2)

$$
\begin{aligned}
D_j^1(u) &= \frac{1}{12}u_{j-2} - \frac{2}{3}u_{j-1} + \frac{2}{3}u_{j+1} - \frac{1}{12}u_{j+2} \\
D_j^2(u) &= -\frac{1}{12}u_{j-2} + \frac{4}{3}u_{j-1} - \frac{5}{2}u_j + \frac{4}{3}u_{j+1} - \frac{1}{12}u_{j+2} \\
D_j^3(u) &= -\frac{1}{2}u_{j-2} + u_{j-1} - u_{j+1} + \frac{1}{2}u_{j+2}
\end{aligned}
$$

Stencil (-1)

$$D_j^1(u) = -\frac{1}{4}u_{j-1} - \frac{5}{6}u_j + \frac{3}{2}u_{j+1} - \frac{1}{2}u_{j+2} + \frac{1}{12}u_{j+3}$$

$$D_j^2(u) = \frac{11}{12}u_{j-1} - \frac{5}{3}u_j + \frac{1}{2}u_{j+2} - \frac{1}{12}u_{j+3}$$

$$D_j^3(u) = -\frac{3}{2}u_{j-1} + 5u_j - 6u_{j+1} + 3u_{j+2} - \frac{1}{2}u_{j+3}$$

Stencil (0)

$$D_j^1(u) = -\frac{25}{12}u_j + 4u_{j+1} - 3u_{j+2} + \frac{4}{3}u_{j+3} - \frac{1}{4}u_{j+4}$$

$$D_j^2(u) = \frac{35}{12}u_j - \frac{26}{3}u_{j+1} + \frac{19}{2}u_{j+2} - \frac{14}{3}u_{j+3} + \frac{11}{12}u_{j+4}$$

$$D_j^3(u) = -\frac{5}{4}u_j + 4u_{j+1} - \frac{9}{2}u_{j+2} + 2u_{j+3} - \frac{1}{4}u_{j+4}$$

From deconvolution we obtain:

$$\tilde{D}_j^0(u) = D^0(\bar{u}) - \frac{1}{24}D_j^2(\bar{u}), \quad \tilde{D}_j^1(u) = D^1(\bar{u}) - \frac{1}{24}D_j^3(\bar{u})$$

$$\tilde{D}_j^2(u) = D^2(\bar{u}), \quad \tilde{D}_j^3(u) = D^3(\bar{u})$$

As in the case of third order schemes, we consider three schemes, namely *Timeux*, RK3-2, and NCERK4. The first is based on a Taylor expansion, the second on two steps of Runge-Kutta 3, and the last uses one step of RK4 with NCE of degree 3.

Only the analysis for scheme NCERK4 is shown in detail. By applying Runge-Kutta 4 with NCE to (6) one has

$$w_j^{n+\beta} = w_j^n + k[b_1(\beta)F(w^n; j) + b_2(\beta)F(K_1; j) + b_3(\beta)F(K_2; j)$$
$$+ b_4(\beta)F(K_3; j)], \qquad \beta = 1/2, 1$$

$$K_{1,j} = w_j^n + \frac{k}{2}F(w^n; j)$$

$$K_{2,j} = w_j^n + \frac{k}{2}F(K_1; j)$$

$$K_{3,j} = w_j^n + kF(K_2; j)$$

where the classical RK4 scheme has been used with $\mathbf{b}^T = (1, 2, 2, 1)/6$; the terms $b_i(\beta), i = 1, \ldots, 4$ are the NCE polynomials, and

$$kF(w^n; j) = -\lambda\tilde{D}_j^1(\bar{w}^n)$$
$$kF(K_s; j) = -\lambda D_j^1(K_s), \qquad s = 1, 2, 3.$$

Third order schemes		
Scheme	Stencil	Stability region
RK2-2	$j-2, j-1, j, j+1$	$\lambda^*=0.35099$
	$j-1, j, j+1, j+2$	$\lambda^*=0.43403$
	$j, j+1, j+2$	unstable
NCERK2	$j-2, j-1, j, j+1$	$\lambda^*=0.348086$
	$j-1, j, j+1, j+2$	$\lambda^*=0.435831$
	$j, j+1, j+2, j+3$	unstable
Timeux	$j-2, j-1, j, j+1$	$\lambda^*=1/2$
	$j-1, j, j+1, j+2$	$\lambda^*=1/2$
	$j, j+1, j+2, j+3$	unstable

TABLE 1. Stable stencils for third order schemes

After substitution, the expression of the predicted values is given by:

$$
\begin{aligned}
w_j^{n+\beta} &= \tilde{D}_j^0(\overline{w}_j^n) - \beta\lambda[b_1(\beta)\tilde{D}_j^1(\overline{w}^n) + b_2(\beta)D_j^1(K_1) \\
&\quad + b_3(\beta)D_j^1(K_2) + b_4(\beta)D_j^1(K_3)], \qquad \beta = 1/2, 1
\end{aligned}
$$

$$
K_{1,j} = \tilde{D}_j^0(\overline{w}^n) - \frac{\lambda}{2}\tilde{D}_j^1(\overline{w}^n)
$$

$$
K_{2,j} = \tilde{D}_j^0(\overline{w}^n) - \frac{\lambda}{2}D_j^1(K_1)
$$

$$
K_{3,j} = \tilde{D}_j^0(\overline{w}^n) - \lambda D_j^1(K_2)
$$

Now we are ready to express (8) in terms of $\overline{w}^n$.

The amplification factor is obtained by looking for solutions of the form

$$
\overline{w}_j^n = \rho^n e^{ij\xi},
$$

where $i^2 = -1$. By substituting such expression in the numerical schemes one obtains

$$
\overline{w}_{j+\frac{1}{2}}^{n+1} = \rho_\lambda(\xi)e^{i\xi/2}\overline{w}_j^n, \qquad \xi \in [0, 2\pi].
$$

Stability is studied by analyzing the function

$$
P_\lambda(\xi) = |\rho_\lambda(\xi)|^2.
$$

Let λ^* be the maximum value of λ for which

$$
\max_{0\le\xi\le2\pi} |P_\lambda(\xi)| \le 1, \tag{10}
$$

We say that the scheme is stable if there exists $\lambda^* > 0$.

From the analysis of $P_\lambda(\xi)$ corresponding to the previous schemes one obtains the stability results summarized in Table 1 and 2, for third and fourth order schemes respectively.

The values of λ^* have been computed by solving an algebraic equation obtained from condition (10).

Fourth order schemes		
Scheme	Stencil	Stability region
RK3-2	$j-3, j-2, j-1, j, j+1$	unstable
	$j-2, j-1, j, j+1, j+2$	$\lambda^*=9/22$
	$j-1, j, j+1, j+2, j+3$	moder. unstable
	$j, j+1, j+2, j+3, j+4$	unstable
NCERK4	$j-3, j-2, j-1, j, j+1$	unstable
	$j-2, j-1, j, j+1, j+2$	$\lambda^*=9/22$
	$j-1, j, j+1, j+2, j+3$	moder. unstable
	$j, j+1, j+2, j+3, j+4$	unstable
Timeux	$j-3, j-2, j-1, j, j+1$	unstable
	$j-2, j-1, j, j+1, j+2$	$\lambda^*=1/2$
	$j-1, j, j+1, j+2, j+3$	moder. unstable
	$j, j+1, j+2, j+3, j+4$	unstable

TABLE 2. Stable stencils for fourth order schemes

The results of the analysis have been confirmed by the numerical results obtained by the above schemes.

For third order schemes based on Runge-Kutta, we find that stencil (-2) and (-1) are stable, and the stability region is slightly smaller than the one corresponding to scheme *Tadmor* and the schemes based on Taylor expansion. Stencil (0) is unstable.

For fourth order schemes based on Runge-Kutta, only the central stencil (-2) is stable. Stencil (-1) is moderately unstable for $\lambda < 0.4251$, and this instability is observed only after long integration time. Stencil (-3) and (0) are unstable.

Note that for the third order schemes based on ENO reconstruction there are two central stencils (i.e. (-2) and (-1)), and those are both stable. If equation $u_t - u_x = 0$ is considered, then the results for stencil (-2) and (-1) for third order schemes are reversed. In case of systems, because one does not want to do upwinding, the most restrictive CFL condition has to be used. For the fourth order schemes based on ENO there is one central stencil, and it is stable. In both cases, therefore, the stability does not depend on the sign of the characteristic velocity.

2.5. Numerical results

Several tests have been performed, and the results are presented in [3]. In particular the scalar equation has been used to test the accuracy of the schemes, and several test problems in gas dynamics have been considered, in order to study the shock capturing and high resolution properties of the schemes. The best performance, in terms of accuracy and efficiency, has been obtained by schemes NCERK2 and NCERK4. The first is a third order scheme based on MC-ENO reconstruction with piecewise cubic polynomials, and flux evaluation obtained by Runge-Kutta 2, with NCE of degree 2. The second is a fourth order scheme based on MC-ENO reconstruction with piecewise polynomials of degree 4, a nd Runge-Kutta 4 with NCE of degree 3. Both schemes show the prescribed accuracy and sharp shock resolution.

References

[1] Anile A.M., Romano V., and Russo G., *Hydrodynamical models for semiconductors based on Extended Thermodynamics*, SIAM J. Appl. Math., submitted.

[2] Arminjon P., Viallon M.-C. and Madrane A., *A Finite Volume Extension of the Lax-Friedrichs and Nessyahu-Tadmor Schemes for Conservation Laws on Unstructured Grids*, International Journal of Computational Fluid Dynamics, **9** (1997), 1–22.

[3] Bianco F., Puppo G., and Russo G., *High order central scheme for hyperbolic system of conservation laws*, in press on SIAM J. Sci. Comput.

[4] Harten A., Engquist B., Osher S., and Chakravarthy S. *Uniformly high order accurate essentially non-oscillatory schemes III*, J. Comp. Phys., **71** (1987), 231–303.

[5] Jiang, G.S., and Tadmor E., *Non-oscillatory Central Schemes for Multidimensional Hyperbolic Conservation Laws*, CAM Report 96-36, UCLA, October 1996.

[6] Levy D., and Tadmor E., *Non-oscillatory central differencing for the Incompressible 2-D Euler Equations*, CAM Report 96-37, UCLA, October 1996.

[7] Liu X.D., and Tadmor E. *Third order non-oscillatory central scheme for hyperbolic conservation laws*, UCLA preprint.

[8] Nessyahu H., and Tadmor E. *Non-oscillatory central differencing for hyperbolic conservation laws*, J. Comp. Phys., **87** (1990), 408–448.

[9] Romano V., and Russo G. *Numerical solution for hydrodynamical models of semiconductors*, Transactions on Computer-Aided Design of Integrated Circuits and Systems, submitted.

[10] Sanders R., and Weiser A. *High Resolution Staggered Mesh Approach for Nonlinear Hyperbolic Systems of Conservation Laws*, J. Comput. Phys., **101** (1992), 314–329.

[11] Shu C.W. *Numerical experiments on the accuracy of ENO and modified ENO schemes*, J. Sci. Comp., **5 (2)** (1990), 127–149.

[12] Shu C.W., and Osher S. *Efficient Implementation of Essentially Non-Oscillatory Shock-Capturing Schemes, II*, J. Comput. Phys. **83** (1989), 32–78.

[13] Zennaro M., *Natural Continuous Extensions of Runge-Kutta Methods*, Math. Comp. **46** (1986), 119–133.

Dipartimento di Matematica,
Università dell'Aquila,
Via Vetoio, loc. Coppito
67100 L'Aquila, Italy
E-mail address: russo@univaq.it

International Series of Numerical Mathematics
Vol. 129, © 1999 Birkhäuser Verlag Basel/Switzerland

A Simple Algorithm to Improve the Accuracy of TVD-MUSCL Schemes

G. Billet and O. Louedin

Abstract. An improvement of the accuracy of TVD-MUSCL scheme is based on the using of AUSM-VL splitting switch and a triad of limiters defined by the behavior of each physical quantity.

1. Introduction

The simulations of severe flow conditions, such as in the reactive supersonic mixing layers, requires robust numerical methods. Liou-Steffen [1] have proposed a remarkably simple upwind FVS called AUSM. AUSM keeps the qualities of FVS and recovers the accuracy attributed to FDS. Radelspiel-Kroll [2] include a switch from AUSM to vanLeer flux splitting (VL) to solve viscous flows correctly. The switch, presented here, is related with the local accuracy of the scheme. When the scheme degenerates into a first-order one (outside a shock wave), it is convenient to use AUSM; and when the scheme is a second- or third-order, VL is better in order to minimize the error terms. To capture strong and (or) rapid physical fluctuations accurately, the local variation of each quantity has to be incorporated as much as possible in the writing of the scheme. ENO schemes choose the stencil which provides the most regular solution in order to minimize numerical over and undershoots. In this paper, we take the stencil which minimizes the numerical error terms (dissipative and dispersive terms). These terms have different expressions following the local evolution of quantities. To improve efficiency, the equivalent system ES needs to be studied, including the expression for the slope limiters. Their expressions are controlled by the environing physical variation of the quantities. For each quantity, six different cases are considered. A triad of limiters is defined which minimizes or cancels the second-order truncation errors. A new explicit scheme is written. Compared with a common TVD-MUSCL scheme, it is no more complicated and it gives a more precise solution. It is applied to the 1D test case proposed by Shu-Osher [4] to simulate the interaction between a moving shock wave and a turbulent flow. The results show that we obtain the same precision as with their ENO scheme. The improved accuracy is also demonstrated by the computations of a 2D supersonic jet.

2. Flux splittings and MUSCL approach

The hyperbolic part of the conservation form of the 1D Navier-Stokes equations is classically written:

$$V_t + f_x = 0 \quad \text{with} \quad V = [\rho,\ \rho u,\ \rho E]^t,$$

$$V(x, 0) = V^0(x), \quad -\infty < x < +\infty, \quad t \geq 0 \tag{1}$$

where ρ, u and E are the density, the velocity and the total energy. In the discrete form, (1) is expressed as:

$$V^{n+1} = V_j - \sigma\left(F_{j+\frac{1}{2}} - F_{j-\frac{1}{2}}\right) \quad \text{with}$$

$$V_j = V_j^n = V(U_j^n), \ \sigma = \Delta t / \Delta x, \ F_{j+\frac{1}{2}} = F(U_{j-1}^n, U_j^n, U_{j+1}^n, U_{j+2}^n). \tag{2}$$

F is a numerical flux which has to verify $F(U, \ldots, U) = f(V)$, and $U_j^n = U_j$ is a set of physical quantities defined at the time $n\Delta t$ and at the grid point $j\Delta x$. Δx is assumed to be constant and Δt is related to Δx by the CFL condition. With the MUSCL approach [5], the backward and forward extrapolated values of $U_{j+\frac{1}{2}}$ at the interface $j + \frac{1}{2}$ can be written as:

$$U_{j+1/2}^L = L(U_j, U_{j+1}, \varphi_1) \ \text{and} \ U_{j+1/2}^R = R(U_j, U_{j+1}, \varphi_2^R).$$

At the interface $j - 1/2$, we have:

$$U_{j-1/2}^L = L(U_{j-1}, U_j, \varphi_2^L) \ \text{and} \ U_{j-1/2}^R = R(U_{j-1}, U_j, \varphi_1),$$

where φ_1 and $\varphi_2^{L,R}$ are non-linear functions of r_j with $r_j = \frac{U_j - U_{j-1}}{U_{j+1} - U_j}$. The non-linear interpolations L and R have to verify the following properties: homogeneity, translation invariance, left-right symmetry, monotonicity and convexity. The flux $F_{j+1/2}$ is written in the general form $F_{j+1/2} = F(U_{j+1/2}^L, U_{j+1/2}^R) - \Phi\Delta G$ where $\Phi\Delta G = \Phi[G(U_{j+1/2}^R) - G(U_{j+1/2}^L)]$ is a dissipation term. We are more particularly interested in FVS_{VL} and AUSM schemes. As in [2] we couple both schemes; but our coupling is different. It is based on an analysis of the ES and takes account of the properties of each scheme at the first and second-order. This coupling is advantageous because the expression for the fluxes is both very similar and yet exhibits different properties. In the case of a perfect gas, with the constant-pressure specific heat Cp and the specific heat ratio γ assumed to be constant, if we define

$$F_M^{FS} = M = F_M^+ + F_M^- = \frac{(M^L + 1)^2}{4} - \frac{(M^R - 1)^2}{4}$$

$$F_a^{FS} = \begin{bmatrix} 0 \\ p \\ 0 \end{bmatrix} = F_a^+ + F_a^- = \begin{bmatrix} 0 \\ p^L \frac{1+M^L}{2} + p^R \frac{1-M^R}{2} \\ 0 \end{bmatrix}, \ F_c^{DS} = \begin{bmatrix} \rho c \\ \rho c u \\ \rho c H \end{bmatrix},$$

then these both splittings can be written, at the grid point $j + \frac{1}{2}$ and for $-1 \leq M \leq 1$, as follows

– FVS$_{VL}$ scheme:

$$F = F_M^{FS} F_c^{DS} + F_a^{FS} = F_M^+ F_c^{DS}(U^L) + F_M^- F_c^{DS}(U^R) + F_a^+ + F_a^-$$
$$\Phi = 0 \tag{3}$$

– AUSM scheme:

$$F = (F_M^+ + F_M^-)\frac{F_c^{DS}(U^L) + F_c^{DS}(U^R)}{2} + F_a^+ + F_a^-$$
$$\Phi = |M| \quad \text{and} \quad \Delta G = \frac{1}{2}\left[F_c^{DS}(U^R) - F_c^{DS}(U^L)\right] \tag{4}$$

where c, p and H represent, respectively, the sound celerity, the static pressure, and the total enthalpy. It is possible to take different expressions for U. With the reactive flows where the temperature T changes rapidly and where the specific heats depend strongly on T, it seems natural to introduce T in the components of U. We have chosen $[\rho, u, T]^t$. The analysis of ES, obtained from Taylor expansions, quantifies the truncation error of the discrete form as Δx and $\Delta t \to 0$. U and F are assumed to be differentiable functions of C^4. For each component U_i, the expansions reflect the environing physical behaviour associated with the specific approach used here. Six different cases are considered for each component:

$$no\ extremum\ at\ j \quad \begin{bmatrix} \text{case 1: } monotonic\ evolution \\ \text{case 2: } extremum\ at\ the\ nodes\ j-1\ and\ j+1 \\ \text{case 3: } extremum\ at\ the\ node\ j-1\ or\ j+1 \end{bmatrix}$$

$$extremum\ at\ j \quad \begin{bmatrix} \text{case 4: } no\ extremum\ at\ the\ nodes\ j-1\ and\ j+1 \\ \text{case 5: } extremum\ at\ the\ nodes\ j-1\ and\ j+1 \\ \text{case 6: } extremum\ at\ the\ node\ j-1\ or\ j+1 \end{bmatrix}$$

3. First order error terms in space

After calculated the Taylor expansions of $r(U)$, $\varphi_\alpha(r(U))$ $(\alpha = 1, 2)$ and of the fluxes $\Psi = \Psi(\varphi_\alpha(r(U)))$ $\quad with \quad \Psi = F_M^-, F_M^+, F^{DS}, \ldots$ at the node j for both cases $U_x \neq 0$ and $U_x = 0$, (1) is transformed into:

$$V_t + F_x + \Delta x\,[A]\,U_{xx} + O(\Delta x^2) = 0 \tag{5}$$

where $[A]$ is a $(3,3)$ matrix. The first order error term in space, for the kth equation $(k = 1, \ldots, 3)$ of the system (5) and for the splitting (3), can be expressed:

$$Er_k = \sum_{i=1}^{3} A_{ki} U_{i_{xx}} = \frac{1}{2} \sum_{i=1}^{3} \left\{ \begin{array}{l} (F^{DS}F_i^{+\prime} + \frac{F^{FS}}{2}F_i^{L\prime} + \frac{\Phi}{2}G_i^{L\prime})\left[g_i^L\right] \\ -(F^{DS}F_i^{-\prime} + \frac{F^{FS}}{2}F_i^{R\prime} - \frac{\Phi}{2}G_i^{R\prime})\left[g_i^R\right] \end{array} \right\} U_{i_{xx}}.$$

For the mixed approach (4), the expression is a little more complicated:

$$Er_k = \frac{1}{2}\sum_{i=1}^{3}\left\{\begin{array}{l} (F_c^{DS}F_{M_i}^{+\prime} + \frac{F_M^{FS}}{2}F_{C_i}^{L\prime} + F_{a_i}^{+} + \frac{\Phi}{2}G_i^{L\prime})\,[g_i^L] \\ -(F_c^{DS}F_{M_i}^{-\prime} + \frac{F_M^{FS}}{2}F_{C_i}^{R\prime} + F_{a_i}^{-} - \frac{\Phi}{2}G_i^{R\prime})\,[g_i^R] \end{array}\right\}U_{i_{xx}} \quad \text{where}$$

$$g^{L,R} = g^{L,R}(\varphi) = \frac{\varphi_1(-1)}{2} + \frac{\varphi_2^{L,R}(3)}{2} - 1 \quad \text{if } U_x = 0,$$

$$g^{L,R} = g^{L,R}(\varphi') = \varphi_2'^{L,R}(1) - \varphi'_1(1) \quad \text{if } U_x \neq 0,$$

$$\varphi' = \frac{d\varphi}{dr},\ F^{+\prime} = \frac{dF^+}{dU^L},\ F^{-\prime} = \frac{dF^-}{dU^R},\ F^{L,R\prime} = \frac{dF^{DS}}{dU^{L,R}},\ G^{L,R\prime} = \frac{dG}{dU^{L,R}},\cdots$$

In order to develop a simple analysis, we assume at the node $j: |U_i^R - U_i^L| \ll Max(|U_i^R|, |U_i^L|)$, $(i = 1,\ldots,3)$. This says the jump at the interface j is considered to be weak (the strong discontinuities are excluded from the proof). The case $|U_i^R - U_i^L| \approx Max(|U_i^R|, |U_i^L|)$ is not considered in this paper, although it may be present in the velocity under certain circumstances, such as when this quantity has strong fluctuations around zero. The expansions are calculated for positive values of M. The expressions for $M < 0$ are obtained by symmetry (g_i^L is replaced by g_i^R and reciprocally). The first-order term cancels if $g_i^{L,R} = 0$. In general, we assume that

$$\text{for } r < 0,\ \varphi_\alpha = \varphi'_\alpha = 0 \quad \text{and} \quad \text{then} \quad \varphi_1 = \varphi_1(-1) = 0. \tag{6}$$

Therefore, the first-order error term cancels if

$$\varphi_2^L = \varphi_2^R = 2 \quad \text{for } r = 3\ (\textit{if extremum at node } j) \quad \text{or if} \tag{7}$$

$$\varphi_2'^L = \varphi_2'^R = \varphi'_1 \text{ for } r = 1\ (\textit{if not extremum at node } j). \tag{8}$$

The Taylor expansions at node j include the presence of one extremum (cases 4–6) or none (cases 1–3) at this point. On the other hand, they do not say whether one extremum exits or not at the neighbors $j - 1$ and $j + 1$. If there is no extremum associated with $j - 1$ and $j + 1$ (cases 1 and 4), any additional constraint appears; but if an extremum is present at these points, then either a different definition of φ (cases 2 and 3) is required in order to preserve the second-order accuracy or the scheme accuracy automatically degenerates (cases 5 and 6). If $U_{i_x} \neq 0$ at node j, condition (8) is easily met if the nodes $j - 1$, j and $j + 1$ have no extremum for component U_i (case 1). In this case, it is sufficient to take the same function in the second-order TVD domain for each point $j - 1$, j, $j + 1$. If one extremum exits for one or both neighbors of node j (case 2 or 3), the condition is more restrictive. Since we have $\varphi_2'^L = 0$ and/or, the second-order accuracy is ensured only if

$$\varphi'_1(1) = 0. \tag{9}$$

If $U_{i_x} = 0$ at point j, condition (7) is met if $j - 1$ and $j + 1$ are not associated with an extremum (case 4). But this condition is no longer met if there exists at least one extremum at one of the neighbors of j. For these cases (cases 5 and 6), the first-order error term does not cancel. Case 5 corresponds to local phenomena of wave length $2\Delta x$, and case 6 to phenomena of wave length $2\Delta x$ or $3\Delta x$. Therefore, when we have physical variations with wave length fluctuations greater than $3\Delta x$, the scheme is second-order in space if the expression of φ is well-defined. When wave length fluctuations are smaller than or equal to $3\Delta x$ (cases 5 and 6) the first-order error term is still present. In this case, the scheme has strong dissipative properties that can eliminate the numerical instabilities. For $M > 1$, the error terms have the same expression whatever the splitting; but for $M \leq 1$, their expression, $[A]_c^{AUSM}$ for Liou-Steffen and $[A]_c^{VL}$ for vanLeer, depends on the splitting chosen. For case 5, where $g_i^{L,R} = -1$, and for case 6 where $g_i^L = -1$ and $g_i^R = 0$, the error terms are written:

$$[A]^{AUSM} = [A]_c^{AUSM} + [A]_a^{VL},$$

$$[A]^{VL} = [A]_c^{VL} + [A]_a^{VL}, \text{ with}$$

$$[A]_c^{AUSM} = [A]_c^{VL} = -\begin{bmatrix} A_{11} & A_{12} & A_{13} \\ uA_{11} & uA_{12} + \rho A_{11} & uA_{13} \\ HA_{11} & HA_{12} + \rho u A_{11} & HA_{13} + \rho C_p A_{11} \end{bmatrix}, \text{ where}$$

$$A_{11} = \frac{c}{2} M, A_{12} = \frac{\rho}{2}(M\delta_{l5} + b\delta_{l6}), A_{13} = \frac{\rho c}{2T} M d\delta \text{ for } [A]_c^{AUSM},$$

$$A_{11} = \frac{c}{2}(a^2\delta_{l5} + b^2\delta_{l6}), A_{12} = \frac{\rho}{2}(M\delta_{l5} + b\delta_{l6}), A_{13} = \frac{\rho c}{2T} bd\delta \text{ for } [A]_c^{VL}.$$

$$a^2 = \frac{1+M^2}{2}, b = \frac{1+M}{2}, d = \frac{1-M}{2}, \delta = \delta_{l5} + \frac{\delta_{l6}}{2},$$

$$[A]_a^{VL} = -\frac{c^2}{2\gamma}\begin{bmatrix} 0 & 0 & 0 \\ \frac{2}{\rho}A_{12} & \frac{\rho}{c}\delta & \frac{1}{T}(A_{12} + \frac{\rho}{4}\delta_{l6}) \\ 0 & 0 & 0 \end{bmatrix}.$$

δ_{l5} and δ_{l6} are the Kronecker symbol, $l = 5$ (case 5) or 6 (case 6). The error term induced by AUSM splitting is always smaller. The differences become greater when $M \to 0$. For case 5, for example, when $M \to 0$, many terms cancel with AUSM splitting. From this study, we deduce

Condition 1: *AUSM splitting (4) is chosen when the scheme degenerates into a first-order scheme (cases 5 and 6).*

4. Second-order error terms in space

When, at node j, the variations of all the components U_i are included in cases 1–4, and φ_α are chosen well, the spatial derivatives are approximated by a second-order scheme in space. The expressions of φ_α are defined by studying the second-order error term in space associated with (2). The equation has the following expression:

$$V_t + F_x + \Delta x^2 (BU_{xxx} + CU_{xx} + DU_{xx} + EU_x) = O(\Delta x^3) \qquad (10)$$

where

$$
\begin{bmatrix}
B = B(\chi_1, U), \\
C = C(\varphi''_1, \varphi''^R_2, \varphi''^L_2, U, U_x) \text{ if } U_{i_x} \neq 0 \text{ at node } j \ (cases\ 1-3), \\
C \equiv 0 \qquad\qquad\qquad \text{if } U_{i_x} = 0 \text{ at node } j \ (case\ 4), \\
D = D(\chi_2, U, U_x), \qquad E = E(U, U_x), \\
\chi_1 = 1 - 3\varphi'_2 \qquad\qquad\qquad \text{and } \chi_2 = 1 - \varphi'_1 - \varphi'_2 \text{ if } U_{ix} \neq 0 \text{ at } j \\
\chi_1 = 2 + \varphi_1 - \varphi_2 + 2\varphi'_1 - 4\varphi'_2 \text{ and } \chi_2 = 2 + \varphi_1 - \varphi_2 \text{ if } U_{ix} = 0 \text{ at } j.
\end{bmatrix}
$$

By homogeneity, the cancellation of C for cases 1-3 gives the following condition on φ'': $\varphi''_1(1) = \varphi''^R_2(1) = \varphi''^L_2(1)$ .The terms $E_i U_x$ ($i = 1$ (conti. eq.), 2 (momen. eq.), 3 (ener. eq.)) come from fluxes that contain products of at least three primary quantities (for example $\rho u^2, \dots$). With AUSM and VL, the error terms are the same only for $E_i U_x$ and the dissipation term in the energy equation:

$$(E_1 U_x)^{AUSM} = \frac{1}{32} \left[-3\rho u_x (LogT)^2_x - cM\rho_x (LogT)^2_x + 2\rho cM (LogT)^3_x \right],$$

$$(E_1 U_x)^{VL} = 0$$

$$(E_2 U_x)^{AUSM} = \frac{1}{32} \left[\begin{array}{l} 16\rho_x u_x u_x + 4cM\rho_x u_x (LogT)_x - 4\rho u_x u_x (LogT)_x + \\ c^2 M^2 \rho_x (LogT)^2_x + 2\rho c^2 M^2 (LogT)^3_x \end{array} \right],$$

$$(E_2 U_x)^{VL} = \frac{1}{12} (\rho_x u_x u_x),$$

$$(E_3 U_x)^{AUSM} = \frac{1}{32} \left[\begin{array}{l} 20cM\rho_x u_x u_x + 16c^2 (C_2 - \frac{M^2}{4})\rho_x u_x (LogT)_x + \\ 2\rho cM u_x u_x (LogT)_x - c^3 M (3C_2 - \frac{M^2}{2})\rho_x (LogT)^2_x - \\ \rho c^2 (3C_2 + \frac{M^2}{2}) u_x (LogT)^2_x + \frac{3}{2} \rho c^3 M^3 (LogT)^3_x \end{array} \right],$$

$$(E_3 U_x)^{VL} = \frac{1}{8} \left[cM\rho_x u_x u_x + 3C_p \rho_x u_x T_x + \rho u_x u_x u_x \right],$$

$$(D_3 U_{xx})^{VL} = (D_3 U_{xx})^{AUSM} + \frac{\chi_2}{4} \left[C_p cM (\rho_x T_x)_x \right],$$

where $C_2 = \frac{C_P}{R} - 1$. In order to avoid the appearance of numerical oscillations corresponding to case 5 or 6, it is better to eliminate the dispersive error term BU_{xxx}. Although these oscillations are damped by the scheme, as we have seen in the previous paragraph, it is harmful to drop the scheme accuracy artificially if this is not necessary. Therefore, for cases 1 and 4, we let $\chi_1 = 0$. Applying conditions (7–9), we have

$$\varphi_2'(1) \;=\; \varphi_1'(1) = \frac{1}{3} \text{ if } U_{ix} \neq 0 \text{ at nodes } j-1,\ j \text{ and } j+1 \ (case\ 1) \quad (11)$$

$$\varphi_2'(3) \;=\; 0 \text{ if } U_{ix} = 0 \text{ at node } j \ (case\ 4) \qquad\qquad\qquad (12)$$

For cases 2 and 3, as the constraints (8–9) are already imposed, we have: $\chi_1 = 1$. Therefore, BU_{xxx} does not cancel for these cases. But in fact, it is possible to eliminate the dispersive term if we use a multi-time stepping scheme and if we apply different expressions of φ at every time step (not presented here). From conditions (7–9) and (11–12), it is possible to define an adequate limiter. In fact, we define *a triad of limiters, each adapted to the local variation of the physical quantities*. If at node j we have:

$$\text{case 1} \quad : \quad \varphi_1 = \varphi_2 = \varphi = \frac{r+2}{3} , \ (3^{rd} \ order\ extrapolation) \quad (13)$$

$$\text{cases 2 and 3} \quad : \quad \varphi_1 = \varphi_2 = \varphi = 1 , \ (centered\ interpolation) \qquad (14)$$

$$\text{case 4} \quad : \quad \varphi_1 = \varphi_2 = \varphi = \varphi_{superbee}, \qquad\qquad (15)$$

for cases 5 and 6, φ has to verify only the constraint (6). It is easy to see that the φ-triad (13–15) verifies this condition. Obviously, for all cases, the limiters φ have to lie in the second-order TVD area. From (13–15), $\chi_2 = \frac{1}{3}$ for case 1, $\chi_2 = 1$ for cases 2 and 3 and $\chi_2 = 0$ for case 4. If all the components U_i have an isolated extremum at j (case 4) at the same time, the second-order dissipative error terms vanish. For this case, $E_i U_x$ vanishes too. The scheme is then a third-order scheme in space. For the second-order error terms, the main difference between AUSM and VL splittings is in the expression of terms $E_i U_x$. As long as the temperature gradients are weak, these terms can be neglected. But for reactive flows, their values become unnegligible and the choice of the splitting becomes important. For this kind of flow, it is better to take VL splitting because, with it, the expressions for $E_i U_x$ are much simpler and remain the same as those associated with the supersonic flow. Therefore

Condition 2: *When the scheme remains second-order or third-order (cases 1–4), it is recommended to use VL splitting (3).*

Although simpler with this splitting, the $E_i U_x$ terms are not automatically negligible in particular when the temperature variations become high. So, it is to our advantage to eliminate these terms, which appear as additional transport terms in the conservation equations:

$$\rho_t + (\rho u)_x + \Delta x^2 \left(B_1 U_{xxx} + D_1 U_{xx} \right) \;=\; O(\Delta x^3)$$

$$(\rho u)_t + (\rho u^2 + p)_x + (\delta\rho\delta u)\frac{u_x}{12} + \Delta x^2 \left(B_2 U_{xxx} + D_2 U_{xx} \right) \;=\; O(\Delta x^3)$$

$$(\rho E)_t + (\rho u H)_x + \delta(\rho H)\frac{u_x}{8} + \Delta x^2 \left(B_3 U_{xxx} + D_3 U_{xx} \right) \;=\; O(\Delta x^3)$$

where $\delta(\rho H) = 3C_P \delta\rho\delta T + u\delta\rho\delta u + \rho\delta u\delta u$ and δU_i represents the variation of U_i on the mesh size Δx. Formally, the residual error terms can be corrected by adding the opposite value to the expression for the fluxes. For example, by defining the flux at the interfaces $j + \frac{1}{2}$ and $j - \frac{1}{2}$ in the following form:

$$F_{j+1/2}^{FS} = F^-(U_{j+1/2}^R) + F^+(U_{j+1/2}^L) - \delta Q_j u_{j+1/2},$$

$$F_{j-1/2}^{FS} = F^-(U_{j-1/2}^R) + F^+(U_{j-1/2}^L) - \delta Q_j u_{j-1/2},$$

$$u_{j+1/2} = cM = \frac{c^L + c^R}{2}(F_M^+ + F_M^-), \quad c^{L,R} = c(U^{L,R}),$$

$$\delta Q_j = \begin{bmatrix} 0 \\ \dfrac{\delta\rho\delta u}{12} \\ \dfrac{3C_P\delta\rho\delta T + u\delta\rho\delta u + \rho\delta u\delta u}{8} \end{bmatrix}_j,$$

where $\delta U_j = U_{j+1/2} - U_{j-1/2}$, the terms $E_i U_x$ disappear. This correction is activated only if the scheme remains a second or third-order scheme in space.

5. Multi-time stepping algorithm

The analysis of the previous sections was based on the hypothesis of a single time step. When we use a multi-time stepping scheme, two questions come to the mind: What is the effect of a multi-stepping scheme on the spatial error terms and on the CFL criterion? What is the minimum number of time steps needed to achieve sufficient accuracy? For the particular second-order scheme in time:

$$\begin{aligned} \tilde{V}_j &= V_j - \sigma(F_{j+1/2} - F_{j-1/2}) \\ V_j^{n+1} &= \frac{1}{2}\left[(V_j + \tilde{V}_j) - \sigma(\tilde{F}_{j+1/2} - \tilde{F}_{j-1/2})\right] \end{aligned}, \tag{16}$$

we show that if we choose the same limiters ($\varphi = \varphi_1 = \varphi_2$ for the predictor stage and $\tilde{\varphi} = \tilde{\varphi}_1 = \tilde{\varphi}_2$ for the corrector one) for each case considered, the spatial error terms are the same as those generated by a single-time stepping scheme. But, a more restrictive condition on the CFL number is introduced in order to the scheme (16) verifies (6) and remains second-order in space at a node j where, for one or several components, $U_{i_x} = 0$ and $U_{i_{xx}} \neq 0$ (condition 7):

$$\tilde{\varphi}(-a) = \tilde{\varphi}(-1/a) = 0 \text{ and } \tilde{\varphi}(b) = \tilde{\varphi}(c) = 2 \text{ with}$$

$$a = 1 + 4\sigma F_{i_{xx}}/(U_{i_{xx}} - 2\sigma F_{i_{xx}}), \quad 1/a = 1 - 4\sigma F_{i_{xx}}/(U_{i_{xx}} + 2\sigma F_{i_{xx}}),$$

$$b = 1 + 2U_{i_{xx}}/(U_{i_{xx}} - 2\sigma F_{i_{xx}}), \quad c = 1 + 2U_{i_{xx}}/(U_{i_{xx}} + 2\sigma F_{i_{xx}}),$$

$$\Rightarrow \sigma \leq \frac{1}{2}\left|\frac{U_{i_{xx}}}{F_{i_{xx}}}\right|.$$

This inequality associated with the classical condition gives a new condition on the local time step Δt_j :

$$\Delta t_j \leq Min \left[\frac{2\Delta x}{3Max\,|\lambda_i|}, \frac{\Delta x}{2Max\,|\frac{F_{i_{xx}}}{U_{i_{xx}}}|} \right]$$

where the values of the second derivatives act. λ_i represent the eigenvalues of the Jacobian A_J of f.

Knowing $V_{ttt} = - \left(f'^2_V (f'_V V_x) \right)_{xx}$ we can write the *ES*:

$$V_t + f_x - \frac{\Delta t^2}{6} (A^3_J V_x)_{xx} + 2^{nd} \ order \ spatial \ error \ terms = O(\Delta x^3, \Delta t^3).$$

A_J is known and therefore we can express the second-order error terms in time. Their expressions show that a higher order scheme in time (at least third-order) has to be applied so that the first-order and second-order terms are controlled by the spatial discretization only. Therefore, we use the third-order scheme in time proposed in [3] to solve (1).

6. Numerical results

In 1D, the example proposed by Shu and Osher [4] is interesting because it uses the Euler equations to simulate the interaction between a moving Mach 3 shock and a turbulent flow represented by sine waves in density. The initial conditions are described as:

$$\rho = 3.857143 \qquad u = 2.629369 \quad p = 10.33333 \quad \text{if } x < -4$$
$$\rho = 1 + 0.2 \ sin \ 5x \quad u = 0. \qquad\qquad p = 1. \qquad\quad \text{if } x \geq -4.$$

CFL number is equal to 0.5 and the final time is $t = 1.8$. The solid line representing the numerical solution with 1600 cells. Figs. 1a, 1b and 1c show the solution of the density field with 400 cells and the limiters *minmod, superbee*, and $\varphi = (r+2)/3$, applied separately. The limiters *minmod* and *superbee* give middling solutions. If AUSM-VL splitting (3–4) with the selected triad (13–15) is applied (Fig. 1d), the solution is comparable with that of the third-order ENO scheme. In particular, the high frequencies are well-represented and the compression waves and the shock are well-captured. The second test case is a 2D axisymmetric supersonic mixing layer. The inlet conditions are: central jet: $M = 1.74$, $p = 8\ 10^3$Pa, $T = 200$K; peripheral jet: $M = 2.$, $p = 8\ 10^3$Pa, $T = 580$K. Fig.2a shows an instantaneous view of the temperature field, with the scheme using AUSM splitting and the limiter $\varphi = (r + 2)/3$, and Fig. 2b shows the same view, with the same time-stepping scheme but using the *conditions 1 and 2* and the triad of limiters. The transitional zone (A) shows a greater sensitivity of the scheme proposed here to the physical instabilities. We can also see the very weak diffusion of the scheme in

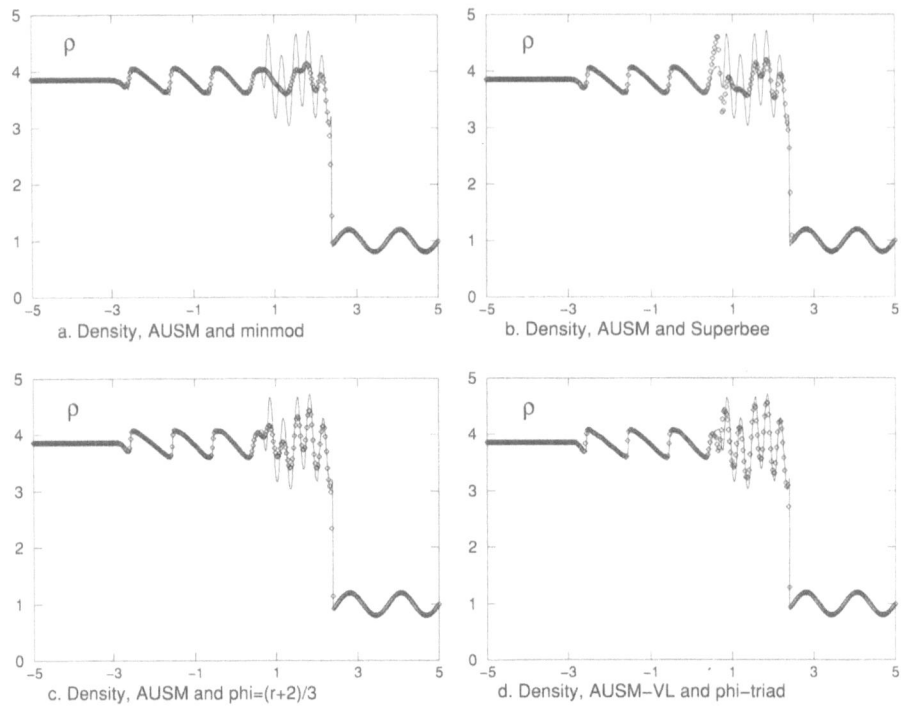

FIGURE 1

the shear layer. In the growth of the large eddies (B), the mixing in the core of
the eddies is more detailed with the method presented here.

7. Conclusion

This paper shows it is possible to improve the accuracy of TVD-MUSCL approach
if: ○ the accuracy in time is greater than the accuracy in space, ○ the non-linear
functions φ are expressed in a triad (13–15) taking into account the local variations
of each quantity, ○ AUSM is used when the scheme degenerates into a first-order
and VL is AUSM is used when the scheme degenerates into a first-order, and VL
is applied when the scheme remains second or third-order. Adding to the basic
well-known avantages of the algorithm proposed herein, the good accuracy of the
numerical solution opens new perspectives for TVD-MUSCL schemes. In particular, Large Eddy Simulations (LES) now seem to come within the field of application
of these schemes. Algorithmically, the correction proposed herein is easily to implement and the additional time consuming is very small. New improvements are

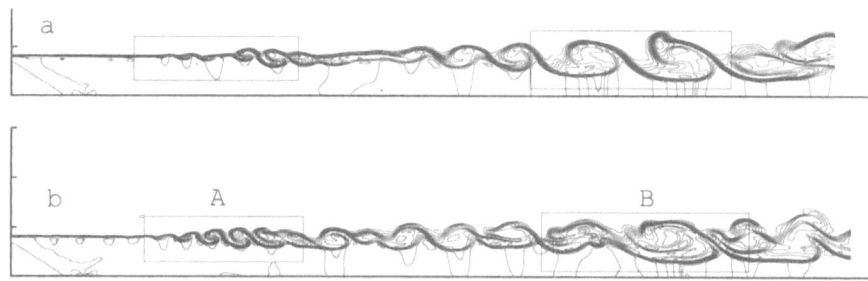

FIGURE 2

possible by using different expressions of φ at each time step of the time integration. Stability study and simulations of freely decaying isotropic turbulence are performing to evaluate the reliability of this scheme in LES.

References

[1] M.S. Liou and Ch. Steffen, *A new flux splitting scheme,* J. Comput. Physics, **107** (1) (1993), 23.

[2] R. Radespiel and N. Kroll, *Accurate Flux Vector Splitting for Shocks and Shear Layers,* J. Comp. Physics, **121** (1995), 66–78 .

[3] C.-W. Shu and S. Osher, *Efficient implementation of essentially non-oscillatory shock-capturing schemes,* J. Comp. Physics, **77** (1988), 439–471.

[4] C.-W. Shu and S. Osher, *Efficient implementation of essentially non-oscillatory shock capturing schemes, II,* CAM Report 88-12, UCLA, April 1988.

[5] B. vanLeer, *Towards the ultimate conservative difference scheme. V. A second order sequel to Godunov's method,* J. Comp. Physics, **32** (1979), 101–136.

ONERA, Chatillon, France
E-mail address: billet@onera.fr

International Series of Numerical Mathematics
Vol. 129, © 1999 Birkhäuser Verlag Basel/Switzerland

Hyperbolic Initial Boundary Value Problems on the Stability of Strong Discontinuities in Continuum Mechanics

Alexander M. Blokhin and Yuri L. Trakhinin

Abstract. A mathematical approach to the investigation of the structural stability of strong discontinuities in hyperbolic models of Continuum Mechanics is presented. Advantages of the described approach are discussed on the example of recent results obtained by the authors for strong discontinuities in Magnetohydrodynamics, Magnetohydrodynamics of Chew, Goldberger and Low with the pressure anisotropy and models of Radiation Hydrodynamics.

1. Introduction

Hyperbolic conservation laws are widely applied in the description of actual processes in Continuum Mechanics. It is known that in these processes strong discontinuities (for example, shock waves) present while a medium is in motion. During the last years such discontinuous flows are advantageously calculated numerically. However, before using computers one should be sure that a surface of strong discontinuity is structurally stable, i.e., it preserves its structure under small perturbations. Note that unstable strong discontinuities don't really exist. Even if the conditions of existence (in the first moment) of unstable discontinuity are created artificially in a physical experiment, then this discontinuity decays immediately.

In this connection, the problem on structural stability of strong discontinuities is of great theoretical and practical importance. It is clear that the solving of this problem is the first and necessary step to investigate discontinuous flows in Continuum Mechanics. In particular, structural stability is necessary condition to prove the convergence of the numerical solution to the weak solution of a corresponding hyperbolic system.

Mathematically the question on the stability of strong discontinuity is reduced to the investigation of some linear initial-boundary value problem (IBVP).

1.1. Standard physical approach

First works devoted to the stability of strong discontinuities were carried out by physicists and relate to Gas Dynamics (see, for example, [1] and [2]). In these works an approach, which can be named now as standard, is suggested.

This approach is as follows. The above-mentioned linear IBVP is formulated and exponential solutions to this problem are sought. A conclusion about the stability or instability of a strong discontinuity is drawn by the behaviour of these solutions. So, in [1] the exponential solution to the linear IBVP on the stability of plane gas dynamic shock wave is sought in the form

$$\mathbf{U} = \mathbf{U}_0 \exp\{i(-\omega t + l x_1 + k x_2)\}, \tag{1}$$

$$t \geq 0, \quad x_1 > 0, \quad x_2 \in \mathbb{R}$$

where $\mathbf{U}_0$ is a constant vector; ω, l, k are constants; t, the time; (x_1, x_2), the Cartesian coordinates. If there is such a solution that

$$\text{Im}\,\omega > 0, \quad \text{Im}\,l > 0, \quad \text{Im}\,k = 0, \tag{2}$$

then the shock wave is unstable. Otherwise, it is stable.

Note that in some works (see, for example, [3]) so-called "neutral stable" strong discontinuities are indicated. It is the case when the stability IBVP doesn't have solutions in form (1) with property (2), but it has the exponential solution with

$$\text{Im}\,\omega = \text{Im}\,l = \text{Im}\,k = 0. \tag{3}$$

Thus, the described standard approach is, in fact, an approach to investigate linear stability. In the present paper we will give arguments which show that in the field of "neutral stability" it is impossible to judge about stability or instability on the linear level of investigation.

1.2. "Equational" approach

We will present another so-called "equational" approach to the investigation of the structural stability of strong discontinuities. This approach was firstly applied in the monograph [4] to the investigation of gas dynamics shock waves. The "equational" approach implies the investigation of well-posedness of the above-mentioned linear IBVP by means of dissipative energy integrals method. Here the term "well-posedness" is used in a classical sense. In the case the IBVP is well-posed the strong discontinuity is structurally stable. Otherwise, the strong discontinuity is unstable.

It will be shown that such an approach gives a connection between linear and nonlinear stability. The term "nonlinear stability" means the well-posedness of IBVP for the initial quasilinear hyperbolic system of conservation laws with boundary conditions on a curvilinear shock front.

In some sense, the main idea of the presented approach is very close to the approach of A. Majda [5] with distinction that he uses as a constructive element the analysis based on a kind of Kreiss-Lopatinsky condition [6], but not theory of dissipative energy integrals.

1.3. Applications of the "equational" approach

Earlier the described "equational" approach proved to be very effective in applications connected with strong discontinuities in the following hyperbolic models of Continuum Mechanics: Gas Dynamics [4], Magnetohydrodynamics [7], Landau equations of superfluid helium [8], relativistic Gas Dynamics [9] etc.

In this paper we present recent results which were obtained with the help of "equational" approach. These results relate to strong discontinuities in two models of Magnetohydrodynamics (MHD) and two models of Radiation Hydrodynamics (RHD).

The first MHD model is classical MHD for an ideal medium. The second one is the Chew, Goldberger and Low (CGL) MHD equations [10] with the pressure anisotropy (with two pressures $p^\|$, $p^\perp$). The MHD CGL equations describe the movement of a collisionless strongly magnetized plasma.

The structural stability of radiation shock waves is considered for the cases of nonrelativistic (immovable continuum) and relativistic RHD. These RHD equations are obtained by Anile, Pennisi and Sammartino [11], [12] and closed with the help of introduction into consideration the Addington factor [13].

A part of the presented results was performed during the fellowship of one of the authors (Yu. L. T.) at the Alexander von Humboldt Foundation.

2. Basic steps of the approach

The announced approach to the investigation of the structural stability of strong discontinuities in hyperbolic models of Continuum Mechanics, which is called by us as "equational" approach, has the following basic steps:

- Symmetrization of initial quasilinear system of conservation laws
- Determination of the equations of strong discontinuity for the initial quasilinear equations of Continuum Mechanics
- Linearization of initial quasilinear equations and relations on the surface of a strong discontinuity. Formulation of the linear IBVP on the structural stability of strong discontinuity
- Separation of ill-posedness domains for the formulated IBVP (Hadamard example);
- Construction of a priori estimates without loss of smoothness for the stability IBVP with the help of dissipative energy integrals techniques in such domains, where Hadamard example is not constructed.

If we succeeded in realizing this approach in full volume, then we have the rigorous mathematical basis of linearization method applied to the investigation of structural stability of strong discontinuities. It is very important that obtaining a priori estimates are without loss of smoothness, because in this case the uniform Kreiss-Lopatinsky condition [6] holds, and we can in principle carry our result to the initial level of quasilinear system of hyperbolic conservation laws with boundary conditions on a curvilinear surface of strong discontinuity. For example, it was performed in [4] for gas dynamic shock waves, and the local theorem (with respect to time) of existence and uniqueness of the classical solution to the quasilinear Gas Dynamics system behind the curvilinear shock wave was proved. Note that the presence of a priori estimate with loss of smoothness (it is the case when only the Kreiss-Lopatinsky condition (LC) holds, and the uniform Kreiss-Lopatinsky condition (ULC) is not fulfilled) does not allow to carry the main results obtained for the case of constant coefficients to the case of variable coefficients (and, so, to a quasilinear case).

It is necessary also to concentrate attention to the following point. In the case when the LC is fulfilled, and the ULC is not fulfilled for the stability IBVP we cannot judge about the stability or instability of strong discontinuity. For example, it is shown in [4] that for the IBVP on the stability of gas dynamic shock waves one can so perturb the system and the boundary conditions that an ill-posedness example of Hadamard type can be constructed for the perturbed IBVP in the domain where the LC is fulfilled, and the ULC is not fulfilled. This domain corresponds to case (3). Thus, so-called "neutral stability" can be found in practice as instability. For this reason, it is necessary to investigate the stability of "neutral stable" strong discontinuities on the initial nonlinear level.

The "equational" approach doesn't also give a solution of the problem on the structural stability of strong discontinuities in the domains where the LC is fulfilled, and the ULC is not fulfilled. But outside of such domains if we have a priori estimates without loss of smoothness for the linear IBVP, then these a priori estimates can be carried to the quasilinear level, and we can conclude of the structural – nonlinear – stability of strong discontinuity.

Let us discuss the mentioned basic steps of the presented approach more in detail. The scheme of symmetrization of quasilinear systems of conservation laws was developed by S.K.Godunov in [14], [15] (see also [16], [17], [18], [19], [20]). After symmetrization the initial system of conservation laws can be presented in the following symmetric form:

$$A^0(\mathbf{Q})\mathbf{Q}_t + \sum_{k=1}^{3} A^k(\mathbf{Q})\mathbf{Q}_{x_k} = 0 \,,$$

where t, the time, $\mathbf{x} = (x_1, x_2, x_3)$, the Cartesian coordinates; A^α are symmetric matrices $(A^\alpha = (A^\alpha)^*)$; $\mathbf{Q} = (q_1, \ldots, q_n)^*$, the vector of canonic variables. This system can also be rewritten as a symmetric system in terms of the initial vector of unknowns $\mathbf{U} = (u_1, \ldots, u_n)^*$ (see [20]):

$$A_0(\mathbf{U})\mathbf{U}_t + \sum_{k=1}^{3} A_k(\mathbf{U})\mathbf{U}_{x_k} = 0 \,, \tag{4}$$

where $A_\alpha = (A_\alpha)^*$. System (4) is symmetric t-hyperbolic (in the sense of Friedrichs [17]) if $A_0 > 0$.

Let the surface of strong discontinuity has the equation

$$f(t, \mathbf{x}') - x_1 = 0, \quad \mathbf{x}' = (x_2, x_3). \tag{5}$$

Performing the usual procedure we can write out jump conditions on surface (5) for the considered equations of Continuum Mechanics. For instance, for the equations of Gas Dynamics the obtained conditions on the surface of strong discontinuity are the well-known Rankine-Hugoniot conditions.

Linearising system (1) and conditions on the surface of strong discontinuity with respect to the piecewise constant solution

$$\mathbf{U}(t, \mathbf{x}) = \begin{cases} \hat{\mathbf{U}} = \text{const}, & x_1 > 0 ; \\ \hat{\mathbf{U}}_\infty = \text{const}, & x_1 < 0 ; \end{cases}$$

which satisfies jump conditions on the surface $x_1 = 0$, we obtain a linear mixed problem to determine the vector of small perturbations $\delta \mathbf{U}$ (in what follows $\delta \mathbf{U} = \mathbf{U}$) and the small disturbance of surface of discontinuity front $F = \delta f = F(t, \mathbf{x}')$.

Thus, the question on the stability of strong discontinuities is reduced to the investigation of IBVP for symmetric t-hyperbolic systems with constant real coefficients in the following form.

Problem 2.1. (Main Problem) *We seek the solutions of the systems*

$$A_0 \mathbf{U}_t + \sum_{k=1}^{3} A_k \mathbf{U}_{x_k} = 0, \quad t > 0, \qquad \mathbf{x} \in \mathbb{R}_+^3 ; \tag{6}$$

$$A_{0\infty} \mathbf{U}_t + \sum_{k=1}^{3} A_{k\infty} \mathbf{U}_{x_k} = 0, \quad t > 0, \quad \mathbf{x} \in \mathbb{R}_-^3 ; \tag{7}$$

satisfying the following boundary conditions

$$M \begin{pmatrix} \mathbf{U}|_{x_1 \to -0} \\ \mathbf{U}|_{x_1 \to +0} \\ F_t \\ F_{x_2} \\ F_{x_3} \end{pmatrix} = 0 \tag{8}$$

at $x_1 = 0$ $(t > 0, \quad \mathbf{x}' \in \mathbb{R}^2)$ *and the initial data*

$$\mathbf{U}|_{t=0} = \mathbf{U}_0(\mathbf{x}), \quad \mathbf{x} \in \mathbb{R}_\pm^3, \quad F|_{t=0} = F_0(\mathbf{x}'), \quad \mathbf{x}' \in \mathbb{R}^2 \tag{9}$$

under $t = 0$.

Here $A_\alpha = A_\alpha(\hat{\mathbf{U}}) = A_\alpha^*$, $A_{\alpha\infty} = A_\alpha(\hat{\mathbf{U}}_\infty) = A_{\alpha\infty}^*$, $A_0 > 0$, $A_{0\infty} > 0$; M, a constant rectangular matrix of order $m \times (2n + 3)$.

Remark 2.2. Before the investigation of well-posedness of the Main problem one should be sure that the geometrical Lax condition [21] is fulfilled, i.e., $m = n^+(A_1) + n^-(A_{1\infty}) + 1$, where n^+ (n^-) is the number of positive (negative) eigenvalues of a matrix. Note that the Lax condition is necessary condition for the well-posedness of the Main problem, i.e., as well for the structural stability of strong discontinuity, which is said evolutionary if the Lax condition is fulfilled.

If boundary conditions (8) are dissipative, i.e.,

$$[(A_1 \mathbf{U}, \mathbf{U})]|_{x_1=0} = (A_1 \mathbf{U}, \mathbf{U})|_{x_1 \to +0} - (A_{1\infty} \mathbf{U}, \mathbf{U})|_{x_1 \to -0} \leq 0, \qquad (10)$$

then, using the standard technique of dissipative energy integrals (see, for example, [22]), we can obtain the following a priori estimate:

$$I(t) \leq I(0) \exp(N_1 t)$$

or

$$\|\mathbf{U}\|_{L_2(\mathbb{R}^3_+)} + \|\mathbf{U}\|_{L_2(\mathbb{R}^3_-)} \leq N_2 \left\{ \|\mathbf{U}_0\|^2_{L_2(\mathbb{R}^3_+)} + \|\mathbf{U}_0\|^2_{L_2(\mathbb{R}^3_-)} \right\}^{1/2}, \qquad 0 < t \leq T < \infty,$$

where $N_1 > 0$, $N_2 = N_2(T) > 0$ are constants;

$$I(t) = \iiint\limits_{\mathbb{R}^3_+} (A_0 \mathbf{U}, \mathbf{U}) \mathrm{d}\mathbf{x} + \iiint\limits_{\mathbb{R}^3_-} (A_{0\infty} \mathbf{U}, \mathbf{U}) \mathrm{d}\mathbf{x}.$$

The a priori estimate for the function $F(t, \mathbf{x}')$ is obtained due to the specific character of boundary conditions (8).

However, this simple idea to obtain a priori estimates, generally saying, does not suit for the most of stability problems because usually condition (10) is not fulfilled. In such a case instead of initial system one can try to construct a so-called expanded system (a system for the vector $\mathbf{U}$ and its derivatives) and dissipative boundary conditions for it. Such expanded systems and dissipative boundary conditions for them for the stability IBVP for gas dynamic shock waves are described in [4], where the stability theorem in the form of the following theorem (2D case) is proved.

Theorem 2.3. (Main Theorem) *Main Problem (6)–(9) is well-posed, and also the following a priori estimates take place for the solutions of this problem:*

$$\|\mathbf{U}(t)\|_{W_2^2(\mathbb{R}^2_+)} \leq K_1 \|\mathbf{U}_0\|_{W_2^2(\mathbb{R}^2_+)}, \qquad 0 < t \leq T < \infty, \qquad (11)$$

$$\|F\|_{W_2^3((0,T)\times\mathbb{R})} \leq K_2, \qquad (12)$$

where $K_{1,2} > 0$ are constants depending on T.

Here a priori estimate (11) is written out in the semiplane $\mathbb{R}^2_+$ (behind the stationary strong discontinuity $x_1 = 0$) because for the IBVP on the stability of gas dynamic shock waves all the eigenvalues of the matrix $A_{0\infty}^{-1} A_{1\infty}$ are positive. So, without loss of generality we can assume that $\mathbf{U}(t, \mathbf{x}) \equiv 0$ under $x_1 < 0$.

In view of previous reasonings, a stability theorem like to the Main theorem gives, in fact, the full answer on the question on the structural stability of strong discontinuity.

3. Structural stability of MHD strong discontinuities

Now we present results on the structural stability of MHD strong discontinuities (in classical MHD and MHD CGL) obtained with the help of "equational" approach.

3.1. Stability of strong discontinuities in MHD for an ideal medium

Unlike the situation in Gas Dynamics [1], [2], [4], the stability problem for MHD strong discontinuities (for example, MHD shock waves) isn't fully investigated. After the publication of classical works [23], [24], [25] only a few works related to this topic can be named (see, for example, [26], [27] and the mentioned monograph [7]).

Applying the "equational" approach the following results are managed to obtain during the last years (see also [7]). It is known there exist two types of evolutionary (see remark 2.2) MHD shock waves: fast and slow shock waves. The structural stability of fast shock waves in the polytropic gas is proved in the asymptotic case of weak magnetic field (see [28]). The stability theorem in the form of Main theorem is proved by means of the dissipative energy integrals method. We can apply this method thanks to the existence of symmetric form of the MHD system which is obtained in [15]. It is shown the rotational discontinuity can be unstable. The instability of rotational discontinuity under the strong magnetic field is proved (see [29]).

3.2. Stability of strong discontinuities in MHD CGL

The system of MHD CGL with the pressure anisotropy was symmetrized in [30] (see also [31]). So, we can apply the energy integrals method to the investigation of the structural stability of strong discontinuities in this model.

As in usual MHD, there are two types of evolutionary shock waves in MHD CGL: fast and slow shock waves. The stability of fast shock waves is proved in the asymptotic case of high pressure $(p^{\|}\gg|\mathbf{H}|^2/(4\pi)$, $p^{\perp}\gg|\mathbf{H}|^2/(4\pi)$; $p^{\|}$, $p^{\perp}$, the longitudinal and transversal pressures; $\mathbf{H}$, the magnetic field) [32]. The instability of slow parallel shock waves and rotational discontinuity is proved in the asymptotic case of cold plasma $(p^{\|}\ll|\mathbf{H}|^2/(4\pi)$, $p^{\perp}\ll|\mathbf{H}|^2/(4\pi))$ [31].

The structural stability of fast parallel and transversal shock waves in MHD CGL is investigated in the general nonasymptotic case too [33]. It is shown that fast parallel shock waves are always stable. The stability domain for fast transversal shock waves is found. Structural stability is proved by means of the test of fulfilment of the ULC (a priori estimates in the form of (11), (12) are not yet obtained).

4. Structural stability of radiation shock waves

The symmetric form of the mentioned models of RHD were obtain in [11], [12] (see also [34]). Concerning the model of relativistic RHD, a natural classification in fast and slow relativistic shock waves is introduced in [35]. The stability problem for relativistic shock waves is completely solved under the assumption of smallness of quantities of a radiation energy density on the surface of a stationary discontinuity. The stability of fast and the instability of slow radiation shock waves are proved [35].

Shock waves in the case of the immovable continuum (nonrelativistic RHD) are stable under general assumptions (see [36], [37] for the 2D and 3D cases). Stability theorems for relativistic and nonrelativistic radiation shock waves have the form of the Main Theorem for the 3D case when estimates are written out in the spaces $W_2^2(\mathbb{R}_+^3)$ and $W_2^3((0,T)\times\mathbb{R}^2)$.

References

[1] S. P. D'iakov, *On stability of shock waves*, Zh. Eksperim. i Teor. Fiz., **27** (1954), 288–296.

[2] J. J. Erpenbeck, *Stability of step shocks*, Phys. Fluids, **5** (1962), 1181–1187.

[3] S. A. Egorushkin and A. G. Kulikovsky, *On the stability of solutions of some boundary value problems for hyperbolic systems*, PMM, **56** (1992), 40–51.

[4] A. M. Blokhin, *Energy integrals and their applications to gas dynamics problems*, Nauka, Novosibirsk, **1986**.

[5] A. Majda, *Compressible Fluid Flow and Systems of Conservation Laws in Several Space Variables*, Springer-Verlag, New York, **1984**.

[6] H.-O. Kreiss, *Initial boundary value problems for hyperbolic systems*, Commun. Pure and Appl. Math., **23** (1970), 277–296.

[7] A. M. Blokhin, *Strong Discontinuities in Magnetohydrodynamics*, Nova Science Publishers, New York, **1993**.

[8] A. M. Blokhin and V. N. Dorovsky, *Mathematical Modelling in the Theory of Multivelocity Continuum*, Nova Science Publishers, New York, **1995**.

[9] A. M. Blokhin and E. V. Mishchenko, *Investigation on shock waves stability in relativistic gas dynamics*, Le Matematiche, **48** (1993), 53–75.

[10] G. Chew, M. Goldberger and F. Low, *The Boltzmann equation and the one-fluid hydromagnetic equations in the absence of particle collisions*, Proc. R. Soc. Lond., **A 236** (1956), 112.

[11] A. M. Anile, S. Pennisi and M. Sammartino, *Covariant radiation hydrodynamics*, Ann. Inst. Henri Poincaré, **56** (1992), 49–74.

[12] A. M. Anile, S. Pennisi and M. Sammartino, *A thermodynamical approach to Eddington factors*, J. Math. Phys., **32** (1991), 544–550.

[13] G. M. Kremer and I. Müller, *Radiation thermodynamics*, J. Math. Phys., **33** (1992), 2265–2268.

[14] S. K. Godunov, *An interesting class of quasilinear systems*, DAN, **39** (1961), 521–523.

[15] S. K. Godunov, *Symmetrization of magnetohydrodynamics equations*, Chislennye Metody Mekhaniki Sploshnoi Sredy, Novosibirsk, **3** (1972), 26–34.

[16] K. O. Friedrichs and P. D. Lax, *System of conservation equations with a convex extension*, Proceedings of the National Academy of Science, U.S.A, **68** (1971), 1686–1688.

[17] K. O. Friedrichs, *Symmetric hyperbolic linear differential equations*, Commun. Pure and Appl. Mathem., **31** (1974), 123–131.

[18] G. Boillat, *Sur l'existence et la recherche d'équations de conservation supplémentaires pour les systèmes hyperboliques*, Comptes Rendues de l'Academie des Sciences, **278A** (1974), 909–912.

[19] T. Ruggeri and A. Strumia, *Mail field and convex covariant density for quasilinear hyperbolic systems*, Ann. Inst. Henri Poincaré, **34** (1981), 65–84.

[20] A. M. Blokhin, *Symmetrization of continuum mechanics equations*, Sib. J. Diff. Eq., **2** (1995), 3–47.

[21] A. Jeffrey, *Quasilinear Hyperbolic Systems and Waves*, Pitman, New York, **1976**.

[22] R. Courant and D. Hilbert, *Methods of Mathematical Physics*, Interscience Publishers, New York, **1962**.

[23] F. Hoffman and E. Teller, *Magnetohydrodynamic shocks*, Phys. Rev., **80** (1950), 696–703.

[24] A. I. Akhiezer, G. Ya. Liubarskii and R. V. Polovin R.V., *On the stability of shock waves in magnetohydrodynamics*, JETP, **8** (1959), 507.

[25] C. S. Gardner and M. D. Kruskal, *Stability of plane magnetohydrodynamic shocks*, Phys. Fluids, **7** (1964), 700–706.

[26] M. Lessen and M. V. Deshpande, *Stability of magnetohydrodynamic shocks waves*, J. Plasma Physics, **1** (1967), 463–472.

[27] O. V. Fillipova, *Stability of plane MHD shock waves in the ideal gas*, Izv. Akad. Nauk SSSR Ser. Mekh. Zhidk. Gaza, No. 6 (1991).

[28] A. M. Blokhin and Yu. L. Trakhinin, *Investigation of the well-posedness of the mixed problem on the stability of a fast shock wave in magnetohydrodynamics*, Le Mathematiche, **49** (1994), 123–141.

[29] A. M. Blokhin and Yu. L. Trakhinin, *Rotational Discontinuity in Magnetohydrodynamics*, Sibirsk. Mat. Zh., **34** (1993), 395–411.

[30] A. M. Blokhin and D. A. Krymskikh, *Symmetrization of equations of magnetohydrodynamics with anisotropic pressure*, in: collected sientific papers "Boundary value problems for partial equations", Institute of Mathematics, Siberian Division of the Academy of Science of USSR, Novosibirsk, 1990, 3–19.

[31] A. M. Blokhin and Yu. L. Trakhinin, *Stability of Strong Discontinuities in Plasma with Anisotropic Pressure*, J. Magnetohydrodynamics and Plasma Res., **4** (1994), 109–207.

[32] A. M. Blokhin and Yu. L. Trakhinin, *Investigation of the fast magnetohydrodynamic shock wave stability in plasma with anisotropic pressure*, Prikl. Mekh. Tekhn. Phys., **36** (1995), 16–35.

[33] A. M. Blokhin and Yu. L. Trakhinin, *Stability of fast parallel and transversal shock waves in plasma with pressure anisotropy*, in preparation.

[34] A. M. Blokhin, V. Romano and Yu. L. Trakhinin, *Some mathematical properties of radiating gas model obtained with a variable Eddington factor*, Z. angew. Math. Phys., **47** (1996), 639–658.

[35] A. M. Blokhin, V. Romano and Yu. L. Trakhinin, *Stability of shock waves in relativistic radiation hydrodynamics*, Ann. Inst. Henri Poincaré, **67** (1997), 145–180.

[36] A. M. Blokhin and Yu. L. Trakhinin, *Stability of shock waves for one model of Radiation Hydrodynamics*, Prikl. Mekh. Tekhn. Phys., **37** (1996), 3–14.

[37] A. M. Anile, A. M. Blokhin and Yu. L. Trakhinin, *Investigation of a mathematical model for Radiation Hydrodynamics*, in preparation.

A. M. Blokhin
Sobolev Institute of Mathematics,
Russian Academy of Sciences,
Universitetsky pr. 4,
630090 Novosibirsk, Russia
E-mail address: blokhin@math.nsc.ru

Yu. L. Trakhinin
Sobolev Institute of Mathematics,
Russian Academy of Sciences,
Universitetsky pr. 4
630090 Novosibirsk, Russia

International Series of Numerical Mathematics
Vol. 129, © 1999 Birkhäuser Verlag Basel/Switzerland

On Hyperbolic Integro-differential von Kármán Equations for Viscoelastic Plates

Igor Bock

Abstract. The nonlinear system of equations describing the dynamics of viscoelastic plates made of short memory material is derived. The system contains the nonlinear hyperbolic equation for great deflections and the 2. kind Volterra integro-differential equation for the Airy stress function. Expressing the Airy stress function through the deflection we obtain the hyperbolic canonical equation with a pseudoparabolic main part and a nonlinear term involving the memory integral term. The existence theorem and the local uniqueness for the corresponding initial-boundary value problem will be verified.

1. Introduction

Nonlinear hyperbolic equations describing vibrations of elastic von Kármán plates were investigated for example in [6], [10], [15], where mainly the existence of a solution was derived. Fully dynamic systems have been studied in [7], [14]. Nowadays the boundary controllability as well as the decay rates of solutions for dynamic von Kármán plates are studied for instance in [5], [8], [9], [13]. The stabilization problem with a memory term has been solved in [11]. Here we consider pseudohyperbolic evolutionary von Kármán equations describing large deflections of thin viscoelastic plates acting under the dynamic load. In addition to previous papers the Airy stress function is a solution of the Volterra type integro-differential equation. The second equation of the system is the hyperbolic equation with a pseudoparabolic main part and the same nonlinear part as in the elastic case. Using the approach from the quasistationary case in [1] the canonical nonlinear hyperbolic equation with a memory term will be derived. The initial-boundary value problem involving this equation will be transformed into the parabolic one and solved using the method of the elliptic regularization. The a priori estimates and the uniqueness conditions for a local time interval will be derived.

This was supported by Grant 1/5094/98 of the Grant Agency of the Slovak Republic.

2. Formulation of the problem

Let us assume a thin viscoelastic plate made of a short memory material of the differential type ([2], [3]). It occupies the domain

$$Q = \{(x, z) \in R^3; \ x = (x_1, x_2) \in \Omega, \ -h/2 < z < h/2, \ h \ll 1\},$$

where Ω is a bounded simply connected domain in R^2 with a Lipschitz boundary $\partial\Omega$. Considering great deflections of the plate we obtain the strain-displacement relations

$$\varepsilon_{ij} = \frac{1}{2}(\partial_i u_j + \partial_j u_i + \partial_i w \partial_j w) - z\partial_{ij}w; \quad i, j = 1, 2 \tag{1}$$

$$\varepsilon_{13} = \varepsilon_{23} = 0, \quad \varepsilon_{33} = \frac{1}{2}[(\partial_1 w)^2 + (\partial_2 w)^2], \tag{2}$$

where (u_1, u_2) is the plane displacement vector, w is the deflection of the middle surface of the plate and

$$\partial_i u_j = \frac{\partial u_j}{\partial x_i}, \quad \partial_{ij}w = \frac{\partial w}{\partial x_i \partial x_j}$$

The viscoelastic stress-strain relations have the form

$$\sigma^{ij}(t) = A_{ijkl}^{(1)}\partial_t \varepsilon_{kl}(t) + A_{ijkl}^{(0)}\varepsilon_{kl}(t), \quad i, j, k, l \in \{1, 2\}; \quad \sigma^{33} = 0. \tag{3}$$

The third order tensors $A_{ijkl}^{(r)}$ are symmetric and positive definite:

$$A_{ijkl}^{(r)} = A_{jikl}^{(r)} = A_{klij}^{(r)} \tag{4}$$

$$A_{klij}^{(r)}\tau_{ij}\tau_{kl} \geq c_r \tau_{ij}\tau_{ij}, \quad c_r > 0; \tag{5}$$

for all $\{\tau_{ij}\} \in R_{sym}^4$ and $r = 0, 1$.

Assuming vibrations in the z-direction only and the plate acting upon a perpendicular load of plane density $f \equiv f(t, x)$ the virtual displacements principle implies the relation

$$\int_\Omega \left\{ -\rho h(\partial_{tt}w + \frac{h^2}{12}\nabla \partial_{tt}w) + f \right\} v dx = \int_\Omega \int_{-h/2}^{h/2} \sigma^{ij}\delta\varepsilon_{ij}dzdx \quad \forall v \in C_0^\infty(\Omega), \tag{6}$$

where v is the virtual displacement in the z-direction.

Let us define plane strains $\varepsilon_{ij0} = \frac{1}{2}(\partial_i u_j + \partial_j u_i + \partial_i w \partial_j w)$ and the stress resultants

$$N^{ij}(t) = \int_{-h/2}^{h/2} \sigma^{ij}dz = h[A_{ijkl}^{(1)}\partial_t\varepsilon_{kl0}(t) + A_{ijkl}^{(0)}\varepsilon_{kl0}(t)], \quad i, j = 1, 2. \tag{7}$$

Simultaneously we introduce the initial stress resultants setting

$$N_0^{ij} = hA_{ijkl}^{(1)}\varepsilon_{kl0}(0), \quad i, j = 1, 2. \tag{8}$$

There exists the Airy stress function Φ fulfilling the relations

$$N^{11} = \partial_{22}\Phi, \quad N^{22} = \partial_{11}\Phi, \quad , N^{12} = -\partial_{12}\Phi.$$

The plane strains fulfil the compatibility condition

$$\partial_{22}\varepsilon_{110} - 2\partial_{12}\varepsilon_{120} + \partial_{11}\varepsilon_{220} = -\frac{1}{2}[w, w],$$

$$[v, w] = \partial_{11}v\partial_{22}w + \partial_{22}v\partial_{11}w - 2\partial_{12}v\partial_{12}w.$$

Let us denote $\mathbb{G}(t) = \exp(-\mathbb{A}_1^{-1}\mathbb{A}_0 t)\mathbb{A}_1^{-1}$ the exponential matrix function defined by the tensors $A_{ijkl}^{(r)}$. Combining the stress-strain relations (7), (8) with compatibility conditions we obtain the integro-differential equation for the Airy stress function Φ :

$$\int_0^t D_{ijkl}(t - s)\partial_{ijkl}\Phi(s)ds = -D_{ijkl}(t)\partial_{ij}\tilde{N}_0^{kl} - \frac{h}{2}[w, w], \tag{9}$$

where

$$D_{ijkl} = D_{jikl} = D_{klij},$$

$$D_{1111}(t) = G_{2222}(t), \quad D_{1112}(t) = -\frac{1}{2}G_{2212}(t), \quad D_{1222}(t) = -\frac{1}{2}G_{1112}(t),$$

$$D_{1122}(t) = \frac{1}{2}G_{1122}(t), \quad D_{1212}(t) = \frac{1}{4}G_{1212}(t), \quad D_{2222}(t) = G_{1111}(t),$$

$$\tilde{N}_0^{11} = N_0^{22}, \quad \tilde{N}_0^{12} = -N_0^{12}, \quad \tilde{N}_0^{22} = N_0^{11}.$$

We assume that the plate is clamped on the boundary and no lateral forces are present. The equation (9) is a first-kind Volterra integro-differential equation. After differentiating it with respect to t we obtain the boundary value problem for the second-kind Volterra equation

$$D_{ijkl}(0)\partial_{ijkl}\Phi(t) + \int_0^t \partial_t D_{ijkl}(t - s)\partial_{ijkl}\Phi(s)ds = -\partial_t D_{ijkl}(t)\partial_{ij}\tilde{N}_0^{kl} - h[\partial_t w, w], \tag{10}$$

$$\Phi(t, \xi) = \frac{\partial\Phi}{\partial\mathbf{n}}(t, \xi) = 0, \quad t \in (0, T], \quad \xi \in \partial\Omega, \tag{11}$$

which together with the hyperbolic-pseudoparabolic equation

$$\rho h(\partial_{tt}w - \frac{h^2}{12}\Delta\partial_{tt}w) + \frac{h^3}{12}(A_{ijkl}^{(1)}\partial_t\partial_{ijkl}w + A_{ijkl}^{(0)}\partial_{ijkl}w) - [\Phi, w] = f(t, x), \tag{12}$$

the initial and boundary conditions

$$w(x, 0) = w_0(x), \quad \partial_t w(x, 0) = w_1(x), \tag{13}$$

$$w(t, \xi) = \frac{\partial w}{\partial\mathbf{n}}(t, \xi) = 0, \quad t \in (0, T], \quad \xi \in \partial\Omega, \tag{14}$$

form the nonlinear initial-boundary value problem for determining the unknown functions Φ and w. It represents the strong formulation of the hyperbolic von Kármán system for a dynamic viscoelastic plate.

3. Weak formulation and the canonical form of the problem

Let us denote by $V = H_0^2(\Omega)$ the Hilbert space with the scalar product and the norm

$$(u,v)_V = \int_\Omega (\partial_{11}u\partial_{11}v + 2\partial_{12}u\partial_{12}v + \partial_{22}u\partial_{22}v)dx, \quad \|u\| = (u,u)_V^{1/2}, \quad u,v \in V,$$

by V^* its dual space with the duality pairing $\langle .,. \rangle$ and the norm $\|.\|_*$. Further we consider $U = H_0^1(\Omega)$, $H = L^2(\Omega)$ the Hilbert spaces with the scalar products and the norms

$$(u,v)_U = \int_\Omega (uv + \partial_1 u\partial_1 v + \partial_2 u\partial_2 v)dx, \quad \|u\|_1 = (u,u)_U^{1/2}, \quad u,v \in U,$$

$$(u,v) = \int_\Omega uv dx, \quad \|u\|_0 = (u,u)^{1/2}, \quad u,v \in H.$$

Let us define the operators $D(t), A_0, A_1 : V \to V^*$ and $A_2 : U \to U^*$ by

$$\langle D(t)u, v \rangle = \int_\Omega D_{ijkl}(t)\partial_{ij}u\partial_{kl}v dx, \tag{15}$$

$$\langle A_r u, v \rangle = \frac{h^3}{12} \int_\Omega A_{ijkl}^{(r)}\partial_{ij}u\partial_{kl}v dx, \quad u,v \in V, \quad r = 0,1, \tag{16}$$

$$\langle A_2 u, v \rangle = \rho h \int_\Omega (uv + \frac{h^2}{12}\nabla u \cdot \nabla v)dx, \quad u,v \in U. \tag{17}$$

The operators A_0, A_1, $D(t)$ are linear, bounded, symmetric, positive definite and fulfil the relations

$$\langle A_r u, u \rangle \geq \frac{h^3}{12}c_r\|u\|^2, \quad c_r > 0, \tag{18}$$

$$|\langle A_r u, v \rangle| \leq \frac{h^3}{12}c_{r+2}\|u\| \, \|v\| \quad r = 0,1, \tag{19}$$

$$\langle D(t)u, u \rangle \geq c_4 \exp c_5 t\|u\|^2 \quad \text{for every } u \in V, \ t \geq 0 \tag{20}$$

$$|\langle D(t)u, v \rangle| \leq c_6 \exp(-\frac{c_0}{c_3}t)\|u\| \, \|v\| \quad \text{for all} \quad u,v \in V, \quad r = 0,1, \tag{21}$$

with the constants $c_4 > 0, c_5, c_6$ depending only on $c_0, \ldots, c_3$.

The operator function $D(.)$ is infinitely many times differentiable, all its derivatives $\partial_t^k D(t) : V \to V^*$ are symmetric and $(-1)^k \partial_t^k D(t) : V \to V^*$ positive definite. First derivatives fulfil

$$\|\partial_t D(t)u\|_* \leq c_7 \exp(-\frac{c_0}{c_3}t)\|u\|. \tag{22}$$

Using the Green's formula we express the initial-boundary value problem (10)–(13) in a weak (operator) form:

$$D(0)\Phi(t) + \int_0^t \partial_t D(t-s)\Phi(s)ds = -\partial_t D_0(t)\tilde{\mathbf{N}}_0 - h[\partial_t w(t), w(t)], \qquad (23)$$

$$A_2\partial_{tt}w(t) + A_1\partial_t w(t) + A_0 w(t) - [\Phi(t), w(t)] = f(t), \qquad (24)$$

$$w(0) = w_0, \quad \partial_t w(0) = w_1, \qquad (25)$$

where a functional $D_0(t)\tilde{\mathbf{N}}_0 \in V^*$ is defined by

$$\langle D_0(t)\tilde{\mathbf{N}}_0, v \rangle = \int_\Omega D_{ijkl}(t)\tilde{N}_0^{ij}\partial_{kl}v(x)dx, \quad v \in V, \qquad (26)$$

In order to obtain the canonical hyperbolic equation for determining the deflection function w we express the Airy stress function Φ from the equation (23). We define the operator $B : V \times V \to V$ solving uniquely the equation

$$\langle D(0)B(u,v), \varphi \rangle = \int_\Omega [u,v]\varphi dx \quad \text{for all} \quad u, v, \varphi \in V. \qquad (27)$$

It can be verified in a same way as in [4], that $B : V \times V \to V$ is bilinear, symmetric and compact and fulfils the estimate

$$\|B(u,v)\| \le c_4^{-1}c_8\|u\| \, \|v\| \quad \text{for all} \quad u, \, v \in V. \qquad (28)$$

The relation (27) enables us to express the integral equation (23) as the second-kind Volterra equation in the Hilbert space V:

$$\Phi(t) - \int_0^t K(t-s)\Phi(s)ds = \Phi_0(t) - hB(\partial_t w(t), w(t)), \quad t > 0 \qquad (29)$$

where

$$K(t) = -D(0)^{-1}\partial_t D(t), \quad \|K(t)\|_{L(V,V)} \le c_4^{-1}c_7 \exp(-\frac{c_0}{c_3}t), \qquad (30)$$

$$\Phi_0(t) = -D(0)^{-1}[\partial_t D_0(t)\tilde{\mathbf{N}}_0], \quad \|\Phi_0(t)\| \le hc_3 c_4^{-1}c_7 \exp(-\frac{c_0}{c_3}t)\|w_0\|_{W_0^{1,4}}^2. \qquad (31)$$

We assumed for simplicity the initial conditions $u_1(0) = u_2(0) = 0$ in the plain strains $\varepsilon_{kl0}(0)$ appearing in $\tilde{\mathbf{N}}_0$.

Applying the theory of Volterra integral equations in Hilbert spaces ([12]) we can express the Airy stress function Φ in a form

$$\Phi(t) = \Phi_0(t) - hB(\partial_t w(t), w(t)) + \int_0^t M(t,s)[\Phi_0(s) - hB(\partial_s w(s), w(s))]ds, \qquad (32)$$

where $M(t,s) : V \to V$ is defined as the series

$$M(t,s) = \sum_{n=1}^\infty K_n(t,s), \quad K_1(t,s) = K(t-s), \quad K_n(t,s) = \int_s^t K(t-\sigma)K_{n-1}(\sigma,s)d\sigma.$$

Inserting the function Φ from (32) into (24) we arrive at the canonical initial value problem for determining the deflection function $w : [0, T] \to V$:

$$A_2 \partial_{tt} w + A_1 \partial_t w(t) + A_0 w(t) + [hB(\partial_t w(t), w(t)) - \Phi_0(t), w(t)]$$

$$+ \left[\int_0^t M(t, s)(hB(\partial_s w(s), w(s)) - \Phi_0(s)ds, w(t)] = f(t), \tag{33}$$

$$w(0) = w_0, \quad \partial_t w(0) = w_1. \tag{34}$$

4. A solution of the hyperbolic canonical problem

We shall solve the problem (33), (34) in the spaces of Bochner integrable functions. For $\tau > 0$ we introduce the Hilbert spaces

$$\mathcal{H} = L^2(0, \tau; H), \quad \mathcal{U} = L^2(0, \tau; U), \quad \mathcal{V} = L^2(0, \tau; V)$$

with the inner products and the norms

$$(u, v)_{\mathcal{H}} = \int_0^\tau (u, v) dt, \quad (u, v)_{\mathcal{U}} = \int_0^\tau (u, v)_U dt, \quad ((u, v)) = \int_0^\tau (u, v)_V dt,$$

$$\|u\|_{\mathcal{H}} = (u, u)_{\mathcal{H}}^{1/2}, \quad \|u\|_{\mathcal{U}} = (u, u)_{\mathcal{U}}^{1/2}, \quad \|u\|_{\mathcal{V}} = ((u, u))1/2.$$

Further we denote $\quad \mathcal{V}_0 = \{v \in \mathcal{V} : \partial_t v \in \mathcal{V}, \ v(0) = 0\} \quad$ and $\langle\langle f, v \rangle\rangle = \int_0^\tau \langle f(t), v(t) \rangle dt, \ f \in \mathcal{V}^*, \ v \in \mathcal{V};$ the duality pairing between $\mathcal{V}^*$ and $\mathcal{V}$.

Before solving the hyperbolic problem (33), (34) we transform it to the first order parabolic problem with the homogeneous initial condition.

Let us define for $\delta > 0$ a function $\gamma_\delta : [0, \infty) \to R$:

$$\gamma_\delta(t) = t - \frac{1}{2\delta} t^2, \quad 0 \le t \le \delta, \quad \gamma_\delta(t) = \frac{\delta}{2}, \quad t > \delta.$$

After the substitution $w = Iu$, $(Iu)(t) = w_0 + w_1 \gamma_\delta(t) + \int_0^t u(s) ds$ and defining the operator $Q : \mathcal{V}_0 \to C([0, \tau]; L^2(\Omega))$ by

$$Q(u)(t) = [hB(u(t) + w_1 \gamma'_\delta(t), (Iu)(t)) - \Phi_0(t), (Iu)(t)]$$

the problem (33), (34) can be expressed in a form

$$A(u) = f, \quad u \in \mathcal{V}_0, \tag{35}$$

where $f \in \mathcal{V}^*$ and $\mathcal{A} : \mathcal{V}_0 \to \mathcal{V}^*$ is a parabolic operator defined by

$$\langle\langle \mathcal{A}(u), v \rangle\rangle = \int_0^\tau \langle A_2 \partial_t u + A_1(u + w_1 \gamma'_\delta(t)) + A_0 Iu, v \rangle dt$$

$$+ \int_0^\tau (Q(u)(t) + \int_0^t \overset{.}{M}(t, s) Q(u)(s) ds, v(t)) dt.$$

Next we introduce the operator $L : \mathcal{V}_0 \to \mathcal{V}$, $Lu = \partial_t u$. Its adjoint L^* : $D(L^*) \to \mathcal{V}$ is defined by

$$L^* v = -\partial_t v, \quad v \in D(L^*); \quad D(L^*) = \{v \in \mathcal{V} : \partial_t v \in \mathcal{V}, \ v(\tau) = 0\}.$$

Both operators are positive. The operator $L : \mathcal{V}_0 \to \mathcal{V}$ is closed and hence $\mathcal{V}_0$ is the Hilbert space with the inner product and the norm induced by L:

$$(u, v)_L = ((u, v)) + ((Lu, Lv)), \quad \|u\|_L = (u, u)_L^{-1/2}, \quad u, v \in \mathcal{V}_0.$$

We denote by $\langle\langle ., . \rangle\rangle_L$ the duality pairing between $\mathcal{V}_0^*$ and $\mathcal{V}_0$.

Next we formulate the regularized elliptic equation corresponding to the original problem (35) in a similar way as in [1], or [10]:

$$\varepsilon J L^* L u_\varepsilon + \mathcal{A}(u_\varepsilon) = f, \tag{36}$$

where $J : \mathcal{V} \to \mathcal{V}^*$ is the canonical isomorphism defined by $\langle\langle Ju, v \rangle\rangle = ((u, v))$.

Lemma 4.1. *There exists $\tau > 0$ and $\varepsilon_0 > 0$ such that for every $\varepsilon \in (0, \varepsilon_0)$ and $f \in \mathcal{V}^*$ there exists a function $u_\varepsilon \in \mathcal{V}_0$ fulfilling $Lu_\varepsilon \in D(L^*)$ and solving (37).*

Proof. Let us define the operator $\mathcal{A}_\varepsilon : \mathcal{V}_0 \to \mathcal{V}_0^*$ by

$$\langle\langle \mathcal{A}_\varepsilon(u), v \rangle\rangle_L = \varepsilon((Lu, Lv)) + \langle\langle \mathcal{A}(u), v \rangle\rangle, \quad u, v \in \mathcal{V}_0.$$

If the operator $\mathcal{A}_\varepsilon$ is coercive and pseudomonotone then due to ([10],ch.2.2.4) there exists for every $f \in \mathcal{V}_0^*$ a solution u_ε of the equation

$$\mathcal{A}_\varepsilon(u_\varepsilon) = f. \tag{37}$$

Using the relations $(Q(u)(t) + \int_0^t M(t, s)Q(u)(s)ds, u(t)) = (-\Phi(t), [Iu(t), u(t)])$ and

$$(-\Phi(t), [Iu(t), u(t) + w_1 \gamma_\delta'(t)]) = \frac{1}{h}\langle D(0)\Phi(t), \Phi(t) \rangle - \int_0^t K(t - s)\Phi(s)ds - \Phi_0(t))$$

we obtain

$$\langle\langle \mathcal{A}_\varepsilon(u), u \rangle\rangle_L = \varepsilon \|Lu\|_\mathcal{V}^2 + \frac{1}{2}\langle A_2 u(\tau), u(\tau) \rangle + \frac{1}{2}(\langle A_0(Iu)\tau, (Iu)\tau \rangle - \langle A_0 w_0, w_0 \rangle)$$

$$+ \int_0^\tau (\langle A_1 u, u \rangle + \frac{1}{h}\langle D(0)\Phi(t), \Phi(t) \rangle - \int_0^t K(t - s)\Phi(s)ds - \Phi_0(t)))dt$$

$$+ \int_0^\delta (\langle A_1 u - A_0 Iu, w_1 \gamma_\delta'(t) \rangle + (\Phi(t), [(Iu)(t), w_1 \gamma_\delta'(t)]))dt.$$

The coercivity of $\mathcal{A}_\varepsilon$ with respect to the norm in $\mathcal{V}_0$ means

$$\lim_{\|u\|_L \to \infty} \langle\langle \mathcal{A}_\varepsilon(u), u \rangle\rangle_L \|u\|_L^{-1} = +\infty \tag{38}$$

If we choose sufficiently small τ and δ in order to estimate the members with convolutive integrals, we obtain (38) as a consequence of the positive definiteness of the operators A_0, A_1, $D(0)$.

The pseudomonotonicity of $\mathcal{A}_\varepsilon = \mathcal{M}_\varepsilon + \mathcal{B}$ expressed as a sum of the linear and nonlinear operator defined by

$$\langle\langle \mathcal{M}_\varepsilon u, v\rangle\rangle_L = \varepsilon((Lu, Lv)) + \int_0^\tau \langle A_2 \partial_t u + A_1(u + w_1\gamma_\delta'(t)) + A_0 Iu, v\rangle dt,$$

$$\langle\langle \mathcal{B}(u), v\rangle\rangle = \int_0^\tau ([hB(u(t) + w_1\gamma_\delta'(t), (Iu)(t)) - \Phi_0(t), (Iu)(t)], v(t)) \, dt$$

$$+ \int_0^\tau \left(\left[\int_0^t (M(t,s)(hB(u(s) + w_1\gamma_\delta'(s), (Iu)(s)) - \Phi_0(\dot{s})) \, ds, (Iu)(t) \right], v(t) \right) dt.$$

can be derived in the same way as in [1]. The linear part $\mathcal{M}_\varepsilon$ is bounded and coercive over $\mathcal{V}_0$ equipped with the norm $\|.\|_L$ and it is monotone. Then it is sufficient to verify the pseudomonotonicity of $\mathcal{B}$. Applying the compactness of the operator $B : V \times V \to V$ we obtain the implication

$$u_n \rightharpoonup u \text{ in } \mathcal{V}_0 \implies \lim_{n\to\infty} \langle\langle \mathcal{B}(u_n), u_n - v\rangle\rangle_L = \langle\langle \mathcal{B}(u), u - v\rangle\rangle_L \; \forall v \in \mathcal{V}_0,$$

which together with the boundedness of $\mathcal{B} : \mathcal{V}_0 \to \mathcal{V}^*$ implies its pseudomonotonicity (see [10]).

Hence there exists a solution u_ε of the equation (37). Expressing this equation in the form

$$((Lu_\varepsilon, Lv)) = \frac{1}{\varepsilon} \langle\langle -\mathcal{A}(u_\varepsilon) + f, v\rangle\rangle \quad \text{for all} \quad v \in \mathcal{V}_0$$

we obtain that $Lu_\varepsilon \in D(L^*)$ and the equation (36) holds what completes the proof. $\qquad\qquad\square$

Using the previous lemma we obtain a solution of the parabolic integro-differential problem (35). We remark that there are stronger regularity assumptions on the function f in order to assure the convergence of the sequence of regularized solutions $\{u_\varepsilon\}$.

Lemma 4.2. *Let $\tau > 0$ be determined from Lemma 4.1. Then for every $f \in \mathcal{U}^*$, or $f \in W^{2,1}(0, \tau; V^*)$ there exists a function $u \in \mathcal{V}$ with $\partial_t u \in \mathcal{U}$ solving the initial value problem*

$$A_2 \partial_t u(t) + A_1 u(t) + A_0(Iu)(t) + [hB(u(t) + w_1\gamma_\delta'(t), (Iu)(t)) - \Phi_0(t),$$

$$(Iu)(t)] + [\int_0^t (M(t,s)(hB(u(s) + w_1\gamma_\delta'(s), (Iu)(s)) - \Phi_0(s))) \, ds, (Iu)(t)]$$

$$= f(t) - A_1 w_1\gamma_\delta'(t) \quad \text{for a.e.} \quad t \in [0, \tau], \quad (39)$$

$$u(0) = 0. \quad (40)$$

Proof. We express the regularized equation (36) in the form

$$\varepsilon(Lu_\varepsilon, Lv)) + \langle\langle \mathcal{A}(u_\varepsilon), v\rangle\rangle = \langle\langle f, v\rangle\rangle \quad \text{for every} \quad v \in \mathcal{V}_0. \quad (41)$$

Using the coercivity of the operator $\mathcal{A}$ we obtain after inserting $v = u_\varepsilon$ into (41) the a priori estimates of the norms $\|u_\varepsilon\|_\mathcal{V}$. After duality pairing of the equation (37) with $\partial_t u_\varepsilon$ we estimate the norms $\{\|u_\varepsilon\|_1\}$ for every $t \in [0, \tau]$ and the derivatives $\{\|\partial_t u_\varepsilon\|_{\mathcal{U}}\}$. Extracting a weakly convergent subsequences $\{u_{\varepsilon_n}\}$ in $\mathcal{V}$, $\{\partial_t u_{\varepsilon_n}\}$ in $\mathcal{U}$ and $\{u_{\varepsilon_n}(t)\}$ in U for $\varepsilon_n > 0$, $\varepsilon_n \to 0$ we obtain a solution u of (39), (40). □

After coming back to the original deflection-function w we obtain the solution of the canonical problem (33), (34) on the interval $[0, \tau]$. Prolonging it to the whole interval $[0, T]$ and recalling the way of deriving the canonical problem we obtain

Theorem 4.3. *i) For every $f \in \mathcal{U}^*$, or $f \in W^{2,1}(0, T; V^*)$, w_0, $w_1 \in V$, $\Phi_0 \in C(0, T; V)$ there exists a solution*

$$w \in C([0, T]; V) \cap C^1([0, T]; U) \cap W^{2,1}(0, T; V) \cap W^{2,2}(0, T; U)$$

of the initial value problem (33), (34).
ii)Let $\Phi_0(t) \in V$, $\Phi \in \mathcal{V}$ be defined by (31), (32). Then a couple $\{\Phi, w\}$, is a solution of the initial value problem (23), (24), (25) or a weak solution of the generalized von Kármán system (10)–(13).

In order to investigate uniqueness conditions we restrict ourselves to small time values. We obtain the results analogous to the elastic and stationary case.

Theorem 4.4. *Let $f \in L^2(0, T; H)$, $T \le T_0 = \frac{c_4}{c_7}\left(\frac{c_0}{2c_3}\right)^{1/2}$.*
Then a solution w of (33), (34) fulfils the a priori estimate

$$\rho\|\partial_t w\|^2_{C(0,T;U)} + c_0\|w\|^2_{C(0,T;V)} + c_1\|\partial_t w\|^2_{L^2(0,T;V)} \le \tag{42}$$

$$\frac{12\rho}{h^2}\|w_1\|_1^2 + +\frac{1}{h^2}c_9\|w_0\|^2 + \frac{12}{\rho h^4}\|f\|^2_{L^1(0,T;H)}, \tag{43}$$

where $c_9 = c_2 + c_0^{-1}c_3^3 c_4^{-2} c_7^2 c_{10}^2$, $\|w_0\|_{W_0^{1,4}} \le c_{10}\|w_0\|$.
If

$$c_9\|w_0\|^2 + \rho\|w_1\|_1^2 + \frac{1}{\rho h^2}\|f\|^2_{L^1(0,T;H)} \le c_{11}h^4, \tag{44}$$

where $c_{11} = (c_0 c_1 c_4)^2[576c_6(c_0^2+c_1^2)(c_0^2+4c_1^2)]^{-1}$, then there exists a unique solution of the initial value problem (33), (34).

Proof. If $T \le T_0$, then $|\int_0^T (\Phi(t), \int_0^t K(t-s)\Phi(s)ds)_V dt| \le \frac{1}{2}\int_0^T \|\Phi\|^2 dt$. This estimate enables us using (27), (28) to estimate $\int_0^T ([\Phi(t), w(t)], \partial_t w(t)) dt$ after duality pairing of the equation (24) with $\partial_t w$ and integrating it. The estimate (43) can then be achieved in a standard way.

Let w_1, w_2 are two solutions of (33), (34). Then their difference $w = w_2 - w_1$ fulfils the estimate

$$\frac{1}{2}c_0\|w\|^2_{C(0,T;V)} + c_1\|\partial_t w\|^2_{L^2(0,T;V)} \le \int_0^T \int_0^1 (DP(w_1 + \xi w) \cdot w(t), \partial_t w)d\xi dt,$$

where $P : V \to H$ is the operator defined by $P(w) = [\Phi(w), w]$ and $DP(u) \cdot w$ is its variation at u in the direction w. The condition (44) rends $w = 0$ after estimating the Airy stress function $\Phi(u)$ through the deflection u and taking into account the estimate (43).   □

Remark 4.5. It is possible to obtain a priori estimates and a uniqueness condition for an arbitrary T but in a more complicated way.

References

[1] I. Bock, *On nonstationary von Kármán equations*, ZAMM, **76** (1996), 559–571.

[2] J. Brilla, *Variational methods in mathematical theory of viscoelasticity*, Proceedings Internat. Conf. on Diff. Equations, Equadiff III (M. Ráb and J. Vosmanský, eds.), University J.E. Purkyně Press, Brno, **1973**, 211–216.

[3] R.V. Christensen, *Theory of viscoelasticity*, Academic Press, New York, 1971.

[4] P.G. Ciarlet and P. Rabier, *Les équations de von Kármán*, Springer Verlag, Berlin, **1980**.

[5] M.A. Horn and I. Lasiecka, *Uniform decay of weak solutions to a von Kármán plate with nonlinear dissipation*, Differential and Integral Equations, **7** (1994), 885–908.

[6] H. Koch and A. Stahel, *Global existence of classical solutions to the dynamical von Kármán equations*, Math. Methods in the applied Sciences **161** (1993), 581-586.

[7] J.E. Lagnese, *Boundary stabilization of thin plates*, SIAM, Philadelphia,**1989**.

[8] J.E. Lagnese, *Uniform asymptotic energy estimates for solutions of the equations of dynamic plane elasticity with nonlinear dissipation at the boundary*, Nonlinear Anal. T.M.A. **16** (1991), 35–54.

[9] I. Lasiecka, *Finite dimensionality of attractors associated with von Kármán plate equations and boundary damping*, Journal Differential Equations **117** (1995), 357–389.

[10] J.L. Lions, *Quelques méthodes de résolution des problèmes aux limites non linéaires*, Dunod, Paris, **1969**.

[11] E. Muñoz Rivera and G. Perla Menzala, *Decay rates of solutions to a von Kármán system for viscoelastic plates with memory*, Journal Quarterly of Applied Math., (to appear).

[12] J. Prüss, *Evolutionary integral equations and applications*, Birkhäuser, Basel, **1993**.

[13] J. Puel and M. Tucsnak, *Boundary stabilization for the von Kármán equations*, SIAM J. on Control **33** 255–273.

[14] J. Puel and M. Tucsnak, *Global existence for the full von Kármán system*, Applied Math. and Optim. **34** (1996), 139–161.

[15] I.I. Vorovič, *O nekotorych priamych metodach v nelinejnoj teorii kolebanij pologich oboloček*, Izvestija Akademii Nauk SSSR **21** (1957), 747–784.

Department of Mathematics
FEI STU, Ilkovičova 3,
812 19 Bratislava, Slovakia
E-mail address: bock@kmat.elf.stuba.sk

International Series of Numerical Mathematics
Vol. 129, © 1999 Birkhäuser Verlag Basel/Switzerland

Courant's Problems and Their Extensions

Oleg I. Bogoyavlenskij

Abstract. New geometric invariants are defined for the hyperbolic systems of first order partial differential equations. Applications of these invariants to the Courant problems are presented along with their applications to the necessary conditions for the existence of Hamiltonian structures. Applications to the gas dynamics equations, to the Benney equations and to their perturbations are developed.

1. Courant's problems

I. The paper is devoted to the investigation of Courant's problems (see [3]) for the first-order hyperbolic systems of partial differential equations:

i) When can a nonlinear system

$$u_t^i = \sum_{j=1}^{n} A_j^i(u^1, \ldots, u^n) \, u_x^j, \qquad i = 1, \ldots, n, \tag{1}$$

be reduced to two non-interacting subsystems?

ii) When is it a system of conservation laws?

A positive solution to problem *(i)* leads to a drastic reduction of the computer time required for a numerical solution. A positive solution to problem *(ii)* allows an application of the theory of conservation laws (see [4], [6]).

To study Courant's problems, we make use of the Nijenhuis (1,2) tensor (see [7])

$$N_{jk}^i = \sum_{\alpha=1}^{n} \left(\frac{\partial A_k^i}{\partial u^\alpha} A_j^\alpha - \frac{\partial A_j^i}{\partial u^\alpha} A_k^\alpha + \frac{\partial A_j^\alpha}{\partial u^k} A_\alpha^i - \frac{\partial A_k^\alpha}{\partial u^j} A_\alpha^i \right), \tag{2}$$

constructed from the (1,1) tensor $A_j^i(u^1, \ldots, u^n)$.

We introduce an invariant polynomial

$$P_N(V) = \det \| N_{jk}^i V^j - \delta_k^i \| \tag{3}$$

on the tangent bundle $T(\mathbb{R}^n)$ which is connected with the Nijenhuis tensor (2). Here V is a tangent vector $V = V^i(u)\partial/\partial u^i \in T_u(\mathbb{R}^n)$. The polynomial $P_N(V)$ has degree $\leq n - 1$ in view of the identity $\det \| N_{jk}^i V^j \| = 0$. The latter follows from the skew-symmetricity of the Nijenhuis tensor $N_{jk}^i = -N_{kj}^i$ that implies $N_{jk}^i V^j V^k = 0$.

The polynomial $P_N(V)$ has applications in the following

Necessary condition for separability. *If a system (1) is separable, then the polynomial $P_N(V)$ must have degree $\leq n - 2$ and must be divisable. That means it can be factored into a product of two polynomials of $V^1, \ldots, V^n$ at all points $u \in \mathbb{R}^n$.*

Proof. Suppose the system (1) is split into two non-interacting subsystems of m and $n - m$ equations. Then the Niienhuis tensor N is a direct sum of Nijenhuis tensors N_1 and N_2 in $\mathbb{R}^m$ and $\mathbb{R}^{n-m}$. Therefore, the polynomial $P_N(V_1 + V_2)$ has the form $P_N(V_1 + V_2) = P_{N_1}(V_1)P_{N_2}(V_2)$. We have $\deg P_{N_1}(V_1) \leq m - 1$ and $\deg P_{N_2}(V_2) \leq n - m - 1$. Hence we get

$$\deg P_N = \deg P_{N_1} + \deg P_{N_2} \leq n - 2.$$

$\square$

Example 1. The Benney system (see [1]) has the form

$$u_{it} = -u_i u_{ix} - \sum_{j=1}^k \eta_{jx}, \qquad \eta_{it} = -\eta_i u_{ix} - u_i \eta_{ix}, \qquad (4)$$

where $i = 1, \ldots, k$. A direct calculation of the Nijenhuis tensor (2) shows that the invariant polynomial $P_N(V)$ for the Benney equations (4) has the form

$$P_N(V) = (V^{k+1} + \cdots + V^{2k} - 1)^{2k-1}, \qquad (5)$$

$$V = V^1 \frac{\partial}{\partial u_1} + \cdots + V^k \frac{\partial}{\partial u_k} + V^{k+1} \frac{\partial}{\partial \eta_1} + \cdots + V^{2k} \frac{\partial}{\partial \eta_k}.$$

The polynomial $P_N(V)$ (5) has degree $n - 1$, $n = 2k$. Hence the Benney system (4) is not separable.

To study invariant properties of system (1) we make use also of the Haantjes (1,2) tensor (see [5]) that has the form

$$H_{jk}^i = \sum_{\alpha,\beta=1}^n \left(A_\alpha^i A_\beta^\alpha N_{jk}^\beta + N_{\alpha\beta}^i A_j^\alpha A_k^\beta - A_\alpha^i N_{\beta k}^\alpha A_j^\beta - A_\alpha^i N_{j\beta}^\alpha A_k^\beta \right). \qquad (6)$$

We introduce the following invariant polynomial

$$P_H(V) = \det \| H_{jk}^i V^j - \delta_k^i \| . \qquad (7)$$

This polynomial has degree $\leq n - 1$ as well as the above polynomial $P_N(V)$.

The second necessary condition for separability. *If a system (1) is separable, then the polynomial $P_H(V)$ must have degree $\leq n - 2$ and must be divisable.*

The proof follows from that for the first necessary condition.

Example 2. Let us consider the one-dimensional gas dynamics equations

$$v_t = -vv_x - \frac{1}{\rho} p_\rho \rho_x - \frac{1}{\rho} p_s s_x, \quad \rho_t = -\rho v_x - v \rho_x, \quad s_t = -v s_x, \tag{8}$$

where $v(x,t)$ is the gas velocity, $\rho(x,t)$ is the mass density, $s(x,t)$ is the density of entropy, and p is the pressure: $p = p(\rho, s)$. A direct calculation shows that the polynomial $P_H(V)$ (7) for the gas dynamics equations (8) has the form

$$P_H(V) = h^2(\rho, s)(V^3)^2 - 1, \tag{9}$$

$$V = V^1 \frac{\partial}{\partial v} + V^2 \frac{\partial}{\partial \rho} + V^3 \frac{\partial}{\partial s}, \quad h(\rho, s) = \rho^2 p_\rho^2 \left(\frac{p_s}{\rho^2 p_\rho} \right)_\rho.$$

The polynomial $P_H(V)$ (9) has degree 2. Hence the gas dynamics equations (8) are not separable if $h(\rho, s) \neq 0$.

2. Necessary condition for the existence of n conservation laws

The system (1) is called a system of conservation laws (see [3], [4], [6]) if it has the form

$$v_t^i = P^i(v)_x, \quad i = 1, \ldots, n \tag{10}$$

in some new coordinates $v^1, \ldots, v^n$. Suppose that a (1,1) tensor $A_j^i(u)$ satisfies an algebraic equation $C(A, u) = 0$ where $C(A, u)$ is a polynomial

$$C(A, u) = \sum_{m=0}^{k} c_m(u) A^m(u). \tag{11}$$

We introduce the following skew-symmetric (1,2) tensor

$$B_C(U, V) = \sum_{m=2}^{k} c_m \sum_{p+q+r=m-2} A^p N(A^q U, A^r V) + \sum_{m=0}^{k} (V(c_m) A^m U - U(c_m) A^m V),$$

where N is the Nijenhuis (1,2) tensor (2). Here U, V are arbitrary tangent vectors, $U, V \in T_u(\mathbb{R}^n)$.

Theorem 2.1. *If a system (1) is a system of conservation laws and the corresponding (1,1) tensor $A_j^i(u)$ satisfies an algebraic equation $C(A, u) = 0$ then the (1,2) tensor $B_C(U, V)$ vanishes.*

Proof. A single conservation law

$$f(u)_t = P(u)_x \tag{12}$$

for the system (1) is equivalent to the equation for the differential 1-forms

$$A df = dP. \tag{13}$$

Here the 1-form Adf has the components

$$(Adf)_i = A_i^\alpha \frac{\partial f}{\partial u^\alpha}.$$

Poincaré's lemma implies that equation (13) locally is equivalent to the equation

$$d(Adf) = 0. \tag{14}$$

The conservation law (12)-(13) yields a countable set of equalities

$$2d(A^m df) = -df\, N_m, \quad m \geq 2, \tag{15}$$

where $N_m(U, V)$ is the skew-symmetric (1,2) tensor

$$N_m(U, V) = \sum_{p+q+r=m-2} A^p N(A^q U, A^r V).$$

Here $df\, N_m$ means the differential 2-form

$$(df\, N_m)(U, V) = df\,(N_m(U, V)).$$

Equation (15) for $m = 2$ ($N_2 = N$) has the form

$$2d(A^2 df) = -df\, N. \tag{16}$$

This equation follows from the main properties of the Nijenhuis tensor $N(U, V)$ (see [7]). Equations (15) for $m > 2$ follow from equation (16) by the induction.
Let us prove that the equation

$$2d(C(A, x)df)(U, V) = -df\,(B_C(U, V)) \tag{17}$$

holds for any polynomial $C(A, u)$ (11). The left hand side L of equation (17) has the form

$$L = 2d\left(\sum_{m=0}^{k} c_m(x) A^m df\right) = 2\sum_{m=0}^{k} c_m(x) d(A^m df) + 2\sum_{m=0}^{k} dc_m \wedge A^m df.$$

Using here equations (14), (15) and equation $d^2 f = 0$, we obtain

$$L = -df \sum_{m=2}^{k} c_m(x) N_m + 2\sum_{m=0}^{k} dc_m \wedge A^m df. \tag{18}$$

The standard properties of the differential forms lead to the equalities

$$2(dc_m \wedge A^m df)(U, V) = dc_m(U)(A^m df)(V) - dc_m(V)(A^m df)(U) =$$
$$= df\,(U(c_m) A^m V - V(c_m) A^m U).$$

Substituting these expressions for $m = 0, 1, \ldots, k$ into equation (18), we obtain equation (17).

Suppose that system (1) has n functionally independent conservation laws

$$f^i(u)_t = P^i(u)_x, \quad i = 1, \dots, n.$$

Then equation (17) and equation $C(A, u) = 0$ yield n equations

$$df^i(B_C(U, V)) = 0$$

for n independent differentials df^i. Hence the equation $B_C(U, V) = 0$ is proved for any tangent vectors U, V.   □

The (1,2) tensor $B_C(U, V)$ leads to a general identity for the Nijenhuis tensor $N(U, V)$. The classical Cayley-Hamilton Theorem implies that any (1,1) tensor $A^i_j(u)$ satisfies its characteristic equation

$$P(A, u) = \sum_{m=0}^{n} a_m(u) A^m(u) = 0,$$

where the characteristic polynomial $P(\lambda, u)$ has the form

$$P(\lambda, u) = \det(A(u) - \lambda) = \sum_{m=0}^{n} a_m(u)\lambda^m.$$

The investigation of the (1,2) tensor $B_C(U, V)$ proves the identity $B_P(U, V) \equiv 0$ where $P(A, u)$ is the characteristic polynomial. The identity $B_P(U, V) \equiv 0$ implies that the Nijenhuis tensor $N(U, V)$ satisfies the identity

$$\sum_{m=2}^{n} a_m(u) \sum_{p+q+r=m-2} A^p N(A^q U, A^r V) = \sum_{m=0}^{n} (U(a_m) A^m V - V(a_m) A^m U).$$

These results show that Theorem 2.1 can be used as a necessary condition for the existence of n functionally independent conservation laws for a system (1) if the (1,1) tensor $A^i_j(u)$ has multiple eigenvalues. In this case the algebraic polynomial $C(A, u)$ (11) is the minimal polynomial for the operator $A^i_j(u)$.

3. Extended Courant's problem

The system (1) possesses a non-degenerate Hamiltonian structure if it has the form

$$w^i_t = c^i \left(\frac{\partial H(w)}{\partial w^i} \right)_x \tag{19}$$

in some variables $w^i = w^i(u^1, \dots, u^n)$, for some constants $c^i \neq 0$ (see [8]). It is obvious that any non-degenerate Hamiltonian system is a system of conservation laws (10) $w^i_t = (P^i(w))_x$, where $P^i(w) = c^i \partial H(w)/\partial w^i$. The notion of Hamiltonian structures leads to the following extension of the Courant problems (i)–(ii):

iii) When does a nonlinear system (1) possess a non-degenerate Hamiltonian structure?

Until recently, there were no mathematical methods to approach this problem. Using the Nijenhuis and Haantjes tensors (2) and (6), we have defined the invariant polynomials $P_N(V)$ (3) and $P_H(V)$ (7) on the tangent bundle $T(\mathbb{R}^n)$. These polynomials lead to fibrations of the algebraic varieties V_N and V_H embedded into $T(\mathbb{R}^n)$ which are defined by the equations $P_N(V) = 0$ and $P_H(V) = 0$ respectively.

In terms of these geometric invariants, we derive the following

Necessary condition for the existence of a non-degenerate Hamiltonian structure. *If a system (1) has a non-degenerate Hamiltonian structure then the polynomial $P_H(V)$ (7) must be even, $P_H(V) = P_H(-V)$, and the algebraic varieties $V_H \subset T(\mathbb{R}^n)$ must be invariant with respect to the involution $V \to -V$.*

The proof follows by a direct calculation of the Haantjes tensor $H(U, V)$ (6) and the polynomial $P_H(V)$ (7) in the coordinates $w^1, \ldots, w^n$ where system (1) has Hamiltonian form (19).

The study of bi-Hamiltonian systems (1) (see [2], Section 7) proves that if a hyperbolic system (1) has two Hamiltonian structures in general position then its Haantjes tensor vanishes and hence the system has n Riemann invariants and is integrable by the generalized holograph transform.

EXAMPLE 3. The Benney equations (4) have the form

$$u_{it} = -\left(\frac{\partial H(u, \eta)}{\partial \eta_i}\right)_x, \quad \eta_{it} = -\left(\frac{\partial H(u, \eta)}{\partial u_i}\right)_x,$$

where

$$H(u, \eta) = \frac{1}{2}\sum_{i=1}^{k} \eta_i u_i^2 + \frac{1}{2}\left(\sum_{i=1}^{k} \eta_i\right)^2.$$

Therefore the Benney system (4) takes Hamiltonian form (19) with constants $c^i = 1$ or -1 in the coordinates $w^i = u_i + \eta_i$, $v^i = u_i - \eta_i$. The above necessary condition is satisfied identically because the Haantjes tensor $H(U, V)$ for the Benney system (4) vanishes, hence $P_H(V) = 1$. Let us consider the following natural perturbations of the Benney equations

$$u_{it} = -u_i u_{ix} - \sum_{j=1}^{k} f_j(\eta_j)\eta_{jx}, \quad \eta_{it} = -\eta_i u_{ix} - u_i \eta_{ix}, \tag{20}$$

where $f_j(\eta_j)$ are arbitrary smooth functions and $i, j = 1, \ldots, k$. System (20) for $f_j(\eta_j) = const \neq 0$ is equivalent to the Benney system (4). A direct calculation proves that the polynomial $P_H(V)$ for the system (20) has degree 3. Hence it is not even. Thus the perturbations (20) of the Benney equations (4) do not satisfy the above necessary condition. Therefore any perturbation (20) of the Benney equations (4) with non-constant functions $f_j(\eta_j)$ has no Hamiltonian structure in spite of the Benney system (4) is Hamiltonian.

4. Algebraic structures defined by the Nijenhuis and Haantjes tensors

For an arbitrary $(1,1)$ tensor $A_j^i(u)$ on a smooth manifold M^n, $u \in M^n$, the Nijenhuis and Haantjes tensors (2) and (6) define skew-symmetric algebraic structures in the tangent spaces $T_u(M^n)$: for example $(N(U,V))^i = N_{j\ell}^i U^j V^\ell$ for $U, V \in T_u(M^n)$. To study these structures, we define the $(1,3)$ tensor

$$B_{Njk\ell}^i = \sum_{\alpha=1}^{n} \left(N_{jk}^\alpha N_{\alpha\ell}^i + N_{k\ell}^\alpha N_{\alpha j}^i + N_{\ell j}^\alpha N_{\alpha k}^i \right),$$

that has the invariant form

$$B_N(U,V,W) = N(N(U,V),W) + N(N(V,W),U) + N(N(W,U),V).$$

It is evident that tensor $B_N(U,V,W)$ is skew-symmetric:

$$B_N(U,V,W) = -B_N(V,U,W), \quad B_N(U,V,W) = B_N(V,W,U) = B_N(W,U,V).$$

If $B_N(U,V,W) = 0$ then the Nijenhuis tensor $N(U,V)$ defines a Lie algebra structure in each tangent space $T_u(M^n)$ because the equality $B_N(U,V,W) = 0$ is equivalent to the Jacobi identity. Hence the $(1,3)$ tensor $B_N(U,V,W)$ characterizes the deviation of the algebraic structures defined by the Nijenhuis tensor from the Lie algebraic structures.

Analogously, for the Haantjes tensor $H(U,V)$, we define the $(1,3)$ tensor

$$B_H(U,V,W) = H(H(U,V),W) + H(H(V,W),U) + H(H(W,U),V).$$

For $n = 2$ the Haantjes $(1,2)$ tensor $H(U,V)$ vanishes identically. A direct calculation proves that for $n = 3$ the $(1,3)$ tensor $B_H(U,V,W)$ vanishes identically for any $(1,1)$ tensor $A_j^i(u)$. Hence the Haantjes tensor for $n = 3$ defines Lie algebraic structures in the tangent spaces $T_u(\mathbb{R}^3)$. If system (1) possesses a non-degenerate Hamiltonian structure then the Lie algebras $H(U,V)$ in $T_u(\mathbb{R}^3)$ are either simple or Abelian. The existence of a Hamiltonian structure for $n = 4$ implies that the Haantjes tensor defines Lie algebraic structures in the tangent spaces $T_u(\mathbb{R}^4)$. For the non-degenerate Hamiltonian systems (1) for $n \geq 4$, the $(1,3)$ tensor

$$B_{Hjk\ell}^i = \sum_{\alpha=1}^{n} \left(H_{jk}^\alpha H_{\alpha\ell}^i + H_{k\ell}^\alpha H_{\alpha j}^i + H_{\ell j}^\alpha H_{\alpha k}^i \right)$$

possesses the same symmetries as the $(1,3)$ Riemann tensor $R_{jk\ell}^i$ in Riemannian geometry.

References

[1] D. J. Benney, *Some properties of long non-linear waves*, Stud. Appl. Math. **52** (1973), 45–50.

[2] O. I. Bogoyavlenskij, *Necessary conditions for existence of non-degenerate Hamiltonian structures*. Commun. Mathem. Phys. **182** (1996), 253–290.

[3] R. Courant and D. Hilbert, *Methods of mathematical physics, II*, Interscience Publishers, New York, **1962**.

[4] J. Glimm, *Solutions in the large for nonlinear hyperbolic systems of equations*. Comm. Pure Appl. Math. **18** (1965), 697–715.

[5] J. Haantjes, *On X_m-forming sets of eigenvectors*. Proc. Kon. Ned. Akad. Amsterdam **58** (1955), 158–162.

[6] P. D. Lax, *Weak solutions of nonlinear hyperbolic equations and their numerical computation*. Comm. Pure Appl. Math. **7** (1954), 159–193.

[7] A. Nijenhuis, *X_{n-1}-forming sets of eigenvectors*. Proc. Kon. Ned. Akad. **54** (1951), 200–212.

[8] S. P. Novikov, *Differential geometry and hydrodynamics of soliton lattices*, in: A. S. Fokas and V. E. Zakharov, Eds., *Important developments in soliton theory*, Springer series in nonlinear dynamics, Springer-Verlag (Berlin) (1993), 242–256.

Department of Mathematics and Statistics,
Queen's University,
Kingston, Canada, K7L 3N6
E-mail address: bogoyavl@oib.mast.Queensu.ca

International Series of Numerical Mathematics
Vol. 129, © 1999 Birkhäuser Verlag Basel/Switzerland

Old and New Hyperbolic Approaches in General Relativity

Carles Bona

Abstract. A brief review of hyperbolic formalisms in General Relativity is presented, stressing the difference between the 'old' ones, where hyperbolicity is achieved by fixing the four coordinate degrees of freedom, and the 'new' ones, where it is achieved by using instead the momentum constraints. The general way of combining the Einstein evolution equations with the constraints in the 3+1 framework is considered with a view on Numerical Relativity applications.

1. The structure of the Einstein field equations

As it is well known, Einstein's field equations relate the energy density and the curvature of the spacetime through a set of ten second order partial differential equations,

$$G_{\mu\nu} = 8\pi \, T_{\mu\nu} \,, \tag{1}$$

where the spacetime geometry is described by the line element coefficients

$$ds^2 = g_{\mu\nu} dx^\mu dx^\nu \quad \mu, \, \nu = 0, 1, 2, 3 \tag{2}$$

which enter into the field equations (1) through the Ricci tensor $R_{\mu\nu}$, namely

$$G_{\mu\nu} = R_{\mu\nu} - \frac{R}{2} \, g_{\mu\nu} \,. \tag{3}$$

The Cauchy problem of the Einstein's field equations is far from being trivial. One can not freely choose for instance the metric coefficients and their first time derivatives as initial data on a given space-like initial ($t = 0$) hypersurface. This is because four of the ten equations do not contain second time derivatives of the metric coefficients and are then to be regarded as constraints, namely

$$G^{0\nu} = \mathcal{C}^\nu(g, \partial g) = 8\pi \, T^{0\nu} \tag{4}$$

(*energy-momentum* constraints). These constraints are actually first integrals of the remaining set of equations (*evolution system*) so that we need only to impose (4) on the initial data. This redundancy means, on the other hand, that the evolution system consists only of six equations, compared with the ten metric coefficients. In order to get a complete evolution system, we must then provide four extra conditions. This is related with the gauge invariance of the equations (1), which are valid in any coordinate system; the gauge is to be fixed by providing a coordinate condition for every one of the four spacetime coordinates.

The mathematical structure of the equations can be studied in more detail by using the following expression for the Ricci tensor [1, 2],

$$2\,R_{\mu\nu} = -\Box g_{\mu\nu} + \partial_\mu\Gamma_\nu + \partial_\nu\Gamma_\mu + H_{\mu\nu}(g,\partial g)\,, \tag{5}$$

where the box stands for the d'Alembert operator acting on functions and we have noted for short

$$\Gamma^\mu \equiv \frac{-1}{\sqrt{|g|}}\partial_\nu(\sqrt{|g|}\,g^{\mu\nu}) = \Box x^\mu\,. \tag{6}$$

Note that in the vacuum case $(T_{\mu\nu} = 0)$ the Einstein field equations (1) are quasi-linear, because the nonlinear terms in the Ricci tensor, collected in the function $H_{\mu\nu}$, involve at most first derivatives of the metric coefficients. We will consider for simplicity only the vacuum case in what follows.

2. The old hyperbolic formalisms

In what we will call 'old' hyperbolic formalisms, the four gauge degrees of freedom were used to prescribe the quantities Γ^μ introduced in Eq. 6 as a given set of functions $f^\mu(x,g)$. This prescription is actually a first order equation and it can be viewed then as a *gauge* constraint. By substituting this constraint into the original evolution system (5), one gets a symmetric hyperbolic system with the following principal part

$$\Box g_{\mu\nu} = \cdots \tag{7}$$

We will call it *relaxed* evolution system because it provides ten evolution equations without any constraint.

Of course, the solution space of the relaxed system is much wider than the solution space of the original evolution system. Even if we restrict ourselves to solutions of the relaxed system obtained from initial data verifying both the energy-momentum constraints and the gauge constraints, they will be solutions of the original evolution system (5) only if all these constraints are shown to be first integrals of the relaxed system. If this happens, this provides a hyperbolic formalism for General Relativity.

The first of such hyperbolic formalisms [3] was obtained by prescribing the four spacetime coordinates to be harmonic functions

$$\Box x^\mu \equiv -\Gamma^\mu = 0 \tag{8}$$

(*Harmonic coordinates*). It is widely used in analytical approximation schemes, but not in Numerical Relativity. This is because, as every 'old' hyperbolic formalism, the four gauge degrees of freedom have been used to get hyperbolicity so that there is no gauge freedom left in order to adapt the coordinates to either the symmetries or the physics of the spacetime one is computing.

Harmonic coordinates have been generalized later [7, 8] by introducing *gauge source* terms in the right-hand-side of Eq. 8. In the most general version [9] the gauge sources can be generic functions of both the spacetime coordinates and the metric components (but not their derivatives). In the meantime, however, a breakthrough in our understanding of the structure of Einstein's field equations pointed into another direction.

3. The 3+1 formalism

This breakthrough was caused by the so called '3+1' or evolution formalism [4] which provides a clear geometric interpretation of the role played by spacetime coordinates. To see this, let us begin with the time coordinate. It can be associated to the slicing of spacetime by the family of constant time hypersurfaces (or, if we take the dual approach, to the congruence of lines normal to these hypersurfaces).

From the geometrical point of view, this means regarding spacetime as a succession of three-dimensional slices, much like when we see real life as a succession of still images in a movie picture. Of course, we are free to choose the three space coordinates in every slice without affecting at all their intrinsic three-dimensional geometry. X-Mozilla-Status: 0000

The four-dimensional line element (2) can be written then

$$ds^2 = -\alpha^2 \, dt^2 + \gamma_{ij}(dx^i + \beta^i \, dt)(dx^j + \beta^j \, dt) \quad i, j = 1, 2, 3 \tag{9}$$

where γ_{ij} is the metric induced on the three-dimensional slices (first fundamental form). The *lapse function* α relates the proper time along the normal lines with coordinate time, and the *shift* β^i measures the deviation of the time lines $(x^i = x_0^i)$ with respect to the normal lines.

The evolution equations in the 3+1 formalism can be expressed in the vacuum case as follows:

$$(\partial_t - \mathcal{L}_\beta)\gamma_{ij} = -2\alpha \, K_{ij} \,, \tag{10}$$

$$(\partial_t - \mathcal{L}_\beta)K_{ij} = -\alpha_{i;j} + \alpha \left[{}^{(3)}R_{ij} - 2K_{ij}^2 + tr \, K \, K_{ij} \right] . \tag{11}$$

where $\mathcal{L}$ stands for the Lie derivative, all index contractions and covariant derivatives are with respect to the induced metric γ_{ij}, and the three-dimensional Ricci tensor constructed from this metric is denoted by ${}^{(3)}R_{ij}$. Note that the extrinsic curvature K_{ij} (second fundamental form) of the three-dimensional slices is introduced here as an independent quantity.

The energy-momentum constraints (4) can also be expressed now as follows:

$$ {}^{(3)}R + (tr \, K)^2 - tr(K^2) = 0 \,, \tag{12}$$

$$ K^k_{i;k} - \partial_i(tr \, K) = 0 \,. \tag{13}$$

where ${}^{(3)}R$ is the trace of the three-dimensional Ricci tensor.

There is no evolution equation for the lapse α nor the shift β^i, which can be prescribed freely. It is then clear transparent that these are purely kinematical quantities related with the choice of time and space coordinates, respectively. The 3+1 formalism was actually discovered [5] for a particular prescription of these quantities ($\alpha = 1$, $\beta^i = 0$) and then extended to the general case [4]. Let us for instance express the harmonic coordinate conditions (8) in the 3+1 language:

$$\partial_t(\frac{\sqrt{\gamma}}{\alpha}) - \partial_i(\frac{\sqrt{\gamma}}{\alpha} \beta^i) = 0 \tag{14}$$

$$\partial_t(\frac{\sqrt{\gamma}}{\alpha} \beta^i) - \partial_j(\frac{\sqrt{\gamma}}{\alpha} \beta^i\beta^j - \alpha\sqrt{\gamma}\,\gamma^{ij}) = 0 \tag{15}$$

The coordinate conditions (15) provide an evolution equation for the shift which contains a strong coupling with the evolving three-dimensional geometry of the space-like slices. However, as far as the choice of the space coordinates on every slice can not affect its intrinsic geometry, a purely kinematical choice of the shift seems to be more appropriate. We will consider for instance taking a zero shift, but keeping the harmonicity condition (14) for the time coordinate (*harmonic slicing*).

Note also that the evolution system (10,11) is first order in time but second order in space. To obtain a system which is also of first order in space, we will follow the standard procedure by introducing the auxiliary variables

$$A_k = \partial_k ln\,\alpha, \quad D_{kij} = 1/2\,\partial_k\gamma_{ij} , \tag{16}$$

so that the principal part of the 3+1 evolution system (10,11) reads

$$\partial_t\gamma_{ij} = \cdots \tag{17}$$

$$\partial_t K_{ij} + \partial_k[\alpha\,\lambda_{ij}^k] = \cdots \tag{18}$$

$$\partial_t D_{kij} + \partial_k[\alpha\,K_{ij}] = 0 , \tag{19}$$

where the flux terms λ_{ij}^k are given by

$$\lambda_{ij}^k = D_{ij}^k + 1/2\,\delta_i^k\,(A_j + 2\,V_j - D_{jr}^{\ \ r}) + 1/2\,\delta_j^k\,(A_i + 2\,V_i - D_{ir}^{\ \ r}) , \tag{20}$$

and we have noted

$$V_k = D_{kr}^{\ \ r} - D_{rk}^r . \tag{21}$$

A straightforward analysis shows that the first order evolution system (17,18), when supplemented with the harmonic slicing condition (14), namely (zero shift)

$$\partial_t(\frac{\sqrt{\gamma}}{\alpha}) = 0 , \quad \partial_t(A_i - D_{ij}^{\ \ j}) = 0 , \tag{22}$$

is not symmetric hyperbolic. Even if all the characteristic speeds are real, the characteristic matrix can not be fully diagonalized due to the combinations V_k appearing in the flux terms (20).

This is quite disappointing both from the theoretical point of view (Cauchy problem) and from the computational one (upwind methods, convergence proofs, etc.). In spite of this fact, the original form (10,11) of the 3+1 formalism has been widely used for years (and is still being used) in Numerical Relativity.

4. The new hyperbolic formalisms

Three decades after the discovery of the 3+1 formalism, a new way of obtaining symmetric hyperbolic system was open [11]. The same 3+1 framework was used with the harmonic slicing condition (22), but an extra time derivative of the equations was considered so that the basic dynamical quantities were not only the three-dimensional metric and the extrinsic curvature but also the time derivative of the second fundamental form.

The key point to get a symmetric hyperbolic system, however, was to use the momentum constraints (13) to kill unwanted terms in the resulting equations. This use of the momentum constraints to ensure hyperbolicity is the point that makes the difference between the old hyperbolic formalisms (where the choice of space coordinates played this role) and the new ones. In the new hyperbolic formalisms, the shift remains an arbitrary kinematical quantity, in keeping with the geometric interpretation of the 3+1 framework. This opens the possibility of adapting the space coordinates to the physics of the problem considered, making these new formalisms more suitable for Numerical Relativity.

All these points have been confirmed much more recently, where the original work [11] has been extended to allow for a generic kinematical prescription of the shift [12] and the resulting formalism has been used in numerical applications by the team of the North Carolina University. It has been shown also that the same idea, but replacing the extra time derivative by a space derivative allows to get again another symmetric hyperbolic formalism [13].

In the meantime, however, we discovered that there is not need to take extra derivatives of any kind to get new hyperbolic formalisms [14]. We realized that the momentum constraint (13) contained space derivatives of the extrinsic curvature which, allowing for (10), is basically a time derivative of the metric. Switching the ordering between space and time derivatives in (13), one gets an evolution equation with a trivial principal part

$$\partial_t V_k = \cdots \tag{23}$$

where V_k are the quantities defined in (21).

Equation (23) is redundant in the original 3+1 formalism, where the space derivatives of the metric are evolved using (19). We will consider instead V_k to be independent quantities, so that equation (23) will be added to the original set of equations (17-19) to get an extended evolution system. The equation (21) will then be considered as an algebraic constraint or, more precisely, the algebraic form of the momentum constraint in the new formalism.

A straightforward calculation using harmonic slicing shows that the extended evolution system is symmetric hyperbolic (remember that the quantities V_k where the ones that prevented the full diagonalization in the original system). This result holds not only for harmonic slicing, but for a generic monotonic functional dependence between the lapse α and the space volume element $\sqrt{\gamma}$ of the form [15]

$$\alpha = F(\sqrt{\gamma}) \qquad F' > 0. \tag{24}$$

In the generic case (24), one of the characteristic cones has a nontrivial characteristic speed (*gauge speed*) which is different from light speed. Only in the harmonic slicing case, where the relationship (24) is linear, gauge speed and light speed do coincide. Condition (24) is general enough to cover most of the algebraic gauge choices that are being used in numerical applications.

In a more recent work [16] these results are extended to allow for a general kinematical prescription for the shift and, on the other hand, to build a one parameter family of symmetric hyperbolic systems by combining the evolution equations with the energy constraint. To explain the last point, let me notice that the principal part of the energy constraint (12) can be written

$$\partial_k V^k = \cdots \tag{25}$$

so that the evolution equation (18) with flux terms given by (20) is equivalent, modulo the energy constraint, to an equation of the same form with the flux terms

$$\lambda_{ij}^k = D_{ij}^k + 1/2\, \delta_i^k \left(A_j + 2 V_j - D_{jr}{}^r \right) + 1/2\, \delta_j^k \left(A_i + 2 V_i - D_{ir}{}^r \right) - \frac{n}{2} V^k \gamma_{ij}, \tag{26}$$

where n is an arbitrary parameter. The choice of this parameter does not affect to the hyperbolicity of the extended evolution system. This indicates that the energy constraint does not play the same central role as the momentum constraint in that respect.

5. The way ahead

Once we have realized that the mathematical properties of the evolution system may be changed quite drastically by mixing it with the energy-momentum constraints, we have opened the Pandora box. Even if we refrain to take extra derivatives or to introduce redundant quantities, we still have a wide choice.

We can consider for instance the more general tensor combination of evolution and constraint equations, namely

$$\partial_t K_{ij} \quad + \quad \partial_k [\alpha\, \lambda_{ij}^k] = \cdots, \tag{27}$$

$$\partial_t D_{kij} + \partial_k [\alpha\, K_{ij}] \quad + \quad a\,\alpha\, [\gamma_{ki}\, \partial_l(K^l{}_j - \delta_j^l\, tr\, K) + \gamma_{kj}\, \partial_l(K^l{}_i - \delta_i^l\, tr\, K)]$$
$$+ \quad b\,\alpha\, \gamma_{ij}\, \partial_l(K^l{}_k - \delta_k^l\, tr\, K) = \cdots, \tag{28}$$

where a and b are arbitrary parameters and the flux terms λ^k_{ij} are still given by (26) but interpreting the symbols V_k no longer as representing additional variables, but again as the shorthand (21).n

The choice $a = b = n = 0$ corresponds to the original 3+1 evolution system (18-20), which is not symmetric hyperbolic. A different choice [17] may lead to a symmetric hyperbolic system when combined with the harmonic slicing condition, the characteristic speeds being either zero or the light speed along the specific direction. Other choices actually lead to symmetric hyperbolic systems with different characteristic speeds: this just means that we are 'propagating' constraint degrees of freedom with quite arbitrary characteristic speeds.

The only degrees of freedom that are not affected by the choice of parameters nor the choice of coordinate gauge are the two characteristic cones that allow gravitational wave propagation. Everything else can be altered and there is no strong theoretical argument that leads to a preferred set of parameters.

There can be, however, pragmatic arguments when considering numerical applications. The choice $n = 1$, for instance, ensures that the evolution system can be obtained directly from the space components of the Einstein tensor, without using at all the energy constraint. These components do not contribute to the Newtonian limit of the Einstein equations, so that we can consider the resulting evolution system to be free of first order (Newtonian) corrections to the flat space-time (Minkowski) metric. A much better accuracy either in weak field scenarios or in far field boundary conditions is then to be expected.

Other arguments may arise when studying the propagation of the errors in the energy-momentum constraints arising from truncation errors in the numerical code [18],[19]. Even if the constraints are first integrals of the exact evolution equations, they are not first integrals of the finite difference versions and the requirement of a stable propagation of the energy-momentum constraints may restrict our parameter space.

In any case, this is an active field of research. The interested reader may find an continuously updated *living review* of these topics in an electronic journal currently published by the Albert-Einstein-Institut [20].

This work is supported by the DGICYT of Spain, under project No. PB-94 1177.

References

[1] T. DeDonder, *La Gravifique Einstenienne* Gauthier-Villars, Paris, 1921.

[2] C. Lanczos, Phys. Z., **23** 537 (1922).

[3] Y. Foures-Bruhat, Acta Math. **88** 141 (1955).

[4] Y. Choquet-Bruhat, C.R. Acad. Sc. Paris **226**, 1071 (1948).
J. Rat. Mec. Analysis, **5** 951 (1956).

[5] A. Lichnerowicz, J. Math. Pures Appl., **23** 37 (1944).

[6] R. Arnowitt, S. Deser and C. W. Misner, in *Gravitation: an Introduction to Current Research*, ed. L. Witten, Wiley, New York, **1962**.

[7] S.W. Hawking, G.F.R. Ellis, *The large scale structure of spacetime*, Cambridge U.P., **1973**.

[8] D. DeTurck, Invent. Math., **65** 179 (1981).

[9] H. Friedrich, Commun. Math. Phys. **107** 587 (1985).

[10] Y. Choquet-Bruhat, in *Gravitation: an Introduction to Current Research*, ed. L. Witten, Wiley, New York, **1962**.

[11] Y. Choquet-Bruhat and T. Ruggeri, Comm. Math. Phys., **89** 269 (1983).

[12] Y. Choquet-Bruhat, J.W. York, C.R. Acad. Sc. Paris **321**, 1089 (1995).
A. Abrahams, A. Anderson, Y. Choquet-Bruhat and J.W. York, Phys. Rev. Lett., **75** 3377 (1995).

[13] H. Friedrich, Class. Quantum Grav., **13** 1451 (1996).

[14] C. Bona and J. Massó, Phys. Rev. Lett., **68** 1097 (1992).

[15] C. Bona, J. Massó, E. Seidel and J. Stela, Phys. Rev. Lett., **75** 600 (1995).

[16] C. Bona, J. Massó, E. Seidel and J. Stela, Phys. Rev. D, **56** 3405 (1997).

[17] S. Fritelli, O.A. Reula, Commun. Math. Phys. **166**, 221 (1994).
Phys. Rev. Lett., **76** 4667 (1996).

[18] M. W. Choptuik, Phys. Rev. D, **44** 3124 (1991).

[19] S. Fritelli, Phys. Rev. D, **55** 5992 (1997).

[20] See the Article by O.A. Reula, in the site http://www.livingreviews.org, maintained by the Albert-Einstein-Institut at Potsdam (Germany).

Departament de Fisica,
Universitat de les Illes Balears,
E-07071 Palma de Mallorca, Spain
E-mail address: dfscbg0@ps.uib.es

International Series of Numerical Mathematics
Vol. 129, © 1999 Birkhäuser Verlag Basel/Switzerland

Differentiability with Respect to Initial Data for a Scalar Conservation Law

François Bouchut and François James

Abstract. We linearize a scalar conservation law around an entropy initial datum. The resulting equation is a linear conservation law with discontinuous coefficient, solved in the context of duality solutions, for which existence and uniqueness hold. We interpret these solutions as weak derivatives with respect to the initial data for the nonlinear equation.

1. Introduction

Consider the one-dimensional scalar conservation law

$$\partial_t u + \partial_x f(u) = 0, \quad 0 < t < T, \quad x \in \mathbb{R}, \tag{1}$$

where f is a C^1 convex function, provided with entropy admissible initial data $u^\circ \in L^\infty(\mathbb{R})$. Kružkov's results [4] assert that the entropy solution u to (1) lies in $L^\infty(]0, T[\times\mathbb{R}) \cap C(0, T; L^1_{loc}(\mathbb{R}))$, and that the following contraction property holds: if u (resp. v) corresponds to the initial data u° (resp. v°), then for all $R > 0$ and any $t > 0$

$$\int_{|x|\leq R} |u(t, x) - v(t, x)| \, dx \leq \int_{|x|\leq R+Mt} |u^\circ(x) - v^\circ(x)| \, dx, \tag{2}$$

where $M = \max\{|f'(s)|, |s| \leq \max(\|u^\circ\|_{L^\infty}, \|v^\circ\|_{L^\infty})\}$. This can be interpreted as a continuity result with respect to initial data in L^1_{loc}. Another kind of stability results was proved by e.g. Majda [5] in the very general setting of a multi-dimensional system of equations: when u° is an admissible shock, then the solution corresponding to a smooth perturbation of the shock is also an admissible shock.

The aim of this paper is to give a few hints about the analysis of a different point of view. We are going to prove some kind of differentiability with respect to initial data, thus generalizing (2). The derivative of the operator $u^\circ \mapsto u$ is solution to the equation obtained by linearizing (1) in the neighborhood of some given u°:

$$\partial_t \mu + \partial_x \left(a(t, x)\mu \right) = 0, \quad a(t, x) = f'(u(t, x)). \tag{3}$$

We shall prove existence and uniqueness for the corresponding Cauchy problem, thus obtaining the first order term of a "Taylor expansion" of the operator. When applied to the perturbation of a shock, we can recover the perturbed shock as well as an approximation of its position. We would like to mention here that a numerical application of this framework is used by Olazabal [6] in fluid dynamics, where a 2-dimensional perturbation of a 1-dimensional shock is computed by simultaneously discretizing in a convenient way both (1) and (3) (see also [3]).

It turns out that the conservation law with discontinuous coefficients (3) has to be solved in the space of measures on $\mathbb{R}$. This leads to the classical problem of defining the product $a\mu$ of a discontinuous function by a measure. Moreover, in order to use (3) for the stability analysis of (1), we need to define this product in such a way that we have weak stability when the coefficient a is perturbed. This was achieved in a preceding paper by the authors [1, 2], where the so-called *duality solutions* are defined. For these solutions, existence and uniqueness hold for the Cauchy problem associated to (3), as well as stability. These results are recalled without proof in Section 2.

In Section 3 we interpret the above results as weak differentiability results with respect to an entropy initial datum, when the flux f is strictly convex. Actually, the solution to (3) behaves like a directional or Gâteaux derivative of the operator which to u° associates the entropy solution to (1). However, this is not rigorously a derivative, since u° lies in $L^1_{loc}(\mathbb{R})$, and the "derivative" is continuous only in the weak sense of measures. We also consider the example of an entropy initial shock, and show how to recover Majda's result.

2. Duality solutions

We recall here briefly the definition and main properties of the so-called duality solutions, which were introduced by Bouchut and James [1, 2]. For detailed results and proofs, we refer to [2]. Duality solutions are measure-valued solutions μ belonging to the space

$$\mathcal{S}_{\mathcal{M}} = C([0,T], \mathcal{M}_{loc}(\mathbb{R}) - \sigma(\mathcal{M}_{loc}(\mathbb{R}), C_c(\mathbb{R}))),$$

where σ denotes the usual weak topology. They are defined as weak solutions, the test functions being Lipschitz solutions to the backward linear transport equation

$$\partial_t p + a(t,x)\partial_x p = 0, \quad p(T,.) = p^T \in \mathrm{Lip}(\mathbb{R}). \tag{4}$$

A formal computation shows that $\partial_t(p\mu) + \partial_x[a(t,x)p\mu] = 0$, and thus

$$\frac{d}{dt}\langle p, \mu \rangle = 0, \tag{5}$$

which defines the duality solutions for suitable p-s. It is well known (see e.g. Oleinik [7]) that the existence of solutions to (4) is ensured by the so-called one-sided

Lipschitz condition

$$\partial_x a \le \alpha(t) \quad \text{in }]0, T[\times\mathbb{R}, \qquad \alpha \in L^1(]0, T[), \tag{6}$$

and that this existence result gives uniqueness for (1) —the same will occur for (3). The point here is that there is no uniqueness for (4) even when (6) holds.

The corner stone in the construction of duality solutions is therefore the introduction of the notion of *reversible* solutions to (4). We shall not need here the precise properties of reversible solutions. We only have to know that they can be characterized in various ways: support properties of $\partial_x p$, monotonicity properties, total variation properties, entropy inequalities. We shall admit that there is existence and uniqueness to the backward Cauchy problem (4) in the class of reversible solutions. We now restrict ourselves to those p-s in (5). More precisely, we state the following definition.

Definition 2.1. *We say that* $\mu \in S_M$ *is a* **duality solution** *to (3) if for any* $0 < \tau \le T$, *and any* **reversible** *solution p to (4) with compact support in x, the function* $t \mapsto \int_{\mathbb{R}} p(x,t)\mu(t, dx)$ *is constant on* $[0, \tau]$.

The main results concerning the Cauchy problem associated to (3) are summarized as follows (see Bouchut & James [2]).

Theorem 2.2. *[Cauchy problem]*

1) Given $\mu^0 \in M_{loc}(\mathbb{R})$, *there exists a unique* $\mu \in S_M$ *duality solution to (3), such that* $\mu(.,0) = \mu^0$.

2) This solution satisfies for any $x_1 < x_2$ *and* $t \in [0, T]$

$$\int_{[x_1, x_2]} |\mu(t, dx)| \le \int_{[x_1 - \|a\|_\infty t, x_2 + \|a\|_\infty t]} |\mu^0(dx)|. \tag{7}$$

3) There exists a bounded Borel function $\hat{a}$, *such that* $\hat{a} = a$ *almost everywhere, and*

$$\partial_t \mu + \partial_x(\hat{a}\mu) = 0 \quad \text{in the distributional sense.}$$

The set of duality solutions is clearly a vector space, but it has to be noted that a duality solution is not defined as a solution in the sense of distributions. The product $\hat{a}\mu$ is defined *a posteriori*, by the equation itself.

Remark 2.3. It is useful to notice that L^∞ distributional solutions to (3) are duality solutions.

We turn now to the most important property of duality solutions, namely weak stability.

Theorem 2.4 (Weak stability). *Let* (a_n) *be a bounded sequence in* $L^\infty(]0, T[\times\mathbb{R})$, *with* $a_n \rightharpoonup a$ *in* $L^\infty(]0, T[\times\mathbb{R}) - w\star$. *Assume* $\partial_x a_n \le \alpha_n(t)$, *where* (α_n) *is bounded* in $L^1(]0,T[)$, $\partial_x a \leq \alpha \in L^1(]0,T[)$. *Consider a sequence* $(\mu_n) \in S_M$ *of duality solutions to*

$$\partial_t \mu_n + \partial_x(a_n \mu_n) = 0 \quad in \quad]0,T[\times\mathbb{R},$$

such that $\mu_n(0,.)$ *is bounded in* $M_{loc}(\mathbb{R})$, *and* $\mu_n(0,.) \rightharpoonup \mu^0 \in M_{loc}(\mathbb{R})$. *Then* $\mu_n \to \mu$ *in* S_M, *where* $\mu \in S_M$ *is the duality solution to*

$$\partial_t \mu + \partial_x(a\mu) = 0 \quad in \quad]0,T[\times\mathbb{R}, \qquad \mu(0,.) = \mu^0.$$

Moreover, $\hat{a}_n \mu_n \rightharpoonup \hat{a}\mu$ *weakly in* $M_{loc}(]0,T[\times\mathbb{R})$.

Remark 2.5. Duality solutions can be defined if $\alpha \in L^1(]0,T[)$ is replaced by $\alpha \in L^1_{loc}(]0,T[)$. The same existence result holds, but uniqueness is lost. A weakened form of Theorem 2.4 is valid, up to a subsequence of (μ_n).

3. Differentiability

Let us now interpret these results as differentiability results. We first compute some kind of directional derivative of the solution with respect to the initial data. Next, we apply this to the stability analysis of a shock.

Theorem 3.1. *Assume that* $f'' \in L^\infty$ *and* $f'' \geq \gamma > 0$. *Consider an entropy admissible initial datum* u°, *i.e.* $\partial_x u^\circ \leq C$ *for some constant* C. *Let* u_λ *be the solution of (1) with initial datum* $u^\circ + \lambda \delta u^\circ$, *where* $\lambda > 0$ *and* $\delta u^\circ \in L^\infty(\mathbb{R})$, *and set* $w_\lambda = (u_\lambda - u)/\lambda$. *Then* $w_\lambda \to \mu$ *in* S_M, *where* μ *is the unique duality solution to*

$$\partial_t \mu + \partial_x(a\mu) = 0, \quad \mu(0,.) = \delta u^\circ. \tag{8}$$

Proof. We first have to check that (8) is well-posed. But this follows immediately since f is strictly convex and smooth: indeed Oleinik's entropy condition and the fact that u° is admissible imply that $a = f'(u)$ satisfies (6) with $\alpha = $ constant and $\partial_x u \leq C$.

Next, we notice that w_λ is a L^∞ distributional solution to the equation

$$\partial_t w_\lambda + \partial_x(a_\lambda w_\lambda) = 0, \quad w_\lambda(x,0) = \delta u^\circ(x), \tag{9}$$

where
$$a_\lambda = \begin{cases} \dfrac{f(u_\lambda) - f(u)}{u_\lambda - u} & if \ u_\lambda \neq u, \\ f'(u) & if \ u_\lambda = u. \end{cases}$$

The key point now is to prove that w_λ is actually a duality solution to (9). Since w_λ is a distributional solution, by Remark 2.3 we only have to verify that a_λ satisfies (6) with some $\alpha \in L^1_{loc}(]0,T[)$. Once again we use Oleinik's entropy condition and the strict convexity of f, to obtain $\partial_x u_\lambda \leq 1/(\gamma t)$. A direct computation and the convexity of f lead to $\partial_x a_\lambda \leq \|f''\|_\infty/(\gamma t)$.

Finally, a_λ is bounded in $L^\infty(]0,T[) \times \mathbb{R}$ and converges almost everywhere to $a = f'(u)$, so that $a_\lambda \rightharpoonup a$ in $L^\infty - w\star$ as λ tends to 0. We are thus in position to apply the weakened version of the stability theorem 2.4 (see Remark 2.5), which gives exactly the expected convergence result, since the limit equation (8) has a unique solution.   □

We actually proved here that μ can be interpreted as a weak directional derivative. Indeed, consider the operator $J : L^\infty \to S_\mathcal{M}$ which to the initial data u° associates the solution u to (1). The above convergence result states exactly that the duality solution μ to (3) is in a weak sense the derivative $DJ(u^\circ; \delta u^\circ)$ of J at u° in the direction δu°. Moreover, (7) in Theorem 2.2 asserts exactly that $DJ(u^\circ; \delta u^\circ)$ is a continuous linear operator on $\mathcal{M}_{loc}(\mathbb{R})$ with respect to δu°, which means that, also in some weak sense, J is Gâteaux differentiable in u°, provided u° is an admissible initial datum, and f is convex.

We illustrate this result on a simple example: consider for u° an admissible shock $u^\circ = u_\ell \mathbb{I}_{x<0} + u_r \mathbb{I}_{x>0}$, with $u_\ell > u_r$, and assume that δu° is a smooth compactly supported function. Denote by u^R the solution to the unperturbed Riemann problem. Our result states that the solution u_ε to this problem satisfies

$$u_\varepsilon = u^R + \varepsilon\mu + \varepsilon\nu, \tag{10}$$

where $\nu \to 0$ in $S_\mathcal{M}$ and μ solves (8). Actually, μ can be computed explicitly by integrating (8). This leads to a transport equation with discontinuous coefficient, which can be solved in the sense of duality solutions (see [2]). Differentiating the solution, we get

$$\begin{aligned}\mu(t,x) \quad = \quad & \delta u^\circ(x - f'(u_\ell)t)\mathbb{I}_{x<\sigma t} + \delta u^\circ(x - f'(u_r)t)\mathbb{I}_{x>\sigma t} + \\ & + \varepsilon\left[v^1((\sigma - f'(u_r))t) - v^1((\sigma - f'(u_\ell))t)\right]\delta_{x-\sigma t},\end{aligned} \tag{11}$$

where σ is the shock velocity, $\sigma = [f(\delta u^\circ)]/[\delta u^\circ]$, and v^1 a primitive of δu°.

Another approach to this problem, followed by Majda [5], is to prove directly that, for ε small enough, the solution u_ε is a function with a smooth discontinuity line $x = \xi_\varepsilon(t)$. One can perform an asymptotic expansion of u_ε and ξ_ε in terms of ε. Quite straightforward computations, which are detailed for instance in [3], lead to the following expressions, up to higher order terms in ε. First, concerning the perturbed shock speed, $\sigma_\varepsilon = d\xi_\varepsilon/dt = \sigma^0 + \varepsilon\sigma^1$, with

$$\begin{aligned}\sigma^0(t) \quad &\equiv \quad \sigma, \\ \sigma^1(t) \quad &= \quad \frac{(f'(u_r) - \sigma)\delta u^\circ((\sigma - f'(u_r))t) - (f'(u_\ell) - \sigma)\delta u^\circ((\sigma - f'(u_\ell))t)}{u_r - u_\ell}.\end{aligned}$$

Next, the solution is given by

$$u_\varepsilon(t,x) = \begin{cases} u_\ell + \varepsilon\delta u^\circ(x - f'(u_\ell)t) & \text{for } x < \sigma t + \varepsilon\xi^1(t), \\ u_r + \varepsilon\delta u^\circ(x - f'(u_r)t) & \text{for } x > \sigma t + \varepsilon\xi^1(t), \end{cases}$$

where $\xi^1(t) = \int_0^t \sigma^1(s)\,ds$.

The last approach thus gives an approximation of the solution as a discontinuous function, as well as an approximation of the position of the perturbed shock. Our framework may seem strange at first sight, since we approximate a function by a measure concentrated on the original shock. However, the information on the perturbed shock position is also contained in formulæ (10)-(11), since the coefficient of the Dirac mass in (11) equals exactly $-(u_r - u_\ell)\xi^1(t)$.

References

[1] F. Bouchut and F. James, *Équations de transport unidimensionnelles à coefficients discontinus*, C.R. Acad. Sci. Paris, Série I, **320** (1995), 1097–1102.

[2] F. Bouchut and F. James, *One-dimensional transport equations with discontinuous coefficients*, Prépublication MAPMO n° 96-13, université d'Orléans, avril 1996, Nonlinear Analysis, TMA, **32** (1998), No 7, 891–933

[3] E. Godlewski, M. Olazabal and P.-A. Raviart, *On the linearization of hyperbolic systems of conservation laws. Application to stability,* Équations aux dérivées partielles et applications, articles dédiés à J.-L. Lions, Gauthier-Villars, Paris, 1998, 549–570

[4] S.N. Kružkov, *First-order quasilinear equations in several independent variables*, Math. USSR Sb., **10** (1970), 217–243.

[5] A. Majda, *The stability of multi-dimensional shock fronts*, Mem. Amer. Math. Soc., **275** (1982).

[6] M. Olazabal *Résolution numérique du système des perturbations linéaires d'un écoulement MHD*, Thèse Université Paris 6, 1998.

[7] O.A. Oleinik, *Discontinuous solutions of nonlinear differential equations*, Amer. Math. Soc. Transl. (2), **26** (1963), 95–172.

Mathématiques, Applications et Physique Mathématique d'Orléans,
UMR CNRS 6628,
Université d'Orléans,
BP 6759, 45067 Orléans Cedex 2, FRANCE
E-mail address: fbouchut@labomath.univ-orleans.fr
james@cmapx.polytechnique.fr

International Series of Numerical Mathematics
Vol. 129, © 1999 Birkhäuser Verlag Basel/Switzerland

Evolution Behavior of Transverse Shocks in a Nonlinear Elastic Layer

Manfred Braun and Sławomir Kosiński

Abstract. The evolution of a plane transverse shock wave propagating in a thin elastic layer is investigated. By averaging the equations of motion over the width of the layer the problem is reduced to a system of partial differential equations in one space dimension and time. The theory of singular surfaces is applied to establish relations between the discontinuities of the strain gradient and its derivatives. Explicit results are developed for a special incompressible elastic material.

1. Introduction

The propagation of a plane transverse shock wave along a thin elastic layer is investigated. The effects of finite lateral dimensions and inertia of the layer are considered by describing the layer as a one-dimensional elastic structure with one scalar variable representing the transverse symmetric motion. The layer has two kinematically independent degrees of freedom which are represented by two independent functions describing the motion within the layer. The transverse shock wave is modelled as a propagating singular surface across which the shear component of the deformation gradient undergoes a discontinuity. The theory of singular surfaces [4] is used to derive the relevant equations for the discontinuous quantities.

To some extent our presentation follows the ideas of Fu and Scott [1], who consider a plane transverse shock wave propagating into a half-space consisting of an incompressible, isotropic elastic material. The main difference is the finite transverse dimension of the layer and its assumed symmetric motion. Unlike in [1] the strength of the transverse shock varies over the width of the layer attaining its maxima at the stress-free lateral surfaces. Nevertheless, some of the equations derived in [1] are regained for the transverse shock propagating in the layer.

Following this introduction the general equations describing the motion of an incompressible, nonlinear elastic medium are recollected in Section 2. The symmetric lateral motion of an elastic layer is presented in Section 3. A special assumption about the lateral motion of the layer makes the problem essentially

The second author (S. K.) gratefully acknowledges the fellowship granted by the DAAD (Deutscher Akademischer Austauschdienst) at Gerhard-Mercator-Universität Duisburg during Oct–Nov 1996.

one-dimensional. As a consequence, the full three-dimensional equations of motion cannot be sustained. The appropriate one-dimensional equations are provided in Section 4 by an averaging procedure. The resulting reduced equations of motion have the form of two coupled PDEs in one space dimension and time.

The analysis of discontinuities is performed by so-called jump conditions, which are presented in Section 5. The evolution behavior of the shock wave is governed, in principle, by an infinite series of ordinary differential equations for the discontinuities of all orders. Finally in Section 6 a special type of incompressible elastic material with three material constants is considered. For this case the propagation speed and the first evolution equation are derived explicitly.

In formulating the general equations we use the index notation for cartesian components. The summation convention has to be invoked where appropriate. A comma followed by one or more subscripts indicates partial differentiation with respect to the corresponding coordinates. Time derivatives are indicated by superposed dots. As in most problems of nonlinear elasticity the material description, using the Piola-Kirchhoff stresses, is preferred.

2. Basic equations

The motion of a three-dimensional continuum is represented by a set of functions

$$x_i = x_i(X_\alpha, t), \qquad i, \alpha = 1, 2, 3 \tag{1}$$

describing the coordinates x_i of the actual position at time t of a material point in terms of its position X_α in the reference configuration. The deformation gradient and the particle velocity have the components

$$x_{i\alpha} = \frac{\partial x_i}{\partial X_\alpha}, \qquad \dot{x}_i = \frac{\partial x_i}{\partial t}. \tag{2}$$

The left Cauchy-Green tensor is defined by

$$B_{ik} = x_{i\alpha} x_{k\alpha}. \tag{3}$$

Its principal invariants I_1, I_2, and I_3 are the deformation invariants.

An incompressible, isotropic elastic material is characterized (i) by the internal constraint of incompressibility, $I_3 = 1$, and (ii) by a strain-energy function

$$W = W(I_1, I_2) \tag{4}$$

expressing the internal energy W per unit undeformed volume as a function of the two free deformation invariants. In this case the first Piola-Kirchhoff stress tensor has the components

$$T_{R i\alpha} = \frac{\partial W}{\partial x_{i\alpha}} + pX_{\alpha i} = \frac{\partial W}{\partial I_k} \frac{\partial I_k}{\partial x_{i\alpha}} + pX_{\alpha i}, \tag{5}$$

where p is a hydrostatic reaction stress, which is not determined by any constitutive equation, and $X_{\alpha i}$ denotes the inverse of the deformation gradient $(2)_1$, i. e. $x_{i\alpha}X_{\alpha k} = \delta_{ik}$.

In the absence of volume forces the balance of momentum can be expressed by the local equation

$$T_{\mathsf{R}i\alpha,\alpha} = \rho_\mathsf{R}\ddot{x}_i, \tag{6}$$

with ρ_R denoting the mass density in the reference configuration. Here the stress and velocity fields are assumed to be differentiable. If the functions $x_i = x_i(X_\alpha, t)$ are continuous everywhere but have discontinuous first derivatives on some propagating surface Σ, the equations of motion (6) have to be supplemented by the jump conditions on this surface,

$$[\![T_{\mathsf{R}i\alpha}]\!]N_\alpha = -\rho_\mathsf{R}U_\mathsf{N}[\![\dot{x}_i]\!], \tag{7}$$

where N_α and U_N denote the normal vector of the wave front and the speed of propagation, respectively, in the reference configuration. The symbol $[\![\psi]\!]$ denotes the jump of any quantity ψ across the wave front Σ.

3. Symmetric motion of the layer

We consider a layer which occupies the material region $X_1 \geq 0$, $-h \leq X_2 \leq +h$ and is infinite in the X_3 direction. The layer is assumed to undergo a symmetric motion of the form

$$x_1 = X_1 + u_1(X_1, t), \qquad x_2 = X_2 + \epsilon_2(X_1, t)X_2, \qquad x_3 = X_3. \tag{8}$$

The deformation is fully determined by the two scalar strain fields $\epsilon_1(X_1, t) = u_{1,1}$ and $\epsilon_2(X_1, t)$. The deformation gradient has the component matrix

$$(x_{i\alpha}) = \begin{bmatrix} 1 + \epsilon_1(X_1, t) & 0 & 0 \\ q(X_1, t)X_2 & 1 + \epsilon_2(X_1, t) & 0 \\ 0 & 0 & 1 \end{bmatrix}, \tag{9}$$

where $q = \epsilon_{2,1}$ denotes the derivative of the transverse strain with respect to X_1. In the same way the inverse deformation gradient $X_{\alpha i}$ and the left Cauchy-Green tensor B_{ik} can be formed, and from the latter one obtains the deformation invariants

$$\begin{aligned} I_1 &= (1 + \epsilon_1)^2 + (1 + \epsilon_2)^2 + (qX_2)^2 + 1, \\ I_2 &= (1 + \epsilon_1)^2 + (1 + \epsilon_2)^2 + (qX_2)^2 + (1 + \epsilon_1)^2(1 + \epsilon_2)^2, \\ I_3 &= (1 + \epsilon_1)^2(1 + \epsilon_2)^2. \end{aligned} \tag{10}$$

Now the condition of incompressibility, $I_3 = 1$, yields

$$(1 + \epsilon_1)(1 + \epsilon_2) = 1 \tag{11}$$

and the first two invariants become identical, $I_1 = I_2$. It follows from the above condition that the functions ϵ_2 and ϵ_1 are no longer kinematically independent but have to satisfy the constraint (11).

We consider the evolution of a shock wave which is generated at the boundary $X_1 = 0$ of the layer by an appropriate sudden loading. To simplify the analysis we assume the discontinuity surface Σ to be a plane surface across the layer with its normal vector N_α pointing in the X_1-direction. All field quantities depend only on the coordinate X_1 and time t.

Since the cross section of the layer cannot change discontinuously, the function $\epsilon_2(X_1, t)$ describing the transversal strain must be continuous according to (8)$_2$ and, due to the incompressibility constraint (11), the longitudinal strain $\epsilon_1(X_1, t)$ must be continuous too. The only component of the deformation gradient (9) that can undergo a discontinuity is x_{21}. We assume that there is a jump

$$[\![x_{21}]\!] = [\![q]\!]X_2, \qquad (12)$$

which means that a transverse shock wave is propagating along the layer. The local strength $[\![x_{21}]\!]$ varies on the shock front between the values $\pm[\![q]\!]h$ attained at the lateral surfaces of the layer. This makes the situation different from that analyzed in [1], where a transverse shock of *uniform* strength propagating in a half-space is considered.

4. Averaged equations of motion

Averaged equations are described by Wright in [6, 7]. For the assumed form of the deformation gradient (9) the equations of motion are reduced to the system of equations for plane strain deformation, viz.

$$
\begin{aligned}
T_{\mathrm{R}11,1} + T_{\mathrm{R}12,2} &= \rho_{\mathrm{R}}\ddot{u}_1, \\
T_{\mathrm{R}21,1} + T_{\mathrm{R}22,2} &= \rho_{\mathrm{R}}X_2\ddot{\epsilon}_2, \\
T_{\mathrm{R}33,3} &= 0.
\end{aligned}
\qquad (13)
$$

The stress vector on the lateral surfaces $X_2 = \pm h$ has the components $T_{\mathrm{R}12}$ and $T_{\mathrm{R}22}$. Since these surfaces are assumed stress-free, we have

$$T_{\mathrm{R}12}(X_1, \pm h, X_3) = T_{\mathrm{R}22}(X_1, \pm h, X_3) = 0. \qquad (14)$$

The stress component $T_{\mathrm{R}12}$ will vanish identically in our case, but the component $T_{\mathrm{R}22}$ is non-zero within the layer, in general.

Now, following Wright's procedure [6, 7], we take the average of the first equation of motion (13)$_1$ over the width of the layer, thus obtaining

$$\frac{1}{2h}\int_{-h}^{+h}\frac{\partial T_{\mathrm{R}11}}{\partial X_1}\,dX_2 + \frac{1}{2h}\int_{-h}^{+h}\frac{\partial T_{\mathrm{R}12}}{\partial X_2}\,dX_2 = \frac{1}{2h}\int_{-h}^{+h}\rho_{\mathrm{R}}\ddot{u}_1\,dX_2. \qquad (15)$$

The second integral on the left-hand side can be integrated explicitly and vanishes due to the boundary condition $(14)_1$. The integration on the right-hand side is trivial, and thus the averaged equation assumes the form

$$\frac{\partial}{\partial X_1}\left(\frac{1}{2h}\int_{-h}^{+h}T_{R11}\,dX_2\right)=\rho_R\ddot{u}_1. \tag{16}$$

In the second equation of motion $(13)_2$ the simple average of all terms vanishes identically due to the assumed symmetry with respect to the plane $X_2=0$. Therefore the first moment of this equation is taken, i. e. the whole equation is multiplied by X_2 before the averaging process. Within the resulting equation

$$\frac{1}{2h}\int_{-h}^{+h}\frac{\partial T_{R21}}{\partial X_1}X_2\,dX_2+\frac{1}{2h}\int_{-h}^{+h}\frac{\partial T_{R22}}{\partial X_2}X_2\,dX_2=\frac{1}{2h}\int_{-h}^{+h}\rho_R\ddot{\epsilon}_2 X_2^2\,dX_2 \tag{17}$$

the second integral on the left-hand side can be integrated by parts, using the boundary condition $(14)_2$. Also the right-hand side can be integrated explicitly. Thus the second equation of motion yields

$$\frac{\partial}{\partial X_1}\left(\frac{1}{2h}\int_{-h}^{+h}X_2 T_{R21}\,dX_2\right)-\frac{1}{2h}\int_{-h}^{+h}T_{R22}\,dX_2=\rho_R\frac{h^2}{3}\ddot{\epsilon}_2. \tag{18}$$

By the averaging process the equations of motion have been reduced to a system of PDEs with only one space coordinate X_1 and time t.

In order to simplify the notations the stress averages occurring in the equations (16) and (18) will be abbreviated by

$$\begin{aligned}S &= \frac{1}{2h}\int_{-h}^{+h}T_{R11}\,dX_2,\\[2mm] P &= \frac{1}{2h}\int_{-h}^{+h}T_{R22}\,dX_2, \\[2mm] Q &= \frac{1}{2h}\int_{-h}^{+h}X_2 T_{R21}\,dX_2.\end{aligned} \tag{19}$$

These quantities are functions of the longitudinal coordinate X_1 and time t. The averaged equations of motion assume the condensed form

$$S'=\rho_R\ddot{u}_1,\qquad Q'-P=\rho_R\frac{h^2}{3}\ddot{\epsilon}_2, \tag{20}$$

where the prime is used to denote the partial derivative $\partial/\partial X_1$ of any function that depends on X_1 and t only.

The averaged equations (20) can also be interpreted as Euler-Lagrange equations of a suitable variational problem. Both the kinetic energy and the strain energy are averaged over the width of the layer, thus giving rise to an average Lagrangian

$$\bar{L} = \frac{\rho_R}{2}\left(\dot{u}_1^2 + \frac{h^2}{3}\dot{\epsilon}_2^2\right) - \bar{W}(u_1', \epsilon_2, \epsilon_2'). \tag{21}$$

Here the average strain energy in the layer,

$$\bar{W} = \frac{1}{2h}\int_{-h}^{+h} W(I_1, I_2)\,\mathrm{d}X_2, \tag{22}$$

is regarded as a function $\bar{W} = \bar{W}(\epsilon_1, \epsilon_2, q)$. Incompressibility imposes a constraint on the possible deformation. Therefore the modified Lagrangian

$$\bar{L}^* = \bar{L} + \bar{p}\left[1 - (1 + \epsilon_1)(1 + \epsilon_2)\right] \tag{23}$$

is formed, in which the incompressibility condition is incorporated with a Lagrange multiplier $\bar{p}$. The Euler-Lagrange equations

$$\frac{\mathrm{d}}{\mathrm{d}t}\left(\frac{\partial \bar{L}^*}{\partial \dot{u}_1}\right) + \frac{\mathrm{d}}{\mathrm{d}X_1}\left(\frac{\partial \bar{L}^*}{\partial u_1'}\right) - \frac{\partial \bar{L}^*}{\partial u_1} = 0,$$

$$\frac{\mathrm{d}}{\mathrm{d}t}\left(\frac{\partial \bar{L}^*}{\partial \dot{\epsilon}_2}\right) + \frac{\mathrm{d}}{\mathrm{d}X_1}\left(\frac{\partial \bar{L}^*}{\partial \epsilon_2'}\right) - \frac{\partial \bar{L}^*}{\partial \epsilon_2} = 0, \tag{24}$$

for the Lagrangian (23) assume the form

$$\rho_R \ddot{u}_1 - \frac{\mathrm{d}}{\mathrm{d}X_1}\left[\frac{\partial \bar{W}}{\partial \epsilon_1} + \bar{p}(1 + \epsilon_2)\right] = 0,$$

$$\rho_R \frac{h^2}{3}\ddot{\epsilon}_2 - \frac{\mathrm{d}}{\mathrm{d}X_1}\left[\frac{\partial \bar{W}}{\partial q} + \bar{p}(1 + \epsilon_1)\right] + \frac{\partial \bar{W}}{\partial \epsilon_2} = 0. \tag{25}$$

By comparing these equations with the averaged equations of motion (20) the following connections can be read off immediately:

$$S = \frac{\partial \bar{W}}{\partial \epsilon_1} + \bar{p}(1 + \epsilon_2), \quad P = \frac{\partial \bar{W}}{\partial \epsilon_2} + \bar{p}(1 + \epsilon_1), \quad Q = \frac{\partial \bar{W}}{\partial q}. \tag{26}$$

From these relations the Lagrange multiplier $\bar{p}$ can be identified as the average of the hydrostatic stress p across the layer.

The strains ϵ_1 and ϵ_2 are kept as the primary variables in all equations. However, due to the incompressibility condition (11) they are not independent. As a consequence the derivative ϵ_1', for instance, can be represented in terms of $q = \epsilon_2'$ by

$$\epsilon_1' = -\frac{1 + \epsilon_1}{1 + \epsilon_2}q = -(1 + \epsilon_1)^2 q, \tag{27}$$

where in the second form the incompressibility condition (11) has been used again.

Whether derived by direct averaging or by the Lagrangian method the averaged equations of motion (20) are valid only, if the required derivatives exist. On the wave front the equations have to be supplemented by corresponding jump conditions for the average motion of the layer. These are derived from the three-dimensional jump condition (7) by the same averaging procedure as in the continuous case. The averaged zero-order jump conditions in X_1- and X_2-directions assume the form

$$[S] = -\rho_R U_N [\dot{u}_1], \qquad [Q] = -\rho_R U_N \frac{h^2}{3}[\dot{\epsilon}_2]. \tag{28}$$

The analysis of propagating discontinuities is mainly concerned with jump conditions of this kind. Even the regular equations of motion (20) are transformed into jump conditions by taking the difference between the states behind and in front of the discontinuity surface. Thus the jump conditions of first order

$$[S'] = \rho_R [\ddot{u}_1], \qquad [Q'] - [P] = \rho_R \frac{h^2}{3}[\ddot{\epsilon}_2] \tag{29}$$

are obtained. There is an infinite series of jump conditions, which can be formed from the original equations of motion (20) by differentiating with respect to X_1 and then taking the jumps. The next pair of this kind,

$$[S''] = \rho_R [\dddot{\epsilon}_1], \qquad [Q''] - [P'] = \rho_R \frac{h^2}{3}[\dddot{q}], \tag{30}$$

constitutes the jump conditions of second order.

The set of equations pertaining to the discontinuities is not complete as yet, without relating the jumps of the space and time derivatives of a quantity to each other. Let $\psi(X_1, t)$ denote an arbitrary quantity with regular behavior on both sides of the wave front. Then the jumps of ψ and its derivatives satisfy the kinematical conditions of compatibility

$$[\psi'] + \frac{1}{U_N}[\dot{\psi}] = \frac{\mathrm{d}}{\mathrm{d}X_1}[\psi]. \tag{31}$$

This is just the one-dimensional version of the more general kinematical conditions of compatibility as given in [4], for instance.

The condition (31) applied to different field quantities makes it possible to express all the discontinuities in terms of the jumps of q and its derivatives. On the right-hand sides of the jump conditions (28) of zeroth order, for instance, the jumps

$$[\dot{u}_1] = -U_N [\epsilon_1] = 0, \qquad [\dot{\epsilon}_2] = -U_N [q] \tag{32}$$

have to be inserted. For the next pair of jump conditions, (29), the expressions

$$[\ddot{u}_1] = -U_N^2 (1 + \epsilon_1)^2 [q], \qquad [\ddot{\epsilon}_2] = U_N^2 ([q'] - 2[q]') - U_N U_N' [q] \tag{33}$$

are needed. (It should be noted that $[\![\epsilon_1']\!]$ can be expressed in $[\![q]\!]$ by exploiting the coupling (27) between ϵ_1' and ϵ_2'.) These examples show that the right-hand sides of the jump conditions of different orders (28)–(30) can be expressed uniquely in terms of the jumps $[\![q]\!]$, $[\![q']\!]$, ... and derivatives thereof.

5. Special type of incompressible elastic material

The analysis will be continued for a special isotropic elastic material characterized by the constitutive equation [8]

$$W = C_1(I_1 - 3) + C_2(I_2 - 3) + C_3(I_1^2 - 9), \tag{34}$$

where C_1, C_2, C_3 are constants. The special case $C_3 = 0$ corresponds to the Mooney-Rivlin material. The relevant components of the Piola-Kirchhoff stress tensor (5) are obtained by use of (9) as

$$
\begin{aligned}
T_{R11} &= 2(C_1 + C_2 + 2C_3 I_1)(1 + \epsilon_1) + (2C_2 + p)(1 + \epsilon_2), \\
T_{R22} &= 2(C_1 + C_2 + 2C_3 I_1)(1 + \epsilon_2) + (2C_2 + p)(1 + \epsilon_1), \\
T_{R21} &= 2(C_1 + C_2 + 2C_3 I_1)q X_2,
\end{aligned}
\tag{35}
$$

with the first Invariant I_1 taken from $(10)_1$.

In order to simplify the expressions we introduce the new material constants

$$c_0^2 = \frac{2}{\rho_R}(C_1 + C_2 + 6C_3), \qquad \eta = \frac{2C_3}{C_1 + C_2 + 6C_3}. \tag{36}$$

Also the function

$$f(\epsilon_1, \epsilon_2) = (1 + \epsilon_1)^2 + (1 + \epsilon_2)^2 - 2 \tag{37}$$

of the strains is used in the following equations. The averaged stresses (19) can then be written in the form

$$
\begin{aligned}
S &= \rho_R c_0^2 \left[1 + \eta \left(f(\epsilon_1, \epsilon_2) + \frac{1}{3}h^2 q^2\right)\right](1 + \epsilon_1) + (2C_2 + \bar{p})(1 + \epsilon_2), \\
P &= \rho_R c_0^2 \left[1 + \eta \left(f(\epsilon_1, \epsilon_2) + \frac{1}{3}h^2 q^2\right)\right](1 + \epsilon_2) + (2C_2 + \bar{p})(1 + \epsilon_1), \\
Q &= \rho_R c_0^2 \left[1 + \eta \left(f(\epsilon_1, \epsilon_2) + \frac{3}{5}h^2 q^2\right)\right]q.
\end{aligned}
\tag{38}
$$

The average hydrostatic stress $\bar{p}$ is eliminated from $(38)_{1,2}$ by forming the linear combination

$$(1 + \epsilon_2)P - (1 + \epsilon_1)S = \rho_R c_0^2 \left[1 + \eta \left(f + \frac{h^2}{3}q^2\right)\right]\left[(1 + \epsilon_2)^2 - (1 + \epsilon_1)^2\right]. \tag{39}$$

It is actually only this connection between P and Q that is needed in the subsequent analysis, and not the explicit expressions $(38)_{1,2}$ themselves. After these preparations the jump conditions for a transverse shock wave propagating along the layer can be formulated.

We consider a shock wave entering the stress-free layer. There are no strains in front of the shock wave and, since the strains are continuous at the wave front, we may set $\epsilon_1 = \epsilon_2 = 0$ in formulating all the jump conditions. First, the jump condition $(28)_2$ with Q taken from $(38)_3$ is formed. Also $[\![\dot{\epsilon}_2]\!]$ on the right-hand side is available from $(32)_2$. The resulting equation yields the propagation speed of the wave in terms of its strength $[\![q]\!]$,

$$U_{\mathsf{N}}^2 = c_0^2 \left(1 + \frac{3}{5} \eta h^2 [\![q]\!]^2 \right). \tag{40}$$

This justifies the notation c_0 introduced above, which is now identified as the propagation speed of a weak wave. In a Mooney-Rivlin material we have $\eta = 0$, and the propagation speed attains the fixed value $U_{\mathsf{N}} = c_0$. As shown in [8] weak nonlinear shocks can exist in this type of material only if $\eta > 0$. Therefore the propagation speed increases with shock strength, in general.

The next jump condition to be exploited is $(28)_1$. Since the right-hand side vanishes according to $(32)_1$, we obtain simply $[\![S]\!] = 0$ and, due to the relation (39), also $[\![P]\!] = 0$. This latter jump appears in $(29)_2$, which is the next condition to be formulated. A lengthy calculation leads to

$$\left(1 + \frac{9}{10} \eta h^2 [\![q]\!]^2 \right) \frac{\mathrm{d}[\![q]\!]}{\mathrm{d}X_1} + \frac{3}{5} \eta h^2 [\![q]\!]^2 [\![q']\!] = 0. \tag{41}$$

In the Mooney-Rivlin case $\eta = 0$ the strength $[\![q]\!]$ of the propagating transverse shock is necessarily constant. For $\eta > 0$ the strength $[\![q]\!]$ may vary, but its growth or decay has to be compensated by a jump $[\![q']\!]$ of opposite sign. The equation (41) is of the same form as the one obtained by Fu and Scott for the corresponding half-space problem.

In principle it is clear how to proceed further, but the explicit calculations become more and more involved. The next conditions should be $(29)_1$ and $(30)_2$, which would end up in an equation relating the jumps $[\![q]\!]$, $[\![q']\!]$, $[\![q'']\!]$ and their derivatives. Eventually one arrives at an infinite system of coupled differential equations for the discontinuities of any order. The system cannot be solved recursively, in general, unless additional closure assumptions are imposed. This makes the behavior of shock waves different from that of acceleration and higher-order waves.

References

[1] Y. B. Fu and H. Scott, *The evolutionary behaviour of plane transverse weak nonlinear shock waves in unstrained incompressible isotropic elastic non-conductors*, Wave Motion, **11** (1989), 351–363.

[2] Y. B. Fu and N. H. Scott, *The evolution law of one-dimensional weak nonlinear shock waves in elastic non-conductors*, Q. J. Mech. Appl. Math., **42** (1989), 23–39.

[3] S. Kosiński, *Evolution law for shock strength in simple elastic structures*, in: K. Malanowski, M. Peszyńska, and Z. Nahorski, Eds., Modelling and Optimization of Distributed Parameter Systems, Application to Engineering, Chapman & Hall (London) **1996**, 339–346.

[4] C.-C. Wang and C. Truesdell, *Introduction to Rational Elasticity*, Noordhoff (Leyden) **1973**, 417ff.

[5] Z. Wesołowski and W. Bürger, *Shock waves in incompressible elastic solids*, Rheol. Acta, **16** (1977), 155–160.

[6] T. W. Wright, *Nonlinear waves in rods*, Proceedings of the IUTAM Symposium on Finite Elasticity held at Lehigh University, August 10–15, 1980, Edited by D. E. Carlson and R. T. Shield, Martinus Nijhoff Publications (Amsterdam) **1981**, 423–443.

[7] T. W. Wright, *Nonlinear waves in rods: results for incompressible elastic materials*, Stud. Appl. Math., **72** (1985), 149–160.

[8] S. Zahorski, *Experimental investigations of certain kinds of rubber*, Engineering Transactions, **10** (1962), 193–207.

Gerhard-Mercator-Universität Duisburg,
Fachbereich Maschinenbau,
47048 Duisburg, Germany
E-mail address: `braun@mechanik.uni.duisburg.de`

Częstochowa University of Technology,
Faculty of Civil Engineering,
ul. Akademicka 3,
42-200 Częstochowa, Poland
E-mail address: `slawek@kmm-lx.p.lodz.pl`

International Series of Numerical Mathematics
Vol. 129, © 1999 Birkhäuser Verlag Basel/Switzerland

A Naive Riemann Solver to Compute
a Non-conservative Hyperbolic System

Thierry Buffard, Thierry Gallouët and Jean-Marc Hérard

Abstract. An approximate Riemann solver is presented in this contribution, which enables to compute shock waves in compressible flows using one or two-equation turbulence model.

1. Introduction

Among the numerous turbulence models, the (one- or two-equation) $K - \epsilon$ closed models, are widely used, in spite of their drawbacks, for industrial applications, mainly in the case of incompressible flows. When focusing on compressible turbulence, they have not been studied very much (except in [5], [18], [11]). We focus here on the numerical approximation of the simplest one, which only contains one (non conservative) equation; this one governs the motion of the turbulent kinetic energy $K =< \rho u_i'' u_i'' >$ using Favre's averaging formalism.

The numerical approach is based on a new approximate Riemann solver proposed recently in [12] in order to solve complex hyperbolic systems in conservative form. This approach was investigated in detail in [17], and a slightly modified version of the scheme was suggested in [3], and used in [4] and [7] in order to compute some single-phase or two-phase flows. One may wonder now whether this approach applies to hyperbolic systems in non conservative form, and this is the main objective of the present contribution. We examine the capability of the proposed scheme to provide as stable results as those obtained using a Godunov-type solver ([11]), and compare efficiencies.

2. Governing equations

Before we introduce the approximate Riemann solver which is used to compute turbulent compressible flows with shocks, we first recall some basic results connected with a simple one equation closure. The set of equations (see [1]) is obtained using Favre's averaging process ([10]):

$$
\begin{cases}
\rho_{,t} + (\rho U_j)_{,j} = 0 \\
(\rho U_i)_{,t} + (\rho U_i U_j)_{,j} + (P\,\delta_{ij} + \Sigma_{ij}^v)_{,j} + (R_{ij})_{,j} = 0 \\
E_{,t} + (U_j\,(E+P))_{,j} + (U_k\,(R_{kj} + \Sigma_{kj}^v))_{,j} = (\sigma_K(\tfrac{K}{\rho})_{,j})_{,j} + (\sigma_E(\tfrac{P}{\rho})_{,j})_{,j} \\
K_{,t} + (K\,U_j)_{,j} - (\sigma_K(\tfrac{K}{\rho})_{,j})_{,j} = -R_{jk}\,U_{j,k} - \epsilon
\end{cases}
\tag{1}
$$

with:

$$
\begin{aligned}
R_{ij} &= \tfrac{2}{3} K\,\delta_{ij} - \mu_t(U_{i,j} + U_{j,i} - \tfrac{2}{3} U_{l,l}\,\delta_{ij}) \\
\Sigma_{ij}^v &= -\mu(U_{i,j} + U_{j,i} - \tfrac{2}{3} U_{l,l}\,\delta_{ij})
\end{aligned}
\tag{2}
$$

Moreover, it is assumed that the perfect gas state law holds; hence:

$$
P = (\gamma - 1)(E - \frac{\rho U_j U_j}{2} - K)
\tag{3}
$$

Admissible states are such as:

$$
\rho(x,t) \geq 0; \quad P(x,t) \geq 0; \quad K(x,t) \geq 0
\tag{4}
$$

Here, ρ stands for the mean density, $\mathbf{U}$ is the mean velocity, E denotes the mean total energy and P is the mean pressure. The Reynolds stress tensor is $\mathbf{R}$, and K is the turbulent kinetic energy ($K = trace(\mathbf{R})/2$); γ is the constant ratio of specific heats; μ is the molecular viscosity. It is assumed that σ_E, σ_K and μ_t are positive bounded functions. The mechanical dissipation ϵ is given through an algebraic relation (see [1]).

We now consider the following entropy-consistent fractional step method which requires solving successively:
Step 1:

$$
\begin{cases}
\rho_{,t} + (\rho U_j)_{,j} = 0 \\
(\rho U_i)_{,t} + (\rho U_i U_j)_{,j} + P_{,i} + \tfrac{2}{3} K_{,i} = -(\Sigma_{ij}^v)_{,j} \\
E_{,t} + (U_j\,(E+P+\tfrac{2}{3} K))_{,j} = -(U_k\,\Sigma_{kj}^v)_{,j} \\
K_{,t} + (K\,U_j)_{,j} + \tfrac{2}{3} K\,U_{j,j} = 0
\end{cases}
\tag{5}
$$

Step 2:

$$
\begin{cases}
\rho_{,t} = 0 \\
(\rho U_i)_{,t} = -(\Sigma_{ij}^t)_{,j} \\
E_{,t} = -(U_k\,\Sigma_{kj}^t)_{,j} + (\sigma_K(K/\rho)_{,j})_{,j} + (\sigma_E(P/\rho)_{,j})_{,j} \\
K_{,t} = -\epsilon + (\sigma_K(\tfrac{K}{\rho})_{,j})_{,j} - \tfrac{1}{2} \Sigma_{kj}^t\,(U_{k,j} + U_{j,k})
\end{cases}
\tag{6}
$$

noting: $\Sigma_{kj}^t = -\mu_t(U_{k,j} + U_{j,k} - \tfrac{2}{3} U_{l,l}\,\delta_{kj})$.

3. Recalling some basic results

In a 2D framework, we introduce the "conservative" state variable:

$$\mathbf{Z}^t = (\rho, \rho U, \rho V, E, K).$$

Given some vector $\mathbf{n}$ in $\mathbb{R}^2$ with unit norm, we also introduce:

$$U_n = \mathbf{U}.\mathbf{n}$$

We know ([11], [16]) that we may derive an entropy inequality for viscous solutions of (1)(2)(3), using the following entropy-entropy flux pair:

$$\eta = -\rho \ln(P\rho^{-\gamma})$$

$$\mathbf{F}_\eta^{nv}(\mathbf{Z}) = \mathbf{U}\,\eta.$$

If $\mathbf{n}$ is normal to the shock which speed is σ, the entropy inequality:

$$-\sigma \left[\eta\right] + \left[\mathbf{F}_\eta^{nv}.\mathbf{n}\right] \leq 0, \tag{7}$$

where $[.]$ represents the jump through the discontinuity, enables to connect states through genuinely non linear fields in a physically meaningful way, when neglecting viscous effects and investigating the one-dimensional Riemann problem. Actually ([11]), the non viscous non conservative first-order differential system (namely the left-hand side of step 1):

$$\mathbf{Z}_{,t} + \sum_{i=1}^{2} (\mathbf{F}_i(\mathbf{Z}))_{,i} + \sum_{i=1}^{2} \mathbf{A}_i^{nc}(\mathbf{Z})\,\mathbf{Z}_{,i} = 0 \tag{8}$$

is a non strictly hyperbolic system, whose eigenvalues are:

$$\lambda_1 = U_n - c_1 ; \quad \lambda_2 = \lambda_3 = \lambda_4 = U_n ; \quad \lambda_5 = U_n + c_1,$$

setting:

$$c_1 = \left(\frac{\gamma P}{\rho} + \frac{10}{9}\frac{K}{\rho}\right)^{1/2}.$$

The 1-wave and the 5-wave are genuinely non linear. If we restrict to sufficiently weak shocks and connect states $(1/\rho, U, V, P, K)$ across some discontinuity travelling in the x-direction with a linear path (see [6], [8], [14], [15]), we get the following approximate jump conditions:

$$\begin{cases} -\sigma\,[\rho] + [\rho U] = 0 \\ -\sigma\,[\rho U] + [\rho U^2 + 2\,K/3 + P] = 0 \\ -\sigma\,[\rho V] + [\rho U V] = 0 \\ -\sigma\,[E] + [U(E + 2\,K/3 + P)] = 0 \\ -\sigma\,[K] + [U\,K] + 2/3\,\bar{K}\,[U] = 0 \end{cases} \tag{9}$$

where σ stands for the speed of the travelling discontinuity, and if L and R subscripts refer to the left- and right-hand side of the discontinuity, we have noted:

$$[\phi] = \phi_R - \phi_L, \quad \bar{\phi} = (\phi_L + \phi_R)/2.$$

Obviously, when the turbulence is set to zero, (9) identifies with exact jump conditions for Gas Dynamics. It may then be proved ([11]) that the 1D-Riemann problem associated with (8) and approximate jump conditions (9) has a unique entropy-consistent solution, provided that the following condition on the data on each side of the initial discontinuity holds

$$U_R - U_L < X_L + X_R$$

setting

$$X_i = \int_0^{\rho_i} \frac{1}{\rho} \left(\frac{\gamma P_i}{\rho_i} (\frac{\rho}{\rho_i})^{\gamma-1} + \frac{10}{9} \frac{K_i}{\rho_i} (\frac{\rho}{\rho_i})^{2/3} \right)^{1/2} d\rho, \quad i = L \text{ or } R.$$

In addition, the approximate solution is in agreement with (4). Based on the latter results, a Godunov-type scheme was previously proposed, which unfortunately requires solving two non linear scalar equations with two unknowns, and the whole results in a rather expensive solver. The main objective here is to derive an approximate Godunov scheme which does not have this drawback.

4. An approximate Riemann solver

Restricting to the two dimensional frame, we present below an extension ([3], [4]) of the approximate Godunov solver which was recently proposed in [12] and [17]. We introduce $\mathbf{B_1}(\mathbf{Z})$ and $\mathbf{B_2}(\mathbf{Z})$:

$$\mathbf{B_1}(\mathbf{Z}) = \frac{d\mathbf{F}_1}{d\mathbf{Z}}(\mathbf{Z}) + \mathbf{A_1^{nc}}(\mathbf{Z}); \quad \mathbf{B_2}(\mathbf{Z}) = \frac{d\mathbf{F}_2}{d\mathbf{Z}}(\mathbf{Z}) + \mathbf{A_2^{nc}}(\mathbf{Z})$$

noting:

$$\mathbf{F_1}(\mathbf{Z}) = \begin{bmatrix} \rho U \\ \rho U^2 + 2K/3 + P \\ \rho U V \\ U(E + 2K/3 + P) \\ U K \end{bmatrix}, \quad \mathbf{F_2}(\mathbf{Z}) = \begin{bmatrix} \rho V \\ \rho U V \\ \rho V^2 + 2K/3 + P \\ V(E + 2K/3 + P) \\ V K \end{bmatrix} \quad (10)$$

We introduce the (explicit) scheme:

$$|\Omega_i|(\mathbf{Z}_i^{n+1} - \mathbf{Z}_i^n) + \Delta t \sum_{j \in V(i)} \Gamma_{ij} \left(\sum_{k=1}^{2} n_k \, \mathbf{F}_k(\mathbf{Z}(\mathbf{Y}^*)) \right)_{ij} + \Delta t \int_{\Omega_i} \mathbf{S}(\mathbf{Z}(\mathbf{Y}^*)) \, d\Omega = 0$$

where:

$$\int_{\Omega_i} \mathbf{S}^t(\mathbf{Z}(\mathbf{Y}^*)) \, d\Omega = (0, 0, 0, 0, \frac{2}{3} \hat{K}_i \sum_{j \in V(i)} \Gamma_{ij}(\mathbf{U}^*.\mathbf{n})_{ij})$$

$$\hat{K}_i = \sum_{j \in V(i)} K_{ij}^* / card(V(i))$$

Δt is a constant time step, $V(i)$ refers to the neighbouring cells of Ω_i, and $\mathbf{n}$ stands for the outward (from Ω_i to Ω_j) normal vector of the interface (which length is

Γ_{ij}) between cells Ω_i and Ω_j. Approximate starred values of states at interface are obtained as follows. Owing to the invariance under frame rotation, and neglecting as usual transverse variations, we may examine the Riemann problem associated with the one-dimensional hyperbolic system in the x-direction in the following fully non conservative form:

$$\mathbf{Z}_{,t} + \mathbf{B}_1(\mathbf{Z})\,\mathbf{Z}_{,x} = 0$$

Restricting to regular solutions, we introduce the following change of variables $\mathbf{Z} \mapsto \mathbf{Y}(\mathbf{Z})$:

$$\mathbf{Y}^t = (\tau, U, V, P, K)$$

(see also [4], [9], [20]), where $\tau = 1/\rho$. Thus comes:

$$\mathbf{Y}_{,t} + (\mathbf{Z}_\mathbf{Y}(\mathbf{Y}))^{-1}\,\mathbf{B}_1(\mathbf{Z}(\mathbf{Y}))\,\mathbf{Z}_\mathbf{Y}(\mathbf{Y})\,\mathbf{Y}_{,x} = 0$$

or:

$$\mathbf{Y}_{,t} + \mathbf{C}_1(\mathbf{Y})\,\mathbf{Y}_{,x} = 0 \tag{11}$$

with:

$$\mathbf{Y} = \begin{pmatrix} \tau \\ U \\ V \\ P \\ K \end{pmatrix} \qquad \mathbf{C}_1(\mathbf{Y}) = \begin{pmatrix} U & -\tau & 0 & 0 & 0 \\ 0 & U & 0 & \tau & 2\tau/3 \\ 0 & 0 & U & 0 & 0 \\ 0 & \gamma P & 0 & U & 0 \\ 0 & 5K/3 & 0 & 0 & U \end{pmatrix} \tag{12}$$

The Riemann problem associated with the one-dimensional non linear hyperbolic system (11), (12) and initial conditions:

$$\mathbf{Y}(x, t = 0) = \begin{cases} \mathbf{Y}_L = \mathbf{Y}(\mathbf{Z}_L) & if\ x < 0 \\ \mathbf{Y}_R = \mathbf{Y}(\mathbf{Z}_R) & if\ x > 0 \end{cases} \tag{13}$$

is linearized as follows:

$$\mathbf{Y}_{,t} + \mathbf{C}_1(\bar{\mathbf{Y}})\,\mathbf{Y}_{,x} = 0 \tag{14}$$

where:

$$\bar{\mathbf{Y}} = (\mathbf{Y}_R + \mathbf{Y}_L)/2.$$

System (14), (12) contains three distinct linearly degenerated fields, and its solution only requires computing five real coefficients noted $\bar{\alpha}_i$ (for: $i=1$ to 5), setting:

$$\mathbf{Y}_R - \mathbf{Y}_L = \sum_{i=1}^{5} \bar{\alpha}_i\,\bar{\mathbf{r}}_i \tag{15}$$

where $\{\bar{\mathbf{r}}_i\}$ represent the basis of right eigenvectors of the matrix $\mathbf{C}_1(\bar{\mathbf{Y}})$. Intermediate states are then uniquely defined, and hence the state $\mathbf{Y}^*$ at the initial location of the data discontinuity. The above mentioned approximate Riemann solver has the following properties:

Remark 4.1. Numerical intermediate states (indexed 1 and 4) are in agreement with the solution of the Riemann problem since:

$$U_1 = U_4$$
$$P_1 + \tfrac{2}{3} K_1 = P_4 + \tfrac{2}{3} K_4 \tag{16}$$

Proof. The right eigenvectors of $\mathbf{C_1}(\bar{\mathbf{Y}})$ associated with the eigenvalue $\lambda_2 = \lambda_3 = \lambda_4 = \bar{U}$ are:

$$\bar{r}_2 = \begin{bmatrix} 1 \\ 0 \\ 0 \\ 0 \\ 0 \end{bmatrix}, \quad \bar{r}_3 = \begin{bmatrix} 0 \\ 0 \\ 1 \\ 0 \\ 0 \end{bmatrix}, \quad \bar{r}_4 = \begin{bmatrix} 0 \\ 0 \\ 0 \\ -2/3 \\ 1 \end{bmatrix}$$

Hence, the jump between the two intermediate states is given by:

$$\mathbf{Y}_4 - \mathbf{Y}_1 = \begin{bmatrix} \bar{\alpha}_2 \\ 0 \\ \bar{\alpha}_3 \\ -2\bar{\alpha}_4/3 \\ \bar{\alpha}_4 \end{bmatrix}$$

We obtain (16) by considering on one hand the second equation, and on the other hand, the two last equations. $\qquad\square$

The expression of vectors or coefficients which occur in the numerical scheme for solving linearized problem (14), (12), (13) is:

$$\bar{r}_1 = \begin{bmatrix} \bar{\tau} \\ \tilde{c}_1 \\ 0 \\ -\gamma\bar{P} \\ -5\bar{K}/3 \end{bmatrix}, \quad \bar{r}_5 = \begin{bmatrix} \bar{\tau} \\ -\tilde{c}_1 \\ 0 \\ -\gamma\bar{P} \\ -5\bar{K}/3 \end{bmatrix}$$

$$\bar{\alpha}_1 = \frac{1}{2\tilde{c}_1}\left([U] - \frac{\bar{\tau}}{\tilde{c}_1}[P + \frac{2}{3}K]\right), \quad \bar{\alpha}_5 = \frac{1}{2\tilde{c}_1}\left(-[U] - \frac{\bar{\tau}}{\tilde{c}_1}[P + \frac{2}{3}K]\right)$$

where: $\tilde{c}_1{}^2 = \bar{\tau}(\gamma\bar{P} + \tfrac{10}{9}\bar{K})$.

Remark 4.2. Consider one travelling (single) discontinuity. Then, approximate jump conditions (9), associated with the non linear system are equivalent to those, associated with the linear hyperbolic system (14), i.e.:

$$-\sigma[\mathbf{Y}] + \mathbf{C_1}(\bar{\mathbf{Y}})[\mathbf{Y}] = 0$$

The proof is similar to the one presented in [4] in the case of Euler system. Let's note that even for zero turbulence, this is not consistency with the integral form of the conservation law.

Remark 4.3. This scheme obviously requires an entropy correction at sonic points in rarefaction waves.

Remark 4.4. When focusing on the Riemann problem associated with:

$$[U] = [P + 2K/3] = 0,$$ (17)

present scheme identifies with Godunov scheme.

Proof. Under condition (17), we have $\bar{\alpha}_1 = \bar{\alpha}_5 = 0$,
hence $\mathbf{Y}_1 = \mathbf{Y}_L$ and $\mathbf{Y}_4 = \mathbf{Y}_R$.
Assuming that $U = U_L = U_R > 0$ provides $\mathbf{Y}^* = \mathbf{Y}_L$. We deduce:

$$\mathbf{F}(\mathbf{Z}(\mathbf{Y}^*)) = \mathbf{F}(\mathbf{Z}_L).$$

In particular, we have also $U^* = U$, and so:

$$\int_{\Omega_i} \mathbf{S}^t(\mathbf{Z}(\mathbf{Y}^*))\, d\Omega = (0,0,0,0,0). \qquad \square$$

$\square$

5. Numerical results

When focusing on conservative hyperbolic systems, it has been underlined in [3], [7], [12], [17] that the behaviour of Godunov scheme and scheme proposed in [12] are quite similar. In particular, restricting to isentropic or non isentropic Euler equations, the measured rates of convergence for all variables are similar for both schemes.

The present numerical scheme has been applicated on several test cases, in particular on the computation of Riemann problems with: two rarefaction waves, two shock waves, one rarefaction wave and one shock wave. Only this last configuration is presented here for which initial conditions are given by:
test 1 :

$$(\rho_L,\ u_L,\ P_L,\ K_L) = (1,\ 0,\ 5.10^5,\ 5000)$$
$$(\rho_R,\ u_R,\ P_R,\ K_R) = (0.125,\ 0,\ 10^4,\ 100)$$

test 2 :

$$(\rho_L,\ u_L,\ P_L,\ K_L) = (1,\ 0,\ 10^5,\ 100)$$
$$(\rho_R,\ u_R,\ P_R,\ K_R) = (0.125,\ 0,\ 10^4,\ 1000)$$

and γ is fixed to 1.4. The first turbulent shock tube experiment presents a supersonic zone, whereas the second remains everywhere subsonic. In both cases, the initial turbulent Mach numbers (namely the square root of the ratio K/P) are rather small.

Regular one dimensional meshes have been used. Several profiles corresponding to test 1 (resp. test 2) are plotted in figure 1 (resp. figure 3) in the case of a mesh containing 20000 points. It is emphasized that the preservation of Riemann invariants U and $P + 2K/3$ through the contact discontinuity is good, even of much coarser grids. Also, Riemann invariants of the 1-wave, namely $P\rho^{-\gamma}$ and $K\rho^{-5/3}$, are rather well predicted (see profiles on figure 2 in the case of test 1). Let's note on figure 3 that the values of P are different on both sides of the contact discontinuity.

6. Conclusion

Present scheme is based on an initial proposition to compute conservative systems (Euler equations with real gas state law). Its stability actually enables to perform computations of strong rarefaction waves, contact discontinuities and shocks, even when the ratio K/P is not negligible, which is much useful for industrial purposes. It is obviously much cheaper than the basic Godunov scheme introduced in [11]. Obviously, derivation of meaningful approximate jump conditions still requires more works (see [2]).

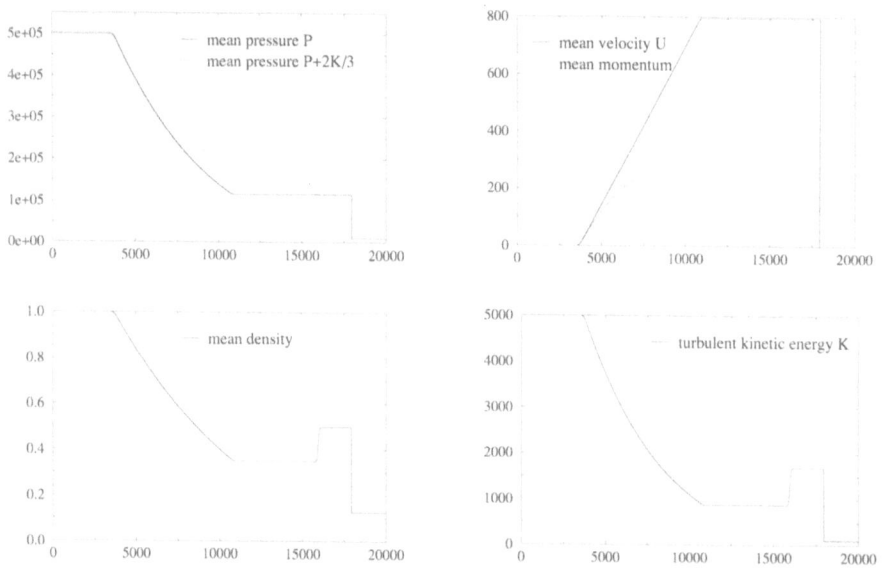

FIGURE 1.   Profiles for a supersonic turbulent shock tube test.

References

[1] B.S. Baldwin, T.J. Barth, *A one equation turbulence transport model for high Reynolds number wall-bounded flows*, NASA TM 102847 (1990).

[2] C. Berthon, F. Coquel, *Travelling wave solutions of a convective-diffusive system with first and second order terms in nonconservative form*, Seventh Int. Conf. on hyperbolic problems, ETH Zürich, February 9–13, 1998.

[3] T. Buffard, T. Gallouët, J.M. Hérard, *Schéma VFRoe en variables caractéristiques: principe de base et application aux gaz réels*, Internal Report EDF-DER HE-41/96/041/A, in french, (1996).

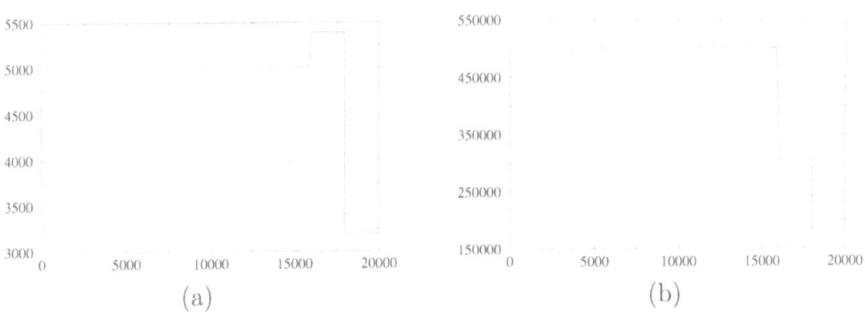

FIGURE 2. Profiles of 1-Riemann invariants for a supersonic turbulent shock tube test: (a) $P\rho^{-\gamma}$, (b) $K\rho^{-5/3}$.

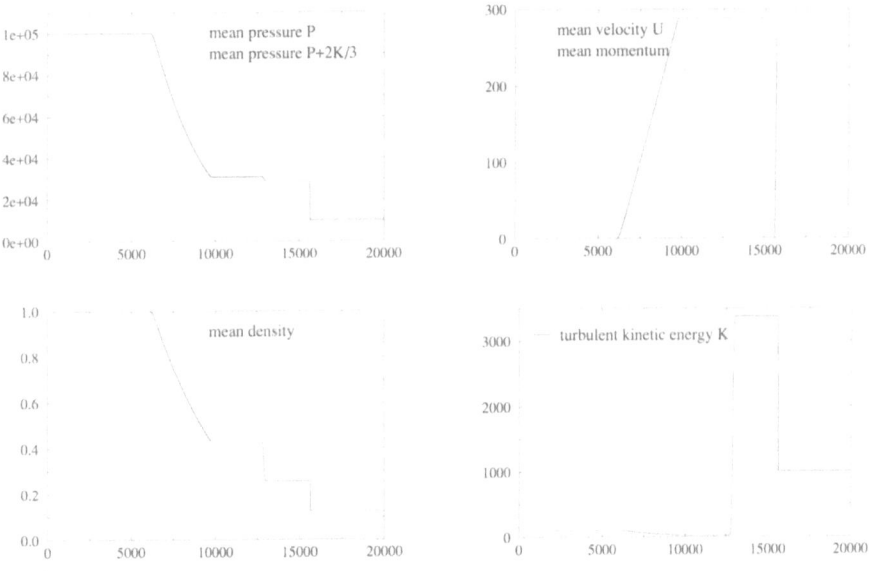

FIGURE 3. Profiles for a subsonic turbulent shock tube test.

[4] T. Buffard, T. Gallouët, J.M. Hérard, *A sequel to a rough Godunov scheme: application to real gases*, Internal Report EDF-DER HE-41/97/053/A, (1997).

[5] B. Cardot, B. Coron, B. Mohammadi, O. Pironneau, *Simulation of turbulence with the $k - \epsilon$ model*, Comput. Methods Appl. Mech. Eng., **87** (1991), 103–116.

[6] J.F. Colombeau, *Multiplication of distributions*, Springer-Verlag, (1992).

[7] L. Combe, *Simulation numérique d'écoulements gaz-particules sur maillage non structuré*, PhD Thesis, I.N.P. Toulouse, (1997).

[8] G. Dal Maso, P.G. Le Floch, F. Murat, *Definition and weak stability of non conservative products,* J. Math. Pures Appl. **74** (1995), 483–548.

[9] F. De Vuyst, *Schémas non conservatifs et schémas cinétiques pour la simulation numérique d'écoulements hypersoniques non visqueux en déséquilibre thermochimique,* PhD Thesis, univ. P. et M. Curie, (1994).

[10] A. Favre, *Equations des gaz turbulents compressibles,* Jour. Mec. **4** (1965), 391–421.

[11] A. Forestier, J.M. Hérard, X. Louis, *A Godunov type solver to compute turbulent compressible flows,* C. R. Ac. Sc. Paris, Série I **324** (1997), 919–926.

[12] T. Gallouët, J.M. Masella, *A rough Godunov scheme,* C. R. Ac. Sc. Paris, Série I **323** (1996), 77–84.

[13] S.K. Godunov, *A difference method for numerical calculation of discontinuous equations of hydrodynamics,* Math. Sb. **47** (1959), 217–300.

[14] P.G. Le Floch, *Entropy weak solutions to non linear hyperbolic systems in non conservative form,* Comm. in Part. Diff. Eq. **13(6)** (1988), 669–727.

[15] P.G. Le Floch, T.P. Liu, *Existence theory for non linear hyperbolic systems in non conservative form,* CMAP report **254** (1992).

[16] X. Louis, PhD Thesis, univ. P. et M. Curie (1995).

[17] J.M. Masella, *Quelques méthodes numériques pour les écoulements diphasiques bi-fluide en conduites pétrolières,* PhD Thesis, univ. P. et M. Curie (1997).

[18] B. Mohammadi, O. Pironneau, *Analysis of the $k - \epsilon$ turbulence model,* Masson-Wiley (1994).

[19] L. Sainsaulieu, , Thèse d'habilitation, univ. P. et M. Curie (1995).

[20] D.M. Salas, A. Iollo, *Entropy jump across an inviscid shock wave,* Theo. Comp. Fluid Dynamics **8** (1996), 365–375.

L.M.A., Université Blaise Pascal,
63177 Aubière cedex, France
E-mail address: `buffard@ucfma.univ-bpclermont.fr`

C.M.I., Université de Provence,
39, rue Joliot Curie
13453 Marseille cedex 13, France
E-mail address: `Thierry.GALLOUET@umpa.ens-lyon.fr`

EDF-DER-Département Laboratoire National d'Hydraulique,
6,quai Watier
78400 Chatou cedex, France
E-mail address: `Jean-Marc.Herard@der.edfgdf.fr`

International Series of Numerical Mathematics
Vol. 129, © 1999 Birkhäuser Verlag Basel/Switzerland

Compactness and Asymptotic Behavior of Entropy Solutions without Locally Bounded Variation for Hyperbolic Conservation Laws

Gui-Qiang Chen

Abstract. We discuss some recent developments and ideas in studying the compactness and asymptotic behavior of entropy solutions without locally bounded variation for nonlinear hyperbolic systems of conservation laws. Several classes of nonlinear hyperbolic systems with resonant or linear degeneracy are analyzed. The relation of the asymptotic problems to other topics such as scale-invariance, compactness of solutions, and singular limits is described.

1. Introduction

We are concerned with the behavior of entropy solutions without locally bounded variation for the Cauchy problem of hyperbolic conservation laws:

$$\partial_t u + \nabla_x \cdot f(u) = 0, \qquad x \in \mathbb{R}^n, \, u \in \mathbb{R}^m, \tag{1}$$

$$u|_{t=0} = u_0(x), \tag{2}$$

where $f : \mathbb{R}^m \to \mathbb{R}^{m \times n}$ is a nonlinear flux function. Such a system arises from many important areas since the conservation law is a fundamental law of nature. The archetypical example is the compressible Euler equations in fluid dynamics. One of the main difficulties in solving (1)–(2) is the development of shock waves, observed in nature, no matter how smooth the initial data are. Therefore, the Sobolev spaces $W^{k,p}, k \geq 1$, are not well-posed for the solutions, for which many powerful techniques are not directly applicable. One expects that the solutions are at most in the space of functions of bounded variations. The Glimm theory indicates that this is indeed the case for strictly hyperbolic systems with initial data of small total variation (see Glimm [21]). The well-posedness in BV is optimal in general since several recent examples indicate that, for the Cauchy problem of strictly hyperbolic systems, the total variation of the solutions blows up in a finite time for the initial data of large variation (e.g. see [25] for example). Another main difficulty is the resonances occurred among different characteristic fields so that the systems become nonstrictly hyperbolic. Such a feature causes extra difficulties even in the linear case. For the multidimensional case, such a situation is generic.

For example, in the 3-D case, if the number of equations $m \equiv 2 \,(mod\,4)$, Lax [28] proved that such systems must be nonstrictly hyperbolic. This is also the case even for $m \equiv \pm2, \pm3, \pm4 \,(mod\,8)$ (see [20]). Many physical systems arising from different areas have such features. Recent existence theories for these systems with L^∞ large initial data indicate that the entropy solutions belong to the following class of functions:

$$u(x,t) \in L^\infty(\mathbb{R}^{n+1}_+), \quad \text{or } L^p(\mathbb{R}^{n+1}_+), \quad \text{for some } p \in [1,\infty), \tag{3}$$

and, for any entropy-entropy flux pair $(\eta, q) \in \mathbb{R} \times \mathbb{R}^n$ with convex $\eta(u)$,

$$\partial_t \eta(u) + \nabla \cdot q(u) \le 0 \tag{4}$$

in the sense of distributions.

In this article we discuss some recent developments and related ideas for studying the behavior of entropy solutions in the class of (3)–(4), especially the asymptotic problems. One of the main difficulties is that the solution class is much larger; and only information is the Lax entropy inequalities (4) in a weak topology: the distributional sense. Another main difficulty is the lack of analytical techniques in L^p, such as the method of generalized characteristics, to follow the characteristics. To explain the ideas more clearly, we will focus mainly on the L^∞ periodic entropy solutions in $x \in \mathbb{R}^n$ with period $P \equiv [0,1]^n$.

2. Asymptotic decay via BV estimates and Glimm-Lax theory

It is well known that, for the linear case with nonzero propagation speeds and periodic initial data, the solution is periodic in time t. One cannot expect the decay as $t \to \infty$. However, this is not true for the genuinely nonlinear case. The observation is that the genuine nonlinearity of equations forces the nonlinear waves of each characteristic family to interact vigorously and to cancel each other. The analysis of Glimm-Lax [22], for scalar equations and 2×2 systems, has indicated that the resulting mutual cancellation of interacting shock and rarefaction waves of the same family induces the decay of periodic solutions. In particular, for 2×2 genuinely nonlinear and strictly hyperbolic systems subject to the following condition: the interaction of two shocks of the same family always produces a shock of the same family plus a rarefaction wave of the opposite family, Glimm-Lax [22] established the decay theory: There exists a constant C_0 and some constant state $\bar{u}$ such that, when $osc(u_0 - \bar{u}) \le C_0$, there exists a global periodic entropy solution $u(x,t)$ in x satisfying $TV(u)([0,1] \times \{t\}) \le \frac{C}{t}$, with C independent of t. This implies that the solution decays like $1/t$ in L^∞. Recently, Dafermos [15] used the method of generalized characteristics to show that any periodic solution with small oscillation and local bounded variation of the 2×2 systems decays asymptotically, with a detailed structure picture as in Lax [27] for the scalar case. Also see Engquist-E [19] for the decay of periodic solutions with local bounded variation for the two-dimensional scalar conservation laws.

3. Asymptotic decay via a new approach

A further problem is whether the decay phenomenon holds for more general cases:
(a) any L^∞ large periodic solutions without restrictions of either small oscillation
or local bounded variation, and (b) more general nonlinear hyperbolic systems,
especially degenerate systems and multidimensional scalar equations. The counterexample of Greenberg-Rascle [23] indicates that this is not always true if the
flux functions are not so smooth; and the asymptotic behavior is very sensitive
with respect to the smoothness of the flux functions. In this section, we discuss a
new approach, developed in Chen-Frid [3, 4], to study the asymptotic behavior of
periodic entropy solutions in a general framework.

3.1. Decay and scale-invariance

One of the main features of (1) is the scale-invariance in the sense that the self-
similar scaling sequence $u^T(x,t) = u(Tx, Tt)$ also satisfies (1) for any $T > 0$.

Definition 3.1. *The periodic solution $u(x,t)$ of the Cauchy problem (1)–(2) in $x \in \mathbb{R}^n$ asymptotically decays to $\bar{u}$, provided that*

$$\|u^T - \bar{u}\|_{L^q_{loc}(\mathbb{R}^{n+1}_+)} \to 0, \quad \text{when } T \to \infty, \quad \text{for some } q \in [1, \infty). \tag{5}$$

*Because of the self-similar structure of $u^T(x,t)$, the limit (5) can be translated
equivalently in terms of the decay geometrically for $u(x,t) = U(x/t,t)$ along the
rays $x/t = \xi$, $\xi \in \mathbb{R}^n$, passing the origin in the (x,t)-plane (also see [3]):*

$$\frac{1}{T}\int_0^T |U(\xi,t) - \bar{u}|^q dt \to 0, \quad \text{in } L^1_{loc}(\mathbb{R}^n_\xi), \quad \text{when } T \to \infty. \tag{6}$$

The limits (5)–(6) for $u(x,t) \in L^\infty(\mathbb{R}^2_+)$ are also equivalent to

$$\frac{n+1}{T^{n+1}}\int_0^T |U(\xi,t) - \bar{u}|^q t^n dt \to 0, \quad \text{in } L^1_{loc}(\mathbb{R}^n_\xi), \quad \text{when } T \to \infty.$$

Remark 3.2. Set $\mu^T(t) = \frac{1}{T}\chi_{[0,T]}(t)dt$. Then the limit (6) means

$$< \mu^T, |U - \bar{u}|^q > (\xi) \longrightarrow 0, \quad \text{in } L^1_{loc}(\mathbb{R}^n_\xi), \quad \text{when } T \to \infty. \tag{7}$$

Such averaging probability measures have been widely used to understand macro-
scopically the asymptotic behavior of physical quantities in statistical mechanics,
kinetic theory, ergodic theory, and probability theory. One may extend these notions to more general settings. A periodic solution $u(x,t)$ asymptotically decays
with respect to a family of probability measures $\{\mu^T(t)\}_{\{T>0\}}$ if $u(x,t) = U(x/t,t)$,
locally integrable in $\mathbb{R}^{n+1}_+$, satisfies (7) and $\mu(t) = w - \lim \mu^T(t)$, in the space of
Radon measures over some compactification for $[0,+\infty)$, satisfies $\mathrm{supp}\, \mu(t) = \{+\infty\}$. It would be natural to develop such notions to understand the asymptotic
behavior of stochastic solutions of nonlinear conservation laws.

3.2. Decay and compactness

One of our main observations is that the compactness of L^∞ solution operator in L^1_{loc}, coupling with the weak convergence of periodic initial data to the mean, yields the decay of the L^∞ periodic solution. More precisely, we have

Theorem 3.3. [3, 4]. *Assume that $u(x,t) \in L^\infty(\mathbb{R}^{n+1}_+)$ is a periodic solution of (1)–(2), and its scaling sequence $\{u^T(x,t)\}$ is compact in $L^1_{loc}(\mathbb{R}^{n+1}_+)$. Then $u(x,t)$ asymptotically decays to $\bar{u} \equiv \int_P u_0(x)dx$ in the sense of (5) or (6).*

In 3.3, we assume that $\{u^T(x,t)\}$ is compact in $L^1_{loc}(\mathbb{R}^{n+1}_+)$, which is a corollary of the compactness of solution operator. Such a compactness can be achieved by the method of compensated compactness, the averaging method, and other analytical techniques. For example, we have

Theorem 3.4. [4, 7]. *Consider a system (1) with a strictly convex entropy pair (η_*, q_*). Assume that the uniformly bounded sequence $u^T(x,t) \in L^\infty(\mathbb{R}^{n+1}_+)$ satisfies (4) for any convex entropy pair $(\eta, q) \in \Lambda$, where Λ is a linear space of entropy pairs of (5) including (η_*, q_*). Then the measure sequence*

$$\partial_t \eta(u^T) + \nabla_x \cdot q(u^T) \qquad \text{is compact in } W^{-1,r}_{loc}(\mathbb{R}^{n+1}_+), \quad r \in (1,\infty), \qquad (8)$$

for any entropy pair $(\eta, q) \in \Lambda$ with $|\nabla^2 \eta(u)| \le C_{\eta,K} \nabla^2 \eta_*(u)$ for $u \in K \Subset \mathbb{R}^m$.

Therefore, the compensated compactness method (e.g. [36, 34]) is one of the efficient methods to achieve the compactness with the aid of Theorem 3.4.

3.3. Decay and entropy

The next question is whether the decay in Theorem 3.3 is achieved in a strong sense. Indeed, the entropy inequalities (4) provide such further information.

Theorem 3.5. [3, 4]. *Let the system (1) be endowed with a strictly convex entropy η_*. Then the asymptotic decay of an L^∞ periodic entropy solution $u(x,t)$ of (1)–(2) to the mean $\bar{u}$ of $u_0(x)$ over the period P in the sense of (5) implies its asymptotic decay in $L^q, 1 \le q < \infty$. That is, there exists a set $\mathcal{T} \subset (0,\infty)$, with $\text{meas}((0,\infty) - \mathcal{T}) = 0$, such that*

$$\int_P |u(x,t) - \bar{u}|^q dx \to 0, \qquad t \to \infty, t \in \mathcal{T}, \qquad \text{for any } 1 \le q < \infty. \qquad (9)$$

4. Applications

We now apply the approach established in §3 to analyze the decay of periodic solutions for various nonlinear conservation laws. The proofs of these theorems can be found in Chen-Frid [4, 7], Chen-LeFloch [10], and Chen-Dafermos [2]. In this section we discuss these results and raise some open problems.

4.1. Scalar conservation laws

For multidimensional scalar conservation laws with C^2 flux function $f(u)$ and initial data $u_0(x) \in L^\infty(\mathbb{R}^n)$ with period $P = [0,1]^n$, the existence of global entropy solutions of the Cauchy problem is due to Kruzkov [26]. Combining the approach in §3 with a theorem by Lions-Perthame-Tadmor [29], we conclude

Theorem 4.1. [4, 7]. *Assume that the flux function $f(u)$ satisfies*

$$\text{meas}\left\{ u \in \mathbb{R} \mid \tau + f'(u) \cdot \vec{k} = 0 \right\} = 0, \quad \text{for any } (\tau, \vec{k}) \in S^n. \tag{10}$$

Let $u(x,t)$ be an entropy solution in $\mathbb{R}_+^{n+1}$ with periodic data $u_0(x)$. Then $u(x,t)$ asymptotically decays to $\bar{u} = \int_P u_0(x)dx$ in the sense of (9).

The condition (10) is implied by the generalized genuine nonlinearity condition: $\text{meas}\{ u \in \mathbb{R} \mid \vec{k} \cdot f''(u) = 0 \} = 0$, for all $\vec{k} \in S^{n-1}$. For the one-dimensional case, the assumption (10) can be further relaxed into the following nonlinearity assumption: There is no subinterval $(\alpha, \beta) \in [\min u_0(x), \max u_0(x)]$ on which $f(u)$ is affine. It would be interesting to generalize the condition (10) under which the decay phenomenon for the multidimensional case is still held.

4.2. 2×2 strictly hyperbolic systems

Consider a 2×2 strictly hyperbolic and genuinely nonlinear system of conservation laws with periodic initial data $u_0(x) \in L^\infty(\mathbb{R}; \mathbb{R}^2)$.

Theorem 4.2. [4] *Let $u(x,t) \in L^\infty(\mathbb{R}_+^2)$ be a periodic entropy solution of the Cauchy problem of the 2×2 system. Then $u(x,t)$ asymptotically decays to $\bar{u} = \int_P u_0(x)dx$ in L^q, $1 \le q < \infty$, in the sense of (9).*

We assume neither the small oscillation and the local bounded variation of the periodic solution nor further conditions on the system in Theorem 4.2. The genuine nonlinearity of the system can be relaxed to allow reflection points. Such a typical example is the equations of elasticity:

$$\partial_t \tau - \partial_x v = 0, \quad \partial_t v - \partial_x \sigma(\tau) = 0, \qquad \sigma(\tau) \in C^2(\mathbb{R}_+), \sigma'(\tau) > 0. \tag{11}$$

The genuine nonlinearity is generally precluded in elasticity in general. One typical case is: $\text{sign}(\tau \sigma''(\tau)) > 0, \tau \neq 0$. A more general situation is:

- There exists an interval $(\alpha_0, 0)$ or $(0, \beta_0)$ in which $\sigma''(\tau) \neq 0$;
- There is no interval $(\alpha, \beta), \alpha > 0$ or $\beta < 0$, in which $\sigma(\tau)$ is affine.

For this case, combining a theorem in [24, 17] and the approach in §3, we have

Theorem 4.3. *Let the initial data $(\tau_0(x), v_0(x)) \in L^\infty(\mathbb{R})$ be periodic with period P. Then there exists a periodic entropy solution of (11), which asymptotically decays to $(\bar{\tau}, \bar{v}) = \int_P (\tau_0(x), v_0(x))dx$ in L^q, $1 \le q < \infty$, in the sense of (9).*

Remark 4.4. For (11) with $\sigma(\tau) \notin C^2$, Greenberg-Rascle [23] constructed an example that the periodic solution $(\tau(x,t), v(x,t))$ does not decay in L^q. This fact indicates that $\sigma(\tau) \in C^2$ is necessary to keep the decay phenomenon. It would be interesting to consider more general $\sigma(\tau) \in C^2$ so that the decay phenomenon still holds.

4.3. Isentropic Euler equations

Consider the isentropic Euler equations for compressible fluids:

$$\partial_t \rho + \partial_x m = 0, \qquad \partial_t m + \partial_x(m^2/\rho + p(\rho)) = 0, \tag{12}$$

where ρ, m, and p are the density, the momentum, and the pressure, respectively. In the non-vacuum state $(\rho > 0)$, $v = m/\rho$ is the velocity. $p = p(\rho) \in C^4(0, \infty)$ is a given function of ρ depending on compressible fluids under consideration. Strict hyperbolicity and genuine nonlinearity away from the vacuum require that

$$p'(\rho) > 0, \quad \rho p''(\rho) + 2p'(\rho) > 0, \qquad \rho > 0. \tag{13}$$

Near the vacuum, $p(\rho)$ is only asymptotic to the γ-law pressure (as real gases):

$$|(p(\rho) - \kappa \rho^\gamma)^{(k)}| \le C \rho^{\gamma+1-k}, \quad 0 \le k \le 4, \ \rho \ll 1. \tag{14}$$

Consider the Cauchy problem for (12) with the initial data

$$(\rho, m)|_{t=0} = (\rho_0(x), m_0(x)), \quad 0 \le \rho_0(x) \le C_0, \ |m_0(x)/\rho_0(x)| \le C_0 < \infty. \tag{15}$$

The main difficulty of this system is that strict hyperbolicity fails, and the flux function is only Lipschitz continuous at the vacuum state $\rho = 0$, which occurs in fluid mechanics. Nevertheless, a compactness theorem has been established by using only weak entropy pairs, a subspace of entropy pairs, consisting of those η vanishing on the vacuum $\rho = 0$ for any fixed $m/\rho \in (-\infty, \infty)$. For example, the mechanical energy-energy flux pair $\eta_* = \frac{1}{2}\frac{m^2}{\rho} + \rho \int_0^\rho \frac{p(r)}{r^2} dr$ and $q_* = \frac{m^3}{2\rho^2} + m \int_0^\rho \frac{p'(r)}{r} dr$ is a convex weak entropy pair. One can prove that, for $0 \le \rho \le C$, $|m/\rho| \le C, |\nabla \eta(u)| \le C_\eta, |\nabla^2 \eta(u)| \le C_\eta \nabla^2 \eta_*(u)$, for any weak entropy η, with C_η independent of u.

Theorem 4.5. [10]. *(a) There exists a global solution $(\rho(x,t), m(x,t))$ of the Cauchy problem (12)–(15), satisfying $0 \le \rho(x,t) \le C, |m(x,t)/\rho(x,t)| \le C$, for some C depending only on C_0 and γ, and $\partial_t \eta(\rho, m) + \partial_x q(\rho, m) \le 0$ in the sense of distributions for any convex weak entropy pairs (η, q).*
(b) The solution operator $(\rho, m)(\cdot, t) = S_t(\rho_0, m_0)(\cdot)$, determined by (a), is compact in $L^1_{loc}(\mathbb{R}^2_+)$ for $t > 0$.

For polytropic perfect gases, the similar results were proved by DiPerna [18] for the case $\gamma = 1 + 1/N, N \ge 5$ odd, for $L^2 \cap L^\infty(\mathbb{R})$ initial data, by Ding-Chen-Luo [16] for $\gamma = 3/2$ and Chen [1] for $1 < \gamma \le 5/3$ for usual gases with general L^∞ initial data. The results are also true for $\gamma \ge 3$ due to Lions-Perthame-Tadmor [30] and for $5/3 < \gamma < 3$ due to Lions-Perthame-Souganidis [31]. Using Theorem 4.5, we conclude

Theorem 4.6. *Let* $(\rho(x,t), m(x,t)), 0 \leq \rho(x,t) \leq C, |m(x,t)/\rho(x,t)| \leq C,$ *be a periodic entropy solution of* (12)–(15) *with period* P. *Then* $(\rho(x,t), m(x,t))$ *asymptotically decays to* $\int_P (\rho_0(x), m_0(x)) dx$ *in the sense of* (9).

4.4. 3×3 Euler equations in thermoelasticity

We now consider hyperbolic systems of conservation laws with linear degeneracy: there is at least one characteristic field is linearly degenerate. A typical example is the 3×3 system of Euler equations in Lagrangian coordinates:

$$\partial_t \tau - \partial_x v = 0, \quad \partial_t v + \partial_x p = 0, \quad \partial_t (e + v^2/2) + \partial_x (vp) = 0, \tag{16}$$

where $\tau, v, p,$ and e denote respectively the deformation gradient (specific volume for fluids, strain for solids), the velocity, the pressure, and the internal energy. Other relevant fields are the entropy s and the temperature θ. The system (16) is complemented by the Clausius inequality $\partial_t s \geq 0$. We consider the following class of constitutive relations for the new state vector (w, v, s) with the form

$$\tau = w + \alpha s, \quad p = h(w) > 0, \quad e = -\int_0^w h(\omega) d\omega + \beta s, \quad \theta = \alpha h(w) + \beta, \tag{17}$$

where α and β are positive constants, and $h(w) \in C^2(\mathbb{R})$ with $h'(w) < 0$ satisfying

$$sign(w - \hat{w}) \left((\alpha h(w) + \beta) h''(w) - 4\alpha h'(w)^2\right) < 0, \quad w \neq \hat{w}. \tag{18}$$

Observe that the equations (17) are compatible with the first law of thermodynamics: $\theta \, ds = de + p \, dv$. The model (17) can be regarded as a "first-order correction" to the general constitutive equations (see [2] for the details).

Consider the Cauchy problem for (16)–(18) with periodic initial data

$$(w, v, s)|_{t=0} = (w_0(x), v_0(x), s_0(x)) \tag{19}$$

satisfying $(w_0(x), v_0(x)) \in \{(w, v) \,|\, |v \pm \int_{\hat{w}}^w \sqrt{-h'(\omega)} d\omega| \leq C_0\}, s_0(x) \in \mathcal{M}_{loc}(\mathbb{R})$.

Theorem 4.7. [2]. *(a) There exists a periodic distributional solution* $(w(x,t), v(x,t), s(x,t)) \in L^\infty(\mathbb{R}_+^2; \mathbb{R}^2) \times \mathcal{M}_{loc}(\mathbb{R}_+^2; \mathbb{R})$ *of* (16)–(18), *satisfying*

$$s_t(x,t) \in \mathcal{M}_{loc}(\mathbb{R}_+^2), \quad \theta(w(x,t)) \geq 0, \quad |s|\{\{|x| \leq cT_0\} \times [0, T_0]\} \leq CT_0^2, \tag{20}$$

for any $c, T_0 > 0$, *with* $C > 0$ *independent of* T_0. *Moreover,* $(w(x,t), v(x,t), s(x,t))$ *satisfies the entropy condition:*

$$\partial_t \eta(w, v) + \partial_x q(w, v) \leq 0, \quad s_t \geq 0, \tag{21}$$

in the sense of distributions for any C^2 *entropy pair* $(\eta(v, w), q(v, w))$ *of* $\partial_t w - \partial_x v = 0, \partial_t v + \partial_x h(w) = 0$, *for which the strong convexity condition holds:*

$$\theta \eta_{ww} - \alpha h'(w) \eta_w \geq 0, \quad \theta \eta_{vv} + \alpha \eta_w \geq 0, \quad (\theta \eta_{ww} - \alpha h'(w) \eta_w)(\theta \eta_{vv} + \alpha \eta_w) - \eta_{ww}^2 \geq 0.$$

(b) Any sequence $(w^T(x,t), v^T(x,t))$, *which is uniformly bounded and satisfies* (21), *is compact in* $L^1_{loc}(\mathbb{R}_+^2)$ *when* $t > 0$.

Theorem 4.8. [4]. *Let $(\tau(x,t), v(x,t), s(x,t))$ be a periodic entropy solution of (16)–(18) with period P satisfying $(v(x,t), \tau(x,t) - as(x,t)) \in L^\infty(\mathbb{R}^2_+)$ and (20)–(21). Then the velocity $v(x,t)$ asymptotically decays to $\bar{v} = \int_P v_0(x)dx$ in L^q, $1 \le q < \infty$. Moreover, the pressure $p(w(x,t))$ and the temperature $\theta(w(x,t))$ decay to $\tilde{p} = p(\Theta^{-1}(\int_P \Theta(w_0(x))dx))$, and $\tilde{\theta} = \theta(\Theta^{-1}(\int_P \Theta(w_0(x))dx))$, in L^q, $1 \le q < \infty$, respectively, where $\Theta(w) = \beta w + \alpha \int_0^w h(\omega)d\omega$.*

5. Final remarks

1. Asymptotic decay of L^p entropy solutions. The approach described here can be easily generalized to the decay problem for L^p entropy solutions for $1 < p < \infty$. For example, In Theorems 3.3–3.5, the assumption $u(x,t) \in L^\infty(\mathbb{R}^{n+1}_+)$ may be replaced by $u(x,t) \in L^p(\mathbb{R}^{n+1}_+), p > 2$. Then the decay of an L^p periodic entropy solution $u(x,t)$ of (1)–(2), with period P, in the sense of (5) or (6) in the L^2 norm implies (9) for any $q \in [2,p)$. It would be interesting to generalize the approach for the entropy solutions in L^p_W (weighted L^p spaces).

2. Hyperbolic conservation laws with relaxation. The approach can be easily extended to the asymptotic problems for the entropy solutions of hyperbolic systems of conservation laws with relaxation. See [4] for the decay results for such systems discussed in [11, 12].

3. Asymptotic behavior of entropy solutions with general initial data. The approach and ideas can be developed to study the asymptotic behavior of entropy solutions without locally bounded variation for general L^p initial data. See [5, 6] for such an approach.

4. Asymptotic behavior of entropy solutions for multidimensional conservation laws. In §4.1 we give the simplest multidimensional example: scalar conservation laws. It would be interesting to find some interesting classes of multidimensional systems of conservation laws so that this approach is applicable.

5. The validity of nonlinear geometric optics for entropy solutions. The ideas and observations discussed here have been applied to the validity of the approximation of weakly nonlinear geometric optics for weak entropy solutions without locally bounded variation for conservation laws (cf. [8]).

6. Generalized characteristics in L^p. It would be interesting to explore some approaches to define generalized characteristics for the entropy solutions in L^p so that one could follow the characteristics to study the asymptotic problems. Again, the Lax entropy inequalities should play an important role in this effort.

Acknowledgement. The author thanks his scientific collaborators, C. M. Dafermos, H. Frid, and P. LeFloch for their explicit and implicit contributions to this article. Gui-Qiang Chen's research was supported in part by NSF grants DMS-9623203 and INT-9726215, and by an Alfred P. Sloan Foundation Fellowship.

References

[1] G.-Q. Chen, *Convergence of the Lax-Friedrichs scheme for isentropic gas dynamics (III)*, Acta Math. Sci. **8** (1988), 243–276 (in Chinese); **6** (1986), 75–120.

[2] G.-Q. Chen and C.M. Dafermos, *The vanishing viscosity method in one-dimensional thermoelasticity*, Trans. Amer. Math. Soc. **347** (1995), 531–541.

[3] G.-Q. Chen and H. Frid, *Asymptotic decay of solutions of conservation laws*, C. R. Acad. Sci. Paris, Série I **323** (1996), 257–262.

[4] G.-Q. Chen and H. Frid, *Decay of entropy solutions of nonlinear conservation laws*, Preprint, Submitted for publication (1997).

[5] G.-Q. Chen and H. Frid, *Asymptotic stability of Riemann waves for conservation laws*, Z. Angew. Math. Phys. (ZAMP), **48** (1997), 30–44.

[6] G.-Q. Chen and H. Frid, *Large time behavior of entropy solutions for conservation laws*, Preprint, Submitted for publication (1997).

[7] G.-Q. Chen and H. Frid, *Asymptotic Stability and Decay of Solutions of Conservation Laws*, Lecture Notes, Northwestern University, 1996.

[8] G.-Q. Chen, S. Junca and M. Rascle, in preparation, 1998.

[9] G.-Q. Chen and P. LeFloch, *Entropies and weak solutions of the compressible isentropic Euler equations*, C. R. Acad. Sci. Paris, Série I, **324** (1997), 1105–1110.

[10] G.-Q. Chen and P. LeFloch, *Compressible Euler equations with general pressure law and related equations*, Preprint, Northwestern University, 1998.

[11] G.-Q. Chen and T.-P. Liu, *Zero relaxation and dissipation limits for hyperbolic conservation laws*, Comm. Pure Appl. Math. **46** (1993), 755–781.

[12] G.-Q. Chen, C.D. Levermore and T.-P. Liu, *Hyperbolic systems of conservation laws with stiff relaxation terms and entropy*, C. P. Ap. Math. **47** (1994), 787–830.

[13] C. Cheverry, *The modulation equations of nonlinear geometric optics*, Commun. Partial Diff. Eqs. **21** (1996), 1119–1140.

[14] C.M. Dafermos, *Applications of the invariance principle for compact processes II: Asymptotic behavior of solutions of a hyperbolic conservation laws*, J. Diff. Eqs. **11** (1972), 416–424.

[15] C.M. Dafermos, *Large time behavior of periodic solutions of hyperbolic systems of conservation laws*, J. Diff. Eqs. **121** (1995), 183–202.

[16] X. Ding, G.-Q. Chen and P. Luo, *Convergence of the Lax-Friedrichs scheme for isentropic gas dynamics (I)–(II)*, Acta Math. Sci. **7** (1987), 467–480, **8** (1988), 61-94 (in Chinese); **5** (1985), 415–432, 433–472.

[17] R. DiPerna, *Convergence of approximate solutions to conservation laws*, Arch. Rational Mech. Anal. **82** (1983), 27–70.

[18] R. DiPerna, *Convergence of the viscosity method for isentropic gas dynamics*, Comm. Math. Phys. **91** (1983), 1–30.

[19] B. Engquist and E, W. *Large time behavior and homogenization of solutions of two-dimensional conservation laws*, Comm. Pure Appl. Math. **46** (1993), 1–26.

[20] S. Friedlands, J.W. Robbin and J. Sylvester, *On the crossing rule*, Comm. Pure Appl. Math. **37** (1984), 19–64.

[21] J. Glimm, *Solutions in the large for nonlinear hyperbolic systems of equations*, Comm. Pure Appl. Math. **18** (1965), 95–105.

[22] J. Glimm and P.D. Lax, *Decay of solutions of nonlinear hyperbolic conservation laws*, Memoirs Amer. Math. Soc. **101**, 1970.

[23] J. Greenberg and M. Rascle *Time-periodic solutions to systems of conservation laws*, Arch. Rational Mech. Anal. **115** (1991), 395–407.

[24] G. Gripenberg *Compensated compactness and one-dimensional elastodynamics*, Ann. Scuola Norm. Sup. Pisa Cl. Sci. (4) **22** (1995), 227–240.

[25] H.K. Jenssen, *Blowup for systems of conservation laws*, Preprint /http:/www.math.ntnu.no/conservation/, 1998.

[26] S.N. Kruzkov, *First-order quasilinear equations in several independent variables*, Math. USSR Sb. **10** (1970), 217–243.

[27] P.D. Lax, *Hyperbolic systems of conservation laws*, Comm. Pure Appl. Math. **10** (1957), 537–566.

[28] P.D Lax, *The multiplicity of eigenvalues*, Bull. Amer. Math. Soc. **6** (1982), 213–214.

[29] P.L. Lions, B. Perthame and E. Tadmor, E. *A kinetic formulation of multidimensional scalar conservation laws and related equations*, J. Amer. Math. Soc. **7** (1994), 169–192.

[30] P.L. Lions, B. Perthame and E. Tadmor, *Kinetic formulation of the isentropic gas dynamics and p-system*, Comm. Math. Phys. **163** (1994), 415–431.

[31] P.L. Lions, B. Perthame and P.E. Souganidis, *Existence of entropy solutions for the hyperbolic systems of isentropic gas dynamics in Eulerian and Lagrangian coordinates*, Comm. Pure Appl. Math. **49** (1996), 599–638.

[32] T.-P. Liu, *Linear and nonlinear large time behavior of solutions of general systems of conservation laws*, Comm. Pure Appl. Math. **30** (1977), 767–796.

[33] F. Murat, *L'injection du cone positif de H^{-1} dans $W^{-1,q}$ est compacte pour tout $q < 2$*, J. Math. Pures Appl. **60** (1981), 309–322.

[34] F. Murat, *Compacité par compensation*, Ann. Scuola Norm. Sup. Pisa Cl. Sci. (4) **5** (1978), 489-507.

[35] D. Serre, *Systémes de Lois de Conservation (I)–(II)*, Diderot Editeur, Paris, 1996.

[36] L. Tartar, *Compensated compactness and applications to partial differential equations*, In: Research Notes in Mathematics, Nonlinear Analysis and Mechanics, ed. R. J. Knops, vol. **4**, Pitman Press, New York, 1979, pp. 136–211.

Department of Mathematics
Northwestern University
Lunt Hall, 2033 Sheridan Road
Evanston, IL 60208-2730, USA
E-mail address: gqchen@math.nwu.edu

International Series of Numerical Mathematics
Vol. 129, © 1999 Birkhäuser Verlag Basel/Switzerland

Non-symmetric Conical Supersonic Flow

Shuxing Chen

Abstract. The inviscid supersonic flow passing a conical body is studied. The potential equation is applied to describe the motion of the flow. The problem is then reduced to a boundary value problem of an elliptic equation with free boundary. If the surface of the body is a perturbed circular cone with its vertex angle less than a critical value, then the existence of the solution for such a boundary value problem with an attached conical shock is proved.

1. Introduction

When a supersonic flow pasts a pointed body, an attached shock will be produced generally, provided the body is sharp in some sense. The mathematical analysis of such problems greatly attracts many people's attention because of its physical background. Certainly, to determine the location of the shock and the supersonic flow field behind the shock is the main problem in studying the flow past the body. In the case when the body is a wedge or a wing, there are many results on the local existence of the solution with shock front even for supersonic flow past a three dimensional wing (see [1], [3–6], [11, 12]). However, since the appearance of the pointed top of the body often causes singularities, the study of the supersonic flow past a pointed body is more complicated than the study of the flow past a wing.

If the pointed body is a circular cone, and the direction of the coming flow is parallel to the axis of the cone, then by symmetry the flow field, as well as the location of the shock, can be determined by a boundary value problem of an ordinary differential equation (see [2]). However, if the coming flow has an attack angle or the pointed body is a non-symmetric cone, the solution of the problem is related to a boundary value problem of a partial differential equation with free boundary.

In this paper we describe the inviscid compressible flow by a potential flow equation, which is a good approximation of inviscid compressible flow, if the strength of the shock in the flow under consideration is small.

The equation is

$$\sum_{i=1}^{3} \partial_{x_i}(\phi_{x_i} H) = 0 \tag{1}$$

where ϕ is the velocity potential, satisfying $\vec{v} = \nabla\phi$, and where H stands for the density, which is a function of $C - \frac{1}{2}|\nabla\phi|^2$. For a perfect gas $H(\tau) = (\frac{\gamma-1}{\gamma}\tau)^{1/(\gamma-1)}$. On any shock front $\zeta(x) = 0$, the potential is continuous, while its derivatives have jumps, satisfying

$$\sum_{j=1}^{3} \zeta_{x_j}[\phi_{x_j} H] = 0 , \tag{2}$$

where $[\ ]$ represents the jump of the quantity inside the bracket. The condition (2) is called Rankine-Hugoniot condition for the second order equation (1). Besides, the solution of (1) should also satisfy the entropy condition, which means that the density behind the shock is greater than that ahead of the shock.

Under our assumptions the problem of supersonic flow past a non-symmetric cone can be reduced to a boundary value problem of (1) in the domain $\Omega : x_3 \leq f(x_1, x_2)$, where $x_3 = f(x_1, x_2)$ is the equation of the surface of the pointed body. If the generators of the cone are straight, then the function $f(x_1, x_2)$ is homogeneous of degree one, and it is a small perturbation of $(x_1^2 + x_2^2)^{1/2} / \tan\alpha_0$. The boundary condition is

$$\phi_{x_1} f_{x_1} + \phi_{x_2} f_{x_2} - \phi_{x_3} = 0 \tag{3}$$

Our main result is

Theorem 1.1. *Assume that the coming flow is constant and supersonic with $u_0 > a_0$, the conical surface $x_3 = f(x_1, x_2)$ is a small perturbation of a circular cone with its vertex angle less than a critical value, then there exists a continuous solution ϕ of (1) in Ω, which is homogeneous of degree one. The solution ϕ satisfies the boundary condition (3) on $\partial\Omega$, and satisfies the Rankine-Hugoniot condition and entropy condition on an attached shock front, where the derivative of ϕ has jumps. Moreover, the shock front is also a conical surface with straight generators.*

2. Reduction and decomposition

The problem (1),(2) is invariant under the dilation $x_i \to \alpha x_i, i = 1, 2, 3$ with $\alpha > 0$. Therefore, we can study its self-similar solutions. Introduce $\xi = x_1/x_3, \eta = x_2/x_3$, and let $\phi(x_1, x_2, x_3)$ have the form $x_3\psi\left(\frac{x_1}{x_3}, \frac{x_2}{x_3}\right)$. Then (1) can be rewritten as

$$a^2 \left((1+\xi^2)\psi_{\xi\xi} + 2\xi\eta\psi_{\xi\eta} + (1+\eta^2)\psi_{\eta\eta}\right) - \left((1+\xi^2)\psi_\xi + \xi\eta\psi_\eta - \xi\psi\right)^2 \psi_{\xi\xi}$$
$$-2\left((1+\xi^2)\psi_\xi + \xi\eta\psi_\eta - \xi\psi\right)\left((1+\eta^2)\psi_\eta + \xi\eta\psi_\xi - \eta\psi\right)\psi_{\xi\eta} \tag{4}$$
$$- \left((1+\eta^2)\psi_\eta + \xi\eta\psi_\xi - \eta\psi\right)^2 \psi_{\eta\eta} = 0.$$

Here we have used the relation $H/H' = a^2$.

Consider (4) in polar coordinates. Let $\xi = r\cos\theta$, $\eta = r\sin\theta$, we have

$$a^2\left((1+r^2)\psi_{rr} + \frac{1}{r}\psi_r + \frac{1}{r^2}\psi_{\theta\theta}\right)$$

$$-((1+r^2)\psi_r - r\psi)^2\psi_{rr} - \frac{2}{r^2}((1+r^2)\psi_r - r\psi)\psi_\theta\psi_{r\theta} \qquad (5)$$

$$-\frac{1}{r^4}\psi_\theta^2\psi_{\theta\theta} - \frac{1}{r^3}\psi_r\psi_\theta^2 + \frac{2}{r^3}\left(\psi_r(1+r^2) - r\psi\right)\psi_\theta^2 = 0.$$

It is not difficult to check that the equation is of elliptic type in the domain between the shock and the surface of the cone. The boundary condition on the surface of the cone is

$$(1+b^2)\psi_r - \frac{b'}{b^2}\psi_\theta - b\psi = 0. \qquad (6)$$

On the shock front $r = s(\theta)$, ψ is continuous and equal to the constant corresponding to the state ahead of the shock front, thus we have

$$\psi = \psi_0. \qquad (7)$$

Besides, the Rankine-Hugoniot condition is

$$((1+s^2)\psi_r - \frac{s'}{s^2}\psi_\theta - s\psi_0)H + s\psi_0\rho_0 = 0, \qquad (8)$$

where ρ_0 is the density ahead of the shock front.

The nonlinear problem (5)–(8) will be called (NL) (or $(NL)\{b(\theta)\}$). Notice that we are looking for the solution of (NL) near the background solution in the symmetric case. Since the background solution $\psi_B(r)$ satisfies the entropy condition – the density behind the shock is greater than the density ahead of the shock – then for the perturbation of the background solution, the entropy condition will also hold.

We emphasize that the boundary $r = s(\theta)$ is unknown, which should be determined together with the potential ψ. To overcome the difficulty caused by the free boundary, we introduce a partial holograph transformation. The character of this transformation is that we take ψ as the new coordinate p, and change the position of ψ and r as unknown function and independent variable. The transformation T is: $(r,\theta) \mapsto (p,\sigma)$,

$$\begin{cases} \sigma = \theta \\ p = \psi(r,\theta). \end{cases} \qquad (9)$$

Its inverse transform is T^{-1},

$$\begin{cases} \theta = \sigma \\ r = u(p,\sigma). \end{cases} \qquad (10)$$

In the new coordinates, the shock front becomes a fixed boundary $p = \psi_0$, because the potential ψ outside the shock equals a constant ψ_0. Meanwhile, the interior domain becomes $\{(p, \sigma) : p < \psi_0\}$, and $u(p, \sigma)$ becomes the new unknown function, which satisfies $u(\psi_0, \sigma) = s(\theta)$ on the shock front.

The function $u(p, \sigma)$ satisfies a second order differential equation:

$$a^2\left(-\frac{u_{pp}}{u_p^3}(1+u^2) + \frac{1}{uu_p} + \frac{2u_\sigma}{u^2u_p^2}u_{p\sigma} - \frac{1}{u^2u_p}u_{\sigma\sigma} - \frac{u_\sigma^2}{u^2u_p^3}u_{\sigma\sigma}\right)$$

$$+\left(\frac{1+u^2}{u_p} - up\right)^2\frac{u_{pp}}{u_p^3} + \frac{2}{u^2}\left(\frac{1+u^2}{u_p} - up\right)\frac{u_\sigma}{u_p}\left(\frac{u_\sigma}{u_p^3}u_{pp} - \frac{u_{p\sigma}}{u_p^2}\right) \qquad (11)$$

$$-\frac{1}{u^4}\frac{u_\sigma^2}{u_p^2}\left(-\frac{u_{\sigma\sigma}}{u_p} + \frac{2u_\sigma}{u_p^2}u_{p\sigma} - \frac{u_\sigma^2}{u_p^3}u_{pp}\right) - \frac{u_\sigma^2}{u^3u_p^3} + \frac{2}{u^3}\left(\frac{1+u^2}{u_p} - up\right)\frac{u_\sigma^2}{u_p^2} = 0$$

and the boundary conditions become

$$u = b(\sigma), \qquad (12)$$

$$(1 + b^2(\sigma)) + \frac{b'(\sigma)}{b^2(\sigma)}u_\sigma - b(\sigma)pu_p = 0. \qquad (13)$$

Besides, on the shock front $p = \psi_0$

$$\left((1+u^2) + \frac{u_\sigma^2}{u^2} - upu_p\right)H + upu_p\rho_0 = 0. \qquad (14)$$

The problem (11)–(14) is called $(NL)^*$, which is essentially equivalent to (NL). Obviously, if one of them is solved, then the solution of the other one is also obtained.

When $b(\theta)$ equals the constant b_0, which is less than a critical value, the problem (5)–(8) has the solution $\psi_B(r)$. In the sequel we always assume that $b(\theta) - b_0$ is sufficiently small, and look for a solution of (NL) near the background solution $\psi_B(r)$. Since the potential ψ is unknown on the surface of the cone, then the location of this boundary will become unknown after the transformation T. To avoid this new trouble we decompose (NL) or $(NL)^*$ into two auxiliary nonlinear boundary value problems with fixed boundary. To this end we introduce two constants r_1 and r_2, satisfying

$$b_0 < r_2 < r_1, \quad \psi_{20} < \psi_{10} < \psi_0, \qquad (15)$$

where $\psi_{10} \overset{\triangle}{=} \psi_B(r_1), \psi_{20} \overset{\triangle}{=} \psi_B(r_2)$. Then we introduce two boundary value problems in the inner ring $\Omega_a : b(\theta) < r < r_1, 0 \le \theta \le 2\pi$ and outer ring $\Omega_b : \psi_{20} < p < \psi_0, 0 \le \theta \le 2\pi$ respectively.

$$(NL)^{(a)} : \begin{cases} \text{equation (5)} & \text{in } \Omega_a \\ \text{boundary condition (6)} & \text{on } r = b(\theta) \\ \psi = d(\theta) & \text{on } r = r_1 \end{cases} \qquad (16)$$

$$(NL)^{(b)} : \begin{cases} \text{equation (11)} & \text{in } \Omega_b \\ \text{boundary condition (14)} & \text{on } p = \psi_0 \\ \psi = q(\sigma) & \text{on } p = \psi_{20} \end{cases} \qquad (17)$$

As it will be seen, the solvability and estimates of the solution of $(NL)^{(a)}$ and $(NL)^{(b)}$ will lead us to the solution of problem (NL) (and $(NL)^*$).

In order to make our main idea clearer we assume $|r_1 - b_0|, |\psi_0 - \psi_{20}|$ are small, so that the maximum principle is available by adding at most a transformation. Otherwise, we have to introduce a set of overlapping intervals $\{(\psi_{a_i}, \psi_{b_i})\}_{1 \le i \le k}$ with $\psi_{a_1} = \psi_{20}, \psi_{a_{i+1}} < \psi_{b_i} < \psi_{a_{i+2}} < \psi_{b_{i+1}}, \psi_{b_k} = \psi_0$, so that $r_1 - b_0$ and each $\psi_{b_i} - \psi_{a_i}$ are small, then a similar argument with corresponding revision will lead to our required conclusion.

3. Problem $(NL)^{(a)}$ and $(NL)^{(b)}$

For the problem $(NL)^{(a)}$ we can use the transformation

$$\begin{cases} \tilde{\theta} = \theta \\ \dfrac{\tilde{r} - b_0}{r_1 - b_0} = \dfrac{r - b(\theta)}{r_1 - b(\theta)} \end{cases} \tag{18}$$

to fix the boundary $r = b(\theta)$ on $\tilde{r} = b_0$, and then make the linearization of the problem $(NL)^{(a)}$ at the state $\psi = \psi_B(r)$ with boundary condition (6) on $r = b_0$ and $\psi = \psi_0$ on $r = r_1$:

$$A_{11}\delta\psi_{rr} + A_{22}\delta\psi_{\theta\theta} + B_1\delta\psi_r + C\delta\psi = f, \tag{19}$$

where

$$A_{11} = a^2(1 + r^2) - ((1 + r^2)\psi_r - r\psi)^2, \quad A_{22} = \frac{a^2}{r^2},$$

$$B_1 = \frac{a^2}{r} - (\gamma - 1)((1 + r^2)\psi_{rr} + \frac{\psi_r}{r})((1 + r^2)\psi_r - r\psi) - 2(1 + r^2)((1 + r^2)\psi_r - r\psi)\psi_{rr},$$

$$C = 2r((1 + r^2)\psi_r - r\psi)\psi_{rr} - ((1 + r^2)\psi_{rr} + \frac{1}{r}\psi_r)(\gamma - 1)(\psi - r\psi_r).$$

Correspondingly, the boundary conditions are

$$(1 + b_0^2)\delta\psi_r - b_0\delta\psi = g \quad \text{on} \quad r = b_0, \tag{20}$$

$$\delta\psi = h \quad \text{on} \quad r = r_1. \tag{21}$$

The problem (19)–(21) is denoted by $L^{(a)}$, for which the following estimate is valid.

Lemma 3.1. *There is $\delta > 0$, such that the solution of $L^{(a)}$ uniquely exists, and*

$$\|\delta\psi\|_{C^{2+\alpha}[b_0,r_1;0,2\pi]} \le C_1(\|f\|_{C^\alpha[b_0,r_1;0,2\pi]} + \|g\|_{C^{1+\alpha}(0,2\pi)} + \|h\|_{C^{2+\alpha}(0,2\pi)}), \tag{22}$$

$$\|\delta\psi\|_{C^{2+\alpha}[b_0,r_1-\frac{\delta}{4};0,2\pi]} \le C_2(\|f\|_{C^\alpha[b_0,r_1;0,2\pi]} + \|g\|_{C^{1+\alpha}(0,2\pi)} + \|h\|_{C^0(0,2\pi)}), \tag{23}$$

provided $|r_1 - b_0| < \delta$.

By using the implicit theorem in Banach spaces we can prove

Lemma 3.2. *Assume* $|r_1 - b_0| < \delta$, $\|b(\theta) - b_0\|_{C^{2+\alpha}(0,2\pi)} < \epsilon$, $\|d(\theta) - \psi_{10}\|_{C^{2+\alpha}(0,2\pi)}$
$< \epsilon$, *then the problem* $(NL)^{(a)}\{b(\theta), d(\theta)\}$ *has a unique solution* $\psi(r, \theta)$. *Moreover,*

$$\|\psi(r, \theta) - \psi_B(r)\|_{C^{2+\alpha}(\Omega_a)} \to 0 \quad when \quad \epsilon \to 0 \tag{24}$$

Since $r_1 - b_0$ is small, we can use the maximum principle, at most by adding a transformation $v = e^{k(r-b_0)^2} \delta\psi$ with large k, to prove that the solution of $(NL)^{(a)}$ has some monotonicity with respect to its boundary value. That is:

Lemma 3.3. *Assume that* δ, ϵ *are sufficiently small,* $|r_1 - b_0| < \delta$, $\|d_i(\theta) - \psi_{10}\|_{C^{2+\alpha}}$
$< \epsilon$ $(i = 1, 2)$, *and* $\psi_i(r, \theta)$ *is the solution of the problem* $(NL)^{(a)}\{b(\theta), d_i(\theta)\}$. *Then* $d_2 \geq d_1$ *implies* $\psi_2 \geq \psi_1$.

The main idea of the proof of this lemma is that the difference $\delta\psi = \psi_2 - \psi_1$ satisfies a linear boundary value problem with coefficients being the perturbation of the corresponding coefficients of (19)–(21). Therefore, when the domain is assumed to be small, the maximum principle leads to the required monotonicity.

Let $\tilde{b}_0$ be a constant slightly greater than b_0, $\tilde{\psi}_B(r)$ be the solution of the boundary value problem in the symmetric case with b_0 replaced by $\tilde{b}_0$. Then from [2] we know that $\tilde{\psi}_B(r)$ is well-defined on $(\tilde{b}_0, \tilde{s}_0)$ with $\tilde{s}_0 > s_0$, and $\tilde{\psi}_B(r) < \psi_B(r)$ holds on $\tilde{b}_0 \leq r \leq s_0$. Furthermore, if we extend $\tilde{\psi}_B(r)$ from $r > \tilde{b}_0$ to $(b_0, \tilde{b}_0)$ by using equation, then the relation $\tilde{\psi}_B(r) \leq \psi_B(r)$ still holds in $(b_0, \tilde{b}_0)$. For any non-symmetric curve $b(\theta)$, we have

Lemma 3.4. *Assume that* $b_0 \leq b(\theta) \leq \tilde{b}_0$, $\|b(\theta) - b_0\|_{C^{2+\alpha}(0,2\pi)} < \epsilon$, $|b_0 - \tilde{b}_0| < \epsilon$, $\psi(r, \theta)$ *is the solution of* $(NL)^{(a)}\{b(\theta), \psi_{10}\}$, $\tilde{\psi}_B(r)$ *is the solution of* $(NL)\{\tilde{b}_0\}$, *then* $\tilde{\psi}_B(r) \leq \psi(r, \theta) \leq \psi_B(r)$ *holds in* Ω_a. *Besides,*

$$\|\psi(r, \theta) - \psi_B(r)\|_{C^{2+\alpha}(\Omega_a^-)} \leq C(\|d(\theta) - \psi_{10}\|_{C(0,2\pi)} + \epsilon). \tag{25}$$

Now consider the nonlinear problem $(NL)^{(b)}$ in the outer ring Ω_b with fixed boundary $p = \psi_0$ and $p = \psi_{20}$. The inverse function of $\psi_B(r)$ is also called background solution for $(NL)^*$, and is denoted by $u_B(p)$. After multiplying (11) by u_p^3 we make the linearization for the problem $(NL)^{(b)}$ at the background solution $u = u_B(p)$, and obtain

$$[N_1^2 - (1 + u^2)a^2]\delta u_{pp} - \frac{u_p^2}{u^2}\delta u_{\sigma\sigma}$$

$$+ \left[\frac{2a^2 u_p}{u} - 2u_{pp}N_1\frac{1 + u^2}{u_p^2} + N_2\left(\frac{1 + u^2}{u_p^3} - \frac{up}{u_p^2}\right)\right]\delta u_p \tag{26}$$

$$+ \left[\frac{2N_1 u_{pp}(2u - pu_p)}{u_p} - \frac{a^2}{u^2}u_p^2 - 2a^2 u u_{pp} + N_2\left(\frac{p}{u_p} - \frac{u}{u_p^2}\right)\right]\delta u = f,$$

where $N_1 = (1 + u^2)/u_p - up$, $N_2 = (\gamma - 1)(u_p^2/u - (1 + u^2)u_{pp})$.

The boundary condition on $p = \psi_0$ is

$$\left[\left(1 - \tfrac{\rho_0}{\rho}\right)pu\frac{1 + u^2 - upu_p}{a^2}\left(\left(p - \frac{u}{u_p}\right)\frac{u}{u_p^2} - \frac{1}{u_p^3}\right)\right]\delta u_p$$

$$+\left[-2u + \left(1 - \tfrac{\rho_0}{\rho}\right)pu_p - \frac{1 + u^2 - upu_p}{a^2}\left(p - \frac{u}{u_p}\right)\frac{1}{u_p}\right]\delta u = g, \tag{27}$$

while the condition on $p = \psi_{20}$ is

$$\delta u = h. \tag{28}$$

Similar to the discussion on the problem $(NL)^{(a)}$ and $L^{(a)}$, we can establish the following propositions under the assumption that $|\psi_0 - \psi_{20}|$ is small.

Lemma 3.5. *The linearized problem (26)–(28) has a unique solution, which satisfies the estimates*

$$\|\delta u\|_{C^{2+\alpha}(\Omega_b)} \le C_1(\|f\|_{C^\alpha(\Omega_b)} + \|g\|_{C^{1+\alpha}(\Omega_b)} + \|h\|_{C^{2+\alpha}(0,2\pi)}), \tag{29}$$

$$\|\delta u\|_{C^{2+\alpha}(\Omega_b^-)} \le C_2(\|f\|_{C^\alpha(\Omega_b)} + \|g\|_{C^{1+\alpha}(\Omega_b)} + \|h\|_{C^0(0,2\pi)}), \tag{30}$$

where $\Omega_b^- = [\psi_{20} + \delta_1, \psi_0; 0, 2\pi]$, and δ_1 is a small number satisfying $0 < \delta_1 < \delta$.

Lemma 3.6. *Assume that $\|h(\sigma) - r_2\|_{C^{2+\alpha}(0,2\pi)} < \epsilon$, r_2 satisfies $b_0 < r_2 < r_1$, then $(NL)^{(b)}\{h(\sigma)\}$ has a unique solution. Moreover,*

$$\|u(p,\sigma) - u_B(p)\|_{C^{2+\alpha}(\Omega_b)} \to 0 \quad \text{when} \quad \epsilon \to 0. \tag{31}$$

Lemma 3.7. *Assume that r_1, ψ_{20} are given as above. For $i = 1, 2$, $\|h_i(\sigma) - r_2\|_{C^{2+\alpha}}$ $< \epsilon_0$, $h_2(\sigma) \ge h_1(\sigma)$, and u_i is the solution of $(NL)^{(b)}\{h_i(\sigma)\}$, then*

$$u_2(\psi_{10}, \sigma) \ge u_1(\psi_{10}, \sigma). \tag{32}$$

Besides,

$$\|u_i(p,\sigma) - u_B(p)\|_{C^{2+\alpha}(\Omega_b^-)} \le C_2\|h_i(\sigma) - r_2\|_{C^0(0,2\pi)}. \tag{33}$$

4. Solution for nonlinear problems (NL) and $(NL)^*$

Based on the discussion of $(NL)^{(a)}$ and $(NL)^{(b)}$, we are able to construct the solution of (NL) or $(NL)^*$ now. After making a first choice of an approximate solution we improve the degree of approximation alternatively by solving $(NL)^{(a)}$ in the inner ring Ω_a and $(NL)^{(b)}$ in the outer ring Ω_b. Namely, we will establish a sequence $\{\psi^{(n)}\}$ of solutions of $(NL)^{(a)}$ in Ω_a and a sequence $\{u^{(n)}\}$ of solutions of $(NL)^{(b)}$ in Ω_b alternatively, and then prove that these sequences are convergent and their limits are the inverse of each other.

In the sequel we also use the following notations. If $p = \psi(r, \theta)$, then r as a function of p, θ is denoted by $\psi^{-1}(p, \theta)$, and if $r = u(p, \sigma)$, then p as a function of

r, θ is denoted by $u^{-1}(r, \sigma)$. The sequences $\{\psi^{(n)}(r, \theta)\}$ and $\{u^{(n)}(p, \sigma)\}$ are established as follows. $\psi^{(1)}(r, \theta)$ is the solution of the problem $(NL)^{(a)}\{b(\theta), \psi_B(r_1))\}$, then inductively, $u^{(n)}(p, \sigma)$ is the solution of problem $(NL)^{(b)}\{(\psi^{(n)})^{-1}(\psi_{20}, \sigma)\}$, and $\psi^{(n+1)}(r, \theta)$ is the solution of problem $(NL)^{(a)}\{b(\theta), (u^{(n)})^{-1}(r_1, \theta)\}$.

Lemma 4.1. *If* $\|b(\theta) - b_0\|_{C^{2+\alpha}} < \epsilon$ *with* ϵ *being sufficiently small, then the sequences* $\{\psi^{(n)}\}, \{u^{(n)}\}$ *established as above are well-defined and satisfy*

$$\tilde{\psi}_B(r) \leq \psi^{(n)}(r, \theta) \leq \psi_B(r) \quad in \quad \Omega_a^-$$
$$\|\psi^{(n)}(r, \theta) - \psi_B(r)\|_{C^{2+\alpha}(\Omega_a^-)} \leq C\epsilon \tag{34}$$

$$\tilde{u}_B(p) \geq u^{(n)}(p, \sigma) \geq u_B(p) \quad in \quad \Omega_b^-$$
$$\|u^{(n)}(p, \sigma) - u_B(p)\|_{C^{2+\alpha}(\Omega_b^-)} \leq C\epsilon \tag{35}$$

where Ω_a^- *is the domain* $b(\theta) < r < r_1 - \delta_1$, $0 \leq \theta \leq 2\pi$, *and* Ω_b^- *is the domain* $\psi_{20} + \delta_1 < p < \psi_0, 0 \leq \sigma \leq 2\pi$. *Moreover,* $\{\psi^{(n)}\}$ *is monotonically decreasing with respect to* n, *and* $\{u^{(n)}\}$ *is monotonically increasing with respect to* n.

Proof. Lemma 3.4 indicates that the solution $\psi^{(1)}(r, \theta)$ of the nonlinear problem $(NL)^{(a)}\{b(\theta), \psi_{10}\}$ exists and satisfies $\tilde{\psi}_B \leq \psi^{(1)} \leq \psi_B$. Therefore, under the assumptions of the lemma we have $\|\tilde{\psi}_B - \psi_B\|_{C^0} < \epsilon$, and then

$$\|\psi^{(1)} - \psi_B\|_{C^0(\Omega_a)} \leq C_3\epsilon, \tag{36}$$

$$\|\psi^{(1)} - \psi_B\|_{C^{2+\alpha}(\Omega_a)} \leq C_4\epsilon. \tag{37}$$

Because $\psi^{(1)}$ is a small perturbation of ψ_B, then $\psi_r^{(1)} > k > 0$, where k is independent of the constant C_1, C_2 given in Lemma 3.1 and 3.5. Hence $(\psi^{(1)})^{-1}$ is well-defined and $(\psi^{(1)})^{-1} \geq (\psi_B)^{-1} = u_B$. Moreover, from (36), (37) we have

$$\|(\psi^{(1)})^{-1}(\psi_{20}, \sigma) - r_2\|_{C^0(0, 2\pi)} < \frac{C_3\epsilon}{k}. \tag{38}$$

Taking $(\psi^{(1)})^{-1}(\psi_{20}, \sigma)$ as the data of the nonlinear problem $(NL)^{(b)}$, we obtain the solution $u^{(1)}(p, \sigma)$ of $(NL)^{(b)}\{(\psi^{(1)})^{-1}(\psi_{20}, \sigma)\}$ by using Lemma 3.6. Besides, Lemma 3.5, 3.7 also indicate the estimate

$$\|u^{(1)} - u_B\|_{C^{2+\alpha}(\Omega_b^-)} \leq C_2\|(\psi^{(1)})^{-1}(\psi_{20}, \sigma) - r_2\|_{C^0(0, 2\pi)} \leq C_2 C_3 \frac{\epsilon}{k}, \tag{39}$$

$$\tilde{u}_B(p) \geq u^{(1)}(p, \sigma) \geq u_B(p) \quad in \quad O_{\delta_1}(\psi_{20}) : \psi_{20} \leq p \leq \psi_{20} + \delta_1, \tag{40}$$

where the domain $O_{\delta_1}(\psi_{20})$ covers the lines $p = \psi_B(r_1)$ and $p = \tilde{\psi}_B(r_1)$ (equivalently, the lines $u_B(p) = r_1$ and $\tilde{u}_B(p) = r_1$). Then from $u_p^{(1)} > 0$ we know $(u^{(1)})^{-1} \leq \psi_B$ on $r = r_1$ and

$$\|(u^{(1)})^{-1}(r_1, \theta) - \psi_B(r_1)\|_{C^0(0, 2\pi)} \leq |\tilde{\psi}_B(r_1) - \psi_{10}| < C_3\epsilon. \tag{41}$$

Furthermore, we will solve the problem $NL^{(a)}\{b(\theta), (u^{(1)})^{-1}(r_1, \theta)\}$ by using Lemma 3.2 to obtain the solution $\psi^{(2)}(r, \theta)$, which satisfies $\psi^{(2)} \leq \psi^{(1)}$ according to Lemma 3.3 and satisfies $\psi^{(2)} \geq \tilde{\psi}_B$ according to Lemma 3.4. Therefore, we have

$$\|\psi^{(2)}(r_2, \theta) - \psi_B(r_2)\|_{C^0(0,2\pi)} < C_3\epsilon \tag{42}$$

$$\begin{aligned}
\|\psi^{(2)}(r, \theta) - \psi_B(r)\|_{C^{2+\alpha}(\Omega_a^-)} &\leq C_2\|(u^{(1)})^{-1}(r_1, \theta) - \psi_B(r_1)\|_{C^0(0,2\pi)} \\
&\leq C_2 C_3 \epsilon \,.
\end{aligned} \tag{43}$$

By the same procedure, we obtain $u^{(2)}(p, \sigma)$, which satisfies

$$\tilde{u}_B(p) \geq u^{(2)}(p, \sigma) \geq u_B(p), \quad \text{in} \quad O_{\delta_1}(\psi_{20}) \tag{44}$$

$$\|u^{(2)} - u_B\|_{C^{2+\alpha}(\Omega_b^-)} \leq C_2 C_3 \frac{\epsilon}{k}, \tag{45}$$

$$\|(u^{(2)})^{-1}(r_1, \theta) - \psi_B(r_1)\|_{C^0(0,2\pi)} < |\tilde{\psi}_B(r_1) - \psi_{10}| < C_3\epsilon, \tag{46}$$

and so on. Inductively we establish (34) and (35).

To prove the monotonicity of the sequences $\{\psi^{(n)}\}$ and $\{u^{(n)}\}$, we start with $\psi^{(1)}(r, \theta) \leq \psi_B(r)$ and $u^{(1)}(p, \sigma) \geq u_B(p)$ in $O_{\delta_1}(\psi_{20})$. Denote $u^{(0)}(p, \sigma) = u_B(p)$, then we verify the inequality

$$(u^{(n)})^{-1}(r_1, \theta) \leq (u^{(n-1)})^{-1}(r_1, \theta) \tag{47}$$

inductively. The validity of (47) for $n = 1$ comes from the construction of $u^{(1)}(p, \theta)$. Now assume that (47) holds for general n, then by using Lemma 4.3 we have $\psi^{(n+1)}(r, \theta) \leq \psi^{(n)}(r, \theta)$. Notice that the derivatives of $\psi^{(n)}$ and $\psi^{(n+1)}$ are positive, then

$$(\psi^{(n+1)})^{-1}(p, \theta) \leq (\psi^{(n)})^{-1}(p, \theta). \tag{48}$$

The inequality (48) on $p = \psi_{20}$ coincides with the requirement of Lemma 3.7, which implies that the solutions of $(NL)^{(b)}$ satisfy

$$u^{(n+1)}(p, \sigma) \geq u^{(n)}(p, \sigma) \quad \text{in} \quad O_{\delta_1}(\psi_{20}). \tag{49}$$

Therefore, we come back to (47) with n replaced by $n+1$. Hence the monotonicity of $\{\psi^{(n)}\}$ and $\{u^{(n)}\}$ is proved by induction.   □

Finally, by passing to the limit we can establish the following lemma, which implies the conclusion of Theorem 1 directly.

Lemma 4.2. *If $\|b(\theta) - b_0\|_{C^{2+\alpha}} \leq \epsilon$ with ϵ being sufficiently small, then the problems* (NL) *and* $(NL)^*$ *are solvable.*

Acknowledgement. This work is partly supported by the NNSF and the Doctoral Programme Foundation of IHE of China. Besides, the author is also grateful to Dening Li for his valuable discussions.

References

[1] S.X. Chen, *Existence of local solutions to supersonic flow past a three-dimensional wing*, Advances in Appl. Math., **13** (1992), 273–304.

[2] R. Courant and K.O. Friedrichs, *Supersonic flow and shock waves*, (1948), Springer-Verlag, New York.

[3] C.H. Gu, *A method for solving the supersonic flow past a curved wedge*, Fudan J. (Natur. Sci.), **7** (1962), 11–14.

[4] T.T. Li, *Une remarque sur un problème à frontière libre*, C. R. Acad. Sc. Paris, Ser. A, **289** (1979), 99–102.

[5] T.T. Li, *On a free boundary problem*, Chinese Ann. Math., **1** (1980), 351–358.

[6] T.T. Li and W.C. Yu, *Boundary value problem for quasi-linear hyperbolic systems*, Duke Univ. Math., **5** (1985).

[7] A. Majda, *One perspective on open problems in multi-dimensional conservation laws*, Multidimensional Hyperbolic Problems and Computations, Edited by J. Glimm and A. Majda, Springer-Verlag, IMA **29** (1991), 217–237.

[8] C. Miranda *Equazioni alle derivate parziali di tipo ellittico*, (1955) Springer, Berlin.

[9] C.S. Morawetz, *Potential theory for regular and Mach reflection of a shock at a wedge*, Comm. Pure Appl. Math., **47** (1994), 593–624.

[10] A. Majda and E. Thomann, *Multi-dimensional shock fronts for second order wave equations*, Comm. P.D.E., **12** (1987), 777–828.

[11] D.G. Schaeffer, *Supersonic flow past a nearly straight wedge*, Duke Math. J., **43** (1976) 637–670.

[12] D.G. Schaeffer, *An application of the Nash-Moser theorem to a free boundary problem*, Lecture Notes in Math., **648** (1978), 129–143.

Institute of Mathematics,
Fudan University ,
Shanghai 200433, China
E-mail address: sxchen@fudan.ac.cn

International Series of Numerical Mathematics
Vol. 129, © 1999 Birkhäuser Verlag Basel/Switzerland

On the Preconditioning of Finite Volume Schemes

Sébastien Clerc

Abstract. Preconditioning is known to enhance the accuracy of finite volume schemes for steady-state subsonic flows. The aim of this paper is to precise the benefits obtained by preconditioning, namely a precision which is independent of the Mach number, a reduced grid sensitivity, and an improved processing of the source terms. Time accurate transient computations are also possible thanks to an implicit time-stepping.

1. Introduction

Some problems of computational fluid dynamics are characterized by a low Mach number but important compressibility effects. This is the case, for instance, in combustion and two-phase flow problems. Unfortunately, finite volume schemes designed for transonic and supersonic applications become inefficient in low Mach number regimes. The precision of these schemes deteriorates when the Mach number goes to zero. This fact was investigated by [12], and accounted for by [9].

The so-called preconditioning technique can restore a Mach number independent convergence. This fact was noted by many authors, see e.g. [10, 9, 11]. Another benefit of preconditioning is the acceleration of explicit time-marching schemes for the solution of steady-state subsonic computations. However, the drawback of this approach is the loss of time-consistency. Recently, Viozat [11] proposed an alternative solution based on a consistent implicit time-stepping which allows transient computations as well.

In this paper, we illustrate other benefits of preconditioning. Usual finite volume schemes tend to create discontinuities in the velocity and pressure field aligned with grid lines on cartesian meshes. This problem becomes severe at low Mach numbers. Noticeable improvements are obtained with preconditioning even at moderately low Mach numbers. It can thus be said that the preconditioned scheme exhibits genuinely multi-dimensional features.

Further, source term processing can be enhanced by preconditioning when used in conjunction with source term upwinding.

For both aspects, we give numerical examples with Saint-Venant's equations of shallow water flow. The effect of the topography is modeled by a source term in the momentum equation.

In Section 1, we present the preconditioning technique applied to finite volume schemes. In Section 2, we give numerical examples. First, the propagation of a two-dimensional acoustic wave in a subsonic flow is computed. Second, a steady-state shallow water flow over a circular bump is shown. We draw some conclusions for this study in Section 3.

2. Preconditioning

Historically, preconditioning was introduced for the computation of steady-state solutions of systems of conservation laws of the form:

$$\partial_t U + \nabla \cdot \mathbf{F}(U) = 0. \tag{1}$$

The idea consists in using a time-marching method for a modified system:

$$\mathbf{P}^{-1} \partial_t U + \nabla \cdot \mathbf{F}(U) = 0, \tag{2}$$

where $\mathbf{P}$ is a non-singular matrix, possibly depending on U (see e.g. [9, 2]). Note that the steady-state solution is not affected by the procedure.

The numerical flux should now be based on one-dimensional Riemann problems for system (2) rather than the original system. To make this assertion more precise, we have to introduce some notations. Consider a cell K of a finite volume mesh, and U_K^n the vector of the unknowns in this cell at time t^n. A standard finite volume scheme for (1) would write:

$$|K|(U_K^{n+1} - U_K^n) + \sum_{e \in \partial K} |e_{KJ}| \Phi(n_{KJ}, U_K, U_J) = 0. \tag{3}$$

The normal vector to an edge e_{KJ}, oriented from K to J is naturally n_{KJ}. The volume of the cell K is denoted by $|K|$ while $|e_{KJ}|$ denotes the length of the edge. We will now use Roe's flux:

$$\Phi(n, U_L, U_R) = \frac{1}{2}\left(\mathbf{F}_n(U_L) + \mathbf{F}_n(U_R)\right) - \frac{1}{2}|\mathbf{A}_n|(U_R - U_L). \tag{4}$$

Here $\mathbf{A}_n$ is the average Jacobian of $\mathbf{F}_n$, and $|\mathbf{A}_n| = \sum_i |\lambda_i| r_i \otimes l_i$ if $\mathbf{A}_n = \sum_i \lambda_i r_i \otimes l_i$ is the eigen-decomposition of $\mathbf{A}_n$. For simplicity, we will now drop the subscript n.

These notations being introduced, we can now express the preconditioned finite volume scheme:

$$|K|\mathbf{P}^{-1}(U_K^{n+1} - U_K^n) + \sum_{e \in \partial K} |e_{KJ}| \Phi_P(n_{KJ}, U_K, U_J) = 0, \tag{5}$$

with the modified numerical flux:

$$\Phi_P(n, U_L, U_R) = \frac{1}{2}\left(\mathbf{F}_n(U_L) + \mathbf{F}_n(U_R)\right) - \frac{1}{2}\mathbf{P}^{-1}|\mathbf{PA}_n|(U_R - U_L). \tag{6}$$

2.1. Time-consistent preconditioning

Time consistency can be recovered by using the original finite volume scheme (1) with the modified numerical flux (6): see [11]. Note that at steady-state, the time-consistent and the pseudo-transient approaches coincide.

It turns out that the resulting explicit scheme is not stable under the usual CFL condition. Indeed, the viscosity matrix of the numerical scheme is $\mathbf{P}^{-1}|\mathbf{PA}_n|$: the spectral radius of this matrix can be different from that of $|\mathbf{A}_n|$. More precisely, it can be proved that the spectral radius must be of order $1/M^2$ to ensure an accurate numerical scheme at low Mach number (see [8]). For a usual finite volume scheme, stable under the CFL condition, the spectral radius is only of order $1/M$. The time step should therefore be much smaller than required by the CFL condition. However, this restriction disappears if an implicit time-stepping is used.

In this work, we have used the linearly implicit version of Roe's scheme, in which the Roe matrices at each interface are frozen. The solution of the resulting linear system is obtained with GMRES and a block-Jacobi preconditioner.

2.2. Source term upwinding

Source term upwinding is known to increase precision and stability in simulations of systems of balance laws with source terms (see for instance [7]). This treatment is natural for the source terms which involve the space derivative of a given scalar, with the general form $g(U)\nabla s$. Classical examples of such source terms are those which account for a geometrical simplification (symmetry, quasi 1-D nozzle, shallow water approximation with topography) and the gravity. For a presentation of the method in the case of the topography source term in Saint-Venant's equations, see [1]. Recently, Jenny [5] extended this technique to more general source terms on a cartesian mesh.

Another way to look at this technique is to consider that the scalar quantity s is in fact discretized in a piece-wise constant fashion. The resulting problem is therefore non-conservative, since both $g(U)$ and s can be discontinuous at an interface. It is however possible to give a mathematical sense to the Riemann problem (in the non-resonant case): see [3, 4].

Practically, the jump of the source term is computed at each interface. If e_{KJ} is the interface between the cells K and J, an average value $\tilde{g}$ of the source $g(U)$ is computed, say for instance $\frac{1}{2}(g(U_K) + g(U_J))$. The jump of the source term is thus $\tilde{g}(s_J - s_K)$. This vector is then split according to the characteristic velocities. The projectors on the left and right going characteristic variables are denoted as Π^- and Π^+. If A_n satisfies $A_n = R^{-1}\Lambda R$, the projectors are given by $\Pi^{\pm} = R^{-1}H(\pm\Lambda)R$, where H is the Heaviside function ($H(\lambda) = 1$ where $\lambda > 0$ and 0 elsewhere). With this notation, the contribution of the source term to the cell K is:

$$\Pi^-\tilde{g}(s_J - s_K),$$

and, naturally, the contribution to the cell J is:

$$\Pi^+ \tilde{g}(s_J - s_K).$$

This methodology can be adapted to retain the advantages of precondition-ing. If a preconditioner $\mathbf{P}$ is used, we still denote by $\Pi^\pm$ the projectors on the left and right going characteristic variables of $\mathbf{PA}$. We then apply the same formulae for the splitting of the source term.

3. Numerical results

3.1. Equations of the model

For simplicity, we give numerical results for Saint-Venant's equations of shallow water flow; preconditioning can naturally be used for the Euler equations of fluid dynamics. Saint-Venant's system reads:

$$\partial_t h + \mathrm{div}(h\mathbf{u}) = 0$$

$$\partial_t h\mathbf{u} + \mathrm{div}(h\mathbf{u} \otimes \mathbf{u}) + \nabla(\frac{1}{2}h^2) = -h\nabla q.$$

Here $q(x,y)$ denotes the (given) topography, h is the water height, $h + q$ is thus the equation of the free surface.

We now give numerical results for Roe's scheme without preconditioning and with the Van-Leer-Lee-Roe (VLLR) preconditioning [10].

3.2. Acoustic wave

The initial data for this test case are specified in table 1. The solution consists of a circular acoustic wave growing in time and slowly advected with the flow.

We perform five implicit time steps with $\delta t/\delta x = 1$. A regular 60×60 mesh is used. The numerical results for Roe's scheme are shown in Fig. 1 & 2. For each scheme, we present the density field and the Mach number (in this case the Froude number) $M = |u|/c$. The solution computed by the Roe scheme presents a squarish wave front with discontinuities aligned with the grid lines. The Mach number field shows an important discontinuity which is aligned with the x grid line originating from the initial perturbation. This discontinuity is much reduced with the preconditioned scheme.

The extension of the wave front computed by the preconditioned scheme is correct: this shows that the time-stepping is consistent. Note however that pre-conditioning accelerates the damping of the density variations.

3.3. Circular bump

For this test case, we seek a steady shallow water flow over a circular bump. The topography is given by:

$$q(x,y) = \max\{0.24 - 12[(x - 1/2)^2 + (y - 1/2)^2], 0\}.$$

We still use a 60×60 mesh. On the boundaries, we impose the values of the incoming characteristics computed for a constant state. Here again, the reference

h : – at point $(1/2, 1/2)$	1.1
– elsewhere	1
u	0.1
v	0

TABLE 1. Initial data for the subsonic acoustic wave.

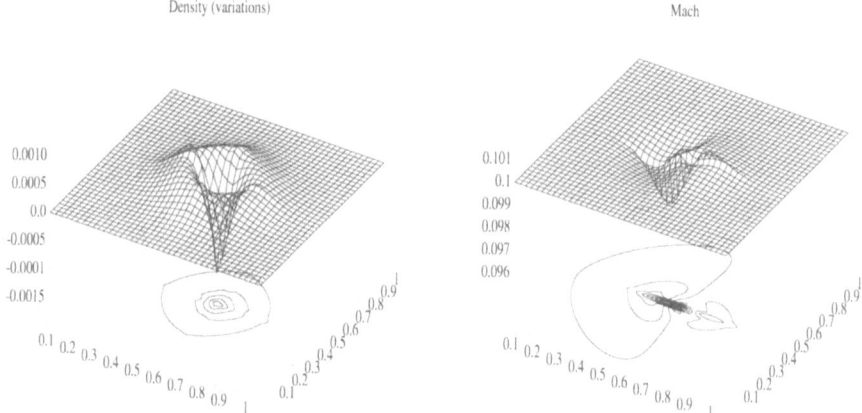

FIGURE 1. Acoustic wave, Roe's scheme: variations of ρ and Mach number. Notice the strong discontinuities in the Mach number field along the grid lines.

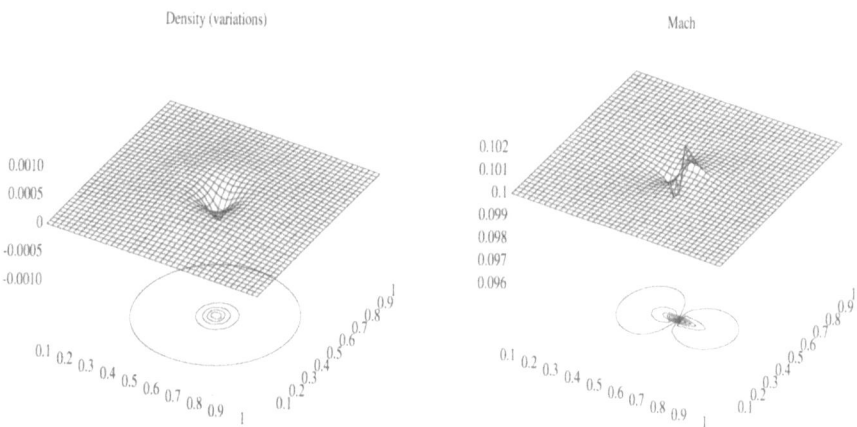

FIGURE 2. Acoustic wave, VLLR preconditioning: variations of ρ and Mach number. The symmetry of the problem is respected.

Total charge

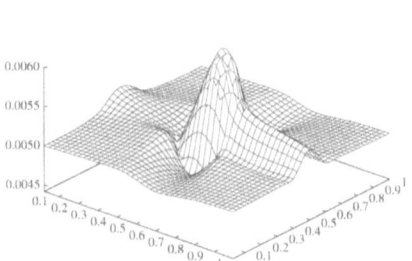

FIGURE 3. Shallow water flow over circular bump, Roe's scheme: total charge. The total charge should remain constant as a consequence of Bernoulli's theorem.

Total charge

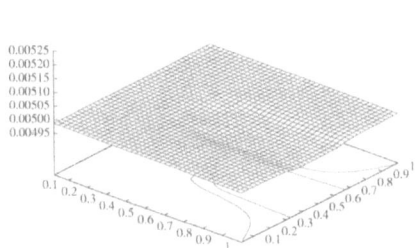

FIGURE 4. Shallow water flow over a circular bump, VLLR preconditioning: total charge. The total charge is almost constant. Notice the blow up on the z axis.

Mach number is 0.1. Since the flow is irrotational, the total charge $h + q + |u|^2/2$ should be constant in the domain. This property is equivalent to the conservation of the total enthalpy for the full Euler equations.

The solution computed with Roe's scheme shows important variations in the total charge, see Fig. 3. On the contrary, the preconditioned scheme preserves an almost constant total charge (Fig. 4).

4. Conclusion

This study shows that preconditioning can improve the results of usual finite volume schemes in the subsonic regime, both for transient and steady-state computations. The improvement is particularly obvious at low Mach number. Among the benefits of preconditioning is a reduced grid sensitivity, and a better treatment of the source terms.

References

[1] A. Bermudez, E. Vazquez, *Upwind methods for hyperbolic conservation laws with source terms*, Computers Fluids, **23 (8)** (1994), 1049–1071.

[2] Y.H.Choi and C.L. Merkle, *The application of preconditioning in viscous flows*, JCP, **105** (1993), 207–223.

[3] S. Clerc, *Ph. D. thesis*, University Paris VI, unpublished (1997).

[4] L. Gosse, *Ph.D. thesis*, University Paris IX, unpublished (1997).

[5] P. Jenny, *Riemann solvers for compressible combustion codes*, in Proceedings of the 1st Int. Symposium on Finite Volumes for Complex Appl., Rouen, France, **1996**.

[6] P.L. Roe and L. Mesaros, *Solving steady mixed conservation laws by elliptic-hyperbolic splitting*, in Proceedings of the 15th Int. Conf. Num. Meth. Fluid Dyn., Lecture Notes in Physics, Springer, **1996**.

[7] P.L. Roe, *Upwind differencing schemes for hyperbolic conservation laws with source terms*, in Proceedings of Nonlinear Hyperbolic Problems, Lecture Notes in Mathematics, **1270**, Springer, (1986).

[8] E. Turkel, *Preconditioned methods for solving the incompressible and low speed compressible equations*, J.C.P., **72** (1987), 277–298.

[9] E. Turkel, *Review of preconditioning methods for fluid dynamics*, Appl. Num. Math., **12** (1993), 257–284.

[10] B. Van Leer, W.-T. Lee and P.L. Roe, *Characteristic time-stepping or local preconditioning of the Euler equations*, A.I.A.A. paper 91-1552, (1991).

[11] C. Viozat, *Implicit upwind schemes for low Mach number compressible flows*, INRIA report 3084, (1997).

[12] G. Volpe, *Performance of compressible flow codes at low Mach number*, AIAA Journal, **31** (1974), 49–56.

SERMA/LETR,
Commissariat à l'Energie Atomique,
CEA SACLAY
91191 GIF-SUR-YVETTE, FRANCE
E-mail address: Sebastien.Clerc@cea.fr

International Series of Numerical Mathematics
Vol. 129, © 1999 Birkhäuser Verlag Basel/Switzerland

A Priori Error Estimates for Nonlinear Scalar Conservation Laws

Bernardo Cockburn, Pierre-Alain Gremaud, and Jimmy X. Yang

Abstract. In this note, we review the recent work of the authors on *a priori* error estimates for nonlinear scalar conservation laws. *A priori* error estimates are important because they shed light into the nature of the corresponding numerical scheme. In this note, we show how to use our *a priori* error estimation technique to study multidimensional monotone schemes defined in non-Cartesian grids. For the so-called *consistent* schemes, we prove that the $L^\infty(0,T;L^1(R^d))$-error goes to zero as $(\Delta x)^{1/2}$ as the discretization parameter Δx goes to zero; all previous results give a rate of convergence of only $(\Delta x)^{1/4}$. The loss of $1/4$ in the exponent is due to the fact that it is not known how to prove that the total variation of the approximate solution is uniformly bounded. We show that such estimate is not necessary to obtain optimal error estimates; only the *structure* of the numerical scheme is important.

1. Introduction

In this note, we discuss the recent work of the authors, [5], [6], and [7], on *a priori* error estimates for the nonlinear scalar conservation law

$$v_t + \nabla \cdot f(v) = 0, \quad \text{in } (0,T) \times \mathbb{R}^d, \tag{1}$$

$$v(0) = v_0, \quad \text{on } \mathbb{R}^d; \tag{2}$$

see [15] for the existence, uniqueness, and regularity properties of the *entropy* solution v. The *a priori* estimates obtained in the above mentioned papers are the first estimates for nonlinear conservation laws written solely in terms of the exact solution, i.e., they are the first estimates of the form

$$e(T) \le \Phi(v,T),$$

where $e(t)$ denotes the L^1-error at time t. *All* the remaining rigorous error estimates for nonlinear conservation laws, e.g., [16], [26], [17], [18], [19],[20], [2], [31], [22], [12], [23], [3], [33], [28], [4], [24], [29], [30],[27], [10], are estimates that depend also on the approximate solution u, i.e., they are of the form

$$e(T) \le \Psi(v,u,T).$$

If Ψ does not depend on v, the above estimate is called an *a posteriori* error estimate. *A posteriori* estimates are very useful for practical computations if they are properly used in the frame of an adaptive strategy; strangely enough, although estimates of this form have been available for a long time, they have not been used for computational purposes until very recently; see, however, [17]. *A priori* error estimates that only depend on the exact solution, on the other hand, are important because they shed light on the nature of the numerical scheme; indeed, the *a priori* error estimates we discuss in this note can be considered to be the rigorous version of the formal local truncation error analysis which is a very popular tool for understanding the nature of the numerical schemes.

The work done [5], [6], and [7] develops a general theory of *a priori* error estimation for scalar nonlinear conservation laws and contains applications to the study of flux-splitting *monotone* schemes. The general theory was developed in [5] where it was applied to obtain optimal *a priori* error estimates for the Engquist-Osher scheme [9], perhaps the simplest flux-splitting monotone scheme, on one-dimensional uniform grids. It was shown that the $L^\infty(0,T;L^1(R^d))$-error goes to zero as $(\Delta x)^{1/2}$ which is the optimal rate of convergence for initial data v_0 in $BV(\mathbb{R})$; see [27]. No regularity property of the approximate solution was used, only its special *structure* of the numerical scheme was exploited.

In [6], the issue of the impact of non-uniform grids on the rate of convergence was studied. For a long time, see, for example, [13], [25], and [32], it has been known that if the numerical fluxes do not properly take into account the irregularity of the grid a *loss of consistency* is generated; however, this lack of consistency does *not* degrade the order of convergence of the numerical scheme. This *supra-convergence* phenomenon remained unexplained for initial data in $BV(\mathbb{R})$ until the publication of [6]; see also [34], [35], and [36], where more regular solutions are considered. In [6], the *a priori* error estimate approach was used to devise flux-splitting monotone schemes that are unaffected by the irregularities of the grids. For those schemes, called *consistent*, optimal *a priori* error estimates where obtained. For *non-consistent* schemes, it was shown that their lack of consistency can be *controlled* by the regularity of the approximate solution thanks to the fact that (i) the scheme is conservative and that (ii) its numerical fluxes are consistent; the nonlinear nature of the equations does not play any role in this mechanism. That was the first explanation of this *supraconvergence* phenomenon for nonlinear equations with low regularity. Supraconvergence for other problems has been analyzed, for example, in [21], [14], [11].

Finally, in [7], the results obtained in [6] were extended to the multidimensional case for flux-splitting monotone schemes of the form

$$u_K^{n+1} = u_K^n - \frac{\Delta t}{|K|} \sum_{e \in \partial K} |e| f_{e,K}(u_K^n, u_{K_e}^n), \qquad (3)$$

where K denotes a finite volume, e denotes a face of K, and K_e represents the finite volume sharing the face e with K. Again, it was shown that for *consistent* schemes, the optimal rate of convergence of $(\Delta x)^{1/2}$ is achieved in the $L^\infty(0, t; L^1(R^d))$-norm. All previous results, [3], [33], [4], and [24], give a rate of convergence of only $(\Delta x)^{1/4}$- only $(\Delta x)^{1/8}$ for the streamline-diffusion method [7]. The loss of $1/4$ (of $3/8$ for the streamline-diffusion method) in the exponent of Δx is due to the fact that in all the above mentioned papers, a bound of the total variation of the approximate solution is needed. This bound was obtained for uniform Cartesian grids [16] and later for non-uniform Cartesian grids [26]. However, when the grid is non-Cartesian, it is not known how to obtain a uniform bound for the total variation of the approximate solution even for simple Lax-Friedrichs-like schemes (3) in equilateral triangles; see [7]. The advantage of the approach used in [7] is that no estimate of the total variation is required to obtain optimal error estimates for the so-called *consistent* schemes. Again, only the *structure* of the scheme is used.

In [8], the work done in [5], [6], and [7] was discussed; the emphasis was on (i) the notion of *consistency* of the numerical scheme that stemmed from the analysis and on (ii) the explanation of the *supraconvergence* phenomenon. In this note, the emphasis will be in the description of our technique. We describe this technique, which is a modification of the Kuznetsov approximation theory [16], in Section 2. In Section 3, we use it to obtain optimal *a priori* error estimates for *consistent* monotone schemes (3) in multidimensions. We end in Section 4 with some concluding remarks.

2. The technique

2.1. Kuznetsov's approximation theory

To describe our approach to *a priori* error estimation, we need to describe the approximation theory of Kuznetsov [16]. This theory is written in terms of a form that we will denote by $E(w_1, w_2; T)$. This form, sometimes called the Kružkov form, is constructed in such a way that $E(w_1, w_2; T) \leq 0$ for all functions w_2 and for all times T if the function w_1 is the *entropy* solution of (1). It is well known [15] that v is the entropy solution of (1) if the following inequality, called the *entropy inequality*, is satisfied in the sense of distributions for any real constant c:

$$|v - c|_t + F(v, c)_x \leq 0,$$

where $F(v, c) = (f(v) - f(c)) \, sign(v - c)$. If we set $c = u(x', t')$, 'multiply' the above inequality by a nonnegative test function φ, and integrate with respect to variables $(x, t) \times (x', t')$ on $\mathbb{Q} \equiv \mathbb{R}^d \times (0, T) \times \mathbb{R}^d \times (0, T)$, we obtain the form $E(v, u; T)$. In this way we have, by construction, that $E(v, u; T) \leq 0$ for any function u and every time T, as wanted.

Note that in the above construction, v is always evaluated at (x, t) and u at (x', t'). It is thus impossible to use $E(v, u; T)$ to compare v and u at the same points unless φ is a smooth approximation of the Dirac with support on $\{t = t'\} \times \{x = x'\}$. Hence, Kuznetsov picked $\varphi = \omega_{\epsilon_t}(t - t') \, \omega_{\epsilon_x}(x_1 - x_1') \ldots \omega_{\epsilon_x}(x_d - x_d')$ where

$\omega_\eta(a) = \omega(a/\eta)/\eta$ and ω is a nonnegative, even function of integral equal to one and support equal to $[-1, 1]$.

The following result is a slight variation of a result by Kuznetsov [16]. In it, the integral from 0 to t of ω_{ϵ_t} is denoted by $W(t)$.

Theorem 2.1. [16]. *We have:*

$$e(T) \le \ e(0) + \left(\epsilon_x + \|f'(v)\| \epsilon_t\right) |v_0|_{TV(\mathbb{R}^d)}$$
$$+\left(\epsilon_x \|\nabla_{x'} u\|_{L^\infty(0,T;L^1(\mathbb{R}^d))} + \epsilon_t \|u_{t'}\|_{L^\infty(0,T;L^1(\mathbb{R}^d))}\right)$$
$$+E(u,v;T)/2\,W(T).$$

Kuznetsov stated this result for $T \ge \epsilon_t$, case in which $2\,W(T) = 1$.

Note that if u is an entropy solution, then $E(u,v;T) \le 0$ and we can take set $\epsilon_t = \epsilon_x = 0$ to obtain the well-known L^1-contraction property of entropy solutions:

$$e(T) \le e(0).$$

Let us see what error estimate does this theorem give us for the model case in which the approximate solution u is the solution of the parabolic regularization:

$$u_{t'} + \nabla_{x'} \cdot f(u) - \nu \Delta_{x'} u = 0, \quad \text{in } (0,T) \times \mathbb{R}^d. \tag{4}$$

In this case, we have:

$$E(u,v;T) = \int_Q \left(|u-v|_{t'} + F(u,v)_{x'}\right) \varphi\, dx\, dt\, dx'\, dt',$$

where $u = u(x',t')$, $v = v(x,t)$ and $\varphi = \varphi(t-t', x-x')$. After a simple algebraic manipulation, we get

$$E(u,v;T) \le \nu \int_Q \varphi \Delta_x |u-v|\, dx\, dt\, dx'\, dt',$$

and, after a simple integration by parts,

$$E(u,v;T) \le -\nu \int_Q \nabla_{x'} |u-v| \cdot \nabla_{x'} \varphi\, dx\, dt\, dx'\, dt' \tag{5}$$

Kuznetsov stopped here and estimated the right hand in terms of u as follows:

$$E(u,v;T) \le \nu \frac{2C}{\epsilon_x}\, W(T)\, \|u\|_{L^1(0,T;TV(\mathbb{R}^d))}, \tag{6}$$

where the constant C is equal to the total variation of the auxiliary function ω. Setting the auxiliary parameter $\epsilon_t = 0$ and minimizing over the parameter ϵ_x Kuznetsov [16] obtained the error estimate

$$e(T) \le e(0) + \sqrt{8C(|v_0|_{TV(\mathbb{R})} + \|u\|_{L^\infty(0,T;TV(\mathbb{R}^d))})\|u\|_{L^1(0,T;TV(\mathbb{R}^d))})} \nu^{1/2}. \tag{7}$$

Note the presence of the approximate solution u in the error estimate. This is an *a posteriori* error estimate that is very useful for practical computation when used with adaptive strategies, but it is not an *a priori* error estimate of the type we are seeking.

2.2. *A priori* **error estimation**

To obtain the *a priori* error estimates we want, we must get rid of the moduli of continuity of u appearing in the right-hand side of Theorem 2.1. This non-trivial step was done in [3] and [4]. However, this is not enough, since the term $E(u, v; T)/2\, W(T)$ must be estimated in terms of the exact solution v only; in [3] and [4] it was estimated in terms of the approximate solution u, following Kuznetsov's original approach.

Fortunately, in [5], a very simple way to estimate $E(u, v; T)/2\,W(T)$ in terms of v was found. The idea is to 'replace' $\nabla_{x'}\,|\,u - v\,|$ appearing in (5) by $\nabla_x\,|\,u - v\,|$. Next, we show how to do that by simple integration by parts and by exploiting the fact that $\varphi = \varphi(t - t', x - x')$. Starting from (5), we get

$$
\begin{aligned}
E(u, v; T) \quad &\leq \nu \int_Q |\,u - v\,|\,\nabla_{x'} \cdot \nabla_{x'}\varphi \, dx \, dt \, dx' \, dt' \\[2mm]
&\leq -\nu \int_Q |\,u - v\,|\,\nabla_x \cdot \nabla_{x'}\varphi \, dx \, dt \, dx' \, dt' \\[2mm]
&\leq \nu \int_Q \nabla_x|\,u - v\,| \, \cdot \nabla_{x'}\varphi \, dx \, dt \, dx' \, dt', \qquad (8)
\end{aligned}
$$

which is what we wanted. Compare with (5): We have 'replaced' $\nabla_{x'}|\,u - v\,|$ by $-\nabla_x|\,u - v\,|$. We can now obtain the estimate

$$
E(u, v; T) \leq \nu \frac{2\,C}{\epsilon_x}\, W(\tau) \,\|\,v\,\|_{L^1(0,T;TV(\mathbb{R}^d))}.
$$

Compare this estimate with (6): We have effectively 'replaced' the approximate solution u by the exact solution v. This was possible is because we could 'transfer' all the derivatives to φ and then use the fact that

$$
\partial_{x_i}\varphi = -\partial_{x_i'}\varphi.
$$

This is the key idea of the approach to *a priori* error estimation!

We could have stopped here to state an *a priori* error estimate in terms of the Kružkov form $E(u, v; T)$. Instead, we will state our result in terms of a form $\hat{E}(u, v; T)$ that contains all the relevant information about the approximate solution u. In our model case, the form $\hat{E}(u, v; T)$ is obtained as follows. Take $u = u(x', t')$, $v = v(x, t)$, and φ as in the Kuznetsov approach. Now, multiply the equation (4) by $sign(u - v)\,\varphi$ and integrate with respect to variables $(x, t) \times (x', t')$ on Q, the result is the form $\hat{E}(u, v; T)$. Next, noting that we can rewrite

$$
sign(u - v)\,\big(u_{t'} + \nabla_{x'} \cdot f(u) - \nu\, \Delta_{x'}\, u\big) \qquad (9)
$$

as the following sum:

$$
\big(|\,u - v\,|_{t'} + \nabla_{x'} \cdot F(u, v) - \nu\,\Delta_{x'}\,|\,u - v\,|\big) + \big(sign'(u - v)\,|\,\nabla_{x'}u\,|^2\big), \qquad (10)
$$

and noting that first term is in *divergence* form and that the second contains the entropy *dissipation*, we rewrite the form $\hat{E}$ as the sum $\hat{E}_{div} + \hat{E}_{diss}$. The very last part of this procedure is to get the form $\hat{E}^\star_{div}$. This form is obtained from $\hat{E}_{div}$ by simply 'transferring' to the the function φ all the derivatives:

$$\hat{E}^\star_{div}(u,v;T) = -\hat{E}_{div}(u,v;T) + \text{(time-boundary terms)}. \tag{11}$$

We are now ready to state the following general result.

Theorem 2.2. [5]. *We have:*

$$e(T) \leq 2\,e(0) + 8\left(\epsilon_x + \epsilon_t\|f'(v)\|\right)|v_0|_{TV(\mathbb{R}^d)}$$

$$+2\sup_{0\leq t\leq T}\left\{\hat{E}^\star_{div}(u,v;t)/W(t)\right\} - 2\inf_{0\leq t\leq T}\left\{\hat{E}_{diss}(u,v;t)/W(t)\right\}.$$

Thus, to obtain *a priori* error estimates we only have to estimate the forms $E^\star_{div}$ and $\hat{E}_{diss}$ in terms of the exact solution v only.

Note that in our model case, we have

$$E^\star_{div}(u,v;T) = \int_Q \left(-|u-v|\varphi_t - F(u,v)\cdot\nabla_x\varphi + \nu|u-v|\Delta_x\varphi\right)dx\,dt\,dx'\,dt', \tag{12}$$

and

$$E_{diss}(u,v;T) = \int_Q sign'(u-v)\,|\nabla_{x'}u|^2\,\varphi\,dx\,dt\,dx'\,dt'. \tag{13}$$

The form $E^\star_{div}(u,v;T)$ can be easily estimated as follows. Since v is the entropy solution,

$$E^\star_{div}(u,v;T) \leq \int_Q \nu\,|u-v|\Delta_{x'}\,\varphi\,dx\,dt\,dx'\,dt'. \tag{14}$$

Integrating by parts once, we get

$$E^\star_{div}(u,v;T) \leq -\nu\int_Q \nabla_x|u-v|\cdot\nabla_x\,\varphi\,dx\,dt\,dx'\,dt'.$$

and so,

$$E^\star_{div}(u,v;T) \leq \nu\frac{2\,C}{\epsilon_x}\,W(\tau)\,\|v\|_{L^1(0,T;TV(\mathbb{R}^d))},$$

as expected.

Since the form $E_{diss}(u,v;T)$ is easily seen to be nonnegative, we obtain from Theorem 2.2 that

$$e(T) \leq 2\,e(0) + 8\left(\epsilon_x + \epsilon_t\|f'(v)\|\right)|v_0|_{TV(\mathbb{R}^d)} + \nu\frac{4\,C}{\epsilon_x}\,\|v\|_{L^1(0,T;TV(\mathbb{R}^d))}.$$

Setting the auxiliary parameter $\epsilon_t = 0$ and minimizing over the parameter ϵ_x, we get the error estimate

$$e(T) \leq 2\,e(0) + \sqrt{128\,C\,|v_0|_{TV(\mathbb{R}^d)}\,\|v\|_{L^1(0,T;TV(\mathbb{R}^d))}}\,\nu^{1/2}. \tag{15}$$

Compare with the estimate (7): The approximate solution u does not appear anymore and the only price we paid was an increase in the *constants* appearing in the estimate.

Finally, let us emphasize here that to obtain the above error estimate, we did not use any regularity property of u; only the *structure* of the equation defining u was used.

3. Application to monotone schemes in multidimensions

To obtain *a priori* error estimates for flux-splitting monotone schemes of the form (3), we simply apply Theorem 2.2 and use a discrete version of the computations we have displayed in the preceding section for the model case.

Let us sketch how to do that; the details can be found in [7]. First, we must define the form $\hat{E}(u, v; T)$, where u is the piecewise constant function equal to u_K^n for $(x, t) \in K \times [t^n, t^{n+1})$ where $t^n = n\Delta t$ and K is a finite volume of the triangulation $\mathbb{T}_{\Delta x}$. Following the model case, we take:

$$\hat{E}(u, v; t^N) = \int_0^{t^N} \int_{\mathbb{R}^d} \sum_{n=0}^{N-1} \sum_{K \in \mathbb{T}_{\Delta x}} \Psi_K^n(v(t, x)) \, \phi(t, x, t^{n+1}, x_K) \, |K| \, \Delta t \, dx \, dt,$$

where Ψ_K^n, the discrete version of (9), is given by

$$\Psi_K^n(c) = sign(u_K^n - c) \left\{ \frac{u_K^{n+1} - u_K^n}{\Delta t} + \frac{1}{|K|} \sum_{e \in \partial K} |e| \, f_{e,K}(u_K^n, u_{K_e}^n) \right\}.$$

The function $\phi(t, x, t^{n+1}, x_K)$ is an averaged test function defined by

$$\phi(t, x, t^{n+1}, x_K) = \frac{1}{|K|} \int_K \varphi(t, x, t^{n+1}, x') \, dx',$$

where x_K denotes the barycenter of the finite volume K.

Next, we must find $\hat{E}_{div}^{\star}(u, v; t^N)$ and $\hat{E}_{diss}(u, v; t^N)$. This is done by rewriting $\Psi_K^n(c)$ as follows:

$$\left\{ \frac{|u_K^{n+1} - c| - |u_K^n - c|}{\Delta t} + \frac{1}{|K|} \sum_{e \in \partial K} |e| \, F_{e,K}(u_K^n, u_{K_e}^n; c) \right\} + LRED_K^n(c),$$

where $F_{e,K}$ is a consistent entropy flux and $LRED_K^n(c)$ is the so-called local rate of entropy dissipation; this is a discrete version of (10). Since we are dealing with a monotone scheme, under a suitable CFL-condition, the quantity $LRED_K^n(c)$ is nonnegative and hence $\hat{E}_{diss}(u, v; t^N) \geq 0$. Moreover, since the first term is in (discrete) *divergence* form and since the numerical flux is a splitting flux, it is possible to 'transfer' all the (discrete) derivatives to ϕ and obtain the form $\hat{E}_{div}^{\star}(u, v; t^N)$ as indicated by (11).

It only remains to obtain an estimate of $\hat{E}_{div}^{\star}(u, v; t^N)$ in terms of the exact solution v. To do that, we write this form as the sum of three terms. The first is associated with the viscosity of the numerical scheme and can be considered to be a discrete version of the right-hand side of (14). The second term is associated with the inconsistency generated by the geometry and nonuniformity of the mesh. The third is a high-order term. The first term can be estimated along the same lines the right-hand side of (14) was estimated. The third term can also be estimated in terms of the exact solution only. However, this is not the case with the second term. The schemes for which the second term is identically equal to zero are called *consistent*. For these schemes, we obtain [7] the following error estimate:

$$e(T) \leq 2\,e(0) + \sqrt{128\,C\,|\,v_0\,|_{TV(\mathbb{R}^d)}\,\|\,v\,\|_{L^1(0,T;TV(\mathbb{R}^d))}}\,\nu^{1/2} + D\,(\Delta x)^{3/4},$$

where ν is an upper bound of the numerical viscosity of the scheme evaluated on the exact solution v and D is a constant in dependent of Δx. Since ν is proportional to Δx, this estimate gives the optimal rate of convergence of $(\Delta x)^{1/2}$. Compare this estimate with the estimate of the model case (15).

For a discussion on the issues related with the estimation of the second term that contains the information about the loss of consistency of the scheme, we refer the reader to the note [8].

4. Concluding remarks

In this note, we have given a simple description of the main ideas of the technique of *a priori* error estimation for nonlinear scalar conservation laws developed in [5], [6], and [7]. The main advantage of this approach is that no regularity property of the approximate solution was needed; only the *structure* of its numerical scheme was used. We believe that this new approach will open a new avenue of research in the area of numerical analysis of nonlinear conservation laws.

Applications of this technique to more general finite volume schemes like the one developed in [1] and to general monotone schemes remain to be done. The application to the streamline-diffusion method and to the discontinuous Galerkin method are currently under way. How to extend this approach to systems of conservation laws is an exciting challenge.

References

[1] T. Boukadida and A.Y. LeRoux, *A new version of the two-dimensional Lax-Friedrichs scheme*, Math. Comp., **63** (1994), 541–553.

[2] B. Cockburn, *The quasi-monotone schemes for scalar conservation laws, Parts I, II, and III*, SIAM J. Numer. Anal., **26** (1989) 1325–1341; **27** (1990) 247–258; and **27** (1990) 259–276, respectively.

[3] B. Cockburn, F. Coquel, and P. LeFloch, *An error estimate for finite volume methods for conservations laws*, Math. Comp., **64** (1994), 77–103.

[4] B. Cockburn and P.-A. Gremaud, *An error estimate for finite element methods for conservations laws*, SIAM J. Numer. Anal., **33** (1996), 522–554.

[5] B. Cockburn and P.-A. Gremaud, *A priori error estimates for numerical methods for scalar conservation laws. Part I: The general approach*, Math. Comp., **65** (1996), 533–573.

[6] B. Cockburn and P.-A. Gremaud, *A priori error estimates for numerical methods for scalar conservation laws. Part II: Flux-splitting monotone schemes on irregular Cartesian grids,*Math. Comp., **66** (1997), 547–572.

[7] B. Cockburn, P.-A. Gremaud, and J.X. Yang *A priori error estimates for numerical methods for scalar conservation laws. Part III: Flux-splitting monotone schemes on non-Cartesian grids,* SIAM J. Numer. Anal., to appear.

[8] B. Cockburn, P.-A. Gremaud, and X. Yang, *New results in conservation laws,* Proceedings of the Interdisciplinary Congress on Free Boundary Problems, Theory and Applications, Herakleion (Crete), Greece, June 1997. Longman, to be published (11 pages).

[9] B. Engquist and S. Osher, *One sided difference approximations for nonlinear conservation laws,* Math. Comp., **36** (1981), 321–351.

[10] B. Engquist and S.-H. Yu, *Convergence of finite difference schemes for piecewise smooth solutions with shocks,* UCLA Preprint (1997).

[11] B. García-Archilla, *A supraconvergent scheme for the Korteweg-de-Vries equation,* Numer. Math., **61** (1992), 291–310.

[12] J. Goodman and Z. Xin, *Viscous limits for piecewise smooth solutions to systems of conservation laws,* Arch. Rational Mech. Anal., **121** (1992), 235–265.

[13] J.D. Hoffman, *Relationship between the truncation error of centered finite difference approximations on uniform and nonuniform meshes,* J. Comput. Phys., **46** (1982), 469–474.

[14] H.-O. Kreiss, T.A. Manteuffel, B. Swartz, B. Wendroff, and A.B. White, Jr., *Supraconvergent schemes on Irregular grids,* Math. Comp., **47** (1986), 537–554.

[15] S.N. Kružkov, *First order quasilinear equations in several independent variables,* Math. USSR Sbornik, **10** (1970), 217–243.

[16] N.N. Kuznetsov, *Accuracy of some approximate methods for computing the weak solutions of a first-order quasi-linear equation,* USSR Comp. Math. and Math. Phys., **16** (1976), 105–119.

[17] B. J. Lucier, *A stable adaptive scheme for hyperbolic conservation laws,* SIAM J. Numer. Anal., **22** (1985), 180–203.

[18] B. J. Lucier, *Error bounds for the methods of Glimm, Godunov and LeVeque,* SIAM J. Numer. Anal., **22** (1985), 1074–1081.

[19] B. J. Lucier, *On nonlocal monotone difference schemes for scalar conservation laws,* Math. Comp., **47** (1986), 19–36.

[20] B. J. Lucier, *A moving mesh numerical method for hyperbolic conservation laws,* Math. Comp., **46** (1986), 59–69.

[21] T.A. Manteuffel and A.B. White, Jr. *The numerical solution of second-order boundary value problems on nonuniform meshes,* Math. Comp., **47** (1986), 511–535.

[22] H. Nessyahu and E. Tadmor, *The convergence rate of approximate solutions for nonlinear scalar conservation laws,* SIAM J. Numer. Anal., **29** (1992), 1505–1519.

[23] H. Nessyahu, E. Tadmor and T. Tassa, *The convergence rate of Godunov type schemes,* SIAM J. Numer. Anal., **31** (1994), 1–16.

[24] S. Nöelle, *A note on entropy inequalities and error estimates for higher-order accurate finite volume schemes on irregular grids,* Math. Comp., **65** (1996), 1155–1163.

[25] J. Pike, *Grid adaptive algorithms for the solution of the Euler equations on irregular grids*, J. Comput. Phys., **71** (1987), 194–223.

[26] R. Sanders, *On convergence of monotone finite difference schemes with variable spatial differencing*, Math. Comp., **40** (1983), 91–106.

[27] Florin Şabac, *The optimal convergence rate of monotone finite difference methods for hyperbolic conservation laws*, SIAM J. Numer. Anal., **34** (1997), 2306–2318.

[28] T. Tang and Z.-H. Teng, *The sharpness of Kuznetsov's $O(\sqrt{\Delta x})$ L^1-error estimate for monotone difference schemes*, Math. Comp., **64** (1995), 581–589.

[29] T. Tang and Z.-H. Teng, *Viscosity methods for piecewise smooth solutions to scalar conservation laws*, Math. Comp., **66** (1997), 495–526.

[30] Z.-H. Teng and P.-W. Zhang, *Optimal L^1-rate of convergence for viscosity methods and monotone schemes to piecewise constant solutions with shocks*, SIAM J. Numer. Anal., **34** (1997), 959–978.

[31] E. Tadmor, *Local error estimates for discontinuous solutions of nonlinear hyperbolic equations*, SIAM J. Numer. Anal., **28** (1991), 891–906.

[32] E. Turkel, *Accuracy of schemes with nonuniform meshes for compressible fluid flow*, App. Numer. Math., **2** (1986), 529–550.

[33] J.-P. Vila, *Convergence and error estimates in finite volume schemes for general multidimensional scalar conservation laws*, Model. Math. Anal. Numer., **28** (1994), 267–295.

[34] B. Wendroff, *Supraconvergence in two dimensions*, Los Alamos National Laboratory report LA-UR-95-3068 (1995).

[35] B. Wendroff and A.B. White, Jr., *Some supraconvergent schemes for hyperbolic equations on irregular grids*, Second International Conference on Hyperbolic Problems, Aachen, Vieweg, (1988), 671–677,

[36] B. Wendroff and A.B. White, Jr., *A supraconvergent scheme for nonlinear hyperbolic systems*, Comput. Math. Appl., **18** (1989) 761–767.

School of Mathematics,
University of Minnesota,
206 Church Street S.E.
Minneapolis, MN 55455
USA
E-mail address: cockburn@math.umn.edu
E-mail address: xyang@math.umn.edu

Center for Research in Scientific Computation,
Department of Mathematics,
North Carolina State University,
Raleigh, NC 27695-8205
USA
E-mail address: gremaud@dali.math.ncsu.edu

International Series of Numerical Mathematics
Vol. 129, © 1999 Birkhäuser Verlag Basel/Switzerland

Traveling Waves for Combustion in Porous Media

J. da Mota, W. Dantas, and D. Marchesin

Abstract. We determine the planar traveling wave solutions for a nonstrictly hyperbolic system of three conservation laws modeling combustion *in-situ* in petroleum reservoirs. The chemical reaction is represented by Arrhenius law. Our analysis shows that the combustion wave is represented by a heteroclinic connection in a non-hyperbolic system of three ordinary differential equations.

1. Introduction

We determine the nonlinear waves in a two-phase flow model that represent combustion of oil and oxygen in a porous medium. Understanding these waves is important to maximize oil recovery in methods such as combustion *in-situ* [1, 2, 3], widely used to extract heavy oil in deep petroleum reservoirs.

Our main result is that combustion waves are represented by connections between two singularities. On one of these singularities the linearized dynamical system has a vanishing eigenvalue.

The system of equations governing the flow we consider here was derived in [4]. It reflects the conservation of mass of each component, as well as the conservation of total energy. It assumes Darcy's law of force. Compressibility and volumetric changes associated with the combustion process are neglected. However, heat conduction, as well as the capillary pressure diffusive effects present in displacement of multiphase fluid in porous media, are all taken into account.

In [4] combustion wave solutions for a similar system were also determined. However, the model employed an unrealistic reaction rate in the equation, which had the effect of preventing the reaction from starting. Here we consider a more physical Arrhenius type equation for the reaction rate.

The flow is described by state quantities depending on (x, t). They are denoted as follows. The saturation of gas is s, and the saturation of oil is $1 - s$; the temperature, which is assumed to be the same for gas, oil, and rock at each point,

Key words and phrases. Porous medium, combustion, traveling waves, multiphase flow.

This work was supported in part by: CNPq under Grant 520725/95-6; CNPq under Grant 300204/83-3; FINEP under Grant 77.97.0315.00; FAPEMIG under Grant CEX153493.

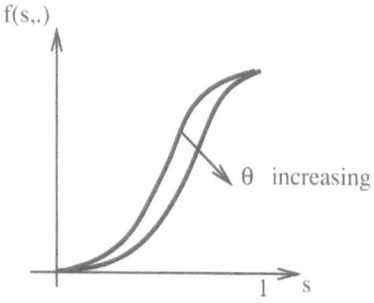

FIGURE 1. Fractional flow function

is θ; and the burnt volume fraction of the gaseous phase is ϵ. The gaseous phase is unburnt when $\epsilon = 0$ and it is burnt when $\epsilon = 1$.

In one spatial dimension, the adimensionalized system may be written as [4]:

$$\frac{\partial s}{\partial t} + \frac{\partial f}{\partial x} = \frac{\partial}{\partial x}(\lambda f \frac{\partial p_c}{\partial x}) \tag{1}$$

$$\frac{\partial}{\partial t}[(s+\alpha)\theta - \eta\epsilon s] + \frac{\partial}{\partial x}[(f+\beta)\theta - \eta\epsilon f] = \frac{\partial}{\partial x}[(\theta - \eta\epsilon)\lambda f \frac{\partial p_c}{\partial x}] + \gamma\frac{\partial^2\theta}{\partial x^2} \tag{2}$$

$$\frac{\partial}{\partial t}(\epsilon s) + \frac{\partial}{\partial x}(\epsilon f) = \frac{\partial}{\partial x}(\epsilon\lambda f \frac{\partial p_c}{\partial x}) + sq. \tag{3}$$

Eq. (1) represents the conservation of total gas mass, Eq. (2) represents the conservation of energy and Eq. (3) represents the conservation of mass of burned gas, and includes a source term representing mass transfer from unburnt to burnt gas. The equation for conservation of energy assumes that heat is released when the combustion occurs, and that the internal energy for gaseous phase, denoted by e, is given by $\rho e = C\theta - \epsilon Q$, where ρ is the density of gas, C is the thermal capacity of gas for unit volume, and Q is the heat released by the combustion per unit volume.

Here f is the so called fractional flow function of the gas, and λ is the mobility of the oil. These are prescribed C^2 functions which depend on s and θ. We assume that f is S-shaped in s for each θ, and that it decreases in θ, see Figure 1. Thus $f(0,\theta) = 0$, $f(1,\theta) = 1$, $\frac{\partial f}{\partial s}$ vanishes for $s = 0$ and $s = 1$ for each θ, $\frac{\partial^2 f}{\partial s^2}$ is first positive and then negative for $0 < s < 1$. We also have $\frac{\partial f}{\partial \theta} < 0$ and $\lambda > 0$ for $0 < s < 1$. These assumptions are used to model two-phase thermal flow in porous media, where oil viscosity is a decreasing function of temperature [5].

The function p_c is the prescribed capillary pressure function, which measures the pressure difference between gas and oil phases. This function depends on s; it will be assumed to be C^2; also $\frac{\partial p_c}{\partial s} > 0$ for $0 < s < 1$.

The function q is the volumetric fraction of burnt gas generated in unit time. It will be assumed to be have the form below [6], a non dimensional version of Arrhenius's formula:

$$q(\epsilon, \theta) = (1 - \epsilon)e^{\frac{-1}{\theta - \theta_0}} \text{ if } \theta > \theta_0, \text{ or } 0 \text{ if } 0 < \theta \le \theta_0, \qquad (4)$$

where θ_0 is a critical temperature at which the reaction starts. In what follows the temperature of any unburnt state will be taken to be θ_0.

The other quantities α, β, η, and γ present in the system are all constant. They are all assumed to be negative, with $\alpha \le \beta < -1$, see [4]. These quantities depend on the thermal capacities of gas, oil, and of the rock forming the porous medium; η depends on heat released by the combustion per unit volume. The value of γ depends on thermal conductivities of the gas, oil, and the rock, it is not really a constant. In the figures presented here, calculations were done with $\alpha = -1.2$, $\beta = -1.1$, $\gamma = \eta = -1$,

$$f(s, \theta) = \frac{s^2}{s^2 + (0.1 + \theta)(1 - s)^2}, \quad \lambda(s, \theta) = (0.1 + \theta)(1 - s)^2.$$

2. Combustion waves

For initial data given by the unburnt state $(s_0, \theta_0, \epsilon_0 = 0)$ and by the burnt state $(s_1, \theta_1, \epsilon_1 = 1)$, we want to know if there is a traveling wave solution of the system (1)–(3) with propagation speed σ. For such solutions, s, θ and ϵ are functions of the single variable $z = x - \sigma t$, with $-\infty < z < \infty$ and

$$\lim_{z \to -\infty} (s(z), \theta(z), \epsilon(z)) = (s_1, \theta_1, \epsilon_1), \quad \lim_{z \to +\infty} (s(z), \theta(z), \epsilon(z)) = (s_0, \theta_0, \epsilon_0), \quad (5)$$

$$\lim_{z \to -\infty} (s'(z), \theta'(z), \epsilon'(z)) = 0, \quad \lim_{z \to +\infty} (s'(z), \theta'(z), \epsilon'(z)) = 0. \qquad (6)$$

In this case we say that the traveling wave solution connects the burnt state on the left to the unburnt state on the right.

Following [4], we can prove the theorem:

Theorem 2.1. *A traveling wave solution of system (1)–(3) connecting the burnt state $(s_1, \theta_1, \epsilon_1 = 1)$ to the unburnt state $(s_0, \theta_0, \epsilon_0 = 0)$ is an orbit satisfying (5) and (6) of the dynamical system*

$$\frac{\partial s}{\partial z} = \frac{a + \sigma s - f(s, \theta)}{h(s, \theta)} \qquad (7)$$

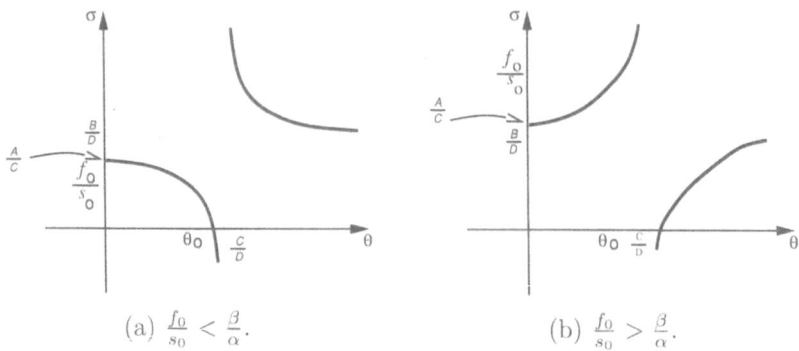

(a) $\frac{f_0}{s_0} < \frac{\beta}{\alpha}$.

(b) $\frac{f_0}{s_0} > \frac{\beta}{\alpha}$.

FIGURE 2. Combustion speed

$$\frac{\partial \theta}{\partial z} = \frac{1}{\gamma}[(a - \sigma\alpha + \beta)\theta - \eta a\epsilon - b] \tag{8}$$

$$\frac{\partial \epsilon}{\partial z} = \frac{1}{a}sq(\theta, \epsilon). \tag{9}$$

Here $h(s, \theta) = -\lambda f(s, \theta)\frac{\partial p_c}{\partial s}(s) < 0$ in $0 < s < 1$, $\theta > 0$, and we have $a = f_0 - \sigma s_0$ and $b = [f_0 + \beta - \sigma(s_0 + \alpha)]\theta_0$, where $f_0 = f(s_0, \theta_0)$.

The speed σ and the states $(s_0, \theta_0, \epsilon_0 = 0)$ and $(s_1, \theta_1, \epsilon_1 = 1)$ are related by the Rankine-Hugoniot shock condition, where $f_1 = f(s_1, \theta_1)$:

$$\sigma = \frac{f_1 - f_0}{s_1 - s_0} = \frac{(f_0 + \beta)\theta_0 + \eta f_0 - (f_0 + \beta)\theta_1}{(s_0 + \alpha)\theta_0 + \eta s_0 - (s_0 + \alpha)\theta_1}. \tag{10}$$

As we will see, this shock condition is not sufficient to guarantee the existence of a combustion wave traveling with speed σ connecting the states $(s_1, \theta_1, \epsilon_1 = 1)$ and $(s_0, \theta_0, \epsilon_0 = 0)$.

Notice that the speed σ of the combustion wave and the constants a and b depend on all unburnt state variables but only on the temperature variable of the burnt state, since

$$\sigma = \frac{A - B\theta_1}{C - D\theta_1}, \tag{11}$$

where $A = (f_0 + \beta)\theta_0 + \eta f_0$, $B = f_0 + \beta$, $C = (s_0 + \alpha)\theta_0 + \eta s_0$ and $D = s_0 + \alpha$.

The plot of $\sigma(\theta_1)$ is shown in Figures 2a and 2b. If $\frac{f_0}{s_0} = \frac{\beta}{\alpha}$, all shocks satisfying (10) are contact discontinuities, and there are no traveling waves.

Now, consider the given unburnt state $I = (s_0, \theta_0, \epsilon_0 = 0)$ and assume $\frac{f_0}{s_0} < \frac{\beta}{\alpha}$. The case $\frac{f_0}{s_0} > \frac{\beta}{\alpha}$ can be treated in a similar way. Given a temperature θ_1, we use equation (11) to find the speed $\sigma(\theta_1)$. Then we use this speed in the first equality in (10) to find all states with this temperature such that condition (10) is satisfied relative to state I. Depending on the temperature we have one, two, or three states in this situation. See Figure 3, which shows the projection on plane $\epsilon = 1$ of the Hugoniot curve through state $I = (s_0, \theta_0, \epsilon_0 = 0)$.

As shown in Figure 3, for $\theta_1 = \theta_a$ we have only one state satisfying the shock condition (10) denoted by II_a. For $\theta_1 = \theta_b$ we have three states, denoted by II_b, IV_b and VI_b. For $\theta_1 = \theta_c$ we have two states denoted by II_c and IV_c.

The question now is the following: which are the states on the Hugoniot curve that can be connected to the unburnt state $I = (s_0, \theta_0, \epsilon_0 = 0)$ by a combustion wave? In the classical Chapman-Jouguet's theory of combustion in gases the same question arises and the answer is that there is only one traveling wave solution, see [6]. The result here is different. For each unburnt state there is a one parameter family of traveling waves. One end of this family corresponds to a distinguished connection, which potentially plays a relevant role in the solution of Riemann problems for combustion.

As we shall see, it is first necessary to study the phase portrait of the dynamical system (7)–(9).

3. The equilibrium points

Here we consider the same unburnt state $I = (s_0, \theta_0, \epsilon_0 = 0)$ as in §2, and we consider a state $(s_1, \theta_1, \epsilon_1 = 1)$ on the Hugoniot curve through I. Now, we introduce the following change of variable in the temperature:

$$T = \frac{\theta - \theta_0}{\theta_1 - \theta_0},$$

which transforms the interval $\theta_0 \leq \theta \leq \theta_1$ into the interval $0 \leq T \leq 1$. This change of variable also transforms the system (7)–(9) into another dynamical system (recall that $0 \leq s \leq 1$, $0 \leq T \leq 1$ and $0 \leq \epsilon \leq 1$):

$$\frac{ds}{dz} = X \equiv \frac{a + \sigma s - f(s, (\theta_1 - \theta_0)T + \theta_0)}{h(s, (\theta_1 - \theta_0)T + \theta_0)} \tag{12}$$

$$\frac{dT}{dz} = Y \equiv \frac{a\eta}{\gamma(\theta_1 - \theta_0)}(T - \epsilon) \tag{13}$$

$$\frac{d\epsilon}{dz} = Z \equiv \frac{1}{a} s(1 - \epsilon) \exp\left(\frac{-1}{(\theta_1 - \theta_0)T}\right) \tag{14}$$

From the definition of C and D, we have that $\theta_0 < \frac{C}{D}$. Studying the phase portrait of the dynamical system (12)–(14), we can see that there is no combustion wave when $\theta_0 \leq \theta_1 \leq \frac{C}{D}$. Depending on this parameter θ_1 with $\theta_1 > C/D$, this system may have from two to six equilibrium points. The component X of the field (X, Y, Z) vanishes only on the cylindrical surface given by

$$\sigma = \frac{f(s, (\theta_1 - \theta_0)T + \theta_0) - f_0}{s - s_0}.$$

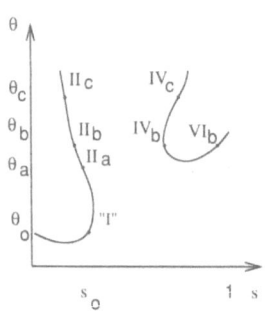

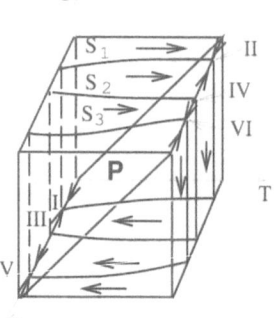

FIGURE 3. Projection
on the plane $\epsilon = 1$
of the Hugoniot curve
through state
$I = (s_0, \theta_0, \epsilon_0 = 0)$

FIGURE 4. Equilibria
I to VI. Surface
branches S_1, S_2 and
S_3 where $X = 0$.
Plane P where $Y = 0$

Depending on θ_1 (see the cases $\theta_1 = \theta_a$, $\theta_1 = \theta_b$, $\theta_1 = \theta_c$ in Figure 4), this surface may have one, two, or three separate branches in (s, T, ϵ) state space. Figure 4 shows three branches denoted by S_1, S_2, and S_3. Arrows on the edges parallel to the s-axis indicate the direction where the component X increases. The component Y vanishes only on the plane $\epsilon = T$, denoted by P in Figures 4, 6a, 6b. Horizontal arrows indicate the direction where the component Y increases. The component Z vanishes on $s = 0$ or $\epsilon = 1$ or $T = 0$. For all other points $Z < 0$, since $a < 0$. Vertical arrows indicate the direction where the component Z increases. Hence, for a parameter θ_1 equal to θ_b in the Figure 3, the field (X, Y, Z) has exactly six equilibrium points, indicated by I, III, and V, for $\epsilon = 0$ and $T = 0$, and by II, IV, and VI for $\epsilon = 1$ and $T = 1$, as in Figures 4, 6a, 6b.

The Jacobian matrix of the system (12)–(14) at each equilibrium point is given by:

$$
M = \begin{pmatrix}
\dfrac{\sigma - f_s}{h} & -\dfrac{(\theta_1 - \theta_0)f_T}{h} & 0 \\[2ex]
0 & \dfrac{b}{\gamma\theta_0} & -\dfrac{b}{\gamma\theta_0} \\[2ex]
0 & 0 & Z_\epsilon
\end{pmatrix}
$$

where $Z_\epsilon = 0$ at points I, III, and V, and $Z_\epsilon < 0$ for all other points in $0 < s \leq 1$. The signs of the eigenvalues of the matrix M at each of the equilibrium points are given in Table 1.

$$
\begin{array}{llll}
I: & -\ -\ 0 & IV: & +\ -\ + \\
II: & -\ -\ + & V: & -\ -\ 0 \\
III: & +\ -\ 0 & VI: & -\ -\ +
\end{array}
$$

Table 1: Signs of the eigenvalues of the Jacobian M at the equilibrium points.

Figure 4 shows that there is no orbit departing from VI and arriving at I. This means that there is no combustion wave connecting VI to I. For each temperature θ_1, there are only two possible cases for which combustion waves can exist. In the first case, II is connected to I. In the other case, IV is connected to I. Here we will consider only the first case. The other may be analyzed in a similar way. Thus, depending on θ_1, we look for an orbit of the dynamical system (12)–(14), which departs from II and enters I. In fact, there is an orbit departing from equilibrium point II, because this point has a one-dimensional unstable manifold (see Table 1).

Considering a fixed s $(0 < s \leq 1)$ in the three-dimensional system (12)–(14), we obtain an associated two-dimensional system, which we study in order to control the orbit departing from II.

4. The two-dimensional dynamical system

We introduce the following system, which is a modification of (12)- (14) with fixed saturation:

$$
\frac{dT}{dz} = Y(T, \epsilon;\, r, \theta) \equiv \frac{a\eta}{\gamma(\theta - \theta_0)}(T - \epsilon) \tag{15}
$$

$$
\frac{d\epsilon}{dz} = Z(T, \epsilon;\, r, \theta) \equiv \frac{1}{a} r(1 - \epsilon) \exp\left(\frac{-1}{(\theta - \theta_0)T}\right), \tag{16}
$$

where we have replaced s by r, to avoid confusion in the notation. In this system, $0 \leq T, \epsilon \leq 1$, and r and θ are parameters with $0 < r \leq 1$, and $\theta > C/D$ is the parameter θ_1 used in Section 3.

For fixed parameters r and θ as above, system (15)–(16) has only two equilibrium points, $(0, 0)$ and $(1, 1)$. Point $(1, 1)$ is a saddle. At point $(0, 0)$ an eigenvalue is negative and the corresponding stable manifold enters $(0, 0)$ tangent to the T-axis. The other eigenvalue is zero and the corresponding eigenvector lies on the straight line $\epsilon = T$. Each orbit which enters $(0, 0)$ tangent to the straight line $\epsilon = T$ is a central manifold.

An orbit of (15)–(16) is denoted by $(T(z), \epsilon(z))$. Analyzing the slope $\epsilon(z)/T(z)$, we can prove the following theorem, see Figures 5a–5c.

Theorem 4.1. *For each fixed $0 < r \leq 1$, there is only one value θ^r of θ, $\frac{C}{D} < \theta^r$, such that:*

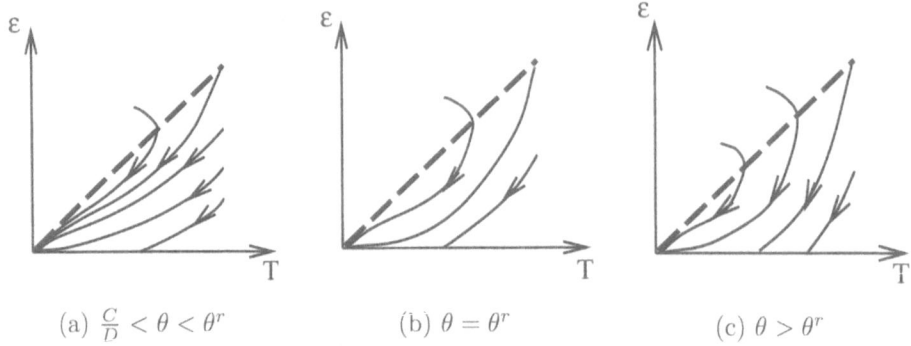

(a) $\frac{C}{D} < \theta < \theta^r$ (b) $\theta = \theta^r$ (c) $\theta > \theta^r$

FIGURE 5. Phase portrait of the two-dimensional system

i) If $\frac{C}{D} < \theta < \theta^r$, the unstable manifold of $(1,1)$ enters $(0,0)$ tangent to the straight line $\epsilon = T$. The stable manifold of $(0,0)$ crosses the straight line $T = 1$ at a point where $0 < \epsilon < 1$. (See Figure 5a.)

ii) If $\theta = \theta^r$, the unstable manifold of $(1,1)$ enters $(0,0)$ tangent to the T-axis. This means that the unstable manifold of $(1,1)$ coincides with the stable manifold of $(0,0)$. (See Figure 5b.)

iii) If $\theta > \theta^r$, the unstable manifold of $(1,1)$ crosses the T-axis at a point where $0 < T < 1$. The stable manifold of $(0,0)$ crosses the straight line $\epsilon = T$ at a point where $0 < T < 1$. (See Figure 5c.)

Given parameters r and θ, $C/D < \theta$, it is easy to see that the part of the stable manifold entering $(0,0)$ which lies in the region $0 \le T \le 1$, $0 \le \epsilon \le T$ is a curve which can be expressed as a function of T. We denote this curve by $\epsilon = \epsilon_\theta^r(T)$, $0 \le T \le 1$, possibly completed by a segment of the diagonal $\epsilon = T$.

Now, we denote by S_θ^r the cylindrical surface $\epsilon = \epsilon_\theta^r(T)$, $0 \le s \le 1$ in the space (s, T, ϵ). We say that this surface is generated by the (r, θ)-stable manifold of $(0,0)$. See Figures 6a–6b, originating from Figures 5a and 5c, respectively.

We denote by A_θ^r the region formed by all points (s, T, ϵ) located above and on the surface S_θ^r, below the plane $\epsilon = T$ and with $0 \le s \le 1$. We also denote by B_θ^r the region located below and on the surface S_θ^r with $\epsilon \ge 0$, $0 \le T \le 1$ and $0 \le s \le 1$. These regions are indicated in Figures 6a and 6b.

We now return to the three-dimensional system (12)–(14).

5. Main results

Let us denote by $(s(z), T(z), \epsilon(z))$, $-\infty < z < +\infty$, an orbit of the system (12)–(14). The components of the equilibrium points are denoted by: $I = (s_I, T_I, \epsilon_I)$, $\ldots$, $VI = (s_{VI}, T_{VI}, \epsilon_{VI})$.

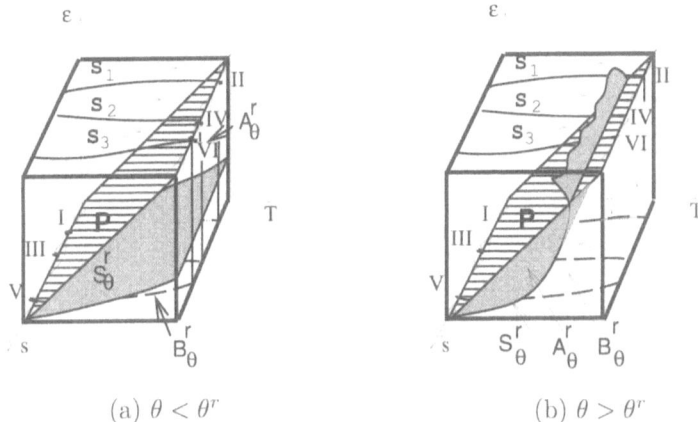

(a) $\theta < \theta^r$ (b) $\theta > \theta^r$

FIGURE 6. Surface generated by the (r, θ)-stable manifold of $(0,0)$.

Analyzing the slope $\epsilon(z)/T(z)$ and the signs of components of the field (X, Y, Z), we can prove the following theorem.

Theorem 5.1. *Let $C/D < \theta$. If the equilibrium point IV does not exist for this value θ, the region A_θ^1 is invariant under the flow (15)-(16). If the equilibrium point IV exists, the part of the region A_θ^1 for which $s \le s_{III}$ is also invariant. In both cases, orbits originating in these regions enter I tangent to the plane $\epsilon = T$.*

We know that the stable manifold of I is two-dimensional and enters I tangent to the plane $\epsilon = 0$. Therefore we have:

Corollary 5.2. *Assuming the same hypothesis as in Theorem (5.1), the points (s, T, ϵ) on the θ-stable manifold of I, with $0 < T < 1$ and $0 < s < s_{III}$ (or $0 < s < 1$ if IV does not exist) lie all in the interior of B_θ^1.*

We now denote by θ^1 the unique temperature given by Theorem (4.1), when $r = 1$. As a consequence of Theorem (5.1) we also have:

Corollary 5.3. *For all θ, $C/D < \theta \le \theta^1$, the orbit which departs from II enters I tangent to the plane $\epsilon = T$.*

It is not difficult to prove that all orbits $(s(z), T(z), \epsilon(z))$ which depart from II (or IV) remain in the region $s \ge r_o$, for some $r_o > 0$, which depends on θ.

The following theorem is similar to Theorem (5.1).

Theorem 5.4. *Let $C/D < \theta$. Consider orbits originating at $B_\theta^{r_o}$, where $s \le s_{III}$ if the equilibrium point IV does not exist. Then these orbits cross the plane $\epsilon = 0$.*

We have immediately the following:

Corollary 5.5. *Assuming the same hypothesis as in Theorem (5.4), then the stable manifold of I must remain in the interior of the region* $A_\theta^{r_o}$.

We denote by θ^{r_o} the unique temperature given by Theorem (4.1), when $r = r_o$. As consequence of Theorem (5.4) we also have:

Corollary 5.6. *For all* θ, $\theta^{r_o} \le \theta$, *there is no orbit which departs from II (or IV) and enters I.*

Now we define two temperatures Θ_1 and Θ_2 as:

$$\Theta_1 = \sup\Omega_1$$

where Ω_1 is the set of Θ with $C/D < \Theta$, such that if $C/D < \theta < \Theta$, then the θ-orbit which departs from II enters I tangent to the plane $\epsilon = T$. Also

$$\Theta_2 = \inf\Omega_2$$

where Ω_2 is the set of Θ with $C/D < \Theta$, such that if $\Theta < \theta < \infty$, then the orbit which departs from II does not enter I.

Clearly, we have

$$\theta^1 \le \Theta_1 \le \Theta_2 \le \theta^{r_o},$$

where the temperatures θ^1 and θ^{r_o} were defined above.

We can prove that the sets Ω_1 and Ω_2 are open. This implies that $\theta^1 < \Theta_1$ and $\Theta_2 < \theta^{r_o}$. Also, if $\theta^1 = \Theta_1$ and $\Theta_2 = \theta^{r_o}$, then the orbit which departs from II enters I tangent to the plane $\epsilon = 0$.

A natural question (which remains unanswered here) is whether $\Theta_1 = \Theta_2$. This should be true under reasonable conditions. If this is true, there is only one temperature θ for which the unstable manifold of II enters I tangent to the plane $\epsilon = 0$. In this case, this unstable manifold lies on the stable manifold of I. For any other temperature θ, if the unstable manifold of II enters I, it is not tangent to the plane $\epsilon = 0$, but it is tangent to the plane $\epsilon = T$.

We conclude that combustion waves in this model are represented by a family of connections between a non hyperbolic saddle and another singularity. This agrees with certain steam injection models [7]. We conjecture that this family ends when the connection becomes distinguished.

References

[1] B.S. Gottfried, *A mathematical model of thermal oil recovery in linear systems*, SPE 1117 (1965), 196–210.

[2] J.W. Grabowski, P.K. Vinsome, R.C. Lin, A. Behie and B. Rubin, *A fully implicit general purpose finite-difference thermal model for in situ combustion and steam*, SPE 8396, (1979), 1–14.

[3] N.K. Baibakov and A.R. Garushev, *Thermal methods of petroleum production*, Elsevier, Amsterdam, **1989**.

[4] J.C. Da-Mota, W.B. Dantas, M.E. Gomes and D. Marchesin, *Combustion fronts in petroleum reservoirs*, Mat. Contemp., **8** (1995), 129–149.

[5] J.C. Da-Mota, *The Riemann problem for a simple thermal model for two phase flow in porous media*, Mat. Aplic. Comp., **11 (2)** (1992), 117–145.

[6] R. Courant and K.O. Friedrichs, *Supersonic flow and shock waves*, Wiley-Interscience, New York, **1948**.

[7] C.J. van Duijn, *Uniqueness conditions in a Hyperbolic model for oil recovery by steam-drive*, Book of Abstracts – Seventh International Conference on Hyperbolic Problems, ETH Zurich, Feb. 9–13, (1998), 35.

Departamento de Matemática,
Universidade Federal de Goiás,
Caixa Postal 131,
64970-001, Goiânia, GO, Brazil
E-mail address: jesus@mat.ufg.br

Departamento de Matemática,
Universidade Federal de Minas Gerais,
Caixa Postal 702,
30161-970, Belo Horizonte, MG, Brazil
E-mail address: wdantas@bhnet.com.br

Instituto de Matemática Pura e Aplicada,
Estrada Dona Castorina 110,
22460-320, Rio de Janeiro, RJ, Brazil
E-mail address: marchesi@impa.br

International Series of Numerical Mathematics
Vol. 129, © 1999 Birkhäuser Verlag Basel/Switzerland

Evolution of a Cusp-like Singularity in a Vortex Patch

Raphaël Danchin

Abstract. We investigate the evolution of cusp-like singularities in the boundary of a vortex patch for two-dimensional Euler equations. According to numerical simulations, cusp singularities appear as limit structures for the evolution of smooth vortex patches when the time goes to infinity. We here present an adaptive scheme that we have used to study the stability of the cusp. We then state a global result of persistence of conormal regularity with respect to vector fields vanishing at a singular point, which generalises the structure of a cusp. This entails the global stability of the cusp with conservation of the order.

1. Introduction

The two-dimensional Euler equations

$$(E) \qquad \begin{cases} (\partial_t + v \cdot \nabla)v = -\nabla p, \\ \operatorname{div} v = 0, \\ v_{|t=0} = v_0. \end{cases}$$

describe the evolution of an inviscid incompressible fluid in the whole plane. $v(t, x)$ and $p(t, x)$ here stand for the velocity and the pressure of the fluid in $x = (x_1, x_2)$ at time t.

We define the vorticity ω as $\omega = \partial_1 v^2 - \partial_2 v^1$. A straightforward computation shows that $(\partial_t + v \cdot \nabla)\omega = 0$. If v has a flow ψ, it entails $\omega(t, x) = \omega_0(\psi_t^{-1}(x))$. We are interested in the evolution of patches of constant vorticity. The initial vorticity ω_0 is assumed to be 1 inside a simply connected bounded domain Ω_0, and 0 outside.

A theorem by V. Yudovitch [11] states that for such an initial datum (without regularity assumptions on the boundary), (E) has a unique solution (v, p) global in time. Moreover the velocity v has a unique continuous flow ψ defined by

$$\psi(t, x) = x + \int_0^t v\big(s, \psi(s, x)\big)\, ds$$

and the vorticity at time t, ω_t, remains the characteristic function of the domain $\Omega_t = \psi_t(\Omega_0)$.

The possible appearance of finite time singularity in an initially regular vortex patch has remained an open question for a long time. Some authors (see for instance [2] and [3]) observed in numerical simulations that singularity in finite time may appear, the singularity being a cusp or a corner.

However, in 1990, J.-Y. Chemin (see [4]) proved that a smooth initial boundary remains smooth for all time. More precisely, $\partial \Omega_t$ remains in the Hölder class C^r for all time if $\partial \Omega_0$ is in C^r and $r \in]1, +\infty] \setminus \mathbb{N}$. Recently, we proved that when the initial boundary is C^r outside a closed subset, it remains C^r outside the closed subset transported by the flow (see [6]). But the evolution of the singularity remains unclear.

Moreover, even though no singularity can appear in a smooth vortex patch, structures very similar to cusps are still observed in numerical simulations (see [12], [2] and [10]). This does not stand in contradiction with double exponential estimates obtained by J.-Y. Chemin for curvature or arclength.

In a certain way, the cusp seems to be "an attractor" for vortex patches. It certainly motivates the study of the stability of cusp-like singularities.

Numerical computations were done to study them and are presented in a joint paper with A. Cohen (see [5]). An adaptive scheme based on multiscale decomposition into interpolatory wavelet bases was used. According to these computations, the cusp remains a cusp. Moreover, the order of the cusp seems to be conserved, at least for small time. We present the principle of the scheme in the first part of that paper.

In the second part, we state a general result of global persistence of conormal regularity with respect to vector fields vanishing at a singular point for two-dimensional incompressible Euler equations, which is proved in [7].

2. An adaptive algorithm for vortex patches

2.1. An equivalent formulation

We first show that our problem is equivalent to a system of non local equations for the evolution of the boundary of the patch. This is the purpose of the following basic computations.

From the Euler equations, we gather

$$\Delta v = \nabla^\perp \omega \quad \text{with} \quad \nabla^\perp = (-\partial_2, \partial_1).$$

Inverting the Laplacian, we thus get

$$v = \frac{1}{2\pi} \int \frac{(x-y)^\perp}{|x-y|^2} \omega(y) dy \quad \text{with} \quad z^\perp = (-z^2, z^1). \tag{1}$$

Let $\gamma_0 : S^1 \to \mathbb{R}^2$ be a parameterisation of $\partial \Omega_0$. Then $\gamma_t = \psi_t(\gamma_0)$ is a parameterisation of $\partial \Omega_t$. Thus we just need to solve the equation

$$\forall s \in S^1, \ \partial_t \gamma_t(s) = v(t, \gamma_t(s)).$$

Using $\omega_t = 1_{\Omega_t}$ and making an integration by parts in (1), we obtain

$$v(t,x) = \frac{1}{2\pi} \int_{\partial\Omega_t} T(\sigma)\log|x - \sigma|\, d\sigma,$$

where $T(\sigma)$ is the vector tangent to $\partial\Omega_t$ at σ. Applying this relation at $x = \gamma_t(s)$, we get

$$\partial_t\gamma_t(s) = \frac{1}{2\pi}\int_0^1 \left(\log|\gamma_t(s) - \gamma_t(s')|\right)\gamma_t'(s')\, ds',$$

where $\gamma_t'(s') = \partial_s\gamma_t(s')$. We make a second integration by parts with respect to s' now, to obtain a bounded integrand. We get $\partial_t\gamma_t(s) = A\gamma_t(s)$ with

$$A\gamma_t(s) = -\frac{1}{2\pi}\int_0^1 \frac{\gamma_t'(s').(\gamma_t(s) - \gamma_t(s'))}{|\gamma_t(s) - \gamma_t(s')|^2}(\gamma_t(s) - \gamma_t(s'))\, ds'.$$

Observe that the integrand tends to $\gamma_t'(s)$ when s' tends to s.

Remark 2.1. The functional A is not local and not linear. But we proved in [6] that $A\gamma$ has the same local regularity as γ. This will justify the adaptive scheme we are going to present.

2.2. Principle of the scheme

Let $G_j = \{2^{-j}k, 0 \le k \le 2^j - 1\}$. Suppose an interpolatory wavelet basis $(\phi_{j,\lambda})_{j\ge0,\lambda\in G_j}$ of $\mathcal{S}^1$ is given (we refer to [9] for an exact definition). We recall that interpolatory wavelets have the same properties as orthogonal wavelets (refinability, localisation, etc), except for orthogonality which is replaced by interpolating properties. The main motivation to use interpolatory wavelets is the existence of fast algorithms to compute coefficients from a function and conversely.

Adaptivity properties of the scheme are obtained thanks to the following thresholding procedure.

We decompose the parameterisation γ into

$$\gamma = \sum_{\lambda\in G_{j_0}} \gamma(\lambda)\phi_{j_0,\lambda} + \sum_{j\ge j_0}\sum_{\lambda\in G_{j+1}\backslash G_j} d_{j,\lambda}\phi_{j+1,\lambda}.$$

The first term of the right hand side is a coarse approximation of γ. For fixed j, each sum in the last term can be considered as details of γ at scale 2^{-j}.

The thresholding procedure consists in keeping only the coefficients bigger than a given threshold ϵ, in the last term. Thanks to the interpolating properties of the basis, this amounts to compute γ on an adaptive grid which is automatically refined near the singularity.

In fact, as our problem is an evolution problem, we need a dynamic procedure which would change the adaptive grid at each time step. This can be done in the following way. Suppose an approximation γ_n of γ at time $n\Delta t$ has been computed on an adaptive grid A_n.

- We "inflate" the grid A_n into a larger grid $\tilde{A}_n$.

- We compute $A\gamma$ on $\tilde{A}_n$. This yields γ_{n+1} on $\tilde{A}_n$.
- We do a thresholding to obtain a new adaptive grid A_{n+1}.

Remark 2.2. This algorithm was also used to compute the evolution of corner-like singularities. Both acute and obtuse angle seem to be unstable. The acute angle evolves instantaneously to a cusp while the obtuse angle evolves to a singular flat structure. The right angle seems to be a limit case.

3. Second section

To state precisely what we proved, let's define exactly the kind of structure we are interested in.

Definition 3.1. Let $x_0 \in \mathbb{R}^2$ and $\epsilon \in]0,1[$. Let $X = (X_\lambda)_{\lambda \in \Lambda}$ be a family of vector fields on the plane and $\phi = (\phi_h)_{h>0}$, a family of functions from $\mathbb{R}^2$ to $\mathbb{R}$. The couple (ϕ, X) will be called "ϵ-sharp geometry at x_0" if it verifies the four following conditions:

i) there exists a bilipschitzian Ψ of $\mathbb{R}^2$ such that $\Psi(0) = x_0$ and $g \in C_0^\infty(\mathbb{R}^2)$ 1-valued near 0 such that $\phi_h(x) = g(\Psi^{-1}(x)/h)$,

ii) each X_λ has coefficients and divergence in C^ϵ,

iii) $0 < \mathcal{I}(X) \stackrel{\text{def}}{=} \inf_{x \in \text{Supp}\,\phi_1} \sup_{\lambda \in \Lambda} \dfrac{|X_\lambda(x)|}{|x - x_0|} < +\infty$,

iv) $0 < \mathcal{J}(X) \stackrel{\text{def}}{=} \inf_{x \notin \text{Supp}\,\phi_1} \sup_{\lambda \in \Lambda} |X_\lambda(x)| < +\infty$.

We then define $\|\|X_\lambda\|\|_\epsilon = \|X_\lambda\|_\epsilon + \|\operatorname{div} X_\lambda\|_\epsilon$ and $I(X) = \min(\mathcal{I}(X), \mathcal{J}(X))$.

Definition 3.2. Let $x_0 \in \mathbb{R}^2$, $\alpha > 0$, $\epsilon \in]0,1[$ and $\sigma \in]0,\epsilon[$. Let $\bar{\alpha} = 2/(2+\alpha)$. Let (ϕ, X) be an ϵ-sharp geometry at x_0. We then define $C_\alpha^{\sigma,\epsilon}(X)$ as the space of bounded scalar functions ω on $\mathbb{R}^2$ such that:

i) $\|\omega\| \stackrel{\text{def}}{=} \|\omega\|_{L^1 \cap L^\infty} < +\infty$,

ii) $\sup_{\lambda \in \Lambda} \|X_\lambda(x, D)\omega\|_{\epsilon-1} < +\infty$,

iii) $\mathcal{N}_\sigma(X, \omega) \stackrel{\text{def}}{=} \sup_{\substack{\lambda \in \Lambda \\ h \in]0,1]}} h^{\frac{\sigma-\epsilon}{\bar{\alpha}}} \|X_\lambda(x, D)(\phi_h \omega)\|_{\sigma-1} < +\infty$,

iv) $\mathcal{N}_\phi^\alpha(\omega) \stackrel{\text{def}}{=} \sup_{h \in]0,1]} h^{-(2+\alpha)} \|\phi_h \omega\|_{L^1} < +\infty$,

where $X(x, D)u$ stands for $\operatorname{div}(X \otimes u) - u \operatorname{div} X$.

We put $N_\sigma(X, \omega) = \mathcal{N}_\sigma(X, \omega) + \sup_{\lambda \in \Lambda} \|X_\lambda(x, D)\omega\|_{\epsilon-1}$, $N_\phi^\alpha(\omega) = \mathcal{N}_\phi^\alpha(\omega) + \|\omega\|$ and

$$\|\omega\|_{\alpha,X}^{\sigma,\epsilon} = \frac{N_\phi^\alpha(\omega) \sup_{\lambda \in \Lambda} \|\|X_\lambda\|\|_\epsilon + N_\sigma(X, \omega)}{I_\gamma(X)}.$$

Remark 3.3. The space $C_\alpha^{\sigma,\epsilon}(X)$ contains the characteristic functions of what we usually expect to be a domain Ω with a cusp ($\alpha = 1/2$ for a "generic" cusp). For example, if in a neighbourhood of the singularity, the boundary can be represented by $x_2 = \pm x_1^{1+\alpha}$, then it is easy to prove that $1_\Omega \in C_\alpha^{\sigma,\epsilon}(X)$. One just has to notice that the vector field $x_1\partial_1 + (1+\alpha)x_2\partial_2$ is tangent to the patch.

We can now state the main result.

Theorem 3.4. *Let $x_0 \in \mathbb{R}^2$, $\epsilon \in]0,1[$, $\alpha > 0$, $\sigma \in]0,\epsilon[$ and (ϕ_0, X_0) be an ϵ-sharp geometry at x_0. Suppose that the initial velocity v_0 is divergence free, that the initial vorticity ω_0 belongs to $C_\alpha^{\sigma,\epsilon}(X_0)$, $\epsilon \in]\bar\alpha, 1[$ and that*

$$\sigma \in \left]0, \epsilon\left(\frac{\epsilon - \bar\alpha}{\epsilon + \bar\alpha}\right)\right[\quad with \quad \bar\alpha = \frac{2}{2+\alpha}.$$

Then (E) has a unique solution v with lipschitzian flow ψ. Moreover, for all time t, (ϕ_t, X_t) remains an ϵ-sharp geometry at x_t with $\phi_{t,h} = \phi_{0,h} \circ \psi_t^{-1}$, $X_{t,\lambda} = X_{0,\lambda}(x, D)\psi_t(x)$ and $x_t = \psi_t(x_0)$. Moreover $\nabla v \in L_{loc}^\infty(\mathbb{R}; L^\infty(\mathbb{R}^2))$ and $\omega(t) \in C_\alpha^{\sigma,\epsilon}(X_t)$. More precisely, there is a constant C depending only on the initial datum and such that

$$\|\nabla v(t)\|_{L^\infty} \leq Ce^{Ct},$$

$$\|\omega_t\|_{\alpha,X_t}^{\sigma,\epsilon} \leq C\|\omega_0\|_{\alpha,X_0}^{\sigma,\epsilon} \exp\left(C(e^{Ct} - 1)\right).$$

Remark 3.5. In the special case where the initial datum is a $C^{1+\epsilon}$ vortex patch except at a point x_0 where it has a cusp, we can find an ϵ-sharp geometry (ϕ, X) at x_0 such that the vector fields X_λ are tangent to the patch. The previous theorem can be applied and we obtain the stability of a sharp structure which is a bit more general than a cusp, for all time together with the conservation of the order of the sharpness. Moreover, as in the smooth case, an exponential estimate of $\|\nabla v\|_{L^\infty}$ is obtained. But the theorem 3.4 does not allow us to recover the existence of right and left tangents at the singularity if these limits exist at time $t = 0$. Recently, N. Depauw proved in [8] that this is indeed the case for small time. Whether this is still true for large time is an open question.

To prove 3.4, the point is to notice that, with the given initial datum, the velocity v is lipschitzian, as for smooth vortex patches. This is of course not true for a more general singularity (see for example [1]). More precisely, we have

Theorem 3.6. *If v verifies the assumptions of 3.4 then there exists a constant C depending only on regularity parameters and such that*

$$\|\nabla v\|_{L^\infty} \leq CN_\phi^\alpha(\omega) \log\left(e + \frac{\|\omega\|_{\alpha,X}^{\sigma,\epsilon}}{N_\phi^\alpha(\omega)}\right).$$

The proof of this last result is inspired by [4] and uses Bony's paradifferential calculus and harmonic analysis. It is just a matter of showing that the contributions of the patch near the singularity in the computation of ∇v remain small, though the boundary is singular. This is due to the fact that, according to condition *iv)* of definition 3.2, the area of the patch intersected with the disc of centre x_0 and radius h, behaves like $h^{2+\alpha}$.

This enables us to propagate the initial singularity for all time with conservation of the order of the cusp (all the details are in [7]).

References

[1] H. Bahouri and J.-Y. Chemin, *Équations de transport relatives à des champs de vecteurs non lipschitziens et mécanique des fluides*, Archiv for Rational Mechanics and Analysis, **127** (1994), 159–182.

[2] T. Buttke, *The observation of singularities in the boundary of patches of constant vorticity*, Physical fluids A, **1 (7)** (1989), 1283–1285.

[3] T. Buttke, *A fast adaptive vortex method for patches of constant vorticity*, Journal of computational physics, **89** (1990), 161–186.

[4] J.-Y. Chemin, *Fluides parfaits incompressibles*, Astérisque, **230** (1995).

[5] A. Cohen and R. Danchin, *Multiscale approximation of vortex patches*, to appear in SIAM Journal on Applied Mathematics.

[6] R. Danchin, *Évolution temporelle d'une poche de tourbillon singulière*, Communications in Partial Differential Equations, **22** (1997), 685–721.

[7] R. Danchin, *Évolution d'une singularité de type cusp dans une poche de tourbillon*, Prépublication du Centre de Mathématiques de l'École Polytechnique, **1997**.

[8] N. Depauw, *Solutions peu régulières des équations d'Euler et de Navier-Stokes incompressibles sur un domaine à bord*, thèse de doctorat de l'Université de Paris Nord, **1998**.

[9] G. Deslauriers et S. Dubuc, *Symmetric iterative interpolation processes*, Constructive Approximation, **5** (1989), 49–68.

[10] D. Dritschel and M. Mac Intyre, *Does contour dynamics go singular*, Physics of Fluids A, **2** (1990), 748–753.

[11] V. Yudovitch, *Non stationary flows of an ideal incompressible fluid*, Zurnal vychislitel'noj matematiki i matematiceskoj fiziki, **3** (1963), 1032–1066.

[12] N. Zabusky, M. Hughes and K. Roberts, *Contour dynamics for the Euler equations in two dimensions*, Journal of computational physics, **30** (1979), 96–106.

Laboratoire d'Analyse Numérique,
Université Pierre et Marie Curie,
4 Place Jussieu
75252 Paris Cedex 05, France
E-mail address: danchin@ann.jussieu.fr

International Series of Numerical Mathematics
Vol. 129, © 1999 Birkhäuser Verlag Basel/Switzerland

A Bow Shock Flow Containing (Almost) All Types of ('Exotic') MHD Discontinuities

H. De Sterck, H. Deconinck, S. Poedts, and D. Roose

Abstract. We present 2D numerical simulations of a planar field-aligned ideal Magnetohydrodynamic (MHD) bow shock flow in a regime where fast MHD switch-on shocks are possible. In this regime, the interaction of the 3 non-linear waves of the non-strictly hyperbolic MHD system leads to a complicated solution with interacting shocks of different MHD shock type. Fast and slow shocks and tangential discontinuities are present in the flow. Moreover, we clearly identify non-classical intermediate shocks and shocks with a compound structure, which are a manifestation of the non-convex nature of the hyperbolic MHD system. We discuss some numerical problems encountered when high-resolution numerical MHD schemes derived from common Computational Fluid Dynamics approaches are used to simulate this kind of flows. This bow shock flow simulation is extremely simple in setup, but contains a wealth of interacting MHD shocks, which makes it a good new test case for ideal MHD codes. Observations suggest that the MHD shock formation effects of our simulations may occur in processes in the corona of the sun.

1. Introduction and setup of the numerical simulation

The equations for the hyperbolic system of ideal one-fluid Magnetohydrodynamics (MHD) in conservative form are given by

$$
\frac{\partial}{\partial t}
\begin{bmatrix}
\rho \\
\rho \vec{v} \\
\vec{B} \\
e
\end{bmatrix}
+ \nabla \cdot
\begin{bmatrix}
\rho \vec{v} \\
\rho \vec{v}\vec{v} + I \left(p + \vec{B} \cdot \vec{B}/2 \right) - \vec{B}\vec{B} \\
\vec{v}\vec{B} - \vec{B}\vec{v} \\
\left(e + p + \vec{B} \cdot \vec{B}/2 \right) \vec{v} - \left(\vec{v} \cdot \vec{B} \right) \vec{B}
\end{bmatrix}
= 0. \tag{1}
$$

Here ρ and p are the density and pressure of the fully ionized plasma, $\vec{v}$ is the plasma velocity, $\vec{B}$ the magnetic field, and

$$
e = \frac{p}{\gamma - 1} + \rho \frac{\vec{v} \cdot \vec{v}}{2} + \frac{\vec{B} \cdot \vec{B}}{2} \tag{2}
$$

is the total energy density of the plasma. I is the unity matrix. The magnetic permeability $\mu = 1$ in our units. We take $\gamma = 5/3$ for the adiabatic index.

The MHD conservation laws describe a hyperbolic system which allows for 3 linear wave modes, the fast, the Alfvén and the slow wave, with (positive) anisotropic wave speeds satisfying $c_f > c_A > c_s$ in standard notation. The wave speeds can coincide, which makes the MHD system non-strictly hyperbolic. The MHD system is non-convex ([2, 5]).

In this paper we present numerical simulation results describing the stationary bow shock formed when a uniform flow of a magnetized fluid falls in on a perfectly conducting cylinder ($r = 0.125$) in 2D. This stationary ideal MHD problem is completely specified by the value of the plasma $\beta = 2p/B^2$ and the fast Mach number $M_f = v/c_f$ of the incoming flow. We take ($\rho = 1, p = 0.2, \vec{B} = (1,0)$, $\vec{v} = (1.5,0)$) for the incoming flow. Thus $\beta = 0.4$ and $M_f = 1.5$. The flow speed is faster than the fast MHD speed, so we expect that a steady bow shock will form. The flow speed lies in the region where fast switch-on shocks can be expected (see Sec. 2). The incoming magnetic field and velocity field lie in the same plane and are parallel. It can be proven that this alignment is conserved everywhere for a steady state solution, resulting in a planar field-aligned flow in the whole domain.

In Fig. 1a we show a general view of the resulting bow shock flow. The leading shock front contains a 'dimpled' concave-outward (from the cylinder) central part. The leading shock front is followed by a second front, and additional discontinuities can be seen. In this paper we give a consistent physical interpretation for all the features present in this complicated flow. (Fig. 2a).

Bow shocks are present in space physics plasmas when the solar wind encounters comets or planets. Many numerical simulations have been done to describe such bow shocks (e.g. [4]). Shock formation in the switch-on regime has been studied in the context of fast solar Coronal Mass Ejections (CMEs) ([12, 13]). However, to our knowledge, the steady bow shock in the switch-on regime has not been studied yet, and the interacting shock solution reported in this paper has not been described before. Much research is going on in the development of shock-capturing numerical schemes for MHD (e.g. [11, 4]) and attempts are being made to extend standard techniques from Computational Fluid Dynamics (CFD) ([5]) to MHD. We will discuss some aspects of how to use these techniques for flows with interacting MHD shocks.

This paper is organized as follows. In Sec. 2 we give a brief overview of the properties of MHD discontinuities, which will help us to explain why the traditional bow shock topology of Fig. 2a is not possible in the flow regime under consideration. In Sec. 3 we present a consistent physical interpretation of all the features present in the simulation results. In Sec. 4 we discuss the numerical technique used. In Sec. 5 we discuss the relevance of our simulation results for the problem of the physical existence of non-classical intermediate shocks and compound structures, which are composed of a shock with an attached rarefaction and are a manifestation of the non-convex nature of the MHD system ([2, 14, 8, 9]). We report on the possible observation of the shock formation effects described in this paper in the propagation of solar CMEs. We conclude in Sec. 6.

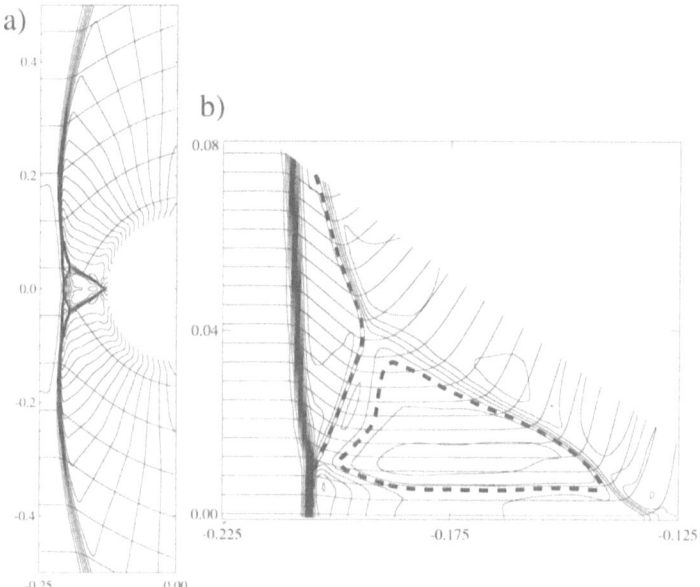

FIGURE 1. (a) Global view of the new bow shock solution. We show density contours and magnetic field lines. The flow comes in from the left. The cylinder fills the space of the white half disc on the right. (b) Detail of the solution upward from the symmetry axis. We show Alfvénic Mach number contours and magnetic field lines. The dashed line is a contour where the Alfvénic Mach number exactly equals 1.

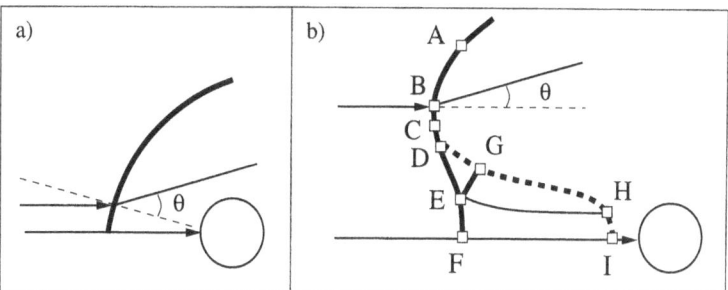

FIGURE 2. Possible bow shock topologies for a uniform flow (magnetic field lines have arrows) falling in from the left on a conducting cylinder. Shock normals are shown as dashed lines. (a) Traditional concave-inward (to the cylinder) geometry. (b) Interacting shock topology of our simulation results.

2. MHD discontinuities and the symmetries of the bow shock flow

The MHD conservation laws allow for many types of discontinuities ([6]). Fast shocks refract $\vec{B}$ away from the normal of the shock surface. The normal plasma velocity (in the shock frame) is greater than the normal fast MHD speed upstream of the fast shock, such that the upstream fast Mach number $M_{f_1} = v_{x_1}/c_{f_1}$ is greater than 1. (The regions upstream and downstream from the shock are traditionally denoted 1 and 2. The x direction is normal to the shock front, and y is tangential.) The flow is sub-fast downstream, so the downstream fast Mach number is smaller than 1. Intermediate shocks change the sign of the tangential component of $\vec{B}$, and the upstream Alfvénic Mach number $M_{A_1} > 1$, whereas downstream $M_{A_2} < 1$. Slow shocks refract $\vec{B}$ towards the normal.

There are limit cases for these three kinds of shocks. Fast switch-on shocks have $B_{y_1} = 0$ but B_{y_2} does not vanish (it is 'switched on'), so the magnetic field is refracted away from the normal over a finite angle θ. The downstream (normal) Alfvénic Mach number M_{A_2} exactly equals 1. Intermediate shocks can have $B_{y_1} = B_{y_2} = 0$, in which case they are called hydrodynamic (or parallel) shocks. Slow switch-off shocks switch off the tangential component of the magnetic field, and thus $B_{y_2} = 0$, and $M_{A_1} = 1$. Switch-on shocks will only occur for plasma speeds in the switch-on region

$$c_{f_1} < v_{x_1} < c_{f_1} \sqrt{\frac{\gamma(1-\beta)+1}{\gamma-1}} \tag{3}$$

where c_f is the (normal) fast speed. Switch-on shocks can only be found in low-β plasmas. Other MHD discontinuities include contact, tangential and rotational discontinuities.

So why do we not find a classical concave-inward bow shock solution (Fig. 2a) in our simulation? The answer can be found in the analysis of the symmetry of an MHD flow in the switch-on regime (as was first done in [12, 13]). If we look at the symmetry of a concave-inward bow shock solution (Fig. 2a), we see that the direction of the $\vec{B}$ field can not change on the stagnation streamline, which is a line of symmetry. When we move downward along the shock front and approach the stagnation streamline, the fast shock will approach a fast switch-on shock ($B_{y_1} = 0$), which will however still turn $\vec{B}$ over a finite angle θ. This is a discontinuity between streamlines, and this simple bow shock topology is thus not possible. In stead, the complex topology of Fig. 2b, satisfying the symmetry constraints by the presence of intermediate shocks, is found as the result of the numerical simulation.

3. Interpretation of the simulation results

In Fig. 1b we show a detail of the flow. We use the lettering of Fig. 2b to identify the shock parts. A-B and D-E are fast shocks because the field is refracted away from the normal. E-F is a hydrodynamic shock because the field lines are not refracted. E-G is an intermediate shock, because it clearly contains the $M_A = 1$ contour (meaning that the flow goes from super-Alfvénic to sub-Alfvénic while passing through the shock), and because the field lines are flipped over the normal. E-F is very reminiscent of a Mach stem as encountered in Hydrodynamic shock reflection. D-G-H-I is a slow switch-off shock, because the upstream Alfvénic Mach number is equal to 1, and because the downstream magnetic field is normal to the shock surface. E-H is a tangential discontinuity. Other tangential discontinuities are stretching out from points G and H along the streamlines to infinity.

In Fig. 3a–d we show plots of the variables along the stagnation streamline. Going from left to right, we see two jumps in the variables, for instance in the pressure. The first jump corresponds to the hydrodynamic shock at point F. The Alfvénic Mach number jumps from above 1 to under 1, indicating an interme-diate shock. The magnetic field does not show a jump, consistent with a purely hydrodynamic shock. Going further to the right, we see that we pass through a rarefaction. The hydrodynamic shock together with the rarefaction make up a compound structure. Then we reach a constant state, before the quantities jump again near the cylinder, where the slow switch-off shock is encountered at point I. We see that the entropy is well conserved on this streamline in the smooth parts of the flow.

In Fig. 3e–f we show a cut along a ray through the center of the cylinder cutting the leading front between points C and D ($\theta = 47.66$ degrees). We cross only one shock, which brings the Alfvénic (and the fast) Mach number from above

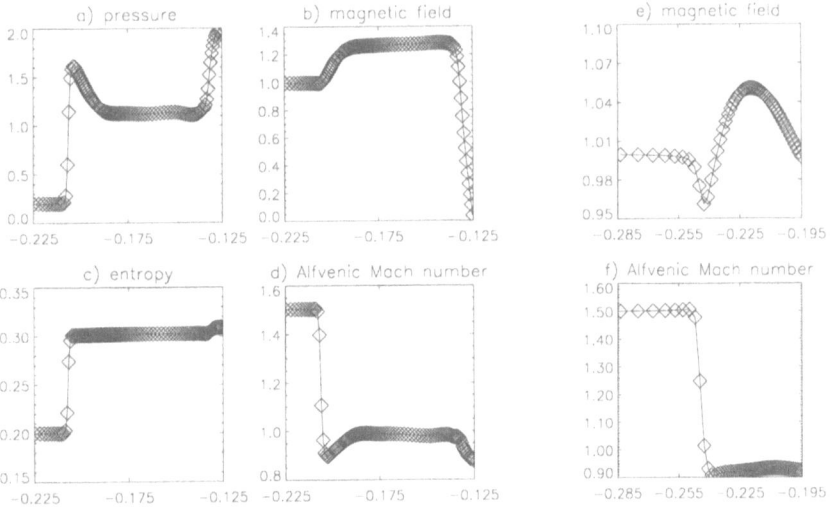

FIGURE 3. (a–d) Plots of some variables along the stagnation streamline. (e–f) Cut along a ray between points C and D.

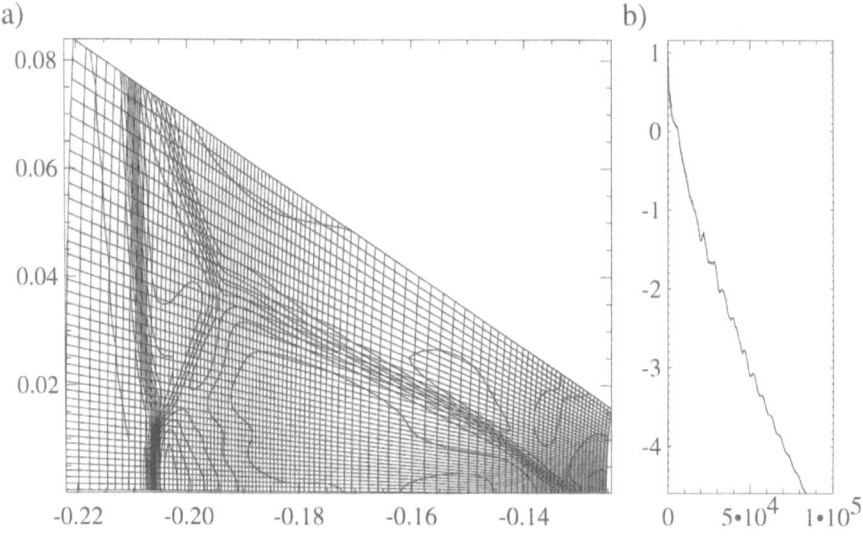

FIGURE 4. (a) Detail of the simulation grid with density contours.
(b) Convergence of the simulation to a steady state. The logarithm
of the root mean square of the density residual is shown as a
function of the number of iterations.

1 to under 1, so B-C-D is clearly a non-classical intermediate shock. Looking at
the magnetic field, we see a clear compound structure (a shock followed by a
rarefaction) for this shock as well.

4. Numerical method

We have simulated the flow in the upper left quadrant, on a 120×120 stretched
elliptic polar-like structured grid, extending to $x = -0.35$ on the x-axis, and to
$y = 1.4$ on the y-axis (see Fig. 4a for a detail of the grid). Starting from a uniform
flow initial condition, the flow evolves until a steady state bow shock solution is
obtained. We solve Eq. (1) using a finite volume high resolution Godunov shock
capturing scheme which is second order in space and time, using a MUSCL approach ([5]) with minmod-limiting on the slopes of the primitive variables. The
time-integration is explicit with a 2-step Runge-Kutta method.

We use the Lax-Friedrichs numerical flux function, which is simple and robust. Contact and tangential discontinuities are not very well resolved due to high
dissipation for these waves, but shocks are well resolved in a steady state calculation. Roe's scheme ([11]) in theory would resolve shocks and especially the

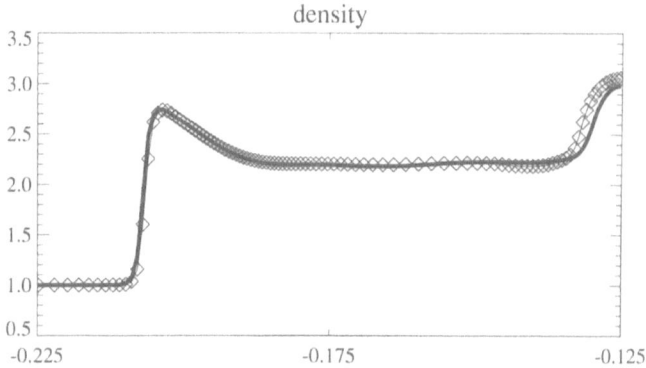

FIGURE 5. Comparison of the density on the stagnation stream-line, using a projection scheme (thin line with diamond symbols) or Powell's source term (thick line). The solutions are almost identical, except near the stagnation point close to the cylinder (on the right).

tangential discontinuities much better, but we found several problems while trying to apply this scheme to our simulation. Roe's scheme suffers from various instabilities, like the carbuncle-instability ([10]) where the shock front intersects the symmetry axis, and this makes that a steady state solution can not be obtained. Adding dissipation to the numerical scheme (for instance using the H-correction proposed in [7]) did seem to cure the carbuncle up to moderate grid resolution, but other instabilities remained. It seems to be difficult to find a good parameter free change to Roe's numerical scheme which solves the problems in a general way. Using the Lax-Friedrichs scheme however, we obtained good convergence of the density residual (Fig. 4b).

We use a projection scheme to keep the $\nabla \cdot \vec{B} = 0$ constraint satisfied up to machine accuracy in the chosen discretization ([1]). As can be seen in Fig. 5, the divergence wave technique proposed by Powell ([4]) gives almost exactly the same solution. However, at the stagnation point close to the cylinder, the two approaches give a slightly different solution.

We carefully checked if the results we obtained do not contain numerical artefacts or features dependent on the numerical resistivity. We did many simulations with grids of different cell sizes and we obtained uniform grid convergence to the solution shown in Fig. 1. We also did simulations on uniform grids and grids with a different accumulation of points than in the grid of Fig. 4a, and refinement studies using these different grids resulted in grid convergence to the same solution of Fig. 1. The fully consistent and complete physical interpretation of all the features in the solution gives further confidence that the features present in our solution are real and physically meaningful.

5. Intermediate and compound shocks and solar coronal mass ejections

Because the MHD system is not convex, compound shocks can exist in principle ([5, 2]). Intermediate shocks satisfy the MHD Rankine-Hugoniot relations. They are called non-classical shocks because more than 1 family of characteristics converges into the shock. MHD intermediate and compound shocks were commonly believed to be unphysical ([6]) or 'exotic'. Intermediate shocks and shocks with a compound structure are however clearly present in our results, and this for the first time in steady 2D ideal MHD simulations. This is a strong indication that those types of discontinuities have to exist, at least in the planar case. Our findings are consistent with earlier 1D ideal MHD simulations, where compound shocks are found ([2]). They are also consistent with recent results on the admissibility of MHD intermediate shocks ([14, 8, 9]). Compound shocks were shown to be necessary and stable ingredients for the solution of some planar MHD Riemann problems ([8, 9]). It is however not clear in how far the steady compound structures of our simulations are related to the time-dependent compound shocks encountered in Riemann problems.

The shock formation effects described in this paper could occur in low-β space plasmas. Solar CMEs are one example. Up to several times a day, large-scale structures in the solar corona disrupt and are ejected out of the corona. Some of these ejecta move away from the sun in the solar coronal plasma with super-fast nearly field-aligned speeds lying in the switch-on region, inducing shock fronts preceding the ejecta, much like the cylinder in our simulations can be thought of as to induce a shock front while moving with a constant speed into a static plasma. Some evidence for the shock formation effects seen in our simulations may

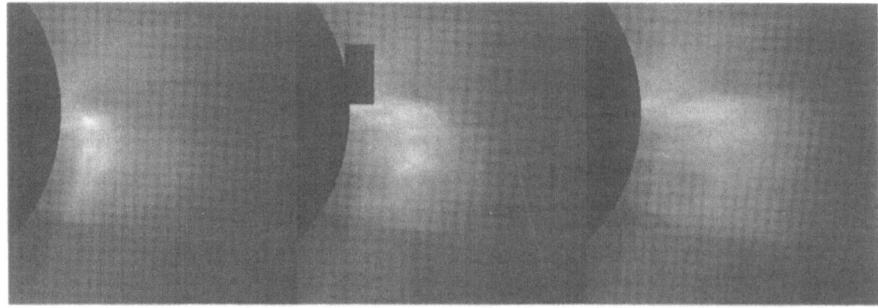

FIGURE 6. Observation of a solar CME with the LASCO C2 coronagraph. The black disc covers the sun and has a radius of approximately $2 \times R_{sun}$. The images, taken on September 28, 1997, at 16:12, 16:31, and 16:53 UT (from left to right), seem to show the evolution of an initially concave-inward (to the sun) CME front to a dimpled front with a double-loop appearance.

be found in coronagraph images of these CMEs. As already noted in ([12, 13]) and as can be seen in Fig. 6, the leading feature of many fast CMEs, which can be interpreted as the signature of a fast shock front, seems to show a dimpled front, consistent with the dimple in the leading front of our simulation results (Fig. 1a). Moreover, the density depletion in the V-shaped region between the two shock fronts in our simulation (Fig. 1a) divides the downstream flow into two distinct lobes with higher density. This structure may explain why some CMEs seem to evolve to a double-loop appearance (Fig. 6 and [3]).

6. Conclusion

In this paper we have discussed simulation results for a model problem describing a planar field-aligned MHD bow shock flow. The interaction of the three types of non-linear waves in the hyperbolic system leads to a complicated flow with a wealth of interacting discontinuities of different types. We have given a clear and consistent physical interpretation of this flow pattern. Non-classical intermediate shocks and shocks with a compound structure, which are a manifestation of the non-convex nature of the hyperbolic MHD system, are clearly identified. Our results are the first clear illustration in 2D of recent findings on the admissibility of non-classical intermediate shocks and compound shocks that were obtained in a 1D framework. High-resolution numerical MHD schemes derived from common CFD approaches are successful in describing this kind of flows, if care is taken about the $\nabla \cdot \vec{B} = 0$ constraint and about the carbuncle phenomenon. Our simulation, very simple in setup, but producing a wealth of interacting discontinuities, is a good new test case for ideal MHD codes. The shock formation effects present in this model problem may occur in the propagation of fast solar CMEs and other phenomena in low-β space plasmas.

Acknowledgement. HDS acknowledges numerous helpful discussions with B. C. Low and A. Hundhausen. The SOHO/LASCO data used here are produced by a consortium of the Naval Research Laboratory (USA), Max-Planck-Institut fuer Aeronomie (Germany)), Laboratoire d'Astronomie (France), and the University of Birmingham (UK). SOHO is a project of international cooperation between ESA and NASA. The National Center for Atmospheric Research is sponsored by the National Science Foundation (USA). HDS is a Research Assistant of the Fund for Scientific Research – Flanders (Belgium), and is also at the Centre for Plasma Astrophysics. SP is a Research Associate of the Fund for Scientific Research – Flanders (Belgium).

References

[1] J. U. Brackbill and D. C. Barnes, J. Comput. Phys. **35**, 426 (1980).

[2] M. Brio and C. C. Wu, J. Comput. Phys. **75**, 400 (1988).

[3] H. De Sterck, *The MHD shock separation effect and the double-loop appearance of fast CMEs*, submitted to Geophys. Res. Lett. (1998).

[4] T. I. Gombosi, K. G. Powell, and D. L. De Zeeuw, J. Geophys. Res. **99**, 21,525 (1994).

[5] R. J. Leveque, *Numerical methods for conservation laws, Lectures in Mathematics ETH Zurich*, Birkhäuser Verlag, Basel, **1992**.

[6] M. A. Liberman and A. L. Velikovich, *Physics of shock waves in gases and plasmas*, Vol. 19 of *Springer Series in Electrophysics*, Springer-Verlag, Berlin, **1985**.

[7] E. Morano, R. Sanders, and M.-C. Druguet, *Multidimensional dissipation for upwind schemes: stability and applications to gas dynamics*, preprint (1997).

[8] R. S. Myong and P. L. Roe, J. Plasma Physics **58**, 485 (1997).

[9] R. S. Myong and P. L. Roe, J. Plasma Physics **58**, 521 (1997).

[10] J. J. Quirk, Int. J. Numer. Methods Fluids **18**, 555 (1994).

[11] P. L. Roe and D. S. Balsara, SIAM J. Appl. Math. **56**, 57 (1996).

[12] R. S. Steinolfson and A. J. Hundhausen, J. Geophys. Res. **95**, 6389 (1990).

[13] R. S. Steinolfson and A. J. Hundhausen, J. Geophys. Res. **95**, 20,693 (1990).

[14] C. C. Wu, J. Geophys. Res. **100**, 5579 (1995).

High Altitude Observatory,
National Center for Atmospheric Research,
P.O. Box 3000
Boulder, CO 80307-3000, USA
E-mail address: `desterck@ucar.edu`

von Karman Institute for Fluid Dynamics,
Waterloose Steenweg 72
1640 Sint-Genesius-Rode, Belgium
E-mail address: `deconinck@vki.ac.be`

Centre for Plasma Astrophysics,
K.U. Leuven,
Celestijnenlaan 200B
3001 Leuven, Belgium
E-mail address: `stefaan.poedts@wis.kuleuven.ac.be`

Department of Cumputer Science,
Celestijnenlaan 200A
3001 Leuven, Belgium
E-mail address: `dirk.roose@cs.kuleuven.ac.be`

International Series of Numerical Mathematics
Vol. 129, © 1999 Birkhäuser Verlag Basel/Switzerland

Application of Kinetic Schemes to All Types of Meshes

S.M. Deshpande

Abstract. Kinetic schemes as pursued in CFD Centre are obtained by taking suitable moments of upwind schemes for Boltzmann equation without collision term. The primary ones among these are KFVS, LSKUM, KFMG and these have been applied successfully to a variety of flow problems using various meshes. These schemes have been found to be very robust.

1. Introduction

The kinetic schemes exploit the connection between the Boltzmann equation of kinetic theory of gases and Euler or Navier-Stokes equations of compressible fluid dynamics. It is well-known that suitable moments of the Boltzmann equation

$$\frac{\partial f}{\partial t} + \vec{v}.\nabla f \;=\; J(f, f) \tag{1}$$

yield the Euler equations when the velocity distribution function f is a Maxwellian F and the Navier-Stokes equations when f is a Chapman-Enskog distribution f_{CE}. Hence a numerical scheme for the solution of

$$\frac{\partial F}{\partial t} + \vec{v}.\nabla F \;=\; 0 \tag{2}$$

maps to a numerical scheme for the solution of Euler equations

$$\frac{\partial U}{\partial t} + \frac{\partial G_x}{\partial x} + \frac{\partial G_y}{\partial y} + \frac{\partial G_z}{\partial z} \;=\; 0 \tag{3}$$

where U is a conserved vector and G_x, G_y and G_z are flux vectors. This is called moment method strategy by Deshpande [1, 2] and has been exploited by Deshpande [3] and Mandal & Deshpande [4] for developing Kinetic Flux Vector Splitting (KFVS) method. We observe that (2) is a linear hyperbolic equation for a scalar F while (3) is a nonlinear vector conservation law. Further, (3) is a classic example of hyperbolic nonlinear PDE which has been the subject of very intense mathematical and numerical study for the past several years. Several possible approaches for studying (3) have been pursued but we will focus our attention on the kinetic schemes, especially those which have been developed and studied in CFD Centre, Bangalore by the author and his co-workers for the past one and half decade.

2. Basic theory of KFVS and some 2-D computations

For illustrating the basic idea behind KFVS, consider 1-D Euler equations

$$\frac{\partial U}{\partial t} + \frac{\partial G}{\partial x} = 0 \tag{4}$$

where $U = \begin{bmatrix} \rho \\ \rho u \\ \rho e \end{bmatrix}, G = \begin{bmatrix} \rho u \\ p + \rho u^2 \\ (\rho e + p) u \end{bmatrix}, e = \frac{p}{\rho(\gamma-1)} + \frac{1}{2}u^2$ and ρ, u, p are respec-

tively mass density, fluid velocity and pressure, e is the specific total energy per unit mass. Define moment function vector and Maxwellian F by

$$\psi = \begin{bmatrix} 1 \\ v \\ I + \frac{v^2}{2} \end{bmatrix}, F = \frac{\rho}{I_0}\sqrt{\frac{\beta}{\pi}} \exp\left[-\beta(v-u)^2 - \frac{I}{I_0}\right]$$

where v = molecular or particle velocity, I_0 = internal energy = $\frac{3-2\gamma}{4(\gamma-1)\beta}, \beta = \frac{1}{2RT}$ and I = internal energy variable. It can be easily verified that (4) can be obtained as

$$\left(\psi, \frac{\partial F}{\partial t} + v\frac{\partial F}{\partial x}\right) = 0 \tag{5}$$

where $(\psi, F) = \psi$-moment of $F = \int_0^\infty dI \int_{-\infty}^{+\infty} \psi F dv$. The relation (5) is the basis of KFVS. For, the flux vector

$$G = (v\psi, F) \tag{6}$$

naturally splits into G^+ and G^- given by

$$G^+ = \left(\frac{v+|v|}{2}\psi, F\right); \quad G^- = \left(\frac{v-|v|}{2}\psi, F\right) \tag{7}$$

After working out the definite integrals in (6) and (7) in terms of exponentials and error functions, we get

$$G^\pm = \begin{bmatrix} \rho u A^\pm(S) \pm \frac{\rho}{2\sqrt{\pi\beta}} B(S) \\ (p + \rho u^2) A^\pm(S) \pm \frac{\rho u}{2\sqrt{\pi\beta}} B(S) \\ (pu + \rho ue) A^\pm(S) \pm \frac{\rho}{2\sqrt{\pi\beta}}\left(\frac{p}{2\rho} + e\right) B(S) \end{bmatrix} \tag{8}$$

where $S = u\sqrt{\beta}, A^\pm(S) = \frac{1\pm erf(S)}{2}, B(S) = \exp\left(-S^2\right)$. The KFVS split Euler equation is therefore given by

$$\frac{\partial U}{\partial t} + \frac{\partial G^+}{\partial x} + \frac{\partial G^-}{\partial x} = 0 \tag{9}$$

and it leads to the KFVS scheme based state update formula

$$U_j^{n+1} = U_j^n - \frac{\Delta t}{\Delta x}\left(G_j^{+n} - G_{j-1}^{+n}\right) - \left(G_{j+1}^{-n} - G_j^{-n}\right) \tag{10}$$

Here we have assumed a uniform mesh of size Δx on the interval $a \le x \le b, U_j^n = U(x_j, t_n)$ = value of U at time level t_n and mesh point x_j. Note the forward differencing of $\frac{\partial G^-}{\partial x}$ and backward differencing of $\frac{\partial G^+}{\partial x}$ for enforcing upwinding. Underlying the KFVS method is an implicit particle model of fluid motion and G_j^+ corresponds to the flux due to particles moving from left to the right and similar interpretation holds good for G_j^-. The update formula (10) can also be written in the cell-centered finite volume form

$$U_j^{n+1} \;=\; U_j^n - \frac{\Delta t}{\Delta x}\left(G_{j+\frac{1}{2}}^n - G_{j-\frac{1}{2}}^n\right) \tag{11}$$

where $G_{j+\frac{1}{2}}^n = G_j^{+n} + G_{j+1}^{-n}$. The flux at cell face $j + \frac{1}{2}$ can be considered as a sum of two parts, one due to G^+ and other due to G^-. The update formula (11) can be made more accurate both in space and time by using reconstruction and more accurate time marching scheme such as RK2 or RK4. Obviously, some kind of limiters are required for getting wiggle-free solution. Mandal & Deshpande [4, 5] applied cell-centered finite volume KFVS method to 1-D shock propagation, bump-in-a-channel problem at transonic speed and shock reflection problem. Mathur [6] and Mathur & Weatherill [7] applied it to many 2-D problems with structured, unstructured meshes with mesh enrichment for solution adaptivity. Two results are shown in Fig. 1 and Fig. 2. The first one is a transonic lifting case with shocks on both upper and bottom surfaces of NACA-0012 airfoil. Fig. 2 shows the C_p−plot for William's airfoil at a low subsonic Mach number $M_\infty = 0.15$. For the latter case, sufficiently fine initial triangular grid, reconstruction for enhancing space accuracy and mesh enrichment were required to obtain pressure distribution close to the exact results.

FIGURE 1. Flow past NACA-0012 airfoil, $M_\infty = 0.85, \alpha = 1.0°$

3. Kinetic treatment of boundary conditions

Mandal & Deshpande [4] developed kinetic treatment of flow tangency boundary condition (called KCBC) using specular reflection model, that is, a particle after impact with a solid wall is reflected with its normal velocity reversed. It is therefore

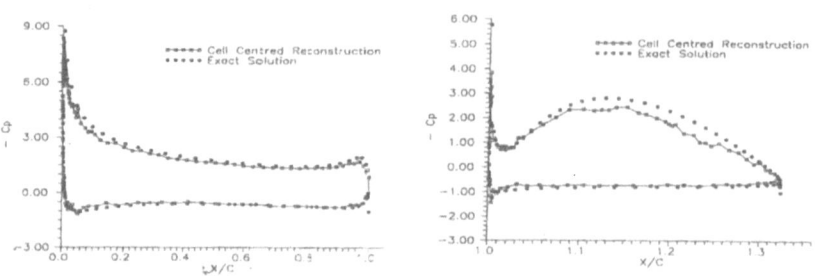

FIGURE 2. C_p plot for William's airfoil, $M_\infty = 0.15, \alpha = 0°$

possible to obtain velocity distribution F_r of reflected particles from the velocity distribution F_i for incident particles. The two are related by the simple relation

$$F_r\left(\vec{v}\right) = F_i\left(\vec{v} - 2\left(\vec{v}.\hat{n}_b\right)\hat{n}_b\right) \tag{12}$$

where $\hat{n}_b$ = unit outward normal at a boundary point b. The outgoing fluxes for reflected particles can be derived by taking ψ-moment of (12). For this purpose consider a cell near the body as shown in Fig. 3(a). The cell face on the body is

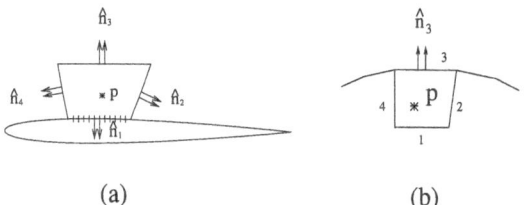

(a)   (b)

FIGURE 3. Finite volume cells near the body surface and the outer boundary

shown hatched, and the problem is to find wall flux G_w using the relation (12). In terms of positive and negative fluxes we obtain

$$G_w^{(l)} \quad = \quad G^{(l)+}\left(S_{n_p}, \beta_p\right) + G^{(l)-}\left(S_{n_p}, \beta_p\right) \tag{13}$$

where $S_{n_p} = u_{n_p}\sqrt{\beta_p}, u_{n_p} = \vec{q}_p.\hat{n}_b, \vec{q}_p$ = fluid velocity at P and $G^{(l)\pm}$ correspond to l-component ($l = 1$ corresponds to mass flux, 2 and 3 to momentum flux etc.) of normal flux, and superscripts + and − are for split fluxes. It is also possible to obtain a kinetic treatment of far field boundary condition [8]. This is different from the usual Riemann invariant method based treatment. Again, we consider a cell near the outer boundary as shown in Fig. 3(b). Dividing the particles into outgoing (obtained from interior data) and incoming (taken as those corresponding to free stream conditions F_∞) we can obtain Kinetic treatment of Outer Boundary Condition (KOBC). The flux on a cell face 3 (refer Fig. 3(b)) for KOBC is given

TABLE 1. Comparison of values obtained by using BHEEMA with KSTBC with experimental data

α	C_N				C_{N_δ} from BHEEMA	C_{N_δ} from experimental data
	$\delta = 0°$	$\delta = 2°$	$\delta = 5°$	$\delta = 10°$		
0.0	0.0	0.0117	0.0297	0.0595	0.341	0.316
2.0	0.1307	0.1421	0.1597	0.1891	0.336	0.316

by

$$G_3^{(l)} = G^{(l)+}\left(S_{n_p}, \beta_p\right) + G^{(l)-}\left(S_{n_\infty}, \beta_\infty\right) \tag{14}$$

where $S_{n_p} = \vec{q}_p.\hat{n}_3$, $S_{n_\infty} = \vec{q}_\infty.\hat{n}_3$, $\beta_\infty =$ value of β in the free stream. We notice that the flux on a cell face lying on the boundary (far field or solid wall) has the same form as the flux on any cell face in the interior of the computational domain, that is, exterior cells and boundary cells are treated similarly in this approach. Also, the formula (14) remains the same regardless of whether the boundary is inflow or outflow, subsonic or supersonic. For a supersonic inflow boundary the second term in (14) becomes negligible compared to the first term. No special care is therefore necessary for various types of the boundaries described above. One of the very important applications of kinetic approach to treatment of boundary conditions is the use of STBC in the computation of flow field around flight vehicles with deflected surfaces. Deshpande et al. [9] have developed Kinetic treatment of Surface Transpiration Boundary Condition (KSTBC) wherein a nonzero transpiration velocity is imposed on the surface of undeflected fins for computing flow with deflected fins.

4. BHEEMA code

Sekar et al. [10] have developed 3-D cell centered finite volume code based on KFVS using local time stepping. This code called BHEEMA(Boltzmann Hyperbolic Euler Equation Solver for Missile Aerodynamics) jointly developed by CFD Centre, IISc and DRDL, Hyderabad. It uses stacked grids for flight vehicles consisting of hemisphere, cones, frustums, flares, cylinders, wire tunnels, nozzles and control surfaces. This code is a standard design & analysis tool routinely used in DRDL and has proven to be very robust. As mentioned before BHEEMA has been used along with KSTBC for obtaining incremental changes in aerodynamic coefficients due to fin deflections. It is not necessary to regrid a configuration everytime fins are deflected, it is only necessary to compute the flow past undeflected configuration twice. First we use BHEEMA with flow tangency boundary condition everywhere, and thereafter we use BHEEMA again with a suitable transpiration velocity to take deflection into account. Table 1 shows the aerodynamic coefficient $C_{N\delta}$ obtained by using BHEEMA with KSTBC and the experimental data obtained from wind tunnels [9]. One more remarkable application of BHEEMA is the prediction of frequency of shock oscillation ahead of spiked intake of a ramjet [11] closed at exit. Fig. 4 shows the schematic diagram of the configuration and the computational domain which has to extend into the far-field to capture the expelled shock. BHEEMA had to be slightly modified to make it time accurate with RK2 method and flow tangency had to be imposed on the outer surface of the intake duct. Several hours of run were required to capture the periodic behavior. Fig. 5 shows several cycles of pressure variation with time at the entry and exit of cowl wall. The computed frequency of 63.5 Hz very closely agrees with the in-flight measured value of 64 Hz.

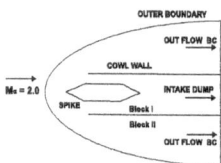

FIGURE 4. Schematic diagram of computational domain over spiked intake of a ramjet

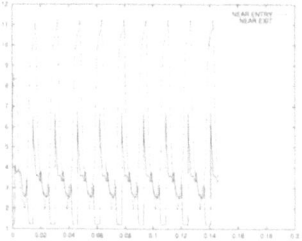

FIGURE 5. Pressure variation with time at entry and exit of the cowl wall

Recently Kulkarni & Deshpande [12, 13] generalized KFVS method to moving grids arising in problems involving fluid-structure interaction. This generalized method called KFMG (Kinetic Flux vector splitting on Moving Grid) method splits the flux vector into positive and negative parts based on $(v_n - w_n) \gtrless 0$ where $\hat{n}$ = normal to a surface, w_n = velocity of surface in the normal direction. Consequently, the new split fluxes $G^{\pm}_{new}$ for a moving surface can be expressed in terms of the old split fluxes given by (8). Krishnamurthy & Deshpande [14] developed 2-D BHEEMA code using KFMG and applied it to an oscillating airfoil problem. Fig. 6 shows $C_L - \alpha$ and $C_m - \alpha$ plots obtained using BHEEMA-KFMG. Work is going on in DRDL and CFD Centre, IISc to extend this code to 3-D geometries to predict aeroelastic behavior of slender configurations.

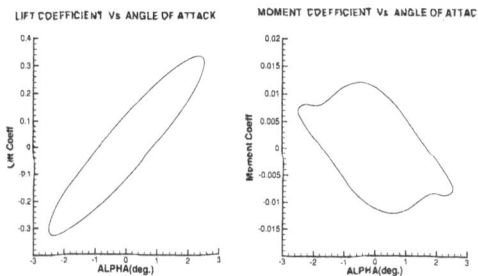

FIGURE 6. Variation of C_l and C_M with α for Oscillating Airfoil

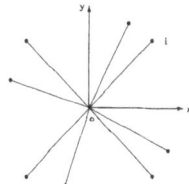

FIGURE 7. Sketch of typical connectivity for LSKUM

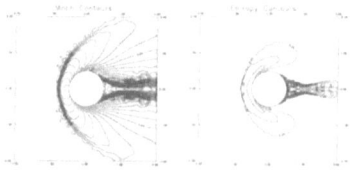

FIGURE 8. Mach and Entropy contours for a supersonic flow past a cylinder, $M_\infty = 3.0$

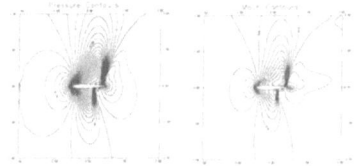

FIGURE 9. Pressure and Mach contours for flow past NACA-0012 airfoil, $M_\infty = 0.85, \alpha = 1°$ obtained from LSKUM

5. LSKUM & SUPERBHEEMA

As the cell-centered finite volume code BHEEMA was applied to various industrial problems, it became clear that there is a need to have a numerical scheme which is capable of operating on any mesh — structured, unstructured, stacked, cartesian, chimera, dragon or just a distribution of points which need <u>not</u> be tessellated. For

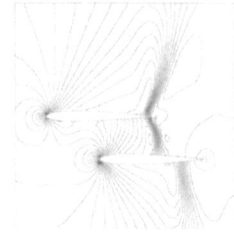

FIGURE 10. Chimera Mesh and Pressure contours for a NACA-
0012 biplane

many practical problems it is almost impossible to generate a grid which is good
everywhere. It is invariably bad at least in some regions, that is, it will be highly
skewed, has rapidly varying volumes and aspect ratios, triangles or quadrilaterals
with very acute or obtuse angles. If cartesian grids are used then one always has
extremely tiny triangles and merging these with neighboring triangles or polygonal
cells is required. The development of SUPERBHEEMA code using LSKUM (Least
Squares Kinetic Upwind Method) avoids many of these problems. Ghosh & Deshpande [15] have solved the problem of discretising spatial derivatives using least
squares. Assuming a node 0 is surrounded by points i (see Fig. 7), the problem
is to determine the derivatives f_{x_0}, f_{y_0} in terms of the data in the neighborhood
$N(0)$ of node 0. This is obviously a classic case of an over-determined problem and
by minimizing e^2 defined by

$$e^2 \;=\; \sum_{i=1}^{n} (\Delta f_i - \Delta x_i f_{x_0} - \Delta y_i f_{y_0})^2, \quad i \in N(0) \tag{15}$$

we get

$$f_{x_0} \;=\; \frac{\left(\sum \Delta y_i^2\right)\left(\sum \Delta f_i \Delta x_i\right) - \left(\sum \Delta x_i \Delta y_i\right)\left(\sum \Delta f_i \Delta y_i\right)}{Det}$$

$$f_{y_0} \;=\; \frac{\left(\sum \Delta x_i^2\right)\left(\sum \Delta f_i \Delta y_i\right) - \left(\sum \Delta x_i \Delta y_i\right)\left(\sum \Delta f_i \Delta x_i\right)}{Det} \tag{16}$$

$$Det \;=\; \left(\sum \Delta x_i^2\right)\left(\sum \Delta y_i^2\right) - \left(\sum \Delta x_i \Delta y_i\right)^2$$

and $\Delta x_i = x_i - x_0, \Delta y_i = y_i - y_0, \Delta f_i = f_i - f_0$. The derivatives are thus known
in terms of the neighboring data and it is clear from the above formulae that we
do not necessarily need a grid. The least squares formulae operate on a structured
grid, unstructured grid, chimera mesh, dragon mesh or for that matter just a
distribution of points. Many aspects of LSKUM such as rotated LSKUM, its connection with Kalman filtering, conservation property and two step formulae for
second order accuracy have been considered by Ghosh [15] and Deshpande et al.
[16]. Here we will show some typical results obtained using LSKUM. Fig. 8 shows
Mach contours for a supersonic flow past a cylinder obtained by using second order
LSKUM with limiters. Fig. 9 shows the pressure and Mach contours obtained by

Ramesh using second order LSKUM for NACA-0012 airfoil at $M_\infty = 0.85$ using unstructured mesh consisting of 4074 vertices, 160 points on the body and outer boundary at a distance of 10 chords. Fig. 10 shows a chimera mesh (consisting of two structured meshes) and pressure contours obtained by preprocessing and running LSKUM for a NACA-0012 biplane. Preprocessing includes blanking out the points of one mesh falling inside the other airfoil and obtaining connectivity for each node via a quadtree search program. Many more computations have been performed by Dhokrikar [17] using 2D SUPERBHEEMA code for cartesian mesh, structured body fitted mesh embedded in a background cartesian mesh and only a few results are shown here for demonstrating the power and flexibility of LSKUM. The development of 3-D SUPERBHEEMA for computing flows around complex configurations is currently underway as a joint project between CFD Centre, National Aerospace Laboratories (NAL) and Defence Research Development Laboratory (DRDL).

6. Conclusions

The kinetic scheme has come a long way after its humble beginning in 1983 and now it is routinely applied to solve many analysis and design problems in aerodynamics. It is very robust and has been applied to unsteady flows using KFMG idea, and its new more powerful version LSKUM holds out a great promise for solving flows with all types of meshes. Also, kinetic treatment of various boundary conditions has turned out to be very fruitful.

Acknowledgement. The author is very thankful to the Indo-French Centre for Promotion of Advanced Research (IFCPAR/CEFIPRA), Delhi for sponsoring this work under the project 1601-1 undertaken jointly with University Pierre et Marie Curie, Paris. Also, travel support given by IFCPAR for presenting this paper in 7th International Conference on Hyperbolic Problems, ETH Zurich is gratefully acknowledged.

References

[1] S. M. Deshpande, *A Second Order Accurate Kinetic Theory Based Method for Inviscid Compressible Flows*, NASA TP-2613, 1986.

[2] S. M. Deshpande, *On the Maxwellian Distribution, Symmetric Form and Entropy Conservation for the Euler Equations*, NASA TP-2583, 1986.

[3] S. M. Deshpande, *Kinetic Theory Based New Upwind Methods for Inviscid Compressible Flows*, AIAA Paper 86-0275, 1986.

[4] J. C. Mandal and S. M. Deshpande, *Kinetic Flux Vector Splitting for Euler Equations*, Computers & Fluids, **23** (1994), 447–478.

[5] J. C. Mandal and S. M. Deshpande, *Higher order accurate Kinetic Flux Vector Splitting method for Euler Equations*, Proceedings of 2nd International Conference on Nonlinear Hyperbolic Problems, **24** (1989), 384–392.

[6] J. S. Mathur, *Application of Kinetic schemes to structured, unstructured, cartesian and hybrid grids,* PhD Thesis, Indian Institute of Science, November 1997.

[7] J. S. Mathur and N. P. Weatherill, *The simulation of inviscid, compressible flows using an upwind kinetic method on unstructured grids,* International Journal of Numerical Methods in Fluids, **15** 1992, 59–82.

[8] V. Ramesh, J. S. Mathur and S. M. Deshpande, *Kinetic treatment of the far-field boundary condition,* Fluid Mechanics Report, 97FM2, Department of Aerospace Engineering, Indian Institute of Science, Bangalore, 1997.

[9] S. M. Deshpande, S. Sekar and M. Nagarathinam, *Kinetic Surface Transpiration with KFVS Euler code for treating control surface deflection,* Sixth Asian Conference on Fluid Mechanics, May 22–26, 1995, Singapore.

[10] S. Sekar, M. Nagarathinam, R. Krishnamurthy, P. S. Kulkarni and S. M. Deshpande, *3-D KFVS Euler code BHEEMA as aerodynamic design and analysis tool for complex configurations,* Proceedings of Fourteenth International Conference on Numerical Methods in Fluid Dynamics, **453** 1995, 525–529.

[11] P. S. Kulkarni, S. M. Deshpande, S. Sekar, M. Nagarathinam *Pressure Oscillation studies for AKASH Air Intake using CFD tools,* JATP-007-97, Internal report of Joint Advanced Technological Programme of Indian Institute of Science, Bangalore.

[12] P. S. Kulkarni and S. M. Deshpande, *KFVS on moving grid for unsteady aerodynamics,* Fluid Mechanics Report, 95FM6, Department of Aerospace Engineering, Indian Institute of Science, Bangalore, June 1995.

[13] P. S. Kulkarni and S. M. Deshpande, *Kinetic Flux Vector Splitting on moving grid for unsteady aerodynamics,* Third ECCOMAS CFD conference, Paris 1996.

[14] R. Krishnamurthy, B. S. Sarma and S. M. Deshpande, *Application of KFMG to an oscillating airfoil problem,* Fluid Mechanics Report (under preparation), Department of Aerospace Engineering, Indian Institute of Science, Bangalore, June 1995.

[15] A. K. Ghosh, *Robust Least Squares Kinetic Upwind Method for inviscid compressible flows,* Ph.D. Thesis, 1996, Department of Aerospace Engineering, Indian Institute of Science, Bangalore, India.

[16] S. M. Deshpande, P. S. Kulkarni and A. K. Ghosh, *New Developments in Kinetic schemes,* Computers in Mathematics Applications, **35** V2, 1998, 75–93.

[17] D. B. Dhokrikar, *Application of Least Squares Kinetic Upwind Method to different grids,* M.E. Thesis, Department of Aerospace Engineering, Indian Institute of Science, Bangalore, January, 1998.

AR & DB Centre of Excellence for Aerospace CFD,
Department of Aerospace Engineering,
Indian Institute of Science,
Bangalore,
India.
E-mail address: `suresh@aero.iisc.ernet.in`

International Series of Numerical Mathematics
Vol. 129, © 1999 Birkhäuser Verlag Basel/Switzerland

On the Regularity of Solutions of the Compressible Isentropic Navier-Stokes Equations

Benoît Desjardins

Abstract. Small time regularity of solutions of the compressible isentropic Navier-Stokes equations is investigated in dimension $N = 2$ or 3 under periodic boundary conditions. The initial density is not required to have a positive lower bound. We prove that weak solutions in $\mathbf{T}^2$ remain smooth as long as the density is bounded in $L^\infty(\mathbf{T}^2)$.

1. Introduction

As mentioned in many papers related to compressible fluid dynamics [8], [9], [17], [18], [19], [21], vacuum is a major difficulty arising in the study of global existence and strong regularity of solutions to the compressible Navier-Stokes equations. As a matter of fact, starting from initial densities that have positive lower bounds, local existence of smooth solutions can be proven by classical means, since lower bounds on the density persist for small enough time. We are interested here in well-posedness results for the compressible isentropic Navier-Stokes equations with periodic boundary conditions in dimension $N = 2$ or $N = 3$ when the initial density is not bounded away from zero. We prove in addition that the maximum norm of the density controls the breakdown of solutions of the 2 dimensional Navier-Stokes equations. In other words, if a solution of the Navier-Stokes equations is initially suitably smooth and loses its regularity at some later time, then the maximum norm of the density grows without bounds as the critical time approaches. Let us first recall the periodic compressible isentropic Navier-Stokes equations in $\mathcal{D}'((0, T) \times \mathbf{T}^N)$ $(N \geq 2)$

$$\partial_t(\rho u) + \operatorname{div}(\rho u \otimes u) - \mu \Delta u - (\lambda + \mu) \nabla \operatorname{div} u + \nabla p(\rho) = \rho f,$$

$$\partial_t \rho + \operatorname{div}(\rho u) = 0, \quad \text{and} \quad p(s) = as^\gamma, \quad \text{where} \quad a > 0 \quad \text{and} \quad \gamma > 1. \tag{1}$$

The unknowns ρ, u respectively correspond to the density of the gas $\rho \geq 0$ and its velocity $u \in \mathbf{R}^N$. The viscosity coefficients λ, μ are assumed to satisfy $\mu > 0$ and $\lambda + 2\mu > 0$, and the external forces f to vanish in order to simplify the presentation (see [5] for the general case). Finally, we complement the above system with initial conditions

$$\rho_{|t=0} = \rho_0 \geq 0, \quad \text{and} \quad \rho u_{|t=0} = m_0. \tag{2}$$

Before the remarkable work of P.-L. Lions, very little was known about global solutions to the compressible isentropic Navier-Stokes equations with finite energy data, at least when $N \geq 2$. In [12], [13], [14], Lions proved a global existence Theorem and weak stability results under the following assumptions on the initial data

$$\rho_0 \in L^1(\mathbf{T}^N) \cap L^\gamma(\mathbf{T}^N), \quad \rho_0 \geq 0, \quad \text{and} \quad |m_0|^2/\rho_0 \in L^1(\mathbf{T}^N), \tag{3}$$

where we agree that $|m_0|^2/\rho_0 = 0$ on $\{x \in \mathbf{T}^N \text{ such that } \rho_0(x) = 0\}$. More precisely, assuming (3) and that $\gamma \geq 3/2$ if $N = 2$, $\gamma \geq 9/5$ if $N = 3$, $\gamma > N/2$ if $N \geq 4$, he proved

Theorem 1.1. *There exists* $(\rho, u) \in L^\infty(0, \infty; L^\gamma(\mathbf{T}^N)) \times L^2(0, \infty; H^1(\mathbf{T}^N))^N$ *solution of* (1) *satisfying in addition:* $\rho \in C([0, \infty); L^p(\mathbf{T}^N))$ *if* $1 \leq p < \gamma$, $\rho|u|^2 \in L^\infty(0, \infty; L^1(\mathbf{T}^2))$, $\rho \in L^q_{loc}([0, \infty); L^q(\mathbf{T}^2))$ *for* $1 \leq q \leq \gamma - 1 + 2\gamma/N$. *Moreover, for almost all* $t \geq 0$, *we have*

$$\int_{\mathbf{T}^N} \left(\frac{1}{2} \rho u^2 + \frac{a}{\gamma - 1} \rho^\gamma \right)(t, x) dx + \int_0^t ds \int_{\mathbf{T}^N} \left(\mu |\nabla u|^2 + (\lambda + \mu)(\operatorname{div} u)^2 \right) dx ds$$
$$\leq \int_{\mathbf{T}^N} \left(\frac{m_0^2}{2\rho_0} + \frac{a}{\gamma - 1} \rho_0^\gamma \right)(x) dx. \tag{4}$$

Local well-posedness issues for the isentropic Navier-Stokes equations have been investigated by Solonnikov in [18]. Considering C^2 pressure laws, initial data (ρ_0, u_0) and external forces f satisfying for some $q > N$

$$0 < m \leq \rho_0(x) < M < +\infty, \quad \rho_0 \in W^{1,q}(\mathbf{T}^N), \quad \text{and} \quad u_0 \in W^{2-\frac{2}{q}, q}(\mathbf{T}^N)^N, \tag{5}$$

$$f \in L^q((0, T) \times \mathbf{T}^N) \quad \text{for all} \quad T > 0, \tag{6}$$

he proved

Theorem 1.2. *There exists a positive time* $T_0 \in (0, \infty]$ *and a unique solution* (ρ, u) *on* $[0, T_0)$ *of* (1) *such that for all* $t \in [0, T_0)$, *the density at time* t *is also bounded and bounded away from zero and for all* $T < T_0$

$$\rho \in L^\infty(0, T; W^{1,\infty}(\mathbf{T}^N)), \quad \partial_t \rho \in L^q((0, T) \times \mathbf{T}^N),$$
$$0 < m(t) \leq \rho(t, .) \leq M(t) < +\infty, \tag{7}$$
$$u \in L^q(0, T; W^{2,q}(\mathbf{T}^N))^N, \quad \text{and} \quad \partial_t u \in L^q((0, T) \times \mathbf{T}^N)^N.$$

A remarkable global well-posedness result was proven by Kazhikhov and Weigant in [10] in the two dimensional case when λ is a suitably increasing function of ρ. In the case of constant λ and dimension $N \geq 1$, Weigant exhibits in [20] a counterexample with spherical symmetry: there exists initial data (ρ_0, u_0) and external forces f satisfying (5) (6) such that for some $T_* < +\infty$, the unique local smooth solution (ρ, u) blows up

$$\lim_{t \to T_*^-} |\rho(t, .)|_{L^\infty(\mathbf{T}^N)} = +\infty.$$

A natural question is to ask whether vacuum plays a role in the development of singularities. Indeed, most of the known local regularity results for solutions of the Navier-Stokes equations require the density to be bounded from below by a positive constant [8] [16] [17]. Moreover, we want to know conversely whether solutions remain smooth if the density ρ has L^∞ bounds.

2. Main results

In order to answer the above two questions, we first deal with existence of smooth solutions in the case when ρ_0 is not assumed to have a positive lower bound. For an interesting investigation in this direction, we refer to the work of Xin [21], who gives estimates for blow-up time of solutions of the full Navier-Stokes equations with compactly supported densities. We require here that $\gamma > \gamma_0$ where $\gamma_0 = 1$ when $N = 2$ and $\gamma_0 = 3$ when $N = 3$. The initial data are supposed to satisfy

$$\rho_0 \in L^\infty(\mathbf{T}^N), \quad \rho_0 \geq 0, \quad \text{and} \quad u_0 \in H^1(\mathbf{T}^N)^N. \tag{8}$$

We want to prove the following Theorem which essentially states that vacuum does not yield singularities and that local strong solutions in dimension 2 remain smooth as long as the density remains bounded.

Theorem 2.1. *(i)   There exists $T_0 \in (0, \infty]$ and a solution (ρ, u) to the Navier-Stokes equations in $(0, T_0)$ such that for all $T < T_0$*

$$\rho \in L^\infty((0, T) \times \mathbf{T}^N) \cap C([0, T]; L^q(\mathbf{T}^N)) \quad \text{for all} \quad q \in [1, \infty), \tag{9}$$

$$\sqrt{\rho}\partial_t u \in L^2((0, T) \times \mathbf{T}^N)^N, \quad Pu \in L^2(0, T; H^2(\mathbf{T}^N))^N, \tag{10}$$

$$\nabla u \in L^\infty(0, T; L^2(\mathbf{T}^N))^{N^2},$$
$$G = (\lambda + 2\mu)\,div\ u - p(\rho) \in L^2(0, T; H^1(\mathbf{T}^N)), \tag{11}$$

where P denotes the projection on the space of divergence-free vector fields.
(ii)   In the case when $N = 2$, the regularity properties (9) (10) and (11) hold as long as $\sup_{t \in [0, T]} |\rho(t, .)|_{L^\infty(\mathbf{T}^2)} < +\infty$.

Notice that the main ingredient for proving (ii) is an a priori estimate stated here without proof (see [5] [6]), which was written in a previous work [4] for incompressible fluids. The main motivation was to study singularities of vacuum bubbles in incompressible multiphase fluids in two dimensions. It turns out that lemma 2.2 allows to prove that vacuum bubbles have essentially the same behavior as fluid bubbles of positive density.

Lemma 2.2. *There exists $C_0 > 0$ such that for all $p > 1$ and (ρ, u) satisfying $u \in H^1(\mathbf{T}^2)^2$, $\fint_{\mathbf{T}^2} u \, dx = 0$, $\rho \in L^\gamma(\mathbf{T}^2)$ for some $\gamma > 1$, we have*

$$
\left| \rho^{\frac{1}{2p}} u \right|_{L^{2p}(\mathbf{T}^2)} \le C_0 \left| \sqrt{\rho} u \right|_{L^2(\mathbf{T}^2)}^{\frac{1}{p}} \left| \nabla u \right|_{L^2(\mathbf{T}^2)}^{1 - \frac{1}{p}} \sqrt{\frac{p\gamma}{\gamma - 1}}
$$

$$
\times \left\{ \log \left(2 + \frac{|\nabla u|_{L^2(\mathbf{T}^2)}^2 |\rho|_{L^\gamma(\mathbf{T}^2)}}{|\sqrt{\rho} u|_{L^2(\mathbf{T}^2)}^2} \right) \right\}^{\frac{1}{2}(1 - \frac{1}{p})}
\tag{12}
$$

Let us also emphasize that the counterexample built by Weigant [20] yields non zero external forces f. The question whether smooth solutions persist in dimension two with vanishing bulk forces f remains open to the author's knowledge.

As far as uniqueness is concerned, "weak-strong" uniqueness results can be shown using the methods introduced by P.-L. Lions in the incompressible case [14]. Recall that a weak solution is called weak-strong unique if and only if it is equal to a strong one, as long as the strong one exists. For a precise statement and detailed proof, we refer to [5].

Finally, following the method we used in [3], we study regularity properties of the integral curves of u when $N = 2$. Roughly speaking, we want to study the Lagrangian point of view by solving $\frac{d}{dt} X(t, s, x) = u(t, X(t, s, x))$ and $X_{|t=s} = x$ in $\mathbf{T}^2$. It turns out that regularity results can be proven for almost every initial point $x \in \mathbf{T}^2$ adapting Di Perna-Lions theory for ordinary differential equations [7] even though the vector field u is not of bounded divergence. More precisely, we have to deal with vector fields u for which there exists $\alpha \in L^2_{loc}([0, \infty))$ such that

$$
u \in L^\infty(0, T; H^1(\mathbf{T}^2)) \quad \text{and} \quad \exp \left(\frac{|\text{div } u(t, x)|^2}{\alpha(t)^2} \right) \in L^\infty(0, T; L^1(\mathbf{T}^2)).
\tag{13}
$$

One of the key argument for the extension of Di Perna Lions theory is a uniqueness lemma for renormalized solutions of linear transport equations that will be detailed in the last section.

3. Proof of Theorem 2.1

The Proof essentially follows 3 steps. First, we have to derive upper a priori bounds for the density ρ, next a priori bounds for the velocity field. Finally, compactness results due to P.-L. Lions [12] [14] allow to conclude rigorously.

3.1. Upper bounds on the density

Applying the div operator to (1) and the inverse Laplacean with zero mean value on $\mathbf{T}^N$, we obtain denoting by $\mathbf{R}_i$ the usual Riecz transforms on $\mathbf{T}^N$

$$
(\partial_t + u.\nabla) \Delta^{-1} \text{div} (\rho u) - (\lambda + 2\mu) \text{div } u + a\rho^\gamma - a \fint_{\mathbf{T}^2} \rho^\gamma = [u_i, R_i R_j](\rho u_j),
\tag{14}
$$

which, combined with

$$
(\partial_t + u.\nabla) \log \rho + \text{div } u = 0,
$$

allows to write for $H = (\lambda + 2\mu) \log \rho + \Delta^{-1} \text{div} (\rho u)$

$$(\partial_t + u.\nabla) H + a\rho^\gamma - a \int_{\mathbf{T}^2} \rho^\gamma = [u_i, R_i R_j](\rho u_j), \tag{15}$$

where G is defined as in Theorem 2.1. In order to derive L^∞ upper bounds on H, we use a result by R. Coifman and Y. Meyer [1] who prove that the commutator at the right hand side of (15) is a smoothing operator

Theorem 3.1. Let (p, q, r) such that $\frac{1}{p} + \frac{1}{q} = \frac{1}{r} < 1$. Then, $(a, b) \mapsto [a_i, R_i R_j]b$ maps continuously $W^{1,p}(\mathbf{T}^N) \times L^q(\mathbf{T}^N)$ into $W^{1,r}(\mathbf{T}^N)$.

Subsequently, upper bounds on ρ are easily obtained in terms of bounds of u stated in Theorem 2.1 (for more details, see [5]).

3.2. Estimates on the velocity

A priori bounds on the velocity are obtained multiplying (1) by $\partial_t u$ and integrating by parts (see [5]). Let us focus on part (ii) related to the two dimensional case. Making use of standard elliptic regularity results, we obtain the following estimate assuming that $\rho \in L^\infty((0, T) \times \mathbf{T}^2)$ for some positive T

$$\int_{\mathbf{T}^2} |\nabla u(t, .)|^2 dx \leq C \int_{\mathbf{T}^2} |\nabla u_0|^2 dx + C + C \int_0^t |\rho^{\frac{1}{2}} u|_{L^4(\mathbf{T}^2)}^4 |\nabla u|_{L^2(\mathbf{T}^2)}^2 ds, \tag{16}$$

thus applying lemma 2.2 with $p = 4$ and estimating $\fint_{\mathbf{T}^2} u$ as in [5], we obtain using the energy bounds (4)

$$\int_{\mathbf{T}^2} |\nabla u(t, .)|^2 dx \leq C + C \int_0^t |\nabla u|_{L^2(\mathbf{T}^2)}^4 \log \left(2 + |\nabla u|_{L^2(\mathbf{T}^2)}^2\right) ds, \tag{17}$$

so that Gronwall's lemma allows to conclude, recalling that (4) provides a priori bounds on u in $L^2(0, T; H^1(\mathbf{T}^2))$ for all $T > 0$.

3.3. Conclusion

The above two subsections provide a priori bounds for solutions of (1) in $(0, T)$. In order to prove that such solutions exist, we can use Theorem 1.2 to construct solutions (ρ_n, u_n) with regularized initial data such that ρ_0 is bounded from below by, for instance, $\frac{1}{n}$. In view of the preceding a priori estimates, there exists $T_0 < \infty$ such that the bounds of Theorem 2.1 hold $(0, T_0)$ for (ρ_n, u_n) uniformly in n. Let ρ and u weak limits of ρ_n et u_n. The compactness in time of ρ_n and $\rho_n u_n$ yield the weak convergences $\rho_n u_n \rightharpoonup \rho u$ and $\rho_n u_n \otimes u_n \rightharpoonup \rho u \otimes u$. In particular, we have $\partial_t \rho + \text{div} (\rho u) = 0$. Next, using the regularization lemma of [7] and the property that $(\rho, u) \in L^\infty((0, T_0) \times \mathbf{T}^2) \times L^2(0, T_0; H^1(\mathbf{T}^2))$, we deduce that for all suitably smooth β, the following equation holds in the sense of distributions

$$\partial_t \beta(\rho) + \text{div} (\beta(\rho)u) + (\rho\beta'(\rho) - \beta(\rho)) \text{div } u = 0, \tag{18}$$

Similarly, denoting $\overline{\beta(\rho)}$ any weak limit of $\beta(\rho_n)$ we obtain using (14)

$$(\lambda + 2\mu)\text{div } u - a\overline{\rho^\gamma} + a \fint_{\mathbf{T}^2} \overline{\rho^\gamma} dx = \partial_t \left(\Delta^{-1}\text{div }(\rho u)\right) + R_i R_j(\rho u_i u_j), \qquad (19)$$

hence multiplying by ρ, and using the fact that $\partial_t \rho + \text{div }(\rho u) = 0$, we can write

$$(\lambda + 2\mu)\rho\text{div } u - a\rho\overline{\rho^\gamma} + a\rho \fint_{\mathbf{T}^2} \overline{\rho^\gamma} dx = \partial_t \left(\rho\Delta^{-1}\text{div }(\rho u)\right)$$
$$+\text{div }\left(\rho u\Delta^{-1}\text{div }(\rho u)\right) + \rho[u_i, R_i R_j](\rho u_j). \qquad (20)$$

On the other hand, multiplying by ρ_n equation (14) considered at a fixed n, we deduce that

$$(\lambda + 2\mu)\overline{\rho\text{div } u} - a\overline{\rho^{\gamma+1}} + a\rho \fint_{\mathbf{T}^2} \overline{\rho^\gamma} dx = \partial_t \left(\rho\Delta^{-1}\text{div }(\rho u)\right)$$
$$+\text{div }\left(\rho u\Delta^{-1}\text{div }(\rho u)\right) + \rho[u_i, R_i R_j](\rho u_j), \qquad (21)$$

using the $L^1(0, T_0; W^{1,2}(\mathbf{T}^2))$ regularity of $[u_i, R_i R_j](\rho u_j)$ provided by Theorem 3.1. Making the difference between (20) and (21), we obtain

$$(\lambda + 2\mu)\left(\overline{\rho\text{div } u} - \rho\text{div } u\right) = a\left(\overline{\rho^{\gamma+1}} - \rho\overline{\rho^\gamma}\right). \qquad (22)$$

Hence denoting $\delta = \overline{\rho\log\rho} - \rho\log\rho$, we conclude that

$$\partial_t\delta + \text{div }(u\delta) + \frac{a}{\lambda + 2\mu}\left(\overline{\rho^{\gamma+1}} - \rho\overline{\rho^\gamma}\right) = 0.$$

Next, convexity arguments yield $\overline{\rho^{\gamma+1}} \geq \rho\overline{\rho^\gamma}$, so that $\partial_t\delta + \text{div }(u\delta) \leq 0$ and $\delta_{|t=0} \equiv 0$. As a consequence, $\delta \equiv 0$ using [7], so that we can easily conclude that ρ_n converges to ρ in $C([0, T_0]; L^p(\mathbf{T}^2))$ for all $p < +\infty$.

4. O.D.E.'s with unbounded divergence

As pointed out in the introduction, the solutions built in Theorem 2.1 have enough regularity to yield existence and uniqueness of a generalized flow X. As a matter of fact, Di Perna Lions theory for ordinary differential equations can be adapted to the case when the divergence of the Eulerian field u is not bounded in $L^1(0, T; L^\infty(\mathbf{T}^2))$. First, we have to introduce the notion of renormalized solution of linear transport equations with coefficients $u \in L^1(0, T; W^{1,1}(\mathbf{T}^N)^N)$

$$\partial_t\phi + u \cdot \nabla\phi = 0, \qquad \phi_{|t=0} = \phi_0. \qquad (23)$$

In the sequel, for all $\phi_0 \in L^1_{loc}(\mathbf{T}^N)$, we define a renormalized solution of (23) ϕ as follows : for all $\beta \in C^1(\mathbf{R}; \mathbf{R})$, the following partial differential equation holds in the sense of distributions

$$\partial_t\beta(\phi) + \text{div }(u\beta(\phi)) + (\phi\beta'(\phi) - \beta(\phi))\text{div } u = 0, \qquad \beta(\phi)_{|t=0} = \beta(\phi_0).$$

We now state the main result of this section. Consider a solution (ρ, u) of (1) in dimension $N = 2$ built as Theorem 2.1 such that (13) holds. Then, we have

Theorem 4.1. *Under the assumptions of Theorem 2.1 when $N = 2$, there exists a unique $X : [0, T_0)^2 \times \mathbf{T}^2 \to \mathbf{T}^2$ defined almost everywhere such that (i) (ii) and (iii) hold*

- (i) *For all $s \in [0, T_0)$, $f(t, x) = X(s, t, x)$ is the unique renormalized solution of the linear transport equation*

$$\partial_t f + u \cdot \nabla f = 0 \ \text{ in } \ \mathcal{D}'((0, T_0) \times \mathbf{T}^2), \quad f_{|t=s} = x.$$

  and $\beta(f(t, x)) \in C([0, T_0); L^p(\mathbf{T}^2))$ for all $p \in [1, \infty)$ and $\beta \in C^1(\mathbf{R}; \mathbf{R})$. Besides, for all $s \in [0, T_0)$, for all $\Phi \in C_0^\infty(\mathbf{R}^2)$, the ordinary differential equation holds in the following sense

$$\partial_t \Phi(X(t, s, x)) = \nabla \Phi(X(t, s, x)) \cdot u(t, X(t, s, x)) \text{ in } \mathcal{D}'((0, T_0) \times \mathbf{T}^2),$$

$$\Phi(X(t, s, x))_{|t=s} = \Phi(x).$$

- (ii) *The image measure of the Lebesgue measure by X satisfies: $\forall p \in (1, +\infty)$, $\exists C = C_{T_0, p} > 0$, $\forall \Phi \in C^0([0, T_0) \times \mathbf{T}^2)$, $\Phi \geq 0$, $\forall (t, t_0) \in [0, T_0)^2$,*

$$\int_{\mathbf{T}^2} \Phi(t_0, X(t_0, t, x)) dx \leq C |\Phi(t_0, .)|_{L^p(\mathbf{T}^2)}. \tag{24}$$

  In addition, $\exists C > 0$ such that $\forall h \in (0, 1)$, $\forall \Phi \in C_c^0([0, T_0) \times \mathbf{T}^2)$,

$$\left| \int_{[0, T_0 - h] \times \mathbf{T}^2} \Phi(t, X(t, t + h, x)) \, dx dt - \int_{[0, T_0] \times \mathbf{T}^2} \Phi(t, x) \, dx dt \right|$$

$$\leq C h |\Phi|_{L^\infty([0, T_0] \times \mathbf{T}^2)}.$$

- (iii) *The following semigroup property holds $\forall (t_2, t_3) \in [0, T_0)^2$, for a.e. $x \in \mathbf{T}^2$, $\forall t_1 \in [0, T_0)$, $X(t_1, t_2, X(t_2, t_3, x)) = X(t_1, t_3, x)$.*

Moreover, the following regularity holds for the integral curves:
- (iv) *$\forall t_0 \in [0, T_0)$, for a.e. x in $\mathbf{T}^2$, $X(., t_0, x) \in C^0([0, T_0))$ and*

$$X(t, t_0, x) = x + \int_{t_0}^t u(s, X(s, t_0, x)) ds \ \ \forall t \in [0, T_0).$$

One of the main argument of the proof relies upon L^p estimates for the Jacobian [2] and bounds on the transported Lebesgue measure on $\mathbf{T}^2$ (24). The above result actually holds for more general vector fields, namely vector fields $u \in L^1(0, T; W^{1,1}(\mathbf{T}^N))$ $(N \geq 1)$ such that div $u \in L^1(0, T; L_\omega^1(\mathbf{T}^N))$, where we define as in [2] the space $L_\omega^1(\mathbf{T}^N)$ as the following subspace of $L^1(\mathbf{T}^N)$

$$L_\omega^1(\mathbf{T}^N) = \left\{ f \in L^1(\mathbf{T}^N) \ / \ \exists C > 0, \text{ for all } A \subset \mathbf{T}^N, \ \int_A |f| dx \leq C \omega(|A|) \right\}$$

where ω denotes a non negative non decreasing function on $\mathbf{R}^+$ such that

$$\int_0^1 \frac{du}{\omega(u)} = +\infty. \tag{25}$$

The space $L_\omega^1(\mathbf{T}^N)$ is naturally embedded with the following norm

$$|f|_{L_\omega^1(\mathbf{T}^2)} = \sup_{|A|\neq 0,\ A\subset \mathbf{T}^N} \frac{1}{\omega(|A|)} \int_A |f|\,dx,$$

In the present case, using Jensen's inequality, we deduce for all $\delta > 0$ and $A \subset \mathbf{T}^N$ of positive measure

$$\frac{1}{|A|}\int_A |\mathrm{div}\ u|dx \leq \delta\Big\{ \log\Big(\frac{1}{|A|}\int_{\mathbf{T}^2} \exp\Big(\frac{|\mathrm{div}\ u|^2}{\delta^2}\Big)dx\Big)\Big\}^{\frac{1}{2}}$$

On the other hand, we have using the expression of G

$$\int_{\mathbf{T}^2} \exp\Big(\frac{|\mathrm{div}\ u|^2}{\delta^2}\Big)dx \leq \int_{\mathbf{T}^2} \exp\Big(C\frac{|G - f_{\mathbf{T}^2}G|}{\delta^2} + \frac{C}{\delta^2}|p(\rho)|_{L^\infty((0,T_0)\times \mathbf{T}^2)}\Big)dx.$$

hence taking $\delta = 1 + \sqrt{\frac{C}{\alpha_0}}|G(t,.)|_{H^1(\mathbf{T}^2)}$ for some small enough α_0, we deduce using Trudinger's inequality that

$$\int_A |\mathrm{div}\ u(t,.)|dx \leq C\left(1 + |G(t,.)|_{H^1(\mathbf{T}^2)}\right)|A|\sqrt{\log\Big(\frac{C}{|A|}\Big)},$$

so that $\mathrm{div}\ u \in L^2(0,T_0; L_{\omega_0}^1(\mathbf{T}^2))$, where $\omega_0(s) = s\sqrt{\log\big(\frac{C}{s}\big)}$. For related results, transport of L^p and Sobolev regularity in linear transport equations, applications to incompressible nonhomogeneous fluids, and extensions, we refer to [2] [3]. Let us point out that most of the above results on ordinary differential equations rely upon a uniqueness argument for bounded solutions of linear transport equations with coefficients $u \in L^1(0,T; W^{1,1}(\mathbf{T}^N))^N$ and $\mathrm{div}\ u \in L^1(0,T; L_\omega^1(\mathbf{T}^N))$ for some ω chosen as in (25).

Lemma 4.2. *Let $\rho \in L^\infty((0,T) \times \mathbf{T}^N)$ be a solution of $\partial_t\rho + \mathrm{div}\ (\rho u) = 0$ with initial data $\rho_{|t=0} \equiv 0$. Then, $\rho \equiv 0$.*

Let us give an idea of the proof. First, using the regularization lemma stated in [7], we obtain $\partial_t\beta(\rho) + u.\nabla\beta(\rho) = 0$ and $\beta(\rho)_{|t=0} \equiv 0$ for all $\beta \in C^1(\mathbf{R};\mathbf{R})$. As a result, we may assume that $\rho \geq 0$. Integrating this equation over $\mathbf{T}^N$, we obtain

$$\int_{\mathbf{T}^N} \rho(t,x)dx \leq \int_0^t \int_{\mathbf{T}^N} \beta(\rho)|\mathrm{div}\ u|dxds.$$

Considering an increasing sequence of smooth function β_k converging to β_λ defined by $\beta_\lambda(s) = 1_{x>\lambda}$ for some given positive λ, we deduce using Beppo-Levi Theorem that

$$|A_\lambda(t)| \leq \int_0^t \int_{A_\lambda(s)} |\text{div } u| dx ds,$$

where we denote $A_\lambda(t) = \{x \in \mathbf{T}^N \ / \ \rho(t,x) > \lambda\}$. Since div $u \in L^1(0,T; L^1_\omega(\mathbf{T}^N))$, we obtain

$$|A_\lambda(t)| \leq \int_0^t |\text{div } u(s,.)|_{L^1_\omega(\mathbf{T}^N)} \omega(|A_\lambda(s)|) ds,$$

so that usual Gronwall type arguments yield $|A_\lambda(t)| = 0$ for all $t \in (0,T)$ and all positive λ, which completes the proof. Notice that similar results have recently been obtained by Kazhikhov and Mamontov, who study the propagation of Orlicz regularity in linear transport equations whose coefficients have divergence in some suitable Orlicz class [11] [15].

References

[1] R. Coifman, P.L. Lions and Y. Meyer, S. Semmes, *Compensated compactness and Hardy spaces*, J. Math. Pures Appl., **72** (1993), 247–286.

[2] B. Desjardins, *A few remarks on ordinary differential equations*, C.P.D.E., **21** (11–12) (1996), 1667–1703.

[3] B. Desjardins, *Linear transport equations with values in Sobolev spaces and application to the Navier-Stokes equations*, Diff. and Int. Equ., **10** (3) (1997), 577–586.

[4] B. Desjardins, *Regularity results for two dimensional viscous flows*, Arch. for Rat. Mech. Anal., **137** (1997), 135–158.

[5] B. Desjardins, *Regularity of Weak Solutions of the Compressible Isentropic Navier-Stokes Equations*, Comm. Partial Differential Equations, **22** (5–6) (1997), 977–1008.

[6] B. Desjardins, *Sur la régularité des solutions faibles de Navier-Stokes isentropiques en dimension deux*, to appear in Séminaire Equations aux Dérivées Partielles, Ecole Polytechnique, Palaiseau, (1997–98).

[7] R.J. Di Perna and P.L. Lions, *Ordinary differential equations, transport theory and Sobolev spaces*, Invent. Math., **98** (1989), 511–547.

[8] D. Hoff, *Global well-posedness of the Cauchy problem for nonisentropic gas dynamics with discontinuous data*, J. Diff. Eq., **95** (1992), 33–73.

[9] A.V. Kazhikov and V.V. Shelukhin, *Unique global solution with respect to time of the initial boundary value problems for one dimensional equations of a viscous gas*, J. Appl. Math. Mech., **41** (1977), 273–282.

[10] A.V. Kazhikhov and V.A. Weigant, *On the existence of global solutions of two-dimensional Navier-Stokes equations of a compressible viscous fluid*, Siberian Math. J., **36** (6) (1995), 1108–1141.

[11] A.V. Kazhikhov and A.E. Mamontov, *One class of convex functions and exact well-posedness of Cauchy problem for the transport equation in Orlicz spaces*, to be published in Siberian Math. Journal, (1998).

[12] P.L. Lions, *Existence globale pour les équations de Navier-Stokes compressibles isentropiques*, C.R. Acad. Sci. Paris, **316** (1993), 1335–1340.

[13] P.L. Lions, *Compacité des solutions des équations de Navier-Stokes compressibles isentropiques*, C.R. Acad. Sci. Paris, **317** (1993), 115–120.

[14] P.L. Lions, *Mathematical Topics in Fluid Mechanics*, Oxford University Press, (1996).

[15] A.E. Mamontov, *Orlicz spaces in the existence problem of global solutions to viscous compressible nonlinear fluid equations*, preprint (1996).

[16] A. Matsumura and T. Nishida, *Initial-boundary value problems for the equations of motion of compressible viscous and heat-conductive fluids*, Comm. Math. Phys., **89 (4)** (1983), 445–464.

[17] D. Serre, *Solutions faibles globales des equations de Navier-Stokes pour un fluide compressible*, C.R. Acad. Sci. Paris, **303** (1986), 629–642.

[18] V.A. Solonnikov, *solvability of the initial boundary value problem for the equation of a viscous compressible fluid*, J. Sov. Math., **14** (1980), 1120–1133.

[19] A. Valli and W.M. Zajaczkowski, *Navier-Stokes equations for compressible fluids : global existence and qualitative properties of the solutions in the general case*, Comm. Math. Phys., **103** (1986), 259–296.

[20] V.A. Weigant, *An example of the nonexistence with respect to time of the global solution of Navier-Stokes equations for a compressible viscous barotropic fluid*, Dokl. Akad. Nauk., **339 (2)** (1994), 155–156.

[21] Xin and Zhouping, *Blow-up of smooth solutions to the compressible Navier-Stokes equations with compact density*, preprint (1997).

Département de Mathématiques et d'Informatique
Ecole Normale Supérieure,
45, rue d'Ulm,
75005 Paris, France
E-mail address: desjardi@dmi.ens.fr

International Series of Numerical Mathematics
Vol. 129, © 1999 Birkhäuser Verlag Basel/Switzerland

Entropy Inequality for High Order Discontinuous Galerkin Approximation of Euler Equations

Bruno Despres

1. Introduction

This paper is devoted to the presentation of a new family of high order numerical schemes in space (first order in time) for the numerical solution of the Euler equations. We restrict the presentation to the 1D case. Full extention to the multi-dimensional case is discussed in [9] with complete discussion of the entropy properties and some numerical simulations. See [8] for an elementary discussion of the entropy inequality.

The model problem is the Euler system in 1D. The unknowns (ρ, u, e) are the density, the velocity and the specific total energy

$$
\begin{cases}
\partial_t \rho + \partial_x \rho u = 0, & \tau = \dfrac{1}{\rho}, \\[2mm]
\partial_t \rho u + \partial_x (\rho u^2 + p) = 0, & p = p(\varepsilon, \tau), \\[2mm]
\partial_t \rho e + \partial_x (\rho u e + p u) = 0, & e = \varepsilon + \dfrac{1}{2} u^2.
\end{cases}
\tag{1}
$$

We suppose that the pressure law p is thermodynamically admissible, that is there exists a strictly concave function called the entropy $S(\varepsilon, \tau)$ and a nonnegative integrating factor called the temperature $T(\varepsilon, \tau)$ such that

$$
T dS = d\varepsilon + p d\tau.
\tag{2}
$$

The discontinuous Galerkin method applied to transport equation is well known since a long time [5]. Many researchers are nowadays interested by the discontinuous Galerkin method for the numerical solution of (1): we refer to [3]–[4] for a presentation of the discontinuous Galerkin method for conservation laws and related problems; see [12] for both a review of the major trends and difficulties in the design of high order entropy schemes (see also [11]–[13]) and a study of an entropy inequality for discontinuous Galerkin methods. However, the entropy considered in [12] is based on the L^2 norm, and is not related to the physical entropy (2). Our main result is precisely that it is possible to obtain a natural entropy inequality for the entropy (2) in the context of a high order discontinuous Galerkin method. To our knowledge this result is new.

The family of schemes is constructed using: a discontinuous Galerkin approximation of the unknowns, a reduction of some exact integration to numerical integration with the use of Gauss-Lobatto points, a splitting of (1) into a Lagrangian phase and a convection phase and the use of a linearized Lagrangian Riemann solver. Gauss-Lobatto points in 1D [1] provide 2 points at the right and left edges of the reference cell: it appears that this property of Gauss-Lobatto points is crucial in order to obtain the entropy inequality of Theorem 2.1.

FIGURE 1. Example of four Gauss-Lobatto points in the reference cell

2. Main result

We present our main result in this section. Suppose that we know the numerical values of the unknowns $(\rho_{k,j}^n, u_{k,j}^n, e_{k,j}^n)$ located at every Gauss point j in every cell Ω_k at step n. Traditionally, one computes $(\rho_{k,j}^{n+1}, u_{k,j}^{n+1}, e_{k,j}^{n+1})$ using some kind of finite volume approximation (which may also be viewed as appropriate integration by parts) and a Riemann solver at the cell edges for the computation of the fluxes. In this paper, we introduce auxiliary unknowns denoted as $(\rho_{k,j}^{n+\frac{1}{2}}, u_{k,j}^{n+\frac{1}{2}}, e_{k,j}^{n+\frac{1}{2}})$ which are calculated in the Lagrangian phase.

These auxiliary unknowns (sometimes denoted as Lagrangian unknowns, or characteristic unknowns) $(\rho_{k,j}^{n+\frac{1}{2}}, u_{k,j}^{n+\frac{1}{2}}, e_{k,j}^{n+\frac{1}{2}})$ are the numerical solutions during a time step Δt of

$$\begin{cases} \rho D_t \tau = \partial_x u, \\ \rho D_t u = -\partial_x p, \\ \rho D_t e = -\partial_x p u. \end{cases} \tag{3}$$

Then we compute $(\rho_{k,j}^{n+1}, u_{k,j}^{n+1}, e_{k,j}^{n+1})$ from $(\rho_{k,j}^n, u_{k,j}^n, e_{k,j}^n)$ and $(\rho_{k,j}^{n+\frac{1}{2}}, u_{k,j}^{n+\frac{1}{2}}, e_{k,j}^{n+\frac{1}{2}})$ solving

$$\begin{cases} \partial_t \rho + \partial_x \rho \tilde{u} = 0, \\ \partial_t \rho u + \partial_x \rho u \tilde{u} = 0, \\ \partial_t \rho e + \partial_x \rho e \tilde{u} = 0, \end{cases} \tag{4}$$

during Δt with the velocity $\tilde{u}$ set to $u_{k,j}^{n+\frac{1}{2}}$. The procedure (3–4) is equivalent to the computation of the fluxes at the edges in 2 steps, and is consistent with (1).

Theorem 2.1. *Under a CFL condition, the scheme (6-7-8) (which is the discrete equivalent of (3)) is entropy consistant. That is at every Gauss-Lobatto point, in every cell, we have* $S_{k,j}^{n+\frac{1}{2}} \geq S_{k,j}^n$. *This result is true for all pressure laws (i.e. for all equations of state) provided these pressure laws satisfy (2) plus the concavity of S.*

We emphasize that this result is true for high order approximations of (3), and with no use of any limitors. It is true also in the multidimensional case. It is a well-known remark that such inequality leads in the discrete case to some nonlinear stability estimate. So the Lagrangian phase is high order accurate and stable. We do not have to use any limitor. In a regular flow the scheme is high order accurate. If there is a shock propagating in the medium the scheme is stable without going down to the 1 order, so that the simulation is not corrupted at first order.

No such strong result is obtained during the convection phase. Nevertheless it is possible to discretize (4) so that the final scheme is stable and high order accurate. The stability in this second phase is just a L^2 stability.

3. Description of the scheme in the Lagrangian phase

In this section we describe how to derive the scheme. We take the first equation of the Lagrangian system $\rho D_t \tau = \partial_x u$ and integrate it by part in an element Ω_k against a test function φ_k. The out-going normal from Ω_k is denoted as ν_k

$$\int_{\Omega_k} \rho D_t \tau . \varphi_k + \int_{\Omega_k} u \partial_x \varphi_k = \int_{\partial \Omega_k} (u.\nu_k) \varphi_k .$$

We now define the velocity at the edge between Ω_k and another neighbouring element $\Omega_{k'}$: the point number j' in $\Omega_{k'}$ is equal to the point number j in Ω_k, even if the degrees of freedom are decoupled at these points. We also define the pressure

$$\begin{cases} u_{kj}^* . \nu_{kj} = \dfrac{1}{2} \left(u_{kj}^n . \nu_{kj} - u_{k'j'}^n . \nu_{k'j'} \right) + \dfrac{1}{2(\rho c)_{kj}^*} \left(p_{kj}^n - p_{k'j'}^n \right) \\[2mm] p_{kj}^* = \dfrac{1}{2} \left(p_{kj}^n + p_{k'j'}^n \right) + \dfrac{(\rho c)_{kj}^*}{2} \left(u_{kj}^n . \nu_{kj} + u_{k'j'}^n . \nu_{k'j'} \right) . \end{cases} \tag{5}$$

The coefficient $(\rho c)_{kj}^* = (\rho c)_{k'j'}^*$ is chosen to be as close as possible to the local value of the density times the sound velocity. So, one recognizes in these edge values the fluxes of a linearized Lagrangian Riemann solver [14]. We emphasize that we re-derive (5) from analogy between (3) and some problems in electromagnetism [6], [7], [2]. However in our context we extend the validity of (5) to the high order case. One obtains

$$\int_{\Omega_k} \rho_k^n \frac{\tau_k^{n+\frac{1}{2}} - \tau_k^n}{\Delta t} \varphi_k + \int_{\Omega_k} u_k^{n+\frac{1}{2}} \partial_x \varphi_k = \int_{\partial\Omega_k} (u_{kj}^* . \nu_{kj}) \varphi_k .$$

We use the family of Gauss-Lobatto points in the element. That is, we replace the exact integral $\int_{\Omega_k} \rho_k^n \frac{\tau_k^{n+\frac{1}{2}} - \tau_k^n}{\Delta t} \varphi_k$ by $\sum_j w_{kj} \rho_{kj}^n \frac{\tau_{kj}^{n+\frac{1}{2}} - \tau_{kj}^n}{\Delta t} \varphi_{kj}$ where the coefficients w_{kj} are the quadrature weights. So

$$\sum_j w_{kj} \rho_{kj}^n \frac{\tau_{kj}^{n+\frac{1}{2}} - \tau_{kj}^n}{\Delta t} \varphi_{kj} + \int_{\Omega_k} u_k^{n+\frac{1}{2}} \partial_x \varphi_k = \int_{\partial\Omega_k} (u_{kj}^* . \nu_{kj}) \varphi_k . \tag{6}$$

We choose a particular test function: $\varphi_k = P_{kj}$ so that the test function equals 1 at the selected Gauss point and 0 at other points. Defining the numerical coefficient

$$\alpha_{kjj'} = -\int_{\Omega_k} P_{kj'} \partial_x P_{kj} dx$$

the previous equation rewrites

$$w_{kj} \rho_{kj}^n \frac{\tau_{kj}^{n+\frac{1}{2}} - \tau_{kj}^n}{\Delta t} - \sum_{j'} \alpha_{kjj'} u_{kj'}^{n+\frac{1}{2}} = z_{kj} (u_{kj}^* . \nu_{kj})$$

where for convenience z_{kj} is set to 0 if the Gauss point is strictly in the element, and z_{kj} is set to 1 if the Gauss point is located on the right or left boundary of the element. The role of z_{kj} is to recall that there is no source term at points inside the element. Doing the same for the second equation one has

$$w_{kj} \rho_{kj}^n \frac{u_{kj}^{n+\frac{1}{2}} - u_{kj}^n}{\Delta t} + \sum_{j'} \alpha_{kjj'} p_{kj'}^{n+\frac{1}{2}} = -z_{kj} p_{kj}^* \nu_{kj} \tag{7}$$

and for the third equation

$$w_{kj} \rho_{kj}^n \frac{e_{kj}^{n+\frac{1}{2}} - e_{kj}^n}{\Delta t} + \left(\sum_{j'\neq j} \alpha_{kjj'} p_{kj'}^{n+\frac{1}{2}} \right) u_{kj}^{n+\frac{1}{2}} + \left(\sum_{j'\neq j} \alpha_{kjj'} u_{kj'}^{n+\frac{1}{2}} \right) p_{kj}^{n+\frac{1}{2}}$$
$$= -z_{kj} p_{kj}^* (u_{kj}^* . \nu_{kj}) . \tag{8}$$

The discrete equivalent of (3) is the nonlinear implicit system (6-7-8). One may wonder why the unknowns are implicit in the integrals, that is why we need $\sum_{j'} \alpha_{kjj'} u_{kj'}^{n+\frac{1}{2}}$ instead of $\sum_{j'} \alpha_{kjj'} u_{kj'}^n$. With $\sum_{j'} \alpha_{kjj'} u_{kj'}^n$ the scheme would be explicit: nevertheless we expect that it will not be stable. On the other hand with the implicit value $\sum_{j'} \alpha_{kjj'} u_{kj'}^{n+\frac{1}{2}}$ we obtain the fundamental result described in Theorem 2.1: all the numerical experiments show that the corresponding numerical scheme is very stable. A drawback is of course that the numerical solution needs to solve a local nonlinear system. It is possible to solve that system using a Newton algorithm: many numerical experiments show that this works very well.

However, it is also possible to linearize this discrete equation in order to simplify the solution algorithm [9].

4. Description of the scheme in the convection phase

The convection phase is based on the numerical approximation of (3) which is just a transport equation. Using a simple integration by part, we replace $\partial_t \rho + \partial_x \rho \tilde{u} = 0$ by

$$\int_{\Omega_k} \partial_t \rho . \psi_k - \int_{\Omega_k} \rho \tilde{u} \partial_x \psi_k = - \int_{\partial \Omega_k} (u.\nu_k) \rho \psi_k$$

and then by

$$\sum_j w_{kj} \frac{\rho_{kj}^{n+1} - \rho_{kj}^n}{\Delta t} \psi_k - \int_{\Omega_k} (\rho_k^{n+1} \times u_k^{n+\frac{1}{2}}) \partial_x \psi_k = - \int_{\partial \Omega_k} (u_{kj}^* . \nu_{kj}) \rho_U^{n+\frac{1}{2}} \psi_k .$$

Here the edge value $\rho_U^{n+\frac{1}{2}}$ is the upwind value of the density evaluated at the end of the Lagrangian phase. Classical considerations lead then to the numerical solution of a linear system. The procedure is the same for both the impulse and the energy. This convection phase is of high order. Nevertheless, it shows less entropy properties than the Lagrangian phase.

5. A simple numerical example

We present the classical Sod shock tube at a time $t = 0.14$, computed with the linearized version of (6-7-8) and 4 Gauss-Lobatto points in every cell (100 cells, 400 degrees of freedom). We emphasize that the solution provides every features of a high order solution, and that no limitor of any kind was used in this simulation. For such a flow the convection phase (which is just stable in the L^2 norm) is stable enough.

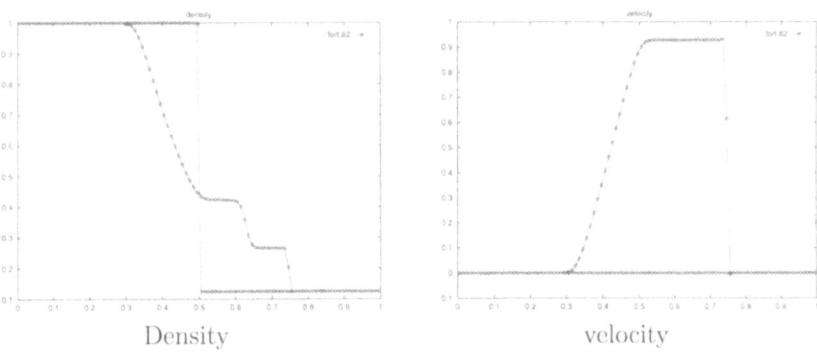

Density   velocity

FIGURE 2. Sod shock tube at $t = .14$

We will also present numerical results concerning other standard cases, including the Harten tube, the Noh tube, the Lax tube, a pure advection case and new results in 2D. For some of these flows, it may be important to build another convection phase (see [9]).

6. Conclusion

In this paper, we have described a new family of entropy schemes for the numerical solution of Euler equations. These schemes are based on discontinuous Galerkin methods, quadrature formulas with Gauss-Lobatto points and the use of the linearized Lagrangian Riemann solver (5) in the Lagrangian phase.

The fundamental result of Theorem 2.1 together with many numerical experiments show that in order to have a stable high order scheme for (3), it is sufficient to satisfy the entropy inequality in the Lagrangian phase. This can be achieved without any limitors of any kind and without the use of artificial viscosity.

We emphasize that we succeeded to add extra physics in these schemes: see [10] for the application to a model of plasma physics with ionic and electronic temperatures. We will report further on the 2D case with application to the cylindrical case.

References

[1] C. Bernardi and Y. Maday, *Approximations spectrales de problèmes aux limites elliptiques.* Springer-Verlag, **1992**.

[2] O. Cessenat and B. Despres, *Application of an ultra weak variational formulation of elliptic PDEs to the 2-dimensional Helmholtz problem.* SIAM J. of Numer. Anal., (1998), to appear.

[3] B. Cockburn and C.W. Shu, *The Runge-Kutta local projection P^1 discontinuous Galerkin method for scalar conservation laws*, M^2AN., **25** (1991), 337–361.

[4] B. Cockburn and C.W. Shu, *The local discontinuous Galerkin method for time-dependent convection-diffusion systems*, Technical Report 9732, ICASE, (1997).

[5] R. Dautray and J.L. Lions, *Analyse mathématique et calcul numérique*, **9**, Evolution: numérique, transport, Masson, Paris, **1987**.

[6] B. Despres, *Sur une formulation variationnelle de type ultra-faible*, Comptes Rendus de l'Académie des Sciences, Série I **318** (1994), 939–944.

[7] B. Despres, *Fonctionnelle quadratique et équations intégrales pour les problèmes d'onde harmonique en domaine extérieur*, M^2AN, **31 (3)** (1997), 1–57.

[8] B. Despres, *Inégalité entropique pour un solveur conservatif du système de la dynamique des gaz en coordonnées de Lagrange*, Comptes Rendus de l'Académie des Sciences, PARIS, Série I **324** (1997), 1301–1306.

[9] B. Despres, *Inégalités entropiques pour un solveur de type Lagrange + convection des équations de l'hydrodynamique*, Technical Report 2822, CEA, (1997).

[10] B. Despres and S. Jaouen, *Solveur entropique d'ordre élevé pour les équations de l'hydrodynamique à deux températures,* in preparation.

[11] E. Godlewski and P.A. Raviart, *Numerical approximation of hyperbolic systems of conservation laws*, Springer-Verlag, **1996**.

[12] G. Jiang and C.W. Shu, *On a cell entropy inequality for discontinuous Galerkin methods*, Math. of Comp., **62 (206)** (1994), 531–538.

[13] B. Perthame, *Boltzmann type schemes for gas dynamics and the entropy property*, Siam J. Numer. Anal., **27** (1990), 1405–1420.

[14] P. L. Roe, *Modern numerical methods applicable to stellar pulsation,* in The numerical modelling of nonlinear stellar pulsations, Kluwe Academic Publishers, **1990**.

CEA,
Bruyeres le Chatel,
91 216 Bruyeres le Chatel BP12, France
E-mail address: despres@bruyeres.cea.fr

International Series of Numerical Mathematics
Vol. 129, © 1999 Birkhäuser Verlag Basel/Switzerland

Asymptotic Equations for Weakly Nonlinear Elastic Waves in a Cubic Crystal

Włodzimierz Domański

Abstract. Using the method of weakly nonlinear geometric optics, we derive asymptotic transport evolution equations for high frequency small amplitude nonlinear elastic waves in an anisotropic medium. Wave propagations for three canonical directions, in the simplest nonlinear anisotropic medium – a cubic crystal, are studied. We demonstrate how a change of the wave direction influences the form of the obtained asymptotic equations. Inviscid Burgers equations are derived for longitudinal waves which are locally genuinely nonlinear in all cases. Modified inviscid Burgers equations are derived for locally linearly degenerate shear waves. Coupled nonlinear systems are obtained for pairs of shear weaves, one of which is linearly degenerate and the other is genuinely nonlinear in the nonstrictly hyperbolic case.

1. Introduction

Weakly nonlinear geometric optics (*WNGO*) is an effective asymptotic method for studying nonlinear hyperbolic wave type problems (see e.g. [1], [2]). The method provides simplified models for the evolution of wave amplitudes (profiles) of non-interacting as well as resonantly interacting waves. A number of papers devoted to both theoretical aspects of *WNGO* and its applications were published in recent years. One can mention just a few of them: [3], [4], [5], [6],[7], (see also the plenary lecture of Guy Métivier [8] at this conference and the references therein).

Here we consider applications of *WNGO* to nonlinear anisotropic elasticity. We have chosen the most symmetric of anisotropic elastic solids – a cubic crystal. The propagation of nonlinear elastic plane waves for three basic directions in the non-resonant case is investigated. The purpose is to show how the wave direction influences not only the form of the elasticity equations in the anisotropic medium but also the choice of the asymptotics.

2. Nonlinear anisotropic elasticity

2.1. General equations in three space dimensions

We begin with a general formulation of nonlinear, dynamical, isentropic, elasticity equations [9] written in Lagrangian coordinates in three space dimensions. The system consists of *the equations of motion*:

$$\rho_0 \frac{\partial \boldsymbol{v}}{\partial t} = \nabla \cdot \boldsymbol{T} \tag{1}$$

together with *the compatibility relations*:

$$\frac{\partial \boldsymbol{F}}{\partial t} = \nabla \boldsymbol{v} \tag{2}$$

Here $\boldsymbol{v}$ – velocity, $\boldsymbol{T}$ – first Piola-Kirchhoff 'stress tensor', $\boldsymbol{F}$ – deformation gradient and ρ_0 – density in the reference configuration. In (1) the body forces are disregarded.

To close this system we need to specify constitutive relations. Both geometrical and physical nonlinearities are taken into account. We assume first that the solid is *hyperelastic* which means that there exists a strain energy function $W = W(\boldsymbol{E})$ such that the first Piola Kirchhoff stress tensor $\boldsymbol{T}$ is expressed as (see [10])

$$\boldsymbol{T} = \boldsymbol{F} \frac{\partial W}{\partial \boldsymbol{E}} \tag{3}$$

where $\boldsymbol{E}$ is the strain tensor defined as

$$\boldsymbol{E} = \frac{1}{2} \left(\boldsymbol{F}^T \boldsymbol{F} - \boldsymbol{I} \right) \tag{4}$$

2.2. Cubic crystal

Next we assume that the medium is *anisotropic*. The general anisotropic medium is characterised by 216 third order constants. However, from the symmetry requirement this number may be greatly reduced. We consider the most symmetric class of cubic crystals, which is characterised by 3 second order and 6 third order elastic constants. The explicit form of the strain energy function $W = W(\boldsymbol{E})$ for such crystal class with third order terms was first given by Birch [11] as

$$
\begin{aligned}
W \;=\; & \tfrac{1}{2} c_{11}(E_{11}^2 + E_{22}^2 + E_{33}^2) + c_{12}(E_{11}E_{22} + E_{22}E_{33} + E_{11}E_{33}) + \\
& c_{44}(E_{12}^2 + E_{21}^2 + E_{23}^2 + E_{32}^2 + E_{31}^2 + E_{13}^2) + \\
& \tfrac{1}{6} c_{111}(E_{11}^3 + E_{22}^3 + E_{33}^3) + \\
& \tfrac{1}{2} c_{112}\{E_{11}^2(E_{22} + E_{33}) + E_{22}^2(E_{11} + E_{33}) + E_{33}^2(E_{11} + E_{22})\} + \\
& c_{144}\{E_{11}(E_{23}^2 + E_{32}^2) + E_{22}(E_{13}^2 + E_{31}^2) + E_{33}(E_{12}^2 + E_{21}^2)\} + \\
& c_{166}\{(E_{11} + E_{22})(E_{12}^2 + E_{21}^2) + (E_{22} + E_{33})(E_{23}^2 + E_{32}^2) + \\
& \quad (E_{11} + E_{33})(E_{13}^2 + E_{31}^2)\} + \\
& c_{123}E_{11}E_{22}E_{33} + 4c_{456}(E_{12}E_{23}E_{31} + E_{21}E_{32}E_{13})
\end{aligned}
\tag{5}
$$

Here the E_{ij} are the components of Lagrangian strain tensor $\boldsymbol{E}$, c_{ij} are the second order elastic constants, and c_{ijk} are the third order elastic constants. We have used more common Brugger's [12] notation for the elastic constants instead of the original Birch's [11] notation.

Given the other direction k of wave propagation, we transform the formula (5) for the energy $W = W(E)$ according to the rule:

$$W(E) \longrightarrow W(Q_k E Q_k{}^T) \tag{6}$$

where Q_k is the unitary matrix of the rotation that transforms the vector $k_1 = [1,0,0]$ into the vector k (the superscript T denotes matrix transposition).

The rotation matrix corresponding to the direction k_2 is chosen as

$$Q_{k_2} = \frac{1}{\sqrt{2}} \begin{bmatrix} 1 & -1 & 0 \\ 1 & 1 & 0 \\ 0 & 0 & \sqrt{2} \end{bmatrix} \tag{7}$$

and to the direction k_3 as

$$Q_{k_3} = \frac{1}{\sqrt{3}} \begin{bmatrix} 1 & -\sqrt{3//2} & -\sqrt{1//2} \\ 1 & \sqrt{3//2} & -\sqrt{1//2} \\ 1 & 0 & \sqrt{2} \end{bmatrix} \tag{8}$$

2.3. First order system in one space dimension

After introducing the constitutive relations into the equations of motion, we obtain a quasilinear first order system of partial differential equations as our basic model of nonlinear elasticity. Instead of the deformation gradient F, it is convenient to introduce *gradient of displacement* ∇u as the dependent variable. Then the deformation gradient F equals to

$$F = I + \nabla u \tag{9}$$

and the strain E takes the form:

$$E = \frac{1}{2}(\nabla u + (\nabla u)^T + (\nabla u)^T \nabla u) \tag{10}$$

Since we are interested in propagation of plane waves, hence we restrict ourselves to the one space dimensional case. Therefore, from now on we assume that all dependent variables are functions of time $- t$, and single Lagrangian space variable $- x$. In the one dimensional setting the displacement gradient becomes a vector. We will denote it by m:

$$m = \frac{\partial u}{\partial x} \tag{11}$$

We assume that equations (1), (2) are properly nondimensionalised so that $\rho_0 = 1$. In the one-dimensional case making use of (9), (10) and (11), our nonlinear elasticity equations (1) and (2) become 6×6 quasilinear system of the form:

$$\frac{\partial w}{\partial t} + A(w) \frac{\partial w}{\partial x} = 0. \tag{12}$$

with the matrix $A(w)$:

$$A(w) = - \begin{bmatrix} 0 & B(w) \\ I & 0 \end{bmatrix} \qquad (13)$$

where I is the 3×3 identity matrix and

$$w = [\,v(x,t), m(x,t)\,] = [\,w_1, w_2, w_3, w_4, w_5, w_6\,]$$

The explicit forms of the 3×3 matrices $B(w) = B(m)$, for three directions of wave propagation, are given in the appendix.

2.4. Eigensystem of matrix A

In this section we analyse the algebraic structure of matrix $A(w)$ from (13), evaluated at zero constant state $w_0 = 0$. Unlike as it is in isotropic elasticity, the form of the equations depends on the chosen direction of propagation. Hence for three directions of wave propagation we obtain three different matrices $A(\mathbf{w})$. However at zero constant state in all three cases there are common features in the structure of the eigensystem of matrix $A(0)$ obtained from (13) with

$$B(0) = \begin{pmatrix} \alpha_1 & 0 & 0 \\ 0 & \alpha_2 & 0 \\ 0 & 0 & \alpha_3 \end{pmatrix}$$

The eigensystem of the matrix $A(0)$ looks as follows.

Eigenvalues:

$$\lambda_1 = -\sqrt{\alpha_1} = -\lambda_2, \quad \lambda_3 = -\sqrt{\alpha_2} = -\lambda_4, \quad \lambda_5 = -\sqrt{\alpha_3} = -\lambda_6$$

Eigenvectors:

$$
\begin{aligned}
r_1 &= [\lambda_1, 0, 0, 1, 0, 0], & l_1 &= \tfrac{1}{2}[\lambda_1^{-1}, 0, 0, 1, 0, 0], \\
r_2 &= [\lambda_2, 0, 0, 1, 0, 0], & l_2 &= \tfrac{1}{2}[\lambda_2^{-1}, 0, 0, 1, 0, 0], \\
r_3 &= [0, \lambda_3, 0, 0, 1, 0], & l_3 &= \tfrac{1}{2}[0, \lambda_3^{-1}, 0, 0, 1, 0], \\
r_4 &= [0, \lambda_4, 0, 0, 1, 0], & l_4 &= \tfrac{1}{2}[0, \lambda_4^{-1}, 0, 0, 1, 0], \\
r_5 &= [0, 0, \lambda_5, 0, 0, 1], & l_5 &= \tfrac{1}{2}[0, 0, \lambda_5^{-1}, 0, 0, 1], \\
r_6 &= [0, 0, \lambda_6, 0, 0, 1], & l_6 &= \tfrac{1}{2}[0, 0, \lambda_6^{-1}, 0, 0, 1].
\end{aligned}
$$

The explicit values of α_j expressed in terms of the elastic constants are given in the appendix. We assume that always $\alpha_j > 0$, so in all three cases we always have three pairs of real eigenvalues. The linearised system (12) is therefore *hyperbolic*, however since we have $\alpha_2 = \alpha_3$ in case of directions **a)** $k_1 = [1, 0, 0]$ and **c)** $k_3 = [1, 1, 1]$ (see the appendix), so the system is *not strictly hyperbolic* in these cases. One pair of eigenvalues corresponds to the velocities of longitudinal waves (subscripts $1, 2$), and the remaining two pairs to the velocities of transverse waves (subscripts $3, 4, 5, 6$). Besides, we always have a complete set of eigenvectors, expressed in terms of the corresponding eigenvalues.

3. Asymptotics

In this part we derive the asymptotic evolution equations for small amplitude weakly nonlinear elastic waves in a cubic crystal. The form of the evolution equations depends heavily on the eigenstructure of the linearised systems presented in the last section. We consider an initial-value problem (IVP) for a quasilinear system (12):

$$
\begin{cases}
\dfrac{\partial w^\epsilon}{\partial t} + A(w^\epsilon)\dfrac{\partial w^\epsilon}{\partial x} = 0 \\[2mm]
w^\epsilon(x,0) = w_0 + \epsilon\, w_1(x, x/\!/\epsilon)
\end{cases}
\tag{14}
$$

The initial data are assumed to have a compact support. This implies the absence of resonant interactions. A *WNGO* asymptotic solution to this IVP around a constant state w_0, is sought for in the form (see [1], [2]):

$$
w^\epsilon(x,t) = w_0 + \epsilon \sum_j \sigma_j\left(x, t, \frac{x - \lambda_j t}{\epsilon}\right) r_j + \mathcal{O}(\epsilon^2)
\tag{15}
$$

with the unknown amplitudes σ_j. In this paper we assume that $w_0 = 0$.

3.1. Asymptotic equations for longitudinal waves

The longitudinal waves correspond to eigensystem of the matrix $A(w_0)$ with subscripts 1 and 2, (see Section 2.4). In all cases considered here, these are locally genuinely nonlinear waves at $w_0 = 0$:

$$
\left.(\nabla \lambda_j(w) \cdot r_j)\right|_{w=0} \neq 0
\tag{16}
$$

for $j = 1, 2$. The canonical transport evolution equations for these waves, derived with the help of classical WNGO expansion (15), in all cases considered here, have the same form of the inviscid Burgers type:

$$
\frac{\partial \sigma_j}{\partial t} + \lambda_j \frac{\partial \sigma_j}{\partial x} + \frac{1}{2}\, \Gamma_j \frac{\partial \sigma_j^2}{\partial \eta} = 0
\tag{17}
$$

for $j = 1, 2$. Here and below η denotes a phase variable.

$$
\lambda_1 = -\sqrt{\alpha_1} = -\lambda_2, \quad \Gamma_1 = -\Gamma_2 = -\frac{\beta_1}{\sqrt{\alpha_1}}
$$

The values of constants α_1 and β_1, which differ for different directions of wave propagation, are displayed in the appendix.

3.2. Asymptotic equations for shear waves

The shear waves correspond to eigensystem of the matrix $A(w_0)$ with subscripts 3, 4, 5 and 6 (see Section 2.4). For these waves a local loss of genuine nonlinearity is observed at the zero constant state (see [14] or [7] in a similar system of equations). Applications of the classical *WNGO* in case of local linear degeneracy gives *linear* transport equations for such waves ([14], [7]). In order to have nonlinear transport evolutions equations, we need to modify the asymptotics. The loss of the local genuine nonlinearity forces us to choose a different scaling of the small parameter in the asymptotics and consider a longer time scale. Uncoupled single *modified Burgers equations* are obtained as the transport equations for shear waves in the nonresonant case. On the other hand, the local loss of strict hyperbolicity results in systems of transport evolution equations for shear waves (in [1,1,1] case). Due to the lack of space we do not show the details of the derivation of these equations here. The methods are essentially similar to those presented by Brio and Hunter ([13]) and the details can be found in ([14]).

a) Direction $k_1 = [1, 0, 0]$

Modified asymptotic expansion:

$$u^\epsilon(x, t) = \epsilon \left(\sigma_s\left(x, t, \frac{\phi_s}{\epsilon^2}\right) r_s + \sigma_{s+2}\left(x, t, \frac{\phi_{s+2}}{\epsilon^2}\right) r_{s+2} \right) \quad s = 3, 4$$

with $\phi_s = x - \lambda_s t$.

Asymptotic equations obtained:

$$\frac{\partial \sigma_s}{\partial t} + \lambda_s \frac{\partial \sigma_s}{\partial x} + \frac{1}{3} G_s \frac{\partial \sigma_s^3}{\partial \eta} = 0 \tag{18}$$

where

$$\lambda_3 = -\sqrt{\alpha_2} = \lambda_5 = -\lambda_4 = -\lambda_6$$

$$G_s = -\frac{1}{4\lambda_s}\left(\frac{3\beta_2^2}{\alpha_2 - \alpha_1} + \gamma_5\right) \quad s = 3, 4, 5, 6 \tag{19}$$

b) Direction $k_2 = [1, 1, 0]$

Modified asymptotic expansion:

$$u^\epsilon(x, t) = \epsilon \sigma_s\left(x, t, \frac{\phi_s}{\epsilon^2}\right) r_s \quad s = 3, 4, 5, 6$$

As the asymptotic equations we get the *modified inviscid Burgers equations* (18) with $\lambda_3 = -\sqrt{\alpha_2} = -\lambda_4$, $\lambda_5 = -\sqrt{\alpha_3} = -\lambda_6$ and G_s equals to (19) for $s = 3, 4$, with different, appropriate for $[1, 1, 0]$ direction constants, and

$$G_s = -\frac{1}{4\lambda_s}\left(\frac{3\beta_3^2}{\alpha_2 - \alpha_1} + \gamma_8\right) \quad s = 5, 6$$

c) Direction $k_3 = [1,1,1]$

Asymptotic expansion:

$$u^\epsilon(x, t) = \epsilon \left(\sigma_s\left(x, t, \frac{\phi_s}{\epsilon}\right) r_s + \sigma_{s+2}\left(x, t, \frac{\phi_{s+2}}{\epsilon}\right) r_{s+2} \right), \quad s = 3, 4$$

Asymptotic systems obtained:

$$\frac{\partial \sigma_s}{\partial t} + \lambda_s \frac{\partial \sigma_s}{\partial x} + \frac{1}{2} \Gamma^s_{s,s+2} \frac{\partial(\sigma_s \sigma_{s+2})}{\partial \eta} = 0$$

$$\frac{\partial \sigma_{s+2}}{\partial t} + \lambda_{s+2} \frac{\partial \sigma_{s+2}}{\partial x} + \frac{1}{2}\left(\Gamma^{s+2}_{s,s} \frac{\partial \sigma_s^2}{\partial \eta} + \Gamma^{s+2}_{s+2,s+2} \frac{\partial \sigma_{s+2}^2}{\partial \eta} \right) = 0$$

where

$$\lambda_3 = -\sqrt{\alpha_2} = \lambda_5 = -\lambda_4 = -\lambda_6$$

$$\Gamma^4_{46} = \frac{\beta_3}{\sqrt{\alpha_2}} = \Gamma^4_{64} = \Gamma^6_{44} = \Gamma^5_{55} = -\Gamma^3_{35} = -\Gamma^3_{53} = -\Gamma^5_{33} = -\Gamma^6_{66}$$

4. Concluding remarks

Our main result is a derivation of the asymptotic canonical evolution equations for the amplitudes of all the weakly nonlinear elastic waves in a cubic crystal for three directions of wave propagation in the non-resonant case. All coefficients in these asymptotic equations are calculated explicitly in the analytical forms and are expressed in terms of the material constants. Derivation of the asymptotic equations for shear waves requires a modification of a classical expansion of *WNGO* used to strictly hyperbolic and genuinely nonlinear waves. The key idea is a proper scaling of a small parameter according to the structure of the eigensystem of the linearised matrix $A(0)$. To obtain this result a modification of a classical approach is required and higher order effects have to be taken into account. We have to include higher order terms both in the Taylor expansion of the fluxes as well as in the expression for the energy. This together with the right scaling of a small parameter allows us to derive modified inviscid Burgers equations as nonlinear evolution transport equations for shear waves in [1,0,0], and [1,1,0] cases. In [1,1,1] case, coupled systems were obtained with the classical scaling. However other expansions may be used here (see [14]).

In this presentation we have focused on the non-resonant case. The problem of resonant interactions of nonlinear elastic waves in a cubic crystal is treated in our separate paper [15].

5. Appendix

In the appendix we display the explicit forms of matrix $B(w)$ from (13) for three directions of wave propagation. We have

$$B(w) = \begin{bmatrix} b_{11} & b_{12} & b_{13} \\ b_{12} & b_{22} & b_{23} \\ b_{13} & b_{23} & b_{33} \end{bmatrix}$$

where

$$
\begin{aligned}
b_{11} &= \alpha_1 + \beta_1\, w_4 + \tfrac{1}{2}\left(\gamma_1\, w_4{}^2 + \gamma_2\, w_5{}^2 + \gamma_3\, w_6{}^2\right)\\
b_{12} &= \beta_2\, w_5 + \gamma_2\, w_4\, w_5\\
b_{13} &= \beta_3\, w_6 + \gamma_4\, w_4\, w_6\\
b_{22} &= \alpha_2 + \beta_2\, w_4 + \beta_4\, w_6 + \tfrac{1}{2}\left(\gamma_2\, w_4{}^2 + \gamma_5\, w_5{}^2 + \gamma_6\, w_6{}^2\right)\\
b_{23} &= \beta_4\, w_5 + \gamma_6\, w_5\, w_6\\
b_{33} &= \alpha_3 + \beta_3\, w_4 - \beta_4\, w_6 + \tfrac{1}{2}\left(\gamma_7\, w_4{}^2 + \gamma_6\, w_5{}^2 + \gamma_8\, w_6{}^2\right)
\end{aligned}
$$

The constants α_j, β_j and γ_j are different for each direction.

a) Direction $k_1 = [1,0,0]$:

$$
\begin{aligned}
\alpha_1 &= c_{11}, \quad \alpha_2 = c_{44} = \alpha_3\\
\beta_1 &= 3c_{11} + c_{111}, \quad \beta_2 = c_{11} + c_{166} = \beta_3, \quad \beta_4 = 0\\
\gamma_1 &= 3(c_{11} + 2c_{111}), \quad \gamma_2 = c_{11} + c_{111} + c_{166} = \gamma_3 = \gamma_4 = \gamma_7\\
\gamma_5 &= c_{11} + 2c_{166} = 3\gamma_6 = \gamma_8
\end{aligned}
$$

b) Direction $k_2 = [1,1,0]$:

$$
\begin{aligned}
\alpha_1 &= \tfrac{1}{2}\left(c_{11} + c_{12} + 2\,c_{44}\right), \quad \alpha_2 = \tfrac{1}{2}\left(c_{11} - c_{12}\right), \quad \alpha_3 = c_{44}\\
\beta_1 &= 3\alpha_1 + \tfrac{1}{4}\left(c_{111} + 3c_{112} + 12\,c_{166}\right), \quad \beta_2 = \alpha_1 + \tfrac{1}{4}\left(c_{111} - c_{112}\right)\\
\beta_3 &= \alpha_1 + \tfrac{1}{2}\left(c_{144} + c_{166} + 2\,c_{456}\right), \quad \beta_4 = 0\\
\gamma_1 &= 3\,\alpha_1 + \tfrac{3}{2}\left(c_{111} + 3c_{112} + 12\,c_{166}\right) \quad \gamma_2 = \alpha_1 + \tfrac{1}{2}\left(c_{111} + c_{112} + 6\,c_{166}\right)\\
\gamma_3 &= \alpha_1 + \tfrac{1}{4}\left(c_{111} + 3c_{112}\right) + \tfrac{1}{4}\left(c_{144} + 7\,c_{166} + 2\,c_{456}\right) = \gamma_4 = \gamma_7\\
\gamma_5 &= 3\alpha_1 + \tfrac{3}{2}\left(c_{111} - c_{112}\right)\\
\gamma_6 &= \alpha_1 + \tfrac{1}{4}\left(c_{111} - c_{112}\right) + \tfrac{1}{2}\left(c_{144} + c_{166} + 2\,c_{456}\right)\\
\gamma_8 &= 3\alpha_1 + 3\left(c_{144} + c_{166} + 2c_{456}\right)
\end{aligned}
$$

c) Direction $k_3 = [1,1,1]$:

$$
\begin{aligned}
\alpha_1 &= \tfrac{1}{3}(c_{11} + 2c_{12} + 4c_{44}), \quad \alpha_2 = \tfrac{1}{3}(c_{11} - c_{12} + c_{44}) = \alpha_3\\
\beta_1 &= 3\alpha_1 + \tfrac{1}{9}(c_{111} + 2c_{123} + 16c_{456}) + \tfrac{2}{3}(c_{112} + 2c_{144} + 4c_{166})\\
\beta_2 &= \alpha_1 + \tfrac{1}{9}(c_{111} - c_{123} - 2c_{456}) + \tfrac{1}{3}(2c_{166} - c_{144})\\
\beta_3 &= \tfrac{\sqrt{2}}{18}(-c_{111} - 2c_{123} + 2c_{456}) + \tfrac{\sqrt{2}}{6}(c_{112} + c_{144} - c_{166}) = \beta_4\\
\gamma_1 &= 3\alpha_1 + \tfrac{2}{3}(c_{111} + 2c_{123} + 16c_{456}) + 4\left(c_{112} + 2c_{144} + 4c_{166}\right)\\
\gamma_2 &= \alpha_1 + \tfrac{1}{9}(2c_{111} + c_{123} + 14c_{456}) + \tfrac{1}{3}(2c_{112} + 3c_{144} + 10c_{166})\\
&= \gamma_3 = \gamma_7\\
\gamma_4 &= 3\alpha_1 + \tfrac{2}{3}(c_{111} - c_{123} - 2c_{456}) + 2\left(2c_{166} - c_{144}\right) = \gamma_5\\
&= 3\gamma_6 = \gamma_8
\end{aligned}
$$

Acknowledgement. The author would like to express his gratitude to the organizers of the Conference for the financial support.

References

[1] J.K. Hunter and J.B. Keller, *Weakly nonlinear high frequency waves*, Comm. Math. Phys., **36** (1983), 547–569 .

[2] A. Majda, *Nonlinear geometric optics for hyperbolic systems of conservation laws*, in *Oscillation Theory, Computation and Methods of Compensated Compactness*, IMA vol. 2, Springer-Verlag, New York (1986), 115–165.

[3] R.J. DiPerna and A. Majda, *The validity of nonlinear geometric optics for weak solutions of conservation laws*, Comm. Math. Phys., **98** (1985), 313–347.

[4] J.-L. Joly, G. Métivier and J.R. Rauch, *Resonant one dimensional geometric optics*, J. Funct. Analy., **114**, (1993), 106–231.

[5] A. Majda and R.R. Rosales, *Resonantly interacting weakly nonlinear hyperbolic waves, I: a single space variable*, Stud. Appl. Math., **71** (1984), 149–179.

[6] J.K. Hunter, *Interaction of elastic waves*, Stud. Appl. Math., **86** (1992), 281–314.

[7] W. Domański, *Weakly nonlinear magnetoelastic waves*, in : *Nonlinear Waves in Solids*, ed. J.L. Wegner and F.R. Norwood, ASME Book No AMR 137, (1995), 241–246.

[8] G. Métivier *Recent Results in Nonlinear Geometric Optics*, A plenary lecture at the Seventh International Conference on Hyperbolic Problems, Theory, Numerics, Applications, ETH Zurich, February 9–13, 1998.

[9] D.R. Bland, *Nonlinear Dynamic Elasticity*, Blaisdel Publishing Company, Woltham, (1969).

[10] R.E. Green, *Ultrasonic Investigation of Mechanical Properties*, v.3 of Treatise on Material Science and Technology, Academic Press, New York, 1973.

[11] F. Birch, "Finite elastic strain of cubic crystals", *Phys. Rev.* **11**, 809–824 (1947).

[12] K.Brugger, *Thermodynamic definition of higher order elastic coefficients*, Phys. Rev., **133** (1964), 1611–1622.

[13] M. Brio and J.K. Hunter, *Asymptotic equations for conservation laws of mixed type*, Wave Motion, **16** (1992), 57–64.

[14] W. Domański, *Propagation of weakly nonlinear elastic waves in a cubic crystal*, to appear.

[15] W. Domański and T.F. Jabłoński, *On resonances of nonlinear elastic waves in a cubic crystal*, submitted to Math. and Mech. of Solids.

Institute of Fundamental Technological Research,
Polish Academy of Sciences,
Świętokrzyska 21,
00-049 Warsaw, Poland
E-mail address: wdoman@ippt.gov.pl

International Series of Numerical Mathematics
Vol. 129, © 1999 Birkhäuser Verlag Basel/Switzerland

Computing Strong Shocks in Ultrarelativistic Flows: A Robust Alternative

Rosa Donat and Antonio Marquina

Abstract. In recent years, shock capturing methods have started to be used in numerical simulations in Relativistic Fluid Dynamics (RFD). These techniques lead to explicit numerical codes that are able to successfully simulate the extreme conditions of the ultrarelativistic regime. After [2], an explicit, ready-to-use description of the *full* spectral decomposition of the Jacobian matrices of the RFD system is available, and this allows us to implement Marquina's scheme [3] in RFD. The scheme is seen to maintain the good behavior shown in [3] with respect to certain numerical pathologies.

1. Introduction

The term Relativistic Fluid Dynamics, (RFD), applies to both those flows in which the velocities (of individual particles or of the fluid as a whole) approach c, the velocity of light in vacuum, or those where the effects of the background gravitational field -or that generated by the matter itself- are so important that a description in terms of Einstein theory of gravity becomes necessary.

Simulations based on the numerical integration of the hydrodynamical equations provide a valuable tool to confront the theoretical models with the observations (as in Astrophysics) or the experimental results (as in Nuclear Physics), which explains the rapid progress, during the last few years towards the development of reliable RFD-codes that work accurately under the extreme conditions of interest.

The first Eulerian code in RFD was developed by Wilson [22], on the basis of explicit finite-differencing techniques and monotonic transport. The code incorporated artificial viscosity techniques based on earlier work of Richtmyer and Morton for the non-relativistic flow equations. Wilson's code and its sequels have been widely used in numerical RFD simulations; however, despite its popularity (almost all codes in numerical relativistic hydrodynamics in the eighties were based in Wilson's procedure) it turned out to be unable to resolve the extremely strong shock structures that appear in the ultrarelativistic regime.

Norman and Winkler analyzed in depth the artificial viscosity approach to RFD in [16]. Their research led them to the conclusion that a fully implicit treatment of the relativistic equations was the only way to increase the accuracy of artificial viscosity formulations in the ultrarelativistic regime.

By the mid-eighties, and fueled by an increasing awareness that the artificial viscosity approach was of limited use in the ultrarelativistic regime, part of the numerical RFD community started to look at other shock capturing techniques that had been successfully applied in classical gas dynamics to obtain accurate numerical approximations in the presence of strong shocks. Although shock capturing techniques for hyperbolic systems of conservation laws were designed with the Euler equations in mind, great care was devoted to the task of formulating them within a systematic mathematical framework that could turn them into general purpose methods for *any* hyperbolic system of conservation laws.

The first explicit shock capturing codes in RFD without artificial viscosity appear in the early nineties [12, 10, 6]. These codes follow the so-called "Godunov approach", and their design is based on two main points: 1) The ability to write the RFD equations as a system of hyperbolic conservation laws, identifying a suitable vector of unknowns. 2) An *approximate Riemann solver* built using the spectral decomposition of the Jacobian matrices of the system.

Nowadays, several research groups in Astrophysics and Nuclear Physics have successfully implemented many of the most successful Godunov-type techniques and their high resolution (higher order) extensions in the RFD context. As a result, accurate numerical simulations in the ultrarelativistic regime have started to appear [1, 5, 4, 7, 14, 13, 19].

HRSC methods are now routinely used in classical gas dynamics to discretize the convective derivatives of a general system of convection-diffusion-reaction equations in any number of spatial dimensions. It is well known, although not particularly well understood, that many of these shock capturing techniques can, on occasions, fail quite spectacularly. An excellent review on the numerical pathologies that can be encountered in gas dynamics simulations is given by Quirk [17]. Usually, the pathological behavior is local, and does not cause the code to crash. However, in complicated situations, the pathological behavior that a particular scheme may display can have disastrous effects on the numerical approximations.

As in Newtonian hydrodynamics, HRSC methods are starting to become part of numerical codes designed to model more complicated situations in RFD. In the references given above, many of the local pathologies observed in Newtonian hydrodynamics can also be observed in relativistic tests.

In [2], an explicit, ready-to-use, formulation of the *full spectral decomposition* of the Jacobian matrices associated with the fluxes in three dimensions is given. This is the essential ingredient in many of the more sophisticated HRSC techniques. The explicit description of the spectral decomposition of the Jacobian matrices makes possible to use a new numerical method described in [3]. The numerical experimentation shown in [3] indicates that this shock capturing technique is less prone to developing numerical pathologies than some well-used methods, (such as Roe-type methods or Shu-Osher characteristic-based ENO schemes).

The scheme described in [3], which will be referred to as *Marquina's scheme* henceforth, and its ability to elude some of the local pathological behavior encountered when using some of the better known HRSC alternatives has been put to work to obtain accurate numerical simulation in highly ultrarelativistic regimes [2]. In this work, we shall present two examples of the behavior of the scheme and compare it to some of the available alternatives.

The paper is organized as follows: In section 2 we describe the scheme in the context of characteristic-based schemes. In section 3 we study a particular shock tube test where special difficulties arise. In section 4 we explore the relativistic extension of a benchmark experiment in Newtonian hydrodynamics: Emery's step test.

The numerical experiments reported here and in [2] show strong numerical evidence that Marquina's scheme is a robust numerical method for the numerical simulation of ultrarelativistic flows. The comparison with other standard Riemann solvers is specially favorable in tests involving critical Riemann problems or special situations in which certain local pathological behavior is known to occur.

2. Characteristic-based Methods and Marquina's scheme

To ensure that discontinuities are captured by the scheme, i.e. they move at the right speed even if they are unresolved, we must write the discrete equations in *conservation form*. That is a form in which the rate of change of conserved quantities is equal to a difference of fluxes. This form guarantees that we conserve the total amount of the states U present, in analogy with the integral form of the system of conservation laws.

A fully discrete *conservative method* has the form

$$U_i^{n+1} - U_i^n + \frac{k}{h}[F_{i+\frac{1}{2}} - F_{i-\frac{1}{2}}] = 0$$

where $F_{i+\frac{1}{2}} = F(U_{i-p}, \ldots, U_{i+q})$ and F is the **numerical flux function** of the scheme.

In the Riemann solver approach, the numerical flux function is computed by solving a Riemann problem at each cell boundary. The main purpose served by introducing a Riemann solver (either exact or approximate) into a finite difference scheme is that of providing physical realism by correctly discriminating between information which should propagate with different speeds.

This is a recurrent theme when solving hyperbolic equations, since the direction in which information propagates is determinant to construct stable upwind finite difference schemes capable of approximating their exact solutions.

The local characteristic structure, and thus the local upwind directions, can also be obtained by diagonalizing the Jacobian matrix rather than by solving directly a Riemann problem. This approach has been used in flux-vector splitting schemes (e.g. [21]) and it is the general technique used in characteristic-based methods.

Let us consider a system of N convective conservation laws in one spatial dimension,

$$U_t + [F(U)]_x = 0$$

the basic idea of characteristic numerical schemes is to transform this non-linear system to a system of (nearly) independent scalar equations of the form

$$u_t + vu_x = 0$$

discretize each scalar equation independently in a v-upwind biased fashion, and then transform the discretized system back into the original variables.

In a smooth region of the flow, we can get a better understanding of the structure of the system by expanding out the derivative as

$$U_t + JU_x = 0$$

where $J = \frac{\partial F}{\partial U}$ is the Jacobian matrix of the system. In a hyperbolic system this matrix is diagonalized by the matrices of left-multiplying and right-multiplying eigenvectors of J. If L is the matrix whose rows are the left eigenvectors of J and R is the matrix whose columns are the right eigenvectors of J we have

$$LJR = Diag(\lambda^p)$$

and the eigenvectors λ^p are all real.

Suppose we want to discretize our equation at the point x_0, where L and R have values L_0 and R_0. To get a locally diagonalized form, we multiply our system of equations by the *constant* matrix L_0 which nearly diagonalizes J over the region near x_0 (we require a constant matrix so that we can move inside all derivatives):

$$[L_0 U]_t + L_0 J R_0 [L_0 U]_x = 0$$

We have inserted $I = R_0 L_0$ to put the equation in a more recognizable form. The spatially varying matrix $L_0 J R_0$ is exactly diagonalized at the point x_0, with eigenvalues λ_0^p, and it should be nearly diagonal at nearby points. The equations should thus be sufficiently decoupled for us to apply upwind-biased discretizations independently to each component, with λ_0^p determining the wind direction for the p-th component equation. Once this system is fully discretized, we multiply the entire system by $L_0^{-1} = R_0$ to return to the original variables.

The Jacobian matrix J is quite important to any characteristic based scheme, as it defines the local linearization of the non-linear problem. It determines the transformation to the local characteristic fields, and thus what the upwind directions are, as well as what quantities are to be upwind differenced.

Recall that in a conservative scheme we require values of the numerical flux at the cell boundaries, i.e. the midpoints between nodes. Thus, in order to transform to characteristic fields to evaluate numerical fluxes, we need the spectral decomposition (eigenvalues and eigenvectors) of the Jacobian at each cell wall. Since only the values of U at grid points are known, the evaluation of the Jacobian at each cell boundary requires some form of interpolation.

The characteristic based approach has been extensively used in the design of Essentially Non Oscillatory (ENO) schemes. In standard ENO schemes it was thought that the precise form of this interpolation was not so important, but recent developments show that in fact it can make a great deal of difference in causing or eliminating certain numerical pathologies.

The standard ENO method uses a single Jacobian evaluated at the linear average of the states at nodes adjacent to the midpoint,

$$J_{i+\frac{1}{2}} = J\left(\frac{U_i + U_{i+1}}{2}\right)$$

In smooth regions, this centered linear approximation is second order accurate; however, in a smooth region it makes little difference whether the derivatives are computed in an upwind biased fashion or in some combination of upwind and downwind. The precise determination of the Jacobian (and the transformation to characteristic fields) is not so important there. It is between nodes in an unresolved steep gradient that the centrally averaged Jacobian (or even the Jacobian evaluated at some other average, like the Roe mean, for example) might cause problems. In this case, an artificially constructed averaged Jacobian can differ significantly from the left and right Jacobian matrices interpolated from left and right nodal state values, and there is no clear reason why any averaged Jacobian should be the right choice for a proper transformation to characteristic variables at a cell boundary.

Near an unresolved steep gradient in the flow, in which the states vary by a large amount from one node to the next, the unambiguous values of the two Jacobian matrices obtained by extrapolation from nodal data at each side of the cell boundary might differ substantially. It, thus, makes sense to try to use these two Jacobian matrices separately, in an upwind fashion, rather than attempt to define a single representative midpoint Jacobian. This is the driving principle of the flux-split formula described in [3].

2.1. Marquina's scheme

For the sake of completeness, we include here the computation of the numerical flux separating the interfaces U_L and U_R in Marquina's scheme.

Compute first the *sided* local characteristic variables and fluxes:

$$\omega_L^p = L^p(U_L) \cdot U_L \qquad \phi_L^p = L^p(U_L) \cdot F(U_L)$$

$$\omega_R^p = L^p(U_R) \cdot U_R \qquad \phi_R^p = L^p(U_R) \cdot F(U_R)$$

Here $L^p(U_L)$, $L^p(U_R)$, for $p = 1, 2 \ldots, m$, are the (normalized) left eigenvectors of the Jacobian matrices $J(U_L), J(U_R)$. Let $\lambda_p(U_L), \lambda_p(U_R)$, $p = 1, 2 \ldots, m$, be their corresponding eigenvalues.

Then proceed as follows:

For $k = 1, \ldots, m$
    If $\lambda_k(U)$ does not change sign in $[U_L, U_R]$,then
        If $\lambda_k(U_L) > 0$ then

$$\phi_+^k = \phi_L^k$$

$$\phi_-^k = 0$$

        else

$$\phi_+^k = 0$$
$$\phi_-^k = \phi_R^k$$

            endif

    else

$$\alpha_k = \max_{U \in \Gamma(U_L, U_R)} |\lambda_k(U)|$$
$$\phi_+^k = .5 \cdot (\phi_L^k + \alpha_k \omega_L^k)$$
$$\phi_-^k = .5 \cdot (\phi_R^k - \alpha_k \omega_R^k)$$

    endif

$\Gamma(U_L, U_R)$ is a curve in the space of states of the system connecting U_L and U_R. For any hyperbolic system where the fields are either genuinely nonlinear or linearly degenerate, we can test the possible sign changes of $\lambda_k(U)$ by checking the sign of $\lambda_k(U_L) \cdot \lambda_k(U_R)$. Also, α_k can be determined as

$$\alpha_k = \max\{|\lambda_k(U_L)|, |\lambda_k(U_R)|\}.$$

The numerical flux that corresponds to the cell-interface separating the states U_L and U_R is then

$$F^M(U_L, U_R) = \sum_{p=1}^{m} \left(\phi_+^p R^p(U_L) + \phi_-^p R^p(U_R) \right) \tag{1}$$

Marquina's scheme can thus be interpreted as a characteristic-based scheme that avoids the use of an *averaged* intermediate state to perform the transformation to the local characteristic fields. The ambiguity in choosing this average is avoided by using directly the unambiguous data on the left and right sides of each cell wall.

To construct higher order versions of the scheme, we follow the *method of lines* approach of [20]. The discretization process is carried out in two steps: First we use an ENO spatial reconstruction for the numerical flux functions. Then we discretize in time using the TVD Runge-Kutta ODE solvers developed in [20]. Our preferred ENO reconstruction is the Piecewise Hyperbolic Method (PHM) developed in [11].

The extension to higher dimensions is accomplished, as in [20], in a dimension by dimension fashion, so that the one dimensional method applies unchanged to higher dimensional problems.

3. A 1D numerical simulation

We concentrate here on the "colliding slabs" Riemann problem. The initial set-up is that of a relativistic shock tube test (see [2] for details) for an ideal gas, $p = (\Gamma - 1)\rho\epsilon$, with $\Gamma = 5/3$, and the following conditions at $t = 0$:

$$\{\epsilon_L = 10^{-4}, \rho_L = 1, v_L = .9995\} \qquad \{\epsilon_R = 10^{-4}, \rho_R = 1, v_R = -.9995\}$$

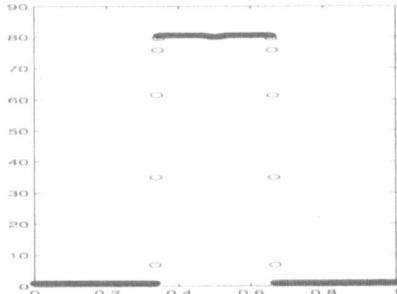 

FIGURE 1. density plots: Marquina-ENO3 left, Shu-Osher-ENO3 right.

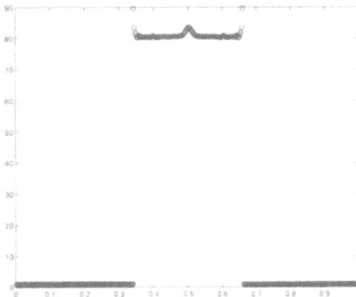

FIGURE 2. density plot: HLLE-ENO3

In the rigid-wall collision of the two slabs of cold gas, a numerical pathology known as 'overheating' ([2, 3] and references therein) occurs on most shock capturing schemes without a heat conduction mechanism.

For our numerical simulations we use a grid of 400 equally spaced points in $[0, 1]$ and $h_t/h_x = .2$. In Figures 1 and 2 we show numerical approximations to the density obtained with a Shu-Osher code, Marquina's scheme and the relativistic extension of the HLLE scheme developed in [19]. The spatial reconstruction procedure is ENO third order polynomial (see [20]), and the time discretization procedure is the third order TVD Runge-Kutta method of [20]. We see that Marquina's scheme is able to dissipate adequately the initial overheating. It is worth mentioning the spurious overshoots obtained by the HLLE scheme. Observe that the velocities of the gas slabs are extremely close to the speed of light (the problem is normalized so that the speed of light is 1). The Lorentz factor is $W \approx 100$, well into the ultrarelativistic regime, and the jumps of the state variables at the shocks are very large; in particular, the pressure jump (not shown) is of 10 orders of magnitude.

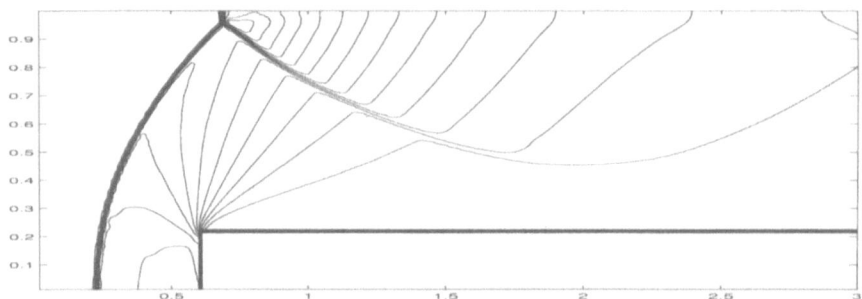

FIGURE 3. density plot: Marquina's scheme (PHM-REF), t=3.0

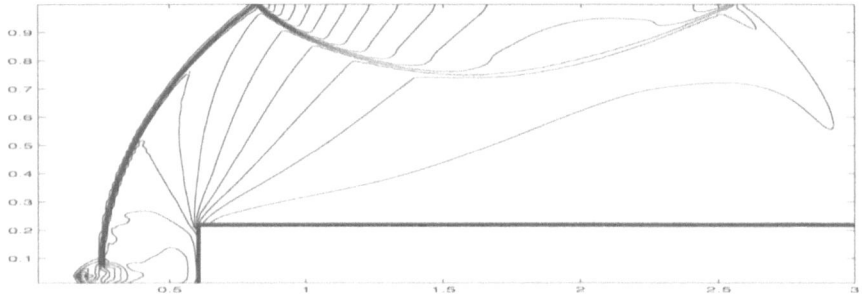

FIGURE 4. density plot: Shu-Osher (PHM-REF), t=2.4

4. A 2D numerical simulation

In [2] we show a numerical simulation of a relativistic extension of Emery's step test, a benchmark test in Newtonian hydrodynamics [23]. We refer to [2] for details on the initial set-up of the test. Here we shall concentrate on an interesting phenomenon observed when the grid is refined. When using the standard Shu-Osher ENO schemes, and a resolution of 240 × 80 cells, a small protuberance starts to form at the base of bow shock. The protuberance grows and causes the code to crash. The numerical pathology can be observed in the density plot of Figure 4. On the other hand, when using our flux-split algorithm, the behavior of the numerical approximation is consistent with the physics of the problem and we observe no numerical pathologies. In Figure 3, we show the numerical approximation obtained with Marquina's scheme and the PHM reconstruction procedure, at time t=3.0; at this time the equivalent Shu-Osher code of Figure 4 has already crashed.

The initial velocity is $v_0^x = .9995$ (we assume a normalization in which the speed of light is 1), that is $W \approx 100$.

Acknowledgement. This work has been supported by the Spanish DGYCIT (grant PB94-0987).

References

[1] D. Balsara, *J. Comput. Phys.*, **114**, 284 (1994).

[2] R. Donat, J.A. Font, J.M. Ibañez and A. Marquina, *J. Comput. Phys. in press*

[3] R. Donat and A. Marquina, *J. Comput. Phys.*, **125**, 42 (1996).

[4] G.C. Duncan and P.A. Hughes, *Astrophys. J.*, **436**, L119 (1994).

[5] A. Dolezal and S.S.M. Wong, *J. Comp. Phys.*, **120**, 266 (1995).

[6] F. Eulderink, Ph.D. Thesis, University of Leiden (1993).

[7] F. Eulderink and G. Mellema, *Astron. Astrophys.*, **284**, 652 (1994).

[8] J.A. Font, J.M$^{\underline{a}}$. Ibáñez, A. Marquina and J.M$^{\underline{a}}$. Martí, *Astron. Astrophys.*, **282**, 304 (1994).

[9] R.J. LeVeque, in *Numerical Methods for Conservation Laws*, Birkhäuser (1991).

[10] A. Marquina, J.M$^{\underline{a}}$. Martí, J.M$^{\underline{a}}$. Ibáñez, J.A. Miralles and R. Donat, *Astron. Astrophys.*, **258**, 566 (1992).

[11] A. Marquina, *SIAM J. Scient. Comp.*, **15**, 892 (1994).

[12] J.M$^{\underline{a}}$. Martí, J.M$^{\underline{a}}$. Ibáñez and J.A. Miralles, *Phys. Rev.*, **D43**, 3794 (1991).

[13] J.M$^{\underline{a}}$. Martí, E. Müller, J.A. Font and J.M$^{\underline{a}}$. Ibáñez, *Astrophys. J.*, **448**, L105 (1995).

[14] J.M$^{\underline{a}}$. Martí and E. Müller, *J. Comp. Phys.*, **123**, 1 (1996).

[15] J.M$^{\underline{a}}$. Martí, E. Müller, J.A. Font, J.M$^{\underline{a}}$. Ibáñez and A. Marquina, *Astrophys. J.*, **479** 151–163 (1997)

[16] M.L. Norman and K-H.A. Winkler, in *Astrophysical Radiation Hydrodynamics*, ed. by M.L. Norman and K-H.A. Winkler (Reidel, 1986).

[17] J. Quirk, *Int. J. Numer. Methods Fluids*, **18** 555 (1994).

[18] P.L. Roe, *J. Comput. Phys.*, **43** 357 (1981).

[19] V. Schneider, V. Katscher, D.H. Rischke, B. Waldhauser, J.A. Marhun and C.-D. Munz, *J. Comput. Phys.*, **105**, 92 (1993).

[20] C. W. Shu and S. J. Osher, *J. Comput. Phys.*, **83**, 32 (1989).

[21] J. Steger and R.F. Warming J. Comput. Phys., v. **40**, (1981), 263–293.

[22] J.R. Wilson, *Astrophys. J.*, **173**, 431 (1972).

[23] P.R. Woodward and P. Colella, *J. Comput. Phys.*, **54** 115 (1984).

Departamento Mathematica Aplicada,

Facultat de Matematiques,

Universitat de Valencia

Dr. Moliner 50

46100 Burjassot (Valencia) Spain

E-mail address: donat@uv.es, marquina@uv.es

International Series of Numerical Mathematics
Vol. 129, © 1999 Birkhäuser Verlag Basel/Switzerland

A Free Boundary Problem for
an Elastic-Plastic Flow Model

Gleb G. Doronin and Aparecido J. Souza

Abstract. The Prandtl-Reuss system of equations with the Mises yield condition is reduced in this paper to a free-boundary problem for a hyperbolic system of conservation laws. A local existence-uniqueness theorem for the linearized problem is proved.

1. Introduction

The problem of non-stationary elastic-plastic deformation has been investigated, as a rule, either computationally, or in special cases, or in general form by the methods of variational inequalities.

In computational investigations, one of the important points is to construct the interface separating the domains of elasticity and plasticity [1]. In other words, one has to describe the evolution of the yield point.

This problem has been resolved for the elastic-plastic impact problem [2] and for some other special model problems with piecewise-constant initial and boundary conditions (see references in [2]). We know of no results concerning mathematical well-posedness of such problems if the initial data are arbitrary smooth functions.

Great mathematical progress in the study of elastic-plastic flow has been obtained using the variational inequalities approach (see [3] and the references). However, this approach only gives weak solutions. This complicates a justification of the time-stepping difference methods, since it is unclear how to describe the interface curve.

Our aim here is to reduce the nonlinear Prandtl-Reuss system of partial differential equations, constrained by the Mises yield condition, to a free-boundary problem for a hyperbolic system of conservation laws. Then we investigate the Cauchy problem for this hyperbolic system, which has discontinuous coefficients along the unknown interface curve. We prove an existence-uniqueness theorem in the class of piecewise smooth functions for the linearized equations. The interface curve is obtained by combining the method of characteristics and the well-known Picard Theorem.

Initial boundary-value problems for hyperbolic systems of equations with dis-continuous coefficients have been studied by several authors (see, for instance, [4], [5] and [6]). Unfortunately the results obtained are not applicable to the problem of elastic-plastic deformation.

2. Statement of the problem

The Prandtl-Reuss system of partial differential equations describing nonstation-ary one-dimensional Elastic-Plastic motion of a continuous medium is written as follows [1]:

$$\rho(u_t + uu_\chi) + (P - S)_\chi = 0; \tag{1a}$$

$$\rho_t + (\rho u)_\chi = 0; \tag{1b}$$

$$S_t + uS_\chi - \frac{4}{3}\mu u_\chi = 0, \text{ in the domain of elasticity;} \tag{1c}$$

$$|S| = Y := \frac{2}{3}Y_0, \text{ in the domain of plasticity.} \tag{1d}$$

Here $\rho(\chi, t)$ and $u(\chi, t)$ are the density and velocity of the medium; $S(\chi, t)$ is the shear stress; P is the pressure; $P = P(\rho)$ is the equation of state satisfying the inequalities $0 < P'(\rho) < \infty$; μ and Y_0 are the positive constants of Lamé and Mises, respectively; and χ and t are independent Eulerian variables, $\chi \in \mathbb{R}$, $t \geq 0$. The domain of elasticity, denoted by D_e, is defined by the inequality $|S| < Y$. The domain of plasticity, denoted by D_p, is defined by the relation $|S| = Y$.

When $t = 0$, we impose the initial conditions

$$(\rho, u, S)(\chi, 0) = (\rho_0, u_0, S_0)(\chi), \ \chi \in \mathbb{R} \tag{2}$$

with sufficiently smooth functions ρ_0, u_0, S_0 and, as usual, $0 < \rho_0(\chi) < \infty$.

If $\rho(\chi, t) > 0$, the characteristic speeds for system (1a)–(1c) are real and distinct:

$$\lambda_1 = u, \ \lambda_{2,3} = u \pm \sqrt{P'(\rho) + 4\mu/(3\rho)}.$$

Thus the system (1a)–(1c) is strictly hyperbolic [7].

However condition (1d) complicates the investigation of system (1a)–(1d) because we do not know "a priori" in what domain one has to find a solution.

The problem we investigate here is to find all unknown physical quantities from (1a)–(1d) in the domain of elasticity and of plasticity as well as the unknown curve that separates these domains.

An advantage of the one-dimensional case is that it is convenient to use Lagrangian coordinates. We recall that the Lagrangian position of the medium located, at time t, at the point χ is the point $x \in \mathbb{R}$ where the medium was at time $t = 0$. The Eulerian and Lagrangian coordinates are connected by the following relation [7]:

$$\chi = x + \int_0^t u(x,t)\, dt\,.$$

In the coordinates (x,t), the system (1a)–(1d) takes the form

$$\rho_0 u_t + (P - S)_x = 0;$$
$$[2ex]\rho_0 (\ln \rho)_t = -\rho u_x;$$
$$\rho_0 S_t = \frac{4}{3}\, \mu\, \rho u_x,\ (x,t) \in D_e;$$
$$[2ex]\,|S| = Y,\ (x,t) \in D_p\,.$$

Without loss of generality, assume that $4\mu/3 = 1$. We also assume that the stress $S(x,t)$ is a nonnegative function everywhere in the domain of flow. The evolution of the plastic domain starts from some yield point, at which the initial stress is maximum. To simplify our study, we take this point to be $(x,t) = (0,0)$. Thus, for the initial shear stress we have the following conditions:

$$0 \le S_0(x) \le Y\;;\; S_0(0) = Y\;;\; S_0'(0) = 0\;;\; S_0''(0) < 0\,.$$

With these assumptions, we can rewrite our system as follows:

$$\rho_0 u_t + (P - s)_x = 0; \tag{3a}$$
$$v_t - u_x = 0; \tag{3b}$$
$$s_t = (\ln v)_t,\ (x,t) \in D_e\,; \tag{3c}$$
$$s = 0,\ (x,t) \in D_p\,. \tag{3d}$$

Here $s = S - Y$, $v = \rho_0/\rho$, $P = P(v)$ and $-\infty < P'(v) < 0$.
   The initial conditions (2) take the form

$$(u,v,s)(x,0) = (u_0,1,s_0)(x),\ x \in \mathbb{R}; \tag{4a}$$
$$-Y \le s_0(x) \le 0,\ s_0(0) = s_0'(0) = 0,\ s_0''(0) < 0\,. \tag{4b}$$

Integrating (3c) with respect to t, substituting the result into (3a), we obtain the following problem:

$$\rho_0 u_t + \sigma_x = 0; \tag{5a}$$
$$v_t - u_x = 0; \tag{5b}$$

$$\sigma = \sigma(v, s_0(x)) = \begin{cases} P(v) - \ln v - s_0(x),\ (x,t) \in D_e; \\ P(v),\ (x,t) \in D_p; \end{cases} \tag{6}$$

$$(u,v)(x,0) = (u_0,1)(x),\ x \in \mathbb{R}\,. \tag{7}$$

The unknown interface curve γ separating D_e and D_p is defined by the yield condition

$$s(x,t)\,|_\gamma = 0\,, \tag{8}$$

where the function $s(x,t)$ is given by the formula

$$s(x,t) = s_0(x) + \ln v(x,t)\,. \tag{9}$$

Therefore from (8) and (9), the yield condition expressed in terms of v and s_0 is

$$(v(x,t) - \exp\{-s_0(x)\})|_\gamma = 0\,. \tag{10}$$

Motivated by physical reasons and practical experiments [2], we expect the curve γ to be a graph of a piecewise smooth function $t = \psi(x)$ starting at $(0,0)$ and having an inverse on each of the intervals $x > 0$ and $x < 0$:

(a) $t = \psi(x)$ for $x > 0$ if and only if $x = \varphi_+(t)$;

(b) $t = \psi(x)$ for $x < 0$ if and only if $x = \varphi_-(t)$;   (11)

(c) $\varphi'_+(t) > 0$; $\varphi'_-(t) < 0$; $\psi(0) = \varphi_\pm(0) = 0$.

Along the curves $(\varphi_\pm(t), t)$, we have the jump conditions [8]

$$\varphi'_\pm(t)[u] = [\sigma]\,;\ \varphi'_\pm(t)[v] = -[u]\,,$$

where $[\cdot]$ denotes the jump in a quantity across the curve.

According to the usual requirement of continuity of the total stress [2], we have $[\sigma] = 0$. Hence $[u] = [v] = 0$ and, therefore, both unknown functions u and v must be continuous across γ.

In the domain D_e we require that $1 \le v(x,t) \le \exp\{-s_0(x)\}$. These inequalities, together with (9), imply that Mises condition $-Y \le s(x,t) \le 0$ holds. Moreover, in order to return to Eulerian coordinates, we require that $v(x,t)$ is bounded from below and from above in both domains D_e and D_p.

Finally, from (10) and (11) we have the nonlinear functional equation for the unknown interface curve γ

$$v(\varphi_\pm(t), t) = \exp\{-s_0(\varphi_\pm(t))\}\,. \tag{12}$$

Thus from the mathematical point of view, our problem is to find functions $(u,v)(x,t)$ and $\psi(x)$ such that

$$(u,v)(x,t) \in C(t \ge 0) \cap C^1(t > 0,\ t \ne \psi(x));$$
$$\psi(x) \in C(\mathbb{R}) \cap C^1(\mathbb{R}\backslash\{0\});$$
$$1 \le v(x,t) \le \exp\{-s_0(x)\},\ t \le \psi(x);$$
$$0 < v(x,t) < \infty\,,\ t > \psi(x)\,.$$

The functions $(u, v)(x, t)$ must satisfy the equations (5) together with (6) for $(t > 0,\ t \neq \psi(x))$ and the initial conditions (7) for $t = 0$; the functions $\varphi_\pm(t)$ defined in (11) must satisfy the condition (12).

Definition 2.1. *The functions $(u, v)(x, t)$ and $\psi(x)$ satisfying the conditions above are called a solution of the problem (5)–(7).*

3. The problem with linearized differential operator

Let us consider the linearization of system (5), (6) around the initial data $(u_0, 1)$. After linearization, the problem (5)–(7) takes the form

$$u_t - a^2 v_x = f(x); \tag{13a}$$

$$v_t - u_x = 0; \tag{13b}$$

$$a^2(x) = \begin{cases} c^2(x) := -(P'(1) - 1)/\rho_0(x), & (x, t) \in D_e\ (t < \psi(x)), \\ c_1^2(x) := -P'(1)/\rho_0(x), & (x, t) \in D_p\ (t > \psi(x)), \end{cases} \tag{14}$$

$$f(x) = \begin{cases} s_0'(x)/\rho_0(x), & t < \psi(x), \\ 0, & t > \psi(x), \end{cases} \tag{15}$$

$$(u, v)(x, 0) = (u_0, 1)(x), \quad x \in \mathbb{R}, \tag{16}$$

$$v(\varphi_\pm(t), t) = \exp\{-s_0(\varphi_\pm(t))\}. \tag{17}$$

We recall that, according to (11), $\varphi_\pm(t)$ are the inverses of $\psi(x)$ when x is positive and negative, respectively.

Notice that although the equations (13) constitute a well-known linear hyperbolic system, the coefficient a^2 is discontinuous and relation (17) is a nonlinear equation connecting the unknown functions $v(x, t)$ and $\psi(x)$. The case $\rho_0 = const$ has been considered in [9].

The main result of this paper is the following.

Theorem 3.1. *For any functions $s_0(x) \in C^3(\mathbb{R})$, $\rho_0(x) \in C^3(\mathbb{R})$, $u_0(x) \in C^2(\mathbb{R})$ with $u_0'(x) > 0$, there exists a positive number T such that the problem (13)–(17) has unique solution in the strip $\{x \in \mathbb{R},\ t \in [0, T]\}$, in the sense of Definition 2.1.*

Proof. Let us introduce the Riemann invariants

$$p = u + av, \quad q = u - av. \tag{18}$$

Then equations (13) and conditions (16) in D_e take the form

$$p_t - c(x)p_x = s_0'/\rho_0 - c'(p-q)/2,$$
$$q_t + c(x)q_x = s_0'/\rho_0 - c'(p-q)/2,$$
$$p(x,0) = p_0(x) := u_0(x) + c(x),$$
$$q(x,0) = q_0(x) := u_0(x) - c(x).$$
(19)

The problem (19) has a unique solution $(p,q)(x,t) \in C^2(D_e)$ (see [7]).

Let us define $A(x,t) = s_0'(x)/\rho_0(x) - c'(x)(p(x,t)-q(x,t))/2$, where $(p,q)(x,t)$ is the solution of (19).

Since the coefficients of p_x and q_x are given functions, we are able to get a representation of the solution of (19) in half-plane $t \geq 0$ as follows.

$$p(x,t) = p_0(x_1(x,t;0)) + \int_0^t A(x_1(x,t;\tau),\tau)\,d\tau,$$

$$q(x,t) = q_0(x_2(x,t;0)) + \int_0^t A(x_2(x,t;\tau),\tau)\,d\tau.$$

Here the function $x_k(x,t;\tau)$, $(k=1,2)$ is solution of the following problem:

$$\begin{cases} \dfrac{\partial x_k}{\partial \tau} = a_k(x_k), \\ x_k(x,t;t) = x, \end{cases} \qquad a_k(x) = \begin{cases} -c(x) & \text{if } k=1, \\ c(x) & \text{if } k=2. \end{cases}$$

Then functions $u(x,t)$ and $v(x,t)$ can be recovered from (18) by formulas

$$
\begin{aligned}
v(x,t) \;=\;& \frac{1}{2c(x)}\,(p-q) = \frac{1}{2c(x)}\Big\{ (u_0+c)(x_1(x,t;0)) \\
& -(u_0-c)(x_2(x,t;0)) + \int_0^t \big[A(x_1(x,t;\tau),\tau) \\
& -A(x_2(x,t;\tau),\tau) \big]d\tau \Big\};
\end{aligned}
\tag{20a}
$$

$$
\begin{aligned}
u(x,t) \;=\;& \frac{1}{2}\Big\{ (u_0+c)(x_1(x,t;0)) \\
& +(u_0-c)(x_2(x,t;0)) + \int_0^t \big[A(x_1(x,t;\tau),\tau) \\
& +A(x_2(x,t;\tau),\tau) \big]d\tau \Big\}.
\end{aligned}
\tag{20b}
$$

Let us differentiate equation (17) with respect to $t > 0$ and obtain ordinary differential equations for functions $x - \varphi(t)$ (from now on we drop the subscripts "$\pm$" whenever it is unambiguous).

After some calculation we find that

$$\varphi'(t) = F(\varphi(t), t) := \frac{-v_t(\varphi(t), t)}{v_x(\varphi(t), t) + s_0'(\varphi(t)) \exp\{-s_0(\varphi(t))\}} \, . \tag{21}$$

Explicit differentiation in (20a) gives us

$$
\begin{aligned}
v_t(x, t) \;=\; & \frac{1}{2c(x)} \Big\{ \Big(\frac{\partial x_1}{\partial t}(x, t; 0)\Big) \big(u_0' + c'\big)(x_1(x, t; 0)) \\
& - \Big(\frac{\partial x_2}{\partial t}(x, t; 0)\Big) \big(u_0' - c'\big)(x_2(x, t; 0)) \\
& + \int_0^t \Big[\frac{\partial A}{\partial x_1}\frac{\partial x_1}{\partial t}(x, t; \tau) - \frac{\partial A}{\partial x_2}\frac{\partial x_2}{\partial t}(x, t; \tau)\Big] d\tau \Big\};
\end{aligned}
\tag{22a}
$$

$$
\begin{aligned}
v_x(x, t) \;=\; & \frac{1}{2c(x)} \Big\{ \Big(\frac{\partial x_1}{\partial x}(x, t; 0)\Big) \big(u_0' + c'\big)(x_1(x, t; 0)) \\
& - \Big(\frac{\partial x_2}{\partial x}(x, t; 0)\Big) \big(u_0' - c'\big)(x_2(x, t; 0)) \\
& + \int_0^t \Big[\frac{\partial A}{\partial x_1}\frac{\partial x_1}{\partial x}(x, t; \tau) - \frac{\partial A}{\partial x_2}\frac{\partial x_2}{\partial x}(x, t; \tau)\Big] d\tau \Big\} \\
& - \frac{c'(x)}{2c^2(x)} \Big\{ \big(u_0 + c\big)(x_1(x, t; 0)) - \big(u_0 - c\big)(x_2(x, t; 0)) \\
& + \int_0^t \big[A(x_1(x, t; \tau), \tau) - A(x_2(x, t; \tau), \tau)\big] d\tau \Big\} \, .
\end{aligned}
\tag{22b}
$$

The derivatives of $x_k(x, t; \tau)$ are given by the formulas below [7]:

$$\frac{\partial x_k}{\partial t}(x, t; \tau) = -a_k(x)\frac{\partial x_k}{\partial x}(x, t; \tau);$$

$$\frac{\partial x_k}{\partial x}(x, t; \tau) = \exp\Big\{\int_t^\tau \frac{\partial a_k}{\partial x_k}(x_k(x, t; \eta))d\eta\Big\}.$$

Therefore

$$\lim_{t \to 0} \big[(v_x(\varphi(t), t) + s_0'(\varphi(t)) \exp\{-s_0(\varphi(t))\}\big] = 0\,;$$

$$\lim_{t \to 0} (v_t(\varphi(t), t)) = u_0'(0) > 0\,.$$

Here we are taking into consideration that $\varphi_\pm(0) = 0$. Consequently we are able to invert the fraction (21) to derive the Cauchy problem for the function $\psi(x)$

$$\frac{d\psi}{dx} = \frac{1}{F(x, \psi(x))}\,, \quad \psi(0) = 0. \tag{23}$$

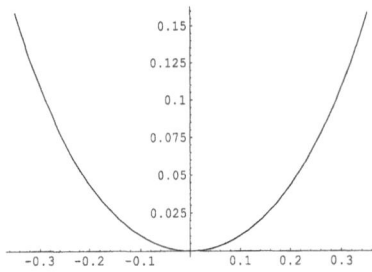

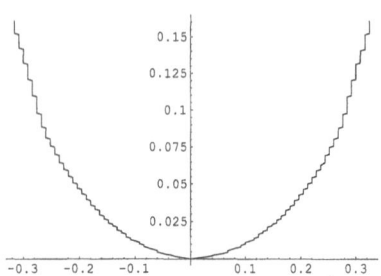

FIGURE 1.   FIGURE 2.
Analytical   Numerical
interface curve   interface curve

Here, by assumption, $F \in C^1$ and for sufficiently small positive t we have $F(x,t) \neq 0$. Hence, by the Picard theorem, there exists a real number $T_1 > 0$ such that the problem (23) has unique differentiable solution for $t \in [0, T_1]$.

Moreover, it is not hard to see by (21)–(23) that

$$\lim_{x \to \pm 0} \left(\frac{d\psi}{dx} \right) = -\frac{s_0'(0)}{u_0'(0)} = 0 \, ;$$

$$\lim_{x \to \pm 0} \left(\frac{d}{dx} \left(\frac{d\psi}{dx} \right) \right) = -\frac{s_0''(0)}{u_0'(0)} > 0 \, . \tag{24}$$

We shall not write out explicit expressions for the derivatives contained in the relations above, since they are obtained in a manner altogether analogous to (22).

Formulas (24) mean that the curve $t = \psi(x)$ is tangent to the line $t = 0$ at the origin and for sufficiently small $t > 0$ there exists an inverse ψ^{-1}. So we have obtained the functions $\varphi_{\pm}(t)$ as defined in (11), (12).

In order to illustrate the interface curve obtained, we show the pictures above. The figure 1 was obtained by using the *Mathematica Symbolic Package* to integrate equation in (23). The figure 2 was obtained by a numerical method, which is based on the Lax-Wendroff scheme, for the problem (13)–(17).

Thus, for $(x,t) \in D_e$, we have constructed the required solution. It is obviously, however, that formulas (20), (23), (24) give us the solution of (13)–(17) in D_e only for $t < T_2$, where the positive number T_2 is the first value of time when the curve $t = \psi(x)$ has a characteristic direction. If such point exists, then the solution above is defined for $t \in [0, \min(T_1, T_2)]$ and for $t \in [0, T_1]$ otherwise.

Now we have to obtain the solution in D_p.

According to (13)–(15) let us consider the following problem:

$$u_t - c_1^2 v_x = 0; \tag{25a}$$

$$v_t - u_x = 0; \tag{25b}$$

$$u(\varphi_\pm(t), t) = u_1(t) := \lim_{x \to \varphi_\pm(t)} u_e(x, t); \tag{25c}$$

$$v(\varphi_\pm(t), t) = v_1(t) := \exp\{-s_0(\varphi_\pm(t))\}. \tag{25d}$$

Here $u_e(x, t)$ is the known function which is defined in D_e by (20b).

The condition $-\infty < P'(v) < 0$ and formulas (24) imply that there exists $T_3 > 0$ such that for $t < T_3$ the curve $t = \psi(x)$ is not a characteristic and has no more than one point of crossing with characteristics of a single family. This means that the Cauchy problem (25) is well-posed. If $u_0 \in C^2(\mathbb{R})$, $s_0 \in C^3(\mathbb{R})$ and $\rho_0 \in C^3(\mathbb{R})$ then the initial data $(u_1, v_1) \in C^1([0, \min(T_1, T_2)])$ and, hence, the Cauchy problem (25) has unique classical solution for $0 \le t \le T = \min(T_1, T_2, T_3)$.

To finish the proof of the Theorem 3.1, notice that formulas (20) and (22) imply that the required inequalities for $v(x, t)$ in D_e and D_p hold for sufficiently small t. Indeed, $v(x, 0) = 1$ and according to (22a), $v_t(x, t) > 0$. Along the curve γ we have, by construction, $v(\varphi(t), t) = \exp\{-s_0(\varphi(t))\} \ge 1$. Therefore in D_e the function $v(x, t)$ is increased from 1 to $\exp\{-s_0(\varphi(t))\}$. In D_p we have that $v(x, t)$ is bounded from below and from above by continuity, as a solution of the Cauchy problem (25). The uniqueness of the solution obtained is followed from the uniqueness of solutions to the Cauchy problems (13)–(16) in D_e and (25) in D_p. This completes the proof. $\qquad \square$

Acknowledgement. We express our gratitude to Nikolai Lar'kin and Bradley Plohr for useful discussions and valuable suggestions.

References

[1] M. L. Wilkins, *Computation of Elastic-Plastic Flows, Methods in Computational Physics, Advances in research and applications*, **3**, Academic Press, 1964.

[2] L. M. Kachanov, *Fundamentals of the theory of plasticity*, Moscow, MIR Publishers, 1969.

[3] G. Duvaut, J. L. Lions, *Les Inéquations en Mécanique et en Physique*, Paris, Dunod, 1972.

[4] T. E. Melnik, *The Stefan type problem for the hyperbolic first order system*, Ukr. Math. Journal, **34 (3)** (1982), 380–384.

[5] K. Yu. Kazakov, S. F. Morozov, *On the definition of unknown curve of discontinuity of solution of initial boundary value problem for the quasi-linear hyperbolic system*, Ukr. Math. Journal, **37 (4)** (1985), 443–450.

[6] C. Denson Hill, *A Hyperbolic Free Boundary Problem*, J. Math. Anal. and Appl., **31** (1970), 117–129.

[7] B. L. Roždestvenskiĭ, N. N. Janenko, *Systems of Quasi-linear Equations and Their Applications to Gas Dynamics,* Providence, AMS, 1983.

[8] J. Smoller, *Shock Waves and Reaction - Diffusion Equations,* Springer-Verlag, New York, 1983.

[9] G. G. Doronin, A. J. Souza, *An Unknown Interface Curve in the Elastic-Plastic Deformation Model,* 46^0 Seminário Brasileiro de Análise, (1997), 690–697.

Permanent address: Institute of Theoretical and Applied Mechanics,
630090, Novosibirsk, Russia;
Current address: Departamento de Matemática e Estatística
Universidade Federal da Paraíba
58109-970, Campina Grande, PB, Brazil
E-mail address: gleb@dme.ufpb.br

Departamento de Matemática e Estatística
Universidade Federal da Paraíba
58109-970, Campina Grande, PB, Brazil
E-mail address: cido@dme.ufpb.br

International Series of Numerical Mathematics
Vol. 129, © 1999 Birkhäuser Verlag Basel/Switzerland

Numerical Errors Downstream of Slightly Viscous Shocks

Gunilla Efraimsson and Gunilla Kreiss

Abstract. Lower order errors downstream of a shock layer have been detected in computations with non-constant solutions when using higher order shock capturing schemes in one and two dimensions, [3]. We analyze the steady state solution of slightly viscous hyperbolic systems of conservation laws using matched asymptotic expansions. The result explains why $O(h)$-errors can appear in smooth regions downstream of a shock layer. The numerical solution of quasi one-dimensional nozzle flow illustrate the analysis.

1. Introduction

Calculations presented in [3] indicate that numerical solutions of conservation laws obtained by a higher order scheme, degenerate in order of accuracy in space, downstream of a shock layer. In one dimension, this effect can only be seen for non-constant initial data or, for stationary solutions, when a lower order term is added to the equations. The degeneracy is only seen for systems of equations. Similar studies are also reported in [1]. Also, in [3], two examples are presented where the degeneracy in accuracy is seen analytically in the discrete approximation.

In many cases a numerical solution of a system of conservation laws can be viewed as a solution of the corresponding slightly viscous system of equations, [10]. In the neighborhood of a shock, the viscous term must be of $O(h)$, to avoid oscillations. Here h is the grid size. In regions where the solution is smooth the viscous term can be of $O(h^2)$ or smaller.

In [4], Goodman and Xin study the relation between the solution of a system of slightly viscous, hyperbolic conservation laws with viscosity ε, here denoted by $\mathbf{v}^v$, and the distributional solution, $\mathbf{v}^i$, of the corresponding inviscid equations. They find, by using matched asymptotic analysis and energy estimates, that outside the shock layer

$$\sup_{\substack{0 \le t \le T, \\ |x - s(t)| \ge \varepsilon^\eta}} |\mathbf{v}^i(x,t) - \mathbf{v}^v(x,t)| \le C\varepsilon, \quad \eta \in (0,1)$$

Here $s(t)$ is the position of the shock.

In this article we study the steady state solution of slightly viscous, one-dimensional hyperbolic systems of conservation laws with a lower order term of the form

$$\mathbf{v}_t^v + \mathbf{f}(\mathbf{v}^v)_x + \mathbf{h}(\mathbf{v}^v) = (\nu \mathbf{v}_x^v)_x \tag{1}$$

and the corresponding inviscid solution, $\mathbf{v}^i$, in order to find the origin of the degeneracy in accuracy. We have here extended the analysis in [2] to systems with nonlinear source terms. In contrast to the work in [4], we let the viscosity term depend on x. In the neighborhood of a shock layer, a viscous term $\varepsilon = O(h)$ is switched on. In the rest of the solution, the viscosity term is of $O(h^2)$. The analysis here is based on the existence of matched asymptotic expansions in ε. In the shock layer an inner solution is valid and an outer solution is valid elsewhere. By considering zeroth and first order terms in the expansions of the inner and outer solutions, it is found that the solution can have an $O(h)$-dependence downstream of the shock layer also when the artificial viscosity term of $O(h)$ is switched off. Also, it is seen that the $O(h)$-dependence downstream of a shock layer is only possible for problems where a characteristic passes through a discontinuity. In shock problems this only occurs for systems of equations. The analysis is applied to a model problem where we explicitly compute the $O(h)$-dependence. Also, we compute the solution for the quasi one-dimensional nozzle flow, that is the Euler equations with a source term, and compute the order of accuracy.

In this paper we consider dissipative schemes, for which we will see that the larger viscosity in the shock layer region does not effect the upstream solution. For non-dissipative, or more generally non-contractive, numerical methods this is not true. For non-contractive schemes it is possible to show that wrong numerical boundary values can influence the numerical solution of the linearized problem upstream, see [8].

It should be stressed that the *order* of accuracy might not always be of main importance, but the accuracy. That is, the constant in front of the $O(h)$-term can be very small, such that the higher order term in h is dominant for reasonable mesh sizes. Hence, a formally higher order scheme can be more efficient than a lower order scheme, although a lower order error is present.

2. Hyperbolic system of conservation laws with a source term

Consider

$$\mathbf{v}_t + \mathbf{f}(\mathbf{v})_x + \mathbf{h}(x, \mathbf{v}) = 0 \quad 0 \le x \le 1 \tag{2}$$

where $\mathbf{v} = (v^{(1)}, v^{(2)}, \dots, v^{(n)})^T$ and $h(x, v)$ is a smoothly varying function. We denote the Jacobian of the flux function by J. The eigenvalues of J are λ_i, $i = 1, 2, \dots, n$. For future reference we also define $H = \partial \mathbf{h}/\partial \mathbf{v}$. In the following we only write out the x-dependence of $\mathbf{h}$ when it is of importance for the analysis.

Assume that $\lambda_i > 0 \ \forall i$ at $x = 0$ and $\lambda_1 < 0$ and $\lambda_i > 0 \ i \neq 1$ at $x = 1$. Hence, suitable boundary conditions are e.g., see [7],

$$\begin{aligned} \mathbf{v} &= \mathbf{v_0} & \text{at } x = 0 \\ R(\mathbf{v}) &= r & \text{at } x = 1 \end{aligned} \tag{3}$$

where $\mathbf{v_0} \in \mathsf{R}^n$ and $r \in \mathsf{R}$ are given and $R(\mathbf{v}) : \mathsf{R}^n \to \mathsf{R}$. The boundary conditions are chosen such that a shock forms at some inner point x_s. We also assume that $\lambda_i > 0 \ \forall i$ for $0 \leq x < x_s$ and $\lambda_1 < 0$ and $\lambda_i > 0 \ i \neq 1$ for $x_s < x \leq 1$

The steady state solution of Equation (2), $\mathbf{v^i}$, satisfies

$$\mathbf{f}(\mathbf{v^i})_x + \mathbf{h}(\mathbf{v^i}) = 0 \tag{4}$$

except at $x = x_s$.

The inviscid problem is closed by the Rankine-Hugoniot condition

$$[\mathbf{f}] = 0 \quad \text{at } x = x_s. \tag{5}$$

We assume that the inviscid solution exists.

Numerical solutions of (2) can be viewed as higher order solutions of the slightly viscous, so called modified, equations

$$\mathbf{v}_t + \mathbf{f}(\mathbf{v})_x + \mathbf{h}(\mathbf{v}) = (\nu \mathbf{v}_x)_x \tag{6}$$

The viscous terms damp oscillations in the numerical solution. In the neighborhood of a shock layer $\nu = O(h)$, where h is the grid size. In regions where the solution is smooth the viscosity term can be of $O(h^2)$ or smaller. A so called limiter determines when to switch on the larger viscous terms. There is a wide variety of limiters and comprehensive work has been performed in order to analyze theoretically and practically the behavior of different limiters. In this article, we will not go into the details in how to construct limiters. More terms can be included in the modified equation. For instance, with a first or second order accurate approximation of the flux derivative a dispersive term $(\nu_2 \mathbf{v}_{xx})_x$ with $\nu_2 = O(h^2)$ is present. See [10]. However, in the present analysis $O(h^2)$-terms in the equation do not effect the existence of $O(h)$-terms in the solution downstream of a shock.

In this paper we consider

$$\mathbf{f}(\mathbf{v}^v)_x + \mathbf{h}(\mathbf{v}^v) = \varepsilon^2 \mathbf{v}_{xx}^v + \varepsilon(\phi(x)\mathbf{v}_x^v)_x \tag{7}$$

with $\varepsilon = O(h)$. That is $\nu = \varepsilon^2 + \phi(x)\varepsilon$. Here ϕ is a smooth function of x/ε satisfying

$$\phi(x) = \begin{cases} 1 & \text{for } |x - x_s| \leq \varepsilon K \\ 0 & \text{for } |x - x_s| \geq \varepsilon(K + 1). \end{cases} \tag{8}$$

Here x_s is the position of the inviscid shock, and K is a sufficiently large constant.

We consider the same boundary conditions as for the inviscid problem (3) together with numerical boundary conditions at $x = 1$. The numerical boundary conditions should be such that no boundary layer is formed at the outflow boundary.

We assume that the solution of Equation (7) can be described by an inner solution, valid in the shock layer, and an outer solution, valid elsewhere. The inner solution is a function of the stretched variable $\tilde{x} = (x - x_s)/\varepsilon$. These solutions can be expanded in ε and matched to sufficient order in a region of overlap. Further, we assume that the position of the shock layer can also be expanded in ε. Thus we have expansions on the form

$$
\begin{aligned}
\text{Inner:} \quad & \mathbf{v}^v \sim \mathbf{g}_0(\tilde{x}) + \varepsilon \mathbf{g}_1(\tilde{x}) + \eta_2(\varepsilon)\mathbf{g}_2(\tilde{x}) + \dots \\
\text{Outer:} \quad & \mathbf{v}^v \sim \mathbf{w}_0(x) + \varepsilon \mathbf{w}_1(x) + \gamma_2(\varepsilon)\mathbf{w}_2(x) + \dots \\
\text{Position:} \quad & x_v \sim x_0 + \varepsilon x_1 + \sigma_2(\varepsilon)x_2 + \dots .
\end{aligned}
\tag{9}
$$

Here $\eta_2(\varepsilon) = \mathrm{o}(\varepsilon)$, $\gamma_2(\varepsilon) = \mathrm{o}(\varepsilon)$ and $\sigma_2(\varepsilon) = \mathrm{o}(\varepsilon)$. We will not prove that these expansions exist. We denote the position of the viscous shock layer as the point where the smallest eigenvalue of the Jacobian, evaluated along the viscous solution, changes sign. We also assume that the matching can be done at points x_m^- and x_m^+ satisfying $x_0 - M \le x_m^- \le x_0 - m$ and $x_0 + m \le x_m^+ \le x_0 + M$, respectively, where $\varepsilon \ll m < M \ll 1$ and $M \to 0$ as $\varepsilon \to 0$. Note that the large viscosity is not present in the matching region.

As is common in asymptotic expansion theory, we also make the assumption that $\mathbf{w}_0 = \mathbf{v}^i$ and $x_0 = x_s$.

In the rest of this section we will show that the above assumptions imply that $\mathbf{w}_1$ can be non-zero in the downstream region. Also, we will obtain the position of the shock layer to order ε.

Firstly, $\mathbf{w}_1$ satisfies

$$
(J(\mathbf{v}^i)\mathbf{w}_1)_x + H(\mathbf{v}^i)\mathbf{w}_1 = 0,
\tag{10}
$$

where $H = \partial \mathbf{h}/\partial \mathbf{v}$. Upstream $\mathbf{w}_1(0) = 0$. Hence, $\mathbf{w}_1(x) \equiv 0$ for $0 \le x \le x_m^-$. In [3] it is proved that $\mathbf{w}_1$ is not present in the upstream region for systems of equations when the numerical scheme is linearly stable and contractive.

Below we will derive the boundary conditions for $\mathbf{w}_1$ at the matching point x_m^+.

Integration of the viscous equation (7) over the shock layer, from the matching point x_m^- to the matching point x_m^+, yields

$$
[\mathbf{f}(\mathbf{v}^v)]_{x_m^-}^{x_m^+} + \int_{x_m^-}^{x_m^+} \mathbf{h}(\mathbf{v}^v)\,dx = \varepsilon^2[\mathbf{v}^v]_{x_m^-}^{x_m^+}.
\tag{11}
$$

We will introduce the expansions into this equality and consider terms of $O(1)$ and $O(\varepsilon)$. At the matching points we use the outer solution and in the shock layer we will use the inner solution.

As expected to zeroth order in ε, Equation (11) is equivalent to the Rankine-Hugoniot relation (5). This is true since

$$
[\mathbf{f}(\mathbf{v}^v)]_{x_m^-}^{x_m^+} = [\mathbf{f}(\mathbf{v}^i)]_{x_m^-}^{x_m^+} + O(\varepsilon) = [\mathbf{f}(\mathbf{v}^i)]_{x_s^-}^{x_s^+} + O(M),
$$

and the other terms in Equation (11) are of higher order in ε.

In the following analysis we need to consider the expansions to the next order, yielding

$$[\mathbf{f}(\mathbf{v}^v)]_{x_m^-}^{x_m^+} = [\mathbf{f}(\mathbf{v}^i)]_{x_m^-}^{x_m^+} + \varepsilon[J(\mathbf{v}^i)\mathbf{w}_1]_{x_m^-}^{x_m^+} + O(\varepsilon^2). \tag{12}$$

By integrating the inviscid equation (4) over the same interval, we obtain

$$[\mathbf{f}(\mathbf{v}^i)]_{x_m^-}^{x_m^+} = -\int_{x_m^-}^{x_m^+} \mathbf{h}(\mathbf{v}^i)dx. \tag{13}$$

After taking into account that $\mathbf{w}_1 \equiv 0$ to the left of the shock layer, and introducing (12) and (13) into Equation (11) we find

$$\varepsilon J(\mathbf{v}^i(x_m^+))\mathbf{w}_1(x_m^+) + \int_{x_m^-}^{x_m^+} (\mathbf{h}(\mathbf{v}^v) - \mathbf{h}(\mathbf{v}^i))dx = O(\varepsilon^2) \tag{14}$$

We denote the integral

$$\mathbf{I}_1 := \int_{x_m^-}^{x_m^+} (\mathbf{h}(\mathbf{v}^v) - \mathbf{h}(\mathbf{v}^i))dx \tag{15}$$

We will now express $\mathbf{I}_1$ using the first term of the inner solution, the inviscid solution at $x = x_s$ and x_1, respectively.

In order to obtain the equation for the inner solution we introduce the stretched variable $\tilde{x} = (x - x_s)/\varepsilon$ into the equation (7), yielding

$$\mathbf{f}(\mathbf{v}^v)_{\tilde{x}} + \varepsilon \mathbf{h}(\varepsilon\tilde{x} + x_s, \mathbf{v}^v) = (\phi(\tilde{x})\mathbf{v}_{\tilde{x}}^v)_{\tilde{x}} + \varepsilon \mathbf{v}_{\tilde{x}\tilde{x}}^v. \tag{16}$$

Thus the first term in the inner expansion satisfies

$$\mathbf{f}(\mathbf{g}_0)_{\tilde{x}} = (\phi(\tilde{x})\mathbf{g}_{0\tilde{x}})_{\tilde{x}}, \tag{17}$$

In accordance with standard matched asymptotic techniques, [5], [9], we use the inviscid solution at the shock position $x = x_s$ as boundary condition at $\tilde{x} = \pm\infty$.

The first order perturbation of the position of the shock layer is x_1. Clearly, x_1 cannot be determined by (17). Let $\mathbf{g}(\tilde{x})$ satisfy

$$\mathbf{f}(\mathbf{g})_{\tilde{x}} = \mathbf{g}_{\tilde{x}\tilde{x}}, \tag{18}$$

together with the same boundary conditions, and the additional condition that the smallest eigenvalue of $J(\mathbf{g})$, vanishes at $\tilde{x} = 0$. Note that $\mathbf{g}$ will approach the boundary values exponentially fast. Then, for sufficiently large K, $\mathbf{g}_0(\tilde{x}) = \mathbf{g}(\tilde{x} + x_1)$, except for exponentially small terms, where K is defined in (8). Below we will determine x_1 by considering the outer problem to $O(\varepsilon)$.

The integral $\mathbf{I}_1$ becomes with the inner variable $\tilde{x} = (x - x_s)/\varepsilon$,

$$\mathbf{I}_1 = \varepsilon \int_{\tilde{x}_m^-}^{\tilde{x}_m^+} (\mathbf{h}(\varepsilon\tilde{x} + x_s, \mathbf{v}^v) - \mathbf{h}(\varepsilon\tilde{x} + x_s, \mathbf{v}^i))d\tilde{x}.$$

Here $\tilde{x}_m^+ = (x_m^+ - x_s)/\varepsilon$ and $\tilde{x}_m^- = (x_m^- - x_s)/\varepsilon$. Introduce the asymptotic expansion of the inner solution,

$$\mathbf{I}_1 = \varepsilon \int_{\tilde{x}_m^-}^{\tilde{x}_m^+} (\mathbf{h}(\varepsilon\tilde{x} + x_s, \mathbf{g}_0(\tilde{x}) + \varepsilon \mathbf{g}_1(\tilde{x}) + \dots) - \mathbf{h}(\varepsilon\tilde{x} + x_s, \mathbf{v}^i(\varepsilon\tilde{x} + x_s)))d\tilde{x} =$$

$$\varepsilon \int_{-\infty}^0 (\mathbf{h}(x_s, \mathbf{g}_0(\tilde{x})) - \mathbf{h}(x_s, \mathbf{v}_-^i))d\tilde{x} + \varepsilon \int_0^\infty (\mathbf{h}(x_s, \mathbf{g}_0(\tilde{x})) - \mathbf{h}(x_s, \mathbf{v}_+^i))d\tilde{x} + \mathbf{o}(\varepsilon)$$

$$:= \varepsilon \mathbf{I}_2 + \mathbf{o}(\varepsilon). \quad (19)$$

Here $\mathbf{v}_\pm^i$ is the inviscid solution of the left and the right branches, respectively, at the inviscid shock position x_s. Since $\mathbf{g}_0(\tilde{x}) = \mathbf{g}(\tilde{x} + x_1)$ we have that

$$\mathbf{I}_2 = \int_{-\infty}^0 (\mathbf{h}(x_s, \mathbf{g}(\tilde{x})) - \mathbf{h}(x_s, \mathbf{v}_-^i))d\tilde{x} +$$

$$\int_0^\infty (\mathbf{h}(x_s, \mathbf{g}(\tilde{x})) - \mathbf{h}(x_s, \mathbf{v}_+^i))d\tilde{x} + x_1[\mathbf{h}(x_s, \mathbf{v}^i)]$$

$$:= \mathbf{I}_3 + x_1[\mathbf{h}(x_s, \mathbf{v}^i)]. \quad (20)$$

Finally, with (19) and (20) in (14) we see that the boundary condition for $\mathbf{w}_1$ at $x = x_m^+$ is

$$\mathbf{w}_1(x_m^+) = -J^{-1}(\mathbf{v}^i(x_m^+))(\mathbf{I}_3 + x_1[\mathbf{h}(x_s, \mathbf{v}^i)]) =$$
$$-J^{-1}(\mathbf{v}_+^i)\mathbf{I}_3 + x_1[\mathbf{h}(x_s, \mathbf{v}^i)]) + O(M). \quad (21)$$

Hence, the equation for $\mathbf{w}_1$ and x_1 is

$$\begin{aligned} (J(\mathbf{v}^i)\mathbf{w}_1)_x + H(\mathbf{v}^i)\mathbf{w}_1 &= 0 \qquad\qquad x_m^+ \le x \le 1 \\ \mathbf{w}_1(x_m^+) &= -J_+^{-1}(\mathbf{I}_3 + x_1[\mathbf{h}(x_s, \mathbf{v}^i)]) \\ \mathbf{L}^T \mathbf{w}_1(1) &= 0 \end{aligned} \quad (22)$$

Here $J_+ = J(\mathbf{v}_+^i)$. Also, $\mathbf{L}^T = (\partial R/\partial v^{(1)}, \partial R/\partial v^{(2)}, \dots, \partial R/\partial v^{(n)})$, evaluated for $\mathbf{v} = \mathbf{v}^i(x = 1)$. To see that $\mathbf{w}_1 \ne 0$ we write Equation (22) on the form

$$(\mathbf{w}_1)_x + D(x)\mathbf{w}_1 = 0$$

where $D(x) = J^{-1}(J_x + H)$. This is a linear ODE with variable coefficients. Since J is nonsingular in the downstream region, the linear problem has a solution and can always be expressed with a solution operator $S(x)$,

$$\mathbf{w}_1(x) = S(x)(-J_+^{-1}\mathbf{I}_3 - x_1 J^{-1}[\mathbf{h}(x_s, \mathbf{v}^i)]))$$

where $S(x_m^+) = I$. Thus, the solution at $x = 1$ is of the form

$$\mathbf{w}_1(1) = \mathbf{y} + x_1 \mathbf{z}$$

where $\mathbf{z} = S(1)J^{-1}[\mathbf{h}(x_s, \mathbf{v}^i)]$. Clearly, x_1 and $\mathbf{w}_1$ can be determined as long as $\mathbf{L}^T S(1) J_+^{-1}[\mathbf{h}(x_s, \mathbf{v}^i)] \ne 0$. In [2] we compute $\mathbf{I}_3$ numerically and solve Equation (22) for a model problem and explicitly see that $\mathbf{w}_1$ and x_1 are non-zero.

For scalar problems the only possible boundary condition at $x = 1$ forces $w_1 \equiv 0$ in the whole region. However, x_1 will be non-zero as long as I_3 is non-zero.

Systems with weak shocks behave in many ways as scalar problems. We expect that from a similar analysis taking the shock strength into account it would follow that $\mathbf{w}_1$ tends to zero as the shock strength tends to zero.

3. Numerical results

We solved the equations for quasi one-dimensional nozzle flow. That is, the Euler equations in $1D$ with a source term

$$\begin{pmatrix} \rho \\ \rho u \\ \rho E \end{pmatrix}_t + \begin{pmatrix} \rho u \\ \frac{(\rho u)^2}{\rho} + p \\ (\rho E + p)\frac{\rho u}{\rho} \end{pmatrix}_x + \frac{\rho u}{\rho}\frac{1}{A}\frac{dA}{dx}\begin{pmatrix} \rho \\ \rho u \\ \rho E + p \end{pmatrix} = 0 \quad 0 \le x \le 1 \quad (23)$$

Here ρ is the density, ρu is the mass flux and ρE the total energy per unit mass. With the assumption of perfect gas, the temperature T and the pressure p are related as $T = p/\rho$. We consider calorically perfect gases and hence

$$p = (\gamma - 1)(\rho E - \frac{(\rho u)^2}{2\rho})$$

with $\gamma = 1.4$. The nozzle cross-section area is denoted by $A(x)$. Here $A(x) = 1.398 + 0.347 \tanh(5(x - 0.4))$.

We solved Equation (23) with the semi-discrete scheme

$$(\mathbf{v}_i)_t + D_0\mathbf{f}(\mathbf{v}_i) + \mathbf{h}(x_i, \mathbf{v}_i) = \varepsilon D_+ \phi_i D_- \mathbf{v}_i + \varepsilon_0 D_+ D_- \mathbf{v}_i \quad (24)$$

Here $\varepsilon_0 = O(\varepsilon^2)$. We discretized in space by introducing $x_i = ih, h = 1/N, i = 0, 1, \ldots, N$ and $\mathbf{v}_i$ is a grid function, where $\mathbf{v}_i \approx \mathbf{v}^v(x_i)$.

We used the following boundary conditions

$$\begin{aligned} M = 1.25 \quad &T = 1.0 \quad p = 0.58 \quad \text{at } x = 0 \\ R_{1,x} = 0 \quad &R_{2,x} = 0 \quad p = 1.1 \quad \text{at } x = 1 \end{aligned}$$

Here

$$R_1 = \frac{p}{\rho^\gamma} \quad R_2 = u + \frac{2c}{\gamma - 1} \quad R_3 = u - \frac{2c}{\gamma - 1}$$

are the three Riemann invariants. The variable c is the speed of sound, $c = \sqrt{(\gamma p/\rho)}$.

The system of ODEs (24) was solved with the classical forth order Runge-Kutta method until the residual was of order 10^{-10}. The time step was $1 \cdot 10^{-4}$. Here the switch ϕ was

$$\phi(x) = \begin{cases} 0.5 \tanh((x - x_s - 5\varepsilon h)/\varepsilon h) + 0.5) & x \le x_s \\ -0.5 \tanh((x - x_s - 5\varepsilon h)/\varepsilon h) + 0.5) & x > x_s \end{cases}$$

This implied that the larger viscosity ε was dominant in the region close to the shock layer, while the smaller viscosity ε_0 was dominant elsewhere. The viscosity coefficient, ε, was chosen such that the boundary layer was well resolved.

	ρ	ρu	ρE
x=0.2	4.1	4.1	4.1
x=0.8	1.9	1.9	1.9

TABLE 1. Euler equations. The quotient (25) for $s = \rho, \rho u, \rho E$ at $x = 0.2$ and $x = 0.8$, respectively. $\varepsilon = 1.5h, \varepsilon_0 = 30h^2$.

We checked the order of accuracy of the solution calculated with the semi-discrete scheme (24), with $\varepsilon = O(h)$ and $\varepsilon_0 = O(h^2)$. The order of accuracy was achieved in the standard way, by calculating the steady state solution of (24) for $N = 200, 400$ and 800 and then compute the quotient

$$\frac{s_{N=200}(x) - s_{N=400}(x)}{s_{N=400}(x) - s_{N=800}(x)} = 2^p \tag{25}$$

where p is the order of accuracy. The results are showed in Table 1 for $s = \rho, \rho u$, and ρE, respectively. Clearly, the solution is second order accurate upstream $(x = 0.2)$ and first order accurate downstream $(x = 0.8)$.

We make the remark that in the nozzle problem the $O(h)$-errors in the steady state solution can be avoided if the equations are formulated in the variables $v = A(\rho, \rho u, \rho E)^T$. This can be seen in computations presented in [1]. To understand this, we note that with this formulation the lower order term is $\mathbf{a}(\mathbf{v}) = (0, p\frac{dA}{dx}, 0)$. Downstream of the shock the $O(h)$-perturbation satisfies a linear system similar to (22). The boundary condition at x_m^+ is

$$\mathbf{w}_1(x_m^+) = -(i_3 + x_1[a_2(\mathbf{v}^i)])J_+^{-1}\begin{pmatrix} 0 \\ 1 \\ 0 \end{pmatrix}$$

where $a_2(\mathbf{v}^i) = p(\mathbf{v}^i)\frac{dA}{dx}|_{x=x_s}$ and the scalar i_3 is an integral determined by the inviscid solution $\mathbf{v}^i$ and the first term of the inner expansion. Clearly, the only possible unique solution is $\mathbf{w}_1 \equiv 0$, $x_1 = -i_3/[a_2(\mathbf{v}^i)]$.

In general, systems of equations cannot be re-formulated in this way. Therefore we have used the nozzle equations in the form (23) in order to study the $O(h)$-errors downstream of a shock.

4. Concluding remarks

In this article we analyze steady state solutions of slightly viscous hyperbolic systems of conservation laws in one dimension. By using matched asymptotic expansions we can explain why the viscous solution can have an $O(h)$-deviation from the inviscid solution in a smooth region, although the viscous term there is of $O(h^2)$. Here h is the space step. Numerical calculations of the quasi one-dimensional nozzle flow illustrate the results.

References

[1] J. Casper , M. H. Carpenter, *Computational Considerations for the Simulation of Shock-induced Sound.* SIAM Journ. of Sci. Comp., **19 (1)**, (1998).

[2] G. Efraimsson, G. Kreiss, *A Remark on Numerical Errors Downstream of Slightly Viscous Shocks* . In Efraimsson G., *A Study of the Influence of Artificial Viscosity Terms on Solutions of Conservation Laws* , PhD Thesis, Royal Institute of Technology, Stockholm, **1996**, ISBN 91-7170-685-2. Submitted.

[3] B. Engquist , B. Sjogreen, *The Convergence Rate of Finite Difference Schemes in the Presence of Shocks*, to appear in SIAM J. Numer. Anal.

[4] J. Goodman, Z. Xin, *Viscous Limits for Piecewise Smooth Solutions to Systems of Conservation Laws*, Arch. for Rat. Mech. and Anal., **121 (3)**, (1992), 235–265.

[5] J. Kevorkian, J. D. Cole, *Perturbation Methods in Applied Mathematics*, Springer-Verlag, **1981**.

[6] G. Kreiss, H.-O. Kreiss, *Nonlinear Stability of Conservation Laws*, Internal Report TRITA-NA-9607, NADA, Royal Institute of Technology, Stockholm, Sweden, **1996**.

[7] H.-O. Kreiss, J. Lorenz, *Initial-Boundary Value Problems and the Navier-Stokes Equations*, Academic Press Inc., **1989**.

[8] H.-O. Kreiss, E. Lundqvist, *On Difference Approximations with Wrong Boundary Values*, Math. Comp., **22 (101)**, 1–12.

[9] P. A. Lagerstrom, *Matched Asymptotic Expansions* , Springer-Verlag **1988**.

[10] R. J. LeVeque, *Numerical Methods for Conservations Laws* , Birkhäuser, **1990**.

Gunilla Efraimsson,
FFA,
Box 11021,
SE-161 11 Bromma
Sweden

Gunilla Kreiss
NADA,
KTH,
SE-100 44 Stockholm,
Sweden
E-mail address: eng@ffa.se, gunillak@nada.kth.se

International Series of Numerical Mathematics
Vol. 129, © 1999 Birkhäuser Verlag Basel/Switzerland

Multiphase Computations
in Geometrical Optics

Björn Engquist and Olof Runborg

Abstract. We propose a new set of partial differential equations which can be seen as a generalization of the classical eikonal and transport equations, to allow for solutions with multiple phases. The traditional geometrical optics pair of equations do not include solutions with multiple phases, corresponding to crossing waves. The new partial differential equations form a hyperbolic system of conservation laws with source terms. They are derived from a closure of the kinetic formulation of geometrical optics. Numerical examples are presented.

1. Introduction

In the direct calculation of wave propagation, the computational effort is larger at higher frequencies. With constant accuracy the work grows algebraically with frequency. For sufficiently high frequencies or short wavelengths it is unrealistic to compute the wave field directly. Fortunately, this is often the regime for which high-frequency asymptotic approximations are quite accurate. These geometrical optics type asymptotic expansions are used in many applications, for example in electromagnetic, elastic and acoustic wave propagation.

Traditionally, ray tracing has been the computational method of choice. Recently, however, the geometrical optics approximations are also being solved by partial differential equation (PDE) techniques. This is e.g. done in [5] and within the framework of seismology in [9], [11] and [12]. The traditional PDEs give only one unique phase at each point in space. In this paper we shall derive equations which allow for multiple phases or crossing rays. The equations are based on the closure assumption of a finite number of crossing rays for the kinetic formulation of geometrical optics.

1.1. High-frequency asymptotics

When high-frequency waves are treated, the computations can be simplified by considering the asymptotic behavior of the solution as the frequency tends to infinity. There are two strongly related ways to formulate this approximation: the PDEs of geometrical optics and ray tracing. Typical wave phenomena, such as diffraction and interference, are lost in the leading terms of the high-frequency approximation.

Classical geometrical optics is based on the scalar wave equation,

$$u_{tt} + c^2 \nabla^2 u = 0. \tag{1}$$

Here $c = c(\boldsymbol{x})$ is the local wave velocity of the medium. We also define the *index of refraction* as $\eta = c_0/c$ with the reference velocity c_0 (e.g. the speed of light in vacuum). Geometrical optics considers the case when the solution to (1) can be written as a series expansion of the form:

$$u = e^{i\omega\phi(\boldsymbol{x},t)} \sum_{k=0}^{\infty} w_k(\boldsymbol{x},t)(i\omega)^{-k}. \tag{2}$$

Entering this expression into (1), we obtain separate equations for the unknown variables in (2). The phase function ϕ will satisfy the *eikonal equation*,

$$\phi_t + c|\nabla\phi| = 0, \tag{3}$$

and the first amplitude coefficient w_0 solves the *transport equation*,

$$(w_0)_t + c\frac{\nabla\phi \cdot \nabla w_0}{|\nabla\phi|} + \frac{c^2\nabla^2\phi - \phi_{tt}}{2c|\nabla\phi|}w_0 = 0. \tag{4}$$

Note that once ϕ is known, the transport equation is a linear equation with variable coefficients.

The problem with the geometrical optics approach is that the class of solutions which justify an expansion of the type (2) is limited. In particular, it does not include solutions with multiple phases, corresponding to crossing waves. In fact, even in the case of a single phase solution, the series does not necessarily converge, for instance when the geometric boundaries create diffraction effects. We shall concentrate on the multiple phase problem.

The eikonal equation is a nonlinear PDE which requires extra conditions to have a unique solution. This solution is known as the *viscosity solution* [3]. At points where the correct solution should have a multi-valued phase, the viscosity solution picks out the phase corresponding to the first arriving wave. In the case of the linear wave equation, clearly, a linear combination of solutions is also a solution. For the nonlinear eikonal equation, this superposition principle does not hold.

Solving the eikonal equation numerically as a PDE instead of using ray tracing has recently been used in seismology. This technique is demonstrated in [9], [11] and [12]. For these applications it is of direct interest to determine the first arrival. A second phase, corresponding to crossing rays was calculated in [5] using two separate eikonal equations. Boundary conditions for the second phase was given at the discontinuity of the first phase or at a geometric reflecting boundary. This boundary could be difficult to determine.

Another way to treat high-frequency waves computationally is through ray tracing, which is based on a kinetic formulation. The waves are postulated to be particles (photons) whose trajectories are rays. The *ray vector*, $\boldsymbol{p}$, is defined as the index of refraction multiplied by the unit vector, $\hat{\boldsymbol{s}}$, in the direction of the ray, i.e. $\boldsymbol{p} = \eta\hat{\boldsymbol{s}}$. For simplicity we will henceforth let $c_0 = 1$, so that the velocity vector $\boldsymbol{v} = c\hat{\boldsymbol{s}} = c^2\boldsymbol{p}$. A transport equation for particles in the space $(\boldsymbol{x}, \boldsymbol{p}, t)$ can then be derived. Denoting the density of particles by $f(\boldsymbol{x}, \boldsymbol{p}, t)$ the evolution of f is described by the Vlasov type equation

$$f_t + \boldsymbol{v} \cdot \nabla_x f + c\nabla_x\eta \cdot \nabla_p f = 0. \tag{5}$$

This follows from the Hamiltonian system with

$$H = \frac{1}{2}\left[\frac{|\boldsymbol{p}|^2}{\eta^2} - 1\right]. \tag{6}$$

Tracing the particle trajectories of (5) corresponds to ray tracing and also to the method of characteristics for (3) and (5). Since (5) is linear the superposition principle is valid.

Because of the large number of independent variables (six in 3D) it is very hard to solve the full equation (5) numerically. If the equation is solved using ray tracing it is difficult to cover the full domain with rays, [11].

1.2. Moment formulation

In this paper we propose a middle way between geometrical optics and the kinetic model. The technique we use to capture multi-valued solutions is based on a closure assumption for a system of equations representing the moments (see [1]). The starting point for this approach is the transport equation (5). Instead of solving the full equation in phase space, we observe that when f is of a simple form in $\boldsymbol{p}$, we can transform (5) to a finite system of moment equations in the reduced space $(\boldsymbol{x}, t)$, analogously to the classical approach of the hydrodynamic limit from a kinetic formulation. We are interested in cases corresponding to a finite number of rays in different directions at each point.

This paper is organized as follows: In Section 2 the moment equations are derived from the kinetic model for high-frequency waves. The system is essentially closed and equivalent to the equations of geometrical optics. We also explore some theoretical issues and find that the resulting hyperbolic equations are not well-posed in the strong sense. Furthermore, we present the solution of the Riemann problem and show that there is no entropy for the one-dimensional system. Finally, in Section 3, we discuss numerical approximations for the equations as well as results of numerical experiments.

2. Moment equations

In this section we will derive and analyze the system of PDEs that follows from the kinetic model and the assumption that a maximum of N rays pass through any given point in space. The analysis is carried out in two-dimensional space.

2.1. Derivation of the moment equations

We start by defining the moments m_{ij}. With $\boldsymbol{p} = (p_1, p_2)$, let

$$m_{ij} = \int_{\mathbb{R}^2} p_1^i p_2^j f d\boldsymbol{p}. \tag{7}$$

Next, we multiply (5) by $p_1^i p_2^j$ and integrate over $\mathbb{R}^2$ with respect to $\boldsymbol{p}$. Using the definition (7) we get the moment equation,

$$(\eta^2 m_{ij})_t + (m_{i+1,j})_x + (m_{i,j+1})_y = i\eta\eta_x m_{i-1,j} + j\eta\eta_y m_{i,j-1}, \tag{8}$$

where we have used the fact that f has compact support in $\boldsymbol{p}$. Since this equation is valid for all $i, j \geq 0$, we have an infinite system of moment equations. For uniformity in notation we have defined $m_{i,-1} = m_{-1,i} = 0$, $\forall i$.

The system (8) is not closed. If truncated at finite i and j, there are more unknowns than equations. To close (8) we use the assumption that for fixed values of $\boldsymbol{x}$ and t, the particle density f can be written

$$f(\boldsymbol{x}, \boldsymbol{p}, t) = \sum_{k=1}^{N} g_k \cdot \delta(|\boldsymbol{p}| - \eta, \arg \boldsymbol{p} - \theta_k). \tag{9}$$

The new variables that we have introduced here are $g_k = g_k(\boldsymbol{x}, t)$, which corresponds to the strength (particle density) of ray k, and $\theta_k = \theta_k(\boldsymbol{x}, t)$ which is the direction of the same ray. Inserting (9) into (7) yields

$$m_{ij} = \sum_{k=1}^{N} \eta^{i+j} g_k \cos^i \theta_k \sin^j \theta_k. \tag{10}$$

A system describing N phases, needs $2N$ equations, corresponding to the N ray strengths g_k and their directions θ_k. We choose here the equations for the moments $m_{2k-1,0}$ and $m_{0,2k-1}$ with $k = 1, \ldots, N$. This system can be essentially closed for all N (see Section 2.3). After scaling the moments, $m_{ij} := \eta^{-i-j} m_{ij}$, and introducing the new variables

$$\boldsymbol{u} = (u_1, \ldots, u_{2N})^T := (g_1 \cos \theta_1, g_1 \sin \theta_1, \ldots, g_N \cos \theta_N, g_N \sin \theta_N)^T. \tag{11}$$

we get a system of nonlinear conservation laws with source terms,

$$\boldsymbol{F}_0(\eta^2 \boldsymbol{u})_t + \boldsymbol{F}_1(\eta \boldsymbol{u})_x + \boldsymbol{F}_2(\eta \boldsymbol{u})_y = \boldsymbol{K}(\boldsymbol{u}, \eta_x, \eta_y), \tag{12}$$

where the functions $\boldsymbol{F}_k$ and $\boldsymbol{K}$ are rather complicated nonlinear functions, which depend on the particular choice of moments above. A different choice of moments gives different functions. Equivalently, with

$$\hat{\boldsymbol{m}} = (m_{10}, m_{01}, m_{30}, m_{03}, \ldots, m_{2N-1,0}, m_{0,2N-1})^T, \tag{13}$$

we could write (12) as

$$(\eta^2 \hat{\boldsymbol{m}})_t + \boldsymbol{F}_1 \circ \boldsymbol{F}_0^{-1}(\eta \hat{\boldsymbol{m}})_x + \boldsymbol{F}_2 \circ \boldsymbol{F}_0^{-1}(\eta \hat{\boldsymbol{m}})_y = \boldsymbol{K}(\boldsymbol{F}_0^{-1}(\hat{\boldsymbol{m}}), \eta_x, \eta_y). \tag{14}$$

In the most simple case, $N = 1$, the function $\boldsymbol{F}_0$ is the identity and

$$\boldsymbol{F}_1 = \frac{u_1}{|\boldsymbol{u}|}\begin{pmatrix} u_1 \\ u_2 \end{pmatrix}, \quad \boldsymbol{F}_2 = \frac{u_2}{|\boldsymbol{u}|}\begin{pmatrix} u_1 \\ u_2 \end{pmatrix}, \quad \boldsymbol{K} = \frac{\eta_x u_2 - \eta_y u_1}{|\boldsymbol{u}|}\begin{pmatrix} u_2 \\ -u_1 \end{pmatrix}. \quad (15)$$

For $N = 2$, let $\boldsymbol{w} = (w_1, w_2)^T$ and

$$\boldsymbol{f}_0 = \begin{pmatrix} w_1 \\ w_2 \\ w_1^3/|\boldsymbol{w}|^2 \\ w_2^3/|\boldsymbol{w}|^2 \end{pmatrix}, \quad \boldsymbol{f}_1 = \frac{w_1}{|\boldsymbol{w}|}\boldsymbol{f}_0, \quad \boldsymbol{f}_2 = \frac{w_2}{|\boldsymbol{w}|}\boldsymbol{f}_0,$$

$$\boldsymbol{k} = \frac{\eta_x w_2 - \eta_y w_1}{|\boldsymbol{w}|}\begin{pmatrix} w_2 \\ -w_1 \\ w_1^2 w_2/|\boldsymbol{w}|^2 \\ -w_1 w_2^2/|\boldsymbol{w}|^2 \end{pmatrix}. \quad (16)$$

Then $\boldsymbol{F}_k = \boldsymbol{f}_k(u_1, u_2) + \boldsymbol{f}_k(u_3, u_4)$ for $k = 0, 1, 2$ and $\boldsymbol{K} = \boldsymbol{k}(u_1, u_2) + \boldsymbol{k}(u_3, u_4)$.

2.2. A comparison with geometrical optics

For smooth solutions, the moment equations (12) are an equivalent formulation of the equations of geometrical optics. Suppose $\{w_{0,k}\}_{k=1}^N$ and $\{\phi_k\}_{k=1}^N$ are the solutions of N pairs of eikonal- and transport equations, (3, 4). Then, if

$$g_k = w_{0,k}^2, \quad \cos\theta_k = \frac{(\phi_k)_x}{|\nabla\phi_k|}, \quad \sin\theta_k = \frac{(\phi_k)_y}{|\nabla\phi_k|}, \quad k = 1, \ldots, N, \quad (17)$$

the corresponding $\boldsymbol{u}$, defined in (11), will solve (12), see [10]. As expected the vector $(u_{2k-1}, u_{2k})^T$ points in the direction of the gradient of ϕ_k. The length of the vector corresponds to the first amplitude coefficient squared.

2.3. Closure of the moment equations

The system (12) is closed if and only if the flux functions $\boldsymbol{F}_1 \circ \boldsymbol{F}_0^{-1}(\hat{\boldsymbol{m}})$ and $\boldsymbol{F}_2 \circ \boldsymbol{F}_0^{-1}(\hat{\boldsymbol{m}})$ are well defined for all solutions, $\hat{\boldsymbol{m}}$, to (14) at all times. This can be proved as long as no two rays meet head-on. To be more precise, introduce the complex variables

$$z_k(\boldsymbol{x}, t) = \cos\theta_k(\boldsymbol{x}, t) + i\sin\theta_k(\boldsymbol{x}, t), \quad k = 1, \ldots, N, \quad (18)$$

so that $g_k z_k = u_{2k-1} + iu_{2k}$. Then the following theorem holds.

Theorem 2.1. *Let $\boldsymbol{F}_k$ be the functions in (12), corresponding to the moment vector $\hat{\boldsymbol{m}}$ defined in (13). Let $\boldsymbol{F}_0|\mathcal{D}$ be the restriction of $\boldsymbol{F}_0$ to the domain*

$$\mathcal{D} = \{\boldsymbol{u} \in \mathbb{R}^{2N} \mid z_k + z_\ell \neq 0, \ \forall k, \ell\}, \quad (19)$$

with z_k defined as in (18). For $\hat{\boldsymbol{m}} \in \boldsymbol{F}_0(\mathcal{D})$ the flux functions

$$\boldsymbol{F}_1 \circ (\boldsymbol{F}_0|\mathcal{D})^{-1}(\hat{\boldsymbol{m}}), \quad \boldsymbol{F}_2 \circ (\boldsymbol{F}_0|\mathcal{D})^{-1}(\hat{\boldsymbol{m}}) \quad (20)$$

are well defined.

The proof can be found in [10]. Note, with a different choice of $\hat{m}$, Theorem 2.1 does not necessarily hold, see [10].

2.4. Analysis of the conservation laws

For simplicity we will mainly deal with the single phase, $N = 1$, one-dimensional case where the medium is vacuum, $\eta = \text{const} = 1$. With the $\boldsymbol{u}$ variables defined in (11) as conserved quantities, the system can be written on the standard form of a conservation law,

$$\boldsymbol{u}_t + \boldsymbol{f}(\boldsymbol{u})_x = 0, \qquad \boldsymbol{f}(\boldsymbol{u}) = \frac{u_1}{|\boldsymbol{u}|}\begin{pmatrix} u_1 \\ u_2 \end{pmatrix}. \tag{21}$$

The Jacobian of $\boldsymbol{f}$ with respect to $\boldsymbol{u}$ has the following form:

$$\frac{\partial \boldsymbol{f}}{\partial \boldsymbol{u}} = G\begin{pmatrix} \cos\theta & -\sin\theta \\ 0 & \cos\theta \end{pmatrix}G^{-1}, \qquad G = \begin{pmatrix} \cos\theta & -\sin\theta \\ \sin\theta & \cos\theta \end{pmatrix}. \tag{22}$$

Thus, the linearized problem has a double real eigenvalue, $\cos\theta$, and an incomplete set of eigenvectors; the system (21) is only *weakly hyperbolic*. In general this means that (21) is not well-posed in the strongly hyperbolic sense. The solution of the linearized problem with frozen coefficients loses one derivative. The L_2 norm of the solution at time $t > t_0$ can be estimated in terms of the H_1 norm of the initial data at time $t = t_0$. There is no bounded variation estimate for (21) and one can expect delta function type measure solutions. Such solutions appear when the correct physical solution contains more phases than the system can describe, cf. Figure 6. See [2] and [4] for analysis of related problems.

It is interesting again to compare the moment equations with the eikonal and transport equations, (3) and (4). The latter also form a weakly hyperbolic system with the same eigenvalue as (21). The viscosity solution may have a jump in $\nabla\phi$. Because of the term $\nabla^2\phi$ in the source term of (4), the first amplitude coefficient w_0 has a measure at these points.

In the general case with N phases, homogeneous medium, the governing equations read

$$\boldsymbol{F}_0(\boldsymbol{u})_t + \boldsymbol{F}_1(\boldsymbol{u})_x = 0. \tag{23}$$

Denoting the Jacobians of $\boldsymbol{F}_k$ with J_k, the following relationship can be derived:

$$J_1 = J_0 \cdot \text{diag}(\frac{\partial \boldsymbol{f}}{\partial \boldsymbol{u}}(\theta_1), \dots, \frac{\partial \boldsymbol{f}}{\partial \boldsymbol{u}}(\theta_N)). \tag{24}$$

This shows that the eigenvalues of the general system are simply the union of the eigenvalues of N systems of the type (21). It also shows that there will not be more than N eigenvectors, for the $2N \times 2N$ system. Hence, we have shown that the system (23) is weakly hyperbolic. This also holds for the full 2D system (12), see [10].

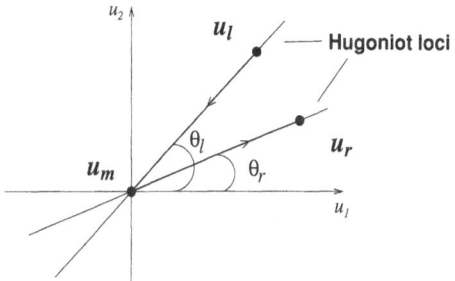

FIGURE 1. Hugoniot loci for left and right state, and solution to the Riemann problem for the first type of discontinuity, plotted in phase space.

2.5. The Riemann problem

We study the one-dimensional Riemann problem for (21),

$$u_t + f(u)_x = 0, \qquad u(x,0) = \begin{cases} u_l & x < 0, \\ u_r & x > 0. \end{cases} \tag{25}$$

At a discontinuity the conservation form gives the Rankine-Hugoniot jump condition,

$$f(u_l) - f(u_r) = s(u_l - u_r), \tag{26}$$

where s represents the propagation speed of the discontinuity. Since $f(u) = \cos\theta u$, this simplifies to

$$\cos\theta_l u_l - \cos\theta_r u_r = s(u_l - u_r). \tag{27}$$

In our case the Hugoniot locus of state u_l is simply αu_l, $\alpha \in \mathbb{R}$. The speed of propagation is $s = \cos\theta_l$. There will be two types of discontinuities. If $\cos\theta_l < \cos\theta_r$, we can use the origin of the phase space as the intermediate state, hence $u_m = 0$. The Hugoniot loci and the solution for this type of discontinuity is illustrated in Figure 1. If $\cos\theta_l > \cos\theta_r$, on the other hand, we do not have a solution in a classical sense. Formally, however, a weak solution to the conservation law with this initial data is given by setting $u_m = \tilde{u}_m \delta(x - st)$. The conservation form gives a slightly modified jump condition for this case,

$$\cos\theta_l u_l - \cos\theta_r u_r = \cos\tilde{\theta}_m(u_l - u_r) + \tilde{u}_m, \tag{28}$$

with the propagation speed $s = \cos\tilde{\theta}_m$.

It is easily verified from (22). that u itself is an eigenvector of the Jacobian of f. Therefore, the Hugoniot locus will coincide with the integral curves of the system's characteristic fields. Since the (double) eigenvalue of (22), $\cos\theta$, is constant along the curves, the fields are linearly degenerate. From this we conclude that the first type of discontinuity is a linear, contact discontinuity; characteristics run parallel to the discontinuity. The linear degeneracy also excludes the possibility of rarefaction wave solutions.

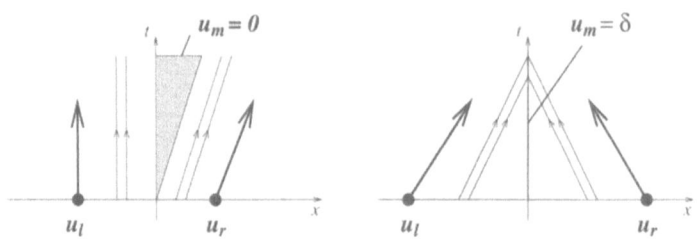

FIGURE 2. The two different types of discontinuities: contact discontinuity (left) and overcompressive shock (right).

The second type of discontinuity will always have two characteristics incident to the discontinuity at each side, because of the double eigenvalue. These discontinuities are thus of overcompressive shock type. The two different discontinuities, plotted in (x, t)-space, are shown in Figure 2.

For the analysis of (21) it would be useful to find an entropy function corresponding to the system. This is, however, not possible.

Theorem 2.2. *There exists no twice continuously differentiable convex entropy pair for the system (21).*

Proof. Let $\boldsymbol{u}$ be a smooth solution to (21) and $U, F \in C^2$ satisfy

$$U(\boldsymbol{u})_t + F(\boldsymbol{u})_x = 0. \tag{29}$$

Then multiplication of (21) by ∇U and identification gives that $\nabla U \frac{\partial \boldsymbol{f}}{\partial \boldsymbol{u}} = \nabla F$. Now, it turns out that

$$\frac{u_1 (u_1^2 U_{u_1,u_1} + 2u_1 u_2 U_{u_1,u_2} u_2^2 U_{u_2,u_2})}{(u_1^2 + u_2^2)^{3/2}} = \nabla \times (\nabla U \frac{\partial \boldsymbol{f}}{\partial \boldsymbol{u}}) = \nabla \times \nabla F = 0. \tag{30}$$

Therefore, with $H(U)$ denoting the Hessian of U,

$$\det H(U) = U_{u_1,u_1} U_{u_2,u_2} - U_{u_1,u_2}^2 = -\frac{1}{4} \left(\frac{u_1}{u_2} U_{u_1,u_1} - \frac{u_2}{u_1} U_{u_2,u_2} \right)^2 \leq 0. \tag{31}$$

3. Numerical treatment

This section includes some results on the numerical treatment of (12). As was discussed in the previous section, the system (12) is only weakly hyperbolic. This makes it sensitive and creates problems for the numerical methods.

The point of departure for our numerical approximations is the basic first order accurate *Lax-Friedrichs* finite difference method. Less smearing of shocks is obtained with the *Godunov* method (see e.g. [7]). For (12) it may, however, produce large L_∞ errors even for smooth problems, cf. Figure 3. The reason for the method's behavior in Figure 3 can be found in the analysis of the Riemann

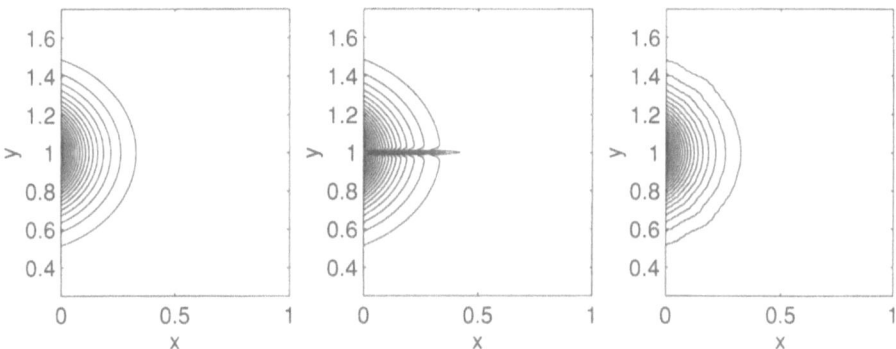

FIGURE 3. Test case with a smooth point source located at $(-0.2, 1)$. The figure shows contour plot of the ray strength $g = m_{00}$ of the single phase system, solved using Lax-Friedrichs (left), Godunov (middle) and splitted Nessyahu-Tadmor (right). Note the problems with using the last two methods.

problem in Section 2.5. Along the line $y = 1$ the Riemann problem in the y-direction corresponds to the situation in Figure 2 (left). Since there is no rarefaction wave solution, there will be no flux in the y direction, and the method reduces to the one-dimensional Godunov method in the x-direction along this line. Hence, along $y = 1$ there will be pure transport, and no damping.

We have also implemented the *Nessyahu-Tadmor* method, [8]. It is a second order TVD method based on the Lax-Friedrichs structure. The results have been very satisfactory when the algorithm was implemented without dimensional splitting. With splitting, the convergence rate is poorer, in particular in L_∞, cf. Figure 3. See [6] for a further discussion.

In the multiple phase case, we have mainly considered the case of two and three phases, and we have only used the Lax-Friedrichs method, see Figures 4–6. A more complete descriptions of the results for the test cases presented here, and others, can be found in [10].

It is more difficult to get reliable calculations when solving (12) with multiple phases, than in the case of a single phase. At each grid point it is necessary to solve a nonlinear system of equations of the type $\boldsymbol{F}_0(\boldsymbol{u}) = \hat{\boldsymbol{m}}$ and in some cases this system had to be solved in the least squares sense.

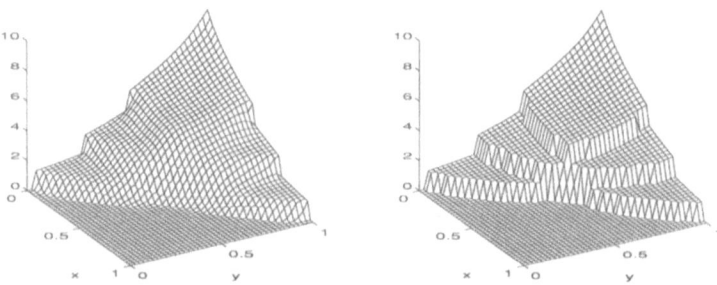

FIGURE 4. Test case with three discontinuous point sources, located at coordinates $(-0.4, 0.8)$, $(0.2, 1.4)$ and $(-0.15, 1.15)$. The figure shows combined strengths of the three waves of the three phase system, $g_1 + g_2 + g_3 = m_{00}$, computed with Lax-Friedrichs method (left) and exact (right).

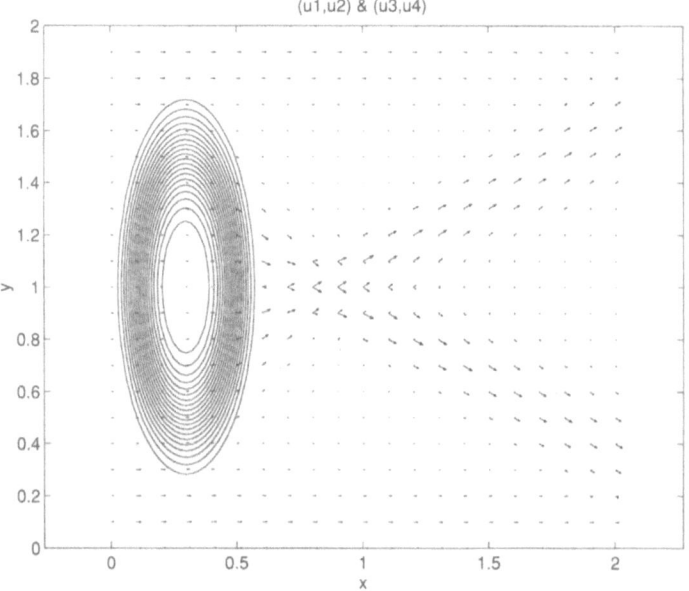

FIGURE 5. Test case with inhomogeneous medium, modeling a smooth convex lens. A plane wave enters from the left. The figure shows vector fields (u_1, u_2), (u_3, u_4) of the two phase system, solved using Lax-Friedrichs, and contour lines of the index of refraction η superimposed.

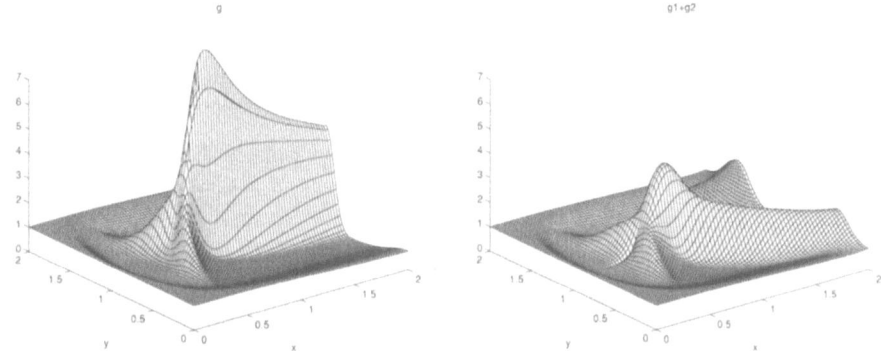

FIGURE 6. Same test case as in Figure 5. The figure shows ray strength g for one phase system (left) and combined ray strengths $g_1 + g_2 = m_{00}$ for two phase system (right). Note how one phase is not sufficient to describe the physically correct solution, so a measure solution is indicated to the left.

References

[1] Y. Brenier and L. Corrias, *Capturing multivalued solutions*, CAM Report 94-46, Department of Mathematics, UCLA, **1994**.

[2] Y. Brenier and E. Grenier, *On the model of pressureless gases with sticky particles*, CAM Report 94-45, Department of Mathematics, UCLA, **1994**.

[3] M. Crandall and P. Lions, *Viscosity solutions of Hamilton-Jacobi equations*, Transactions of the American Mathematical Society, **277** (1983), 1–42.

[4] W. E, Yu. G. Rykov, and Ya. G. Sinai, *Generalized variational principles, global weak solutions and behavior with random initial data for systems of conservation laws arising in adhesion particle dynamics*, to appear.

[5] B. Engquist, E. Fatemi and S. Osher, Numerical solution of the high frequency asymptotic expansion for the scalar wave equation, CAM Report 93-05, Department of Mathematics, UCLA, **1993**.

[6] G.-S. Jiang and E. Tadmor, *Non-oscillatory central schemes for multidimensional hyperbolic conservation laws*, CAM Report 96-36, Department of Mathematics, UCLA, **1996**.

[7] R. J. LeVeque. *Numerical Methods for Conservation Laws*, Birkhäuser, **1992**.

[8] H. Nessyahu and E. Tadmor, *Non-oscillatory central differencing for hyperbolic conservation laws*, Journal of Computational Physics, **87 (2)** (1990), 408–463.

[9] F. Qin et al, *Finite-difference solution of the eikonal equation along expanding wavefronts*, Geophysics, **57 (3)** (1992), 478–487.

[10] O. Runborg, *Multiphase computations in geometrical optics*, Licentiate's Thesis TRITA-NA-9609, Department of Numerical Analysis and Computing Science, KTH, **1996**.

[11] J. van Trier and W. W. Symes, *Upwind finite-difference calculation of traveltimes,* *Geophysics,* **56 (6)** (1991), 812–821,.

[12] J. Vidale, *Finite-difference calculation of traveltimes,* Bulletin of the Seismological Society of America, **78 (6)** (1988), 2062–2076.

Department of Numerical Analysis and Computing Science
Royal Institute of Technology
S-100 44 Stockholm, Sweden
E-mail address: engquist@math.ucla.edu, olofr@nada.kth.se

International Series of Numerical Mathematics
Vol. 129, © 1999 Birkhäuser Verlag Basel/Switzerland

Degenerate Convection-Diffusion Equations and Implicit Monotone Difference Schemes

Steinar Evje and Kenneth Hvistendahl Karlsen

Abstract. We analyse implicit monotone finite difference schemes for nonlinear, possibly strongly degenerate, convection-diffusion equations in one spatial dimension. Since we allow strong degeneracy, solutions can be discontinuous and are in general not uniquely determined by their data. We thus choose to work with weak solutions that belong to the BV (in space and time) class and, in addition, satisfy an entropy condition. The difference schemes are shown to converge to the unique BV entropy weak solution of the problem. This paper complements our previous work [8] on explicit monotone schemes.

1. Degenerate convection-diffusion equations

We are interested in finite difference schemes for nonlinear, possibly strongly degenerate, convection-diffusion problems of the form

$$\partial_t u + \partial_x f(u) = \partial_x(k(u)\partial_x u), \qquad u(x,0) = u_0(x), \tag{1}$$

where $(x,t) \in Q_T = \mathbb{R} \times (0,T)$ and u_0, f, k are given, sufficiently smooth functions. For later use, we need a conservative-form version of (1),

$$\partial_t u + \partial_x f(u) = \partial_x^2 K(u), \qquad K(u) = \int_0^u k(\xi)\, d\xi. \tag{2}$$

By the term 'strongly degenerate' we mean that there are two numbers α and β such that $k(u) = 0$ for all $u \in [\alpha, \beta]$. Hence, the class of equations under consideration is very large and contains, to mention a few, the heat equation, the porous medium equation, the two-phase flow equation and conservation laws. Strongly degenerate equations will in general possess discontinuous – shock wave – solutions. Furthermore, discontinuous weak solutions are not uniquely determined by their data. In fact, an entropy condition is needed to single out the physically relevant weak solution of the problem. We call a bounded measurable function $u(x,t)$ an entropy weak solution if

$$\partial_t |u - c| + \partial_x\big[\mathrm{sgn}(u - c)(f(u) - f(c))\big] + \partial_x^2 |K(u) - K(c)| \leq 0 \quad \text{(weakly)}.$$

It is not difficult to construct an entropy weak solution of (1), even in several space dimensions, see [12]. To the authors knowledge, the main open question

seems to be the uniqueness of such solutions, even in one space dimension. This
has motivated us to seek solutions in the (significantly) smaller class containing the
BV entropy weak solutions. Before introducing this notion of a solution, we recall
that uniqueness of weak solutions for the purely parabolic case (no convection
term) in the class of bounded integrable functions has been proved by Brezis and
Crandall [1], and that uniqueness of entropy weak solutions for hyperbolic problems
(no diffusion term) is a classical result due to Kruzkov [9].

Definition 1.1. *A bounded measurable function $u(x,t)$ is said to be a BV entropy
weak solution of the initial value problem (1) if*

(a) $u(x,t) \in BV(Q_T)$ *and* $K(u) \in C^{1,\frac{1}{2}}(Q_T)$.
(b) *For all non-negative $\phi \in C_0^\infty(Q_T)$ and any $c \in \mathbb{R}$,*

$$\iint_{Q_T} \left(|u - c|\partial_t \phi + \operatorname{sgn}(u - k)(f(u) - f(c) - \partial_x K(u))\partial_x \phi \right) dt \, dx$$

$$+ \int_{\mathbb{R}} |u_0 - c|\phi(x,0) \, dx \geq 0. \tag{3}$$

What makes this definition interesting is that uniqueness of BV entropy
solutions follows from the work of Wu and Yin [13] (actually, instead of (a), they
require $u \in BV(Q_T)$ and only $\partial_x K(u) \in L^1_{\text{loc}}(Q_T)$). We mention here that the
jump conditions proposed by Volpert and Hudjaev [12] are in general not correct,
and thus the uniqueness proof presented there is incomplete, see [13] for more
details. Roughly speaking, entropy weak solutions that are of bounded variation
in both space and time are solutions in the sense of Wu and Yin. One should note
that it is rather restrictive to require BV (in space and time) regularity of solutions
to parabolic equations. In particular, for $\partial_t u$ to be a (locally) finite measure on Q_T,
$\partial_x u$ and $\partial_x^2 K(u)$ need to be (locally) finite measures on Q_T. This fact immediately
implies that the diffusion term $K(u(\cdot, t))$ needs to possess a certain amount of
smoothness, which in turn indicates that it should be harder (than for conservation
laws) to establish the analog of the Crandall and Majda theory [5] for strongly
degenerate parabolic equations. The convergence of a scheme to the desired BV
solution is *not* an immediate consequence of a BV estimate (in space), as is the
case with hyperbolic conservation laws.

It is possible to use the theory developed in [13] to treat strongly degenerate
boundary value problems as well, see Bürger and Wendland [2] (and the references
therein). In [2] the authors analyse their recently proposed model for the settling
and consolidation of a flocculated suspension under the influence of gravity. We
refer to Concha and Bürger [4] for an overview of the activity centring around this
and related sedimentation models. Cockburn and Gripenberg [3] have recently
shown that solutions of degenerate equations also depend continuously on the
nonlinear fluxes of the problem, see [3, 8] for more details.

It is important to realize that solutions of strongly degenerate parabolic equations (1) in general have a more complex structure than solutions of conservation

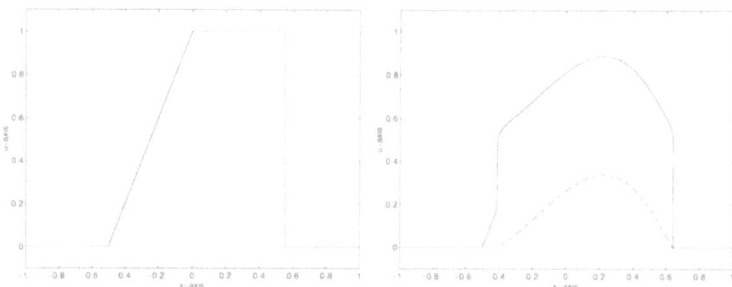

FIGURE 1. Left: The solution (solid) of the inviscid Burgers' equation. Right: The solution (solid) of Burgers' equation with a strongly degenerate diffusion term and the corresponding diffusion function $K(u(\cdot,t))$ (dashed). The initial function is shown as dotted.

laws. The following example demonstrates this. Let $f(u) = u^2$ (referred to as the Burgers flux), and let $k(u) = 0$ for $u \in [0, 0.5]$, $2.5u - 1.25$ for $u \in (0.5, 0.6)$ and 0.25 for $u \in [0.6, 1.0]$. Note that $k(u)$ is continuous and degenerates on the interval $[0, 0.5]$. In Fig. 1 we have plotted the initial function, the solution of the corresponding conservation law, i.e., $k \equiv 0$ in (1), and the solution of (1) at time $T = 0.15$. An interesting observation is that the solution of (1) has a 'new' increasing jump, despite of the fact that f is convex. Thus the solution is not bounded in the so-called Lip^+ norm, as opposed to the solution of the conservation law. Moreover, while the speed of a jump in the conservation law solution is determined solely by f, the speed of a jump in the solution of (1) is in general determined by the jumps in both $f(u)$ and $\partial_x K(u)$, see [13] for precise statements of these jump conditions. Here it suffices to that say the speed s of a jump is

$$s = \frac{f(u^+) - f(u^-) - \left(\lim\limits_{x \to x_0+} \partial_x K(u) - \lim\limits_{x \to x_0-} \partial_x K(u) \right)}{u^+ - u^-}, \tag{4}$$

where u^- and u^+ denote the usual left and right limits (taken along the unit normal to the shock curve) of u respectively. Furthermore, the entropy condition requires that the following inequalities hold for all $c \in int(u^-, u^+)$:

$$\frac{f(u^+) - f(c) - \lim\limits_{x \to x_0+} \partial_x K(u)}{u^+ - c} \le s \le \frac{f(u^-) - f(c) - \lim\limits_{x \to x_0-} \partial_x K(u)}{u^- - c}. \tag{5}$$

See Fig. 2 for an illustration of (5). Finally, we mention here that the techniques developed by Kruzkov (stability) and Kuznetsov (error estimates) for first order equations are not straight on adaptable to second order problems such as (1).

In this paper we are interested in implicit monotone difference schemes for (1). A convergence analysis of explicit monotone schemes was given recently in [8]. In view of the classical monotone theory for conservation laws [5], the main difficulty in obtaining a convergence theory for (1) is to show that the approximations are

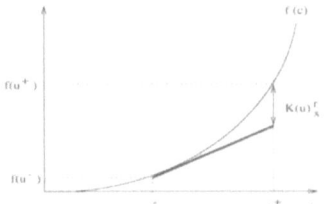 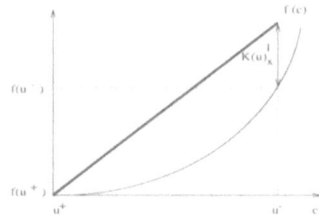

FIGURE 2. Geometric interpretation of the entropy condition (5) for the solution shown in Fig. 1 (right). Left (the left jump): Note that $K(u)^l_x = 0$, $K(u)^r_x > 0$, see Fig. 1 (right). Condition (5) requires that the graph of f restricted to the interval $[u^-, u^+]$ lies above or equals the straight line between $(u^-, f(u^-))$ and $(u^+, f(u^+) - K(u)^r_x)$. Right (the right jump): Note that $K(u)^l_x < 0$, $K(u)^r_x = 0$. Condition (5) requires that the graph of f restricted to $[u^+, u^-]$ lies below or equals the straight line between $(u^+, f(u^+))$ and $(u^-, f(u^-) - K(u)^l_x)$.

L^1 Lipschitz continuous in the time variable, i.e., that they are in BV (in space and time), see Lemmas 2.3 and 2.4 in this paper. To the authors knowledge, there exists no general finite difference theory for strongly degenerate parabolic equations, except for [8]. The main purpose of this paper is show that the theory developed in [8] can be easily extended, using the theory of Crandall and Liggett [6], to implicit schemes as well. An accurate and efficient operator splitting scheme for (1) has been proposed and analysed in [7]. However, for this approximation it is in general impossible to prove L^1 Lipschitz in time regularity, see [7] for details. Finally, we are currently looking into the issue of devising higher order difference schemes for strongly degenerate parabolic equations. In particular, we are investigating to what extent the 'higher order' theory/schemes developed for hyperbolic conservation laws can be taken over to strongly degenerate equations.

2. Implicit monotone difference schemes

We will follow the work [8] closely and refer the reader to it for details not found here. Introduce the difference operators $D_- U_j = \frac{U_j - U_{j-1}}{\Delta x}$ and $D_+ = \frac{U_{j+1} - U_j}{\Delta x}$. We then consider the three-point monotone difference schemes of the form

$$\frac{U_j^{n+1} - U_j^n}{\Delta t} + D_- \left(F(U_j^{n+1}, U_{j+1}^{n+1}) - D_+ K(U_j^{n+1}) \right) = 0, \tag{6}$$

where $F(u, u) = f(u)$ and $\{U_j^0\}$ is some discretization of u_0. In what follows, we assume, without loss of generality, that u_0 has compact support and $f(0) = 0$. The

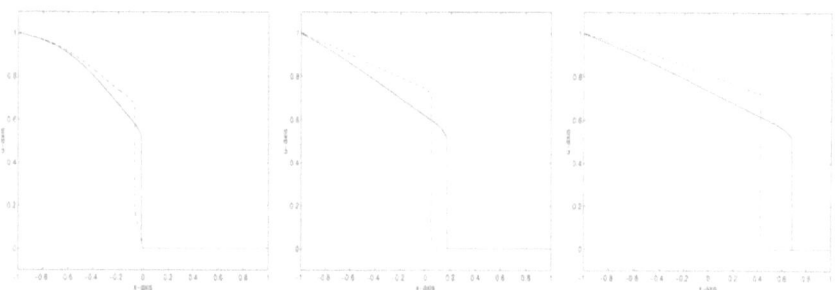

FIGURE 3. The solutions produced by the conservative scheme (6) (solid) and the non-conservative scheme (8) (dashed) plotted at the times $T_1 = 0.0625$, $T_2 = 0.25$ and $T = 1.0$. The initial function is shown as dotted; see the text for a further description of the problem.

assumption of monotonicity in the case of implicit schemes reads

$$F_u(r_1, r_2) + \frac{1}{\Delta x}k(r_3) \geq 0, \qquad \frac{1}{\Delta x}k(r_3) - F_v(r_1, r_2) \geq 0, \qquad \forall(r_1, r_2, r_3). \quad (7)$$

Note that a sufficient condition for (7) to hold is that $F_u \geq 0$ and $F_v \leq 0$. Consequently, any monotone numerical flux F for conservation laws will produce a monotone scheme for (1). To keep the notation simple and making the arguments more transparent, we consider only three-point schemes in this paper.

The monotone schemes (6) are based on differencing the conservative-form equation (2), and not the equation in its original form. One can also devise schemes based on differencing (1) directly, yielding, for example, schemes of the form

$$\frac{U_j^{n+1} - U_j^n}{\Delta t} + D_-\big(F(U_j^{n+1}, U_{j+1}^{n+1}) - k(U_{j+1/2}^{n+1})D_+U_j^{n+1}\big) = 0, \quad (8)$$

where $U_{j+1/2}^{n+1} = \frac{1}{2}(U_j^{n+1} + U_{j+1}^{n+1})$. Although it is possible to prove that (8) converges to a limit, we have not been able to show that this limit satisfies an entropy condition. In fact, we do not believe that (8) will converge to the physically correct solution in the case of strong degeneracy. We now present a simple numerical example intended to support this view. For this purpose we use fluxes $\tilde{f}(u) = \frac{1}{4}u^2$ and $\tilde{k}(u) = 4k(u)$, where k is the one used above. In Fig. 3 we have plotted the solution produced (using small grid parameters) by (6) and (8) at three different times. The convective numerical flux was the upwind flux $F(U_j^{n+1}, U_{j+1}^{n+1}) = f(U_j^{n+1})$ in these calculations. Clearly, the non-conservative scheme (8) produces an incorrect solution. We are currently investigating this phenomenon and will come to back to it in a separate report.

As an aid in the following analysis we shall view the equation (6) in terms of an m-accretive operator and an associated contraction solution operator, i.e., we shall use the Crandall and Liggett theory [6]. A similar treatment of implicit difference schemes for conservation laws has been given earlier by Lucier [11]. If X

is a Banach space, a duality mapping $J : X \to X^*$ has the properties that for all $x \in X$, $\|J(x)\|_{X^*} = \|x\|_X$ and $J(x)(x) = \|x\|_X^2$. A possibly multi-valued operator $\mathcal{A}$, defined on some subset $D(\mathcal{A})$ of X, is said to be accretive if for every pair of elements $(x, \mathcal{A}(x))$ and $(y, \mathcal{A}(y))$ in the graph of $\mathcal{A}$, and for every duality mapping J on X, $J(x - y)(\mathcal{A}(x) - \mathcal{A}(y)) \geq 0$. If, in addition, for all positive λ, $\mathcal{I} + \lambda\mathcal{A}$ is a surjection, then $\mathcal{A}$ is m-accretive. For a fixed n, let us now rewrite the difference equation (6) as (suppressing the Δx dependence)

$$U_j^{n+1} + \Delta t \mathcal{A}(U^{n+1}; j) = U_j^n, \tag{9}$$

where $\mathcal{A}(U; j) = D_- \big(F(U_j^{n+1}, U_{j+1}^{n+1}) - D_+ K(U_j)\big)$. Let $(\Omega, d\mu)$ be a measure space. Recall that, since the dual of $L^1(\Omega)$ is $L^\infty(\Omega)$, any duality mapping J in $L^1(\Omega)$ is of the form $J(u)(v) = \int_\Omega \hat{J}(u)(x)v(x)\,d\mu$, where

$$\hat{J}(u)(x) = \|u\|_{L^1(\Omega)} \begin{cases} 1, & \text{if } u(x) > 0, \\ -1, & \text{if } u(x) < 0, \\ \alpha(x), & \text{if } u(x) = 0, \end{cases} \tag{10}$$

where $\alpha(x)$ is any measurable function with $|\alpha(x)| \leq 1$ for almost every $x \in \Omega$. We shall rely heavily on the following well-known results (see e.g. [6, 11]):

Theorem 2.1. *Let $(\Omega, d\mu)$ be a measure space. Suppose that the nonlinear and possibly multi-valued operator $\mathcal{A} : L^1(\Omega) \to L^1(\Omega)$ is m-accretive. Then for any $\lambda > 0$ and any $u \in L^1(\Omega)$ the equation $T(u) + \lambda\mathcal{A}(T(u)) = u$ has a unique solution $T(u)$. If $\mathcal{A}$ satisfies $\int_\Omega \mathcal{A}(u)\,d\mu = 0$ and commutes with translations, then the solution operator $T : L^1(\Omega) \to L^1(\Omega)$ possesses the following properties: (a) $\int_\Omega T(u)\,d\mu = \int_\Omega u\,d\mu$, (b) $\|T(u) - T(v)\|_{L^1(\Omega)} \leq \|u - v\|_{L^1(\Omega)}$, (c) $|T(u)|_{BV(\Omega)} \leq |u|_{BV(\Omega)}$, (d) $u \leq v \Rightarrow T(u) \leq T(v)$, (e) $\|T(u)\|_{L^\infty(\Omega)} \leq \|u\|_{L^\infty(\Omega)}$.*

The following lemma deals with the question of existence, uniqueness and properties of the solution of the (nonlinear) system (6).

Lemma 2.2. *If (7) is satisfied, then for any U there exists a unique U^* satisfying the equation $\frac{U_j^* - U_j}{\Delta t} + D_- \big(F(U_j^*, U_{j+1}^*) - D_+ K(U_j^*)\big) = 0$, $\forall j \in \mathbb{Z}$. Furthermore, we have the properties: (a) $U_j \leq V_j$ $\forall j \in \mathbb{Z} \Rightarrow U_j^* \leq V_j^*$ $\forall j \in \mathbb{Z}$, (b) $\|U^*\|_{L^\infty(\mathbb{Z})} \leq \|U\|_{L^\infty(\mathbb{Z})}$, (c) $\|U^* - V^*\|_{L^1(\mathbb{Z})} \leq \|U - V\|_{L^1(\mathbb{Z})}$, (d) $|U^*|_{BV(\mathbb{Z})} \leq |U|_{BV(\mathbb{Z})}$.*

Proof. We will first show that the operator $\mathcal{A}$ is accretive. As a first step to achieve this goal, we observe that

$$\sum_{j \in \mathbb{Z}} \text{sgn}(U_j - V_j)\big(\mathcal{A}(U; j) - \mathcal{A}(V; j)\big) \tag{11}$$

$$\geq -\sum_{j \in \mathbb{Z}} \big|cW_j - (\mathcal{A}(U; j) - \mathcal{A}(V; j))\big| + c\sum_{j \in \mathbb{Z}} |W_j|,$$

where W_j denotes $U_j - V_j$ and $c = c(\Delta x) > 0$ is a number chosen so that $c \geq \frac{1}{\Delta x}(F_u(r_1, r_2) - F_v(r_3, r_1)) + \frac{2}{\Delta x^2} k(r_4), \forall (r_1, r_2, r_3, r_4)$. Next, we write

$$\mathcal{A}(U; j) - \mathcal{A}(V; j) = \frac{1}{\Delta x} \left((F_u(\alpha_j, U_{j+1}) W_j + F_v(V_j, \alpha_{j+1}) W_{j+1}) \right.$$
$$- (F_u(\alpha_{j-1}, U_j) W_{j-1} + F_v(V_{j-1}, \alpha_j) W_j))$$
$$- \frac{1}{\Delta x^2} \left(k(\beta_{j-1}) W_{j-1} - 2k(\beta_j) W_j + k(\beta_{j+1}) W_{j+1} \right),$$

for some numbers α_j, β_j between U_j and V_j. Inserting this into (11) yields

$$\sum_{j \in \mathbb{Z}} \mathrm{sgn}(U_j - V_j) \big(\mathcal{A}(U; j) - \mathcal{A}(V; j) \big)$$

$$\geq c \sum_{j \in \mathbb{Z}} |W_j| - \sum_{j \in \mathbb{Z}} \left[\frac{1}{\Delta x} F_u(\alpha_{j-1}, U_j) + \frac{1}{\Delta x^2} k(\beta_{j-1}) \right] |W_{j-1}|$$

$$- \sum_{j \in \mathbb{Z}} \left[c - \frac{1}{\Delta x} (F_u(\alpha_j, U_{j+1}) - F_v(V_{j-1}, \alpha_j)) - \frac{2}{\Delta x^2} k(\beta_j) \right] |W_j|$$

$$- \sum_{j \in \mathbb{Z}} \left[\frac{1}{\Delta x^2} k(\beta_{j+1}) - \frac{1}{\Delta x} F_v(V_j, \alpha_{j+1}) \right] |W_{j+1}| \equiv 0, \qquad (12)$$

which shows that $\mathcal{A}$ is accretive. Observe that $\mathcal{A}$ is Lipschitz continuous, which implies that $\mathcal{A}$ is not only accretive but also m-accretive. We can now invoke Theorem 2.1 to conclude the existence of a unique solution operator $\mathcal{S}$ of (6), i.e., $U_j^* = \mathcal{S}(U; j)$, which proves the first part of the lemma. Since $\sum_{j \in \mathbb{Z}} \mathcal{A}(U; j) = 0$ and $\mathcal{A}$ commutes with translations, the second part of the lemma follows. $\square$

The next lemma plays a key role in our analysis and has no counterpart in the theory of monotone difference approximations for conservation laws.

Lemma 2.3. If (7) is satisfied, we have

$$\left\| F(U_j^{n+1}, U_{j+1}^{n+1}) - D_+ K(U_j^{n+1}) \right\|_{L^\infty(\mathbb{Z})} \leq \left\| F(U_j^0, U_{j+1}^0) - D_+ K(U_j^0) \right\|_{L^\infty(\mathbb{Z})}, \tag{13}$$

$$\left| F(U_j^{n+1}, U_{j+1}^{n+1}) - D_+ K(U_j^{n+1}) \right|_{BV(\mathbb{Z})} \leq \left| F(U_j^0, U_{j+1}^0) - D_+ K(U_j^0) \right|_{BV(\mathbb{Z})}. $$

Proof. From the equation (6), it follows that $V_j^n = \Delta x \sum_{i=-\infty}^{j} \left(\frac{U_i^n - U_i^{n-1}}{\Delta t} \right)$ satisfies

$$V_j^{n+1} = - \left(F(U_j^{n+1}, U_{j+1}^{n+1}) - D_+ K(U_j^n) \right). \tag{14}$$

Next, we derive an equation for the quantity $\{V_j^n\}$. For this purpose consider the difference equation (6) evaluated at $i\Delta x$ and subtract the corresponding equation at time $t = n\Delta t$. Multiplying the resulting equation by Δx and then summing over $i = -\infty, \ldots, j$, yields the following equation

$$\left(V_j^{n+1} - V_j^n \right) + \left(F(U_j^{n+1}, U_{j+1}^{n+1}) - F(U_j^n, U_{j+1}^n) \right) - D_+ \left(K(U_j^{n+1}) - K(U_j^n) \right) = 0.$$

After observing that $\frac{U_j^{n+1}-U_j^n}{\Delta t} = D_- V_j^{n+1}$, we can write

$$F(U_j^{n+1}, U_{j+1}^{n+1}) - F(U_j^n, U_{j+1}^n) = \Delta t\, a_{u,j}^n D_- V_j^{n+1} + \Delta t\, a_j^v D_- V_{j+1}^{n+1}, \qquad (15)$$

where $a_{u,j}^n = F_u(\alpha_j^n, U_{j+1}^{n+1})$, $a_j^v = F_v(U_j^n, \tilde{\alpha}_{j+1}^n)$ and $\alpha_j^n, \tilde{\alpha}_j^n$ are some numbers between U_j^n and U_j^{n+1}. Similarly, we can write

$$K(U_j^{n+1}) - K(U_j^n) = \Delta t\, b_j D_- V_j^{n+1},$$

where $b_j^n = k(\beta_j^n)$ and β_j is a number between U_j^n and U_j^n. Summing up, the sequence $\{V_j^n\}$ satisfies the following linear system of equations

$$\frac{V_j^{n+1} - V_j^n}{\Delta t} + \left(a_{u,j}^n D_- V_j^{n+1} + a_{v,j}^n D_- V_{j+1}^{n+1}\right) = D_+ \left(b_j^n D_- V_j^{n+1}\right). \qquad (16)$$

Observe that this system can be written as

$$A_j^n V_{j-1}^{n+1} + B_j^n V_j^{n+1} + C_j^n V_{j+1}^{n+1} = V_j^n, \qquad (17)$$

where $A_j^n = -[\frac{\Delta t}{\Delta x} a_{u,j}^n + \frac{\Delta t}{\Delta x^2} b_j^n]$, $B_j^n = [1 + \frac{\Delta t}{\Delta x}(a_{u,j}^n - a_{v,j}^n) + \frac{\Delta t}{\Delta x^2}(b_j^n + b_{j+1}^n)]$ and $C_j^n = -[\frac{\Delta t}{\Delta x^2} b_{j+1}^n - \frac{\Delta t}{\Delta x} a_{v,j}^n]$. Because of (7), the linear system (17) is strictly diagonal dominant. Hence, there exists a unique solution V^{n+1} of (17) satisfying $\|V^{n+1}\|_{L^\infty(\mathbb{Z})} \le \|V^n\|_{L^\infty(\mathbb{Z})}$. An argument similar to the one in [8] will also reveal that $|V^{n+1}|_{BV(\mathbb{Z})} \le |V^n|_{BV(\mathbb{Z})}$. The lemma now follows by induction. $\qquad \square$

A direct consequence of (13) is that the approximations are L^1 Lipschitz continuous in the time variable, and thus in BV in space and time.

Lemma 2.4. *If* (7) *is satisfied, we have*

$$\|U^m - U^n\|_{L^1(\mathbb{Z})} \le |F(U_j^0, U_{j+1}^0) - D_+ K(U_j^0)|_{BV(\mathbb{Z})} \frac{\Delta t}{\Delta x} |m - n|.$$

Proof. The result follows directly from (6) and (13), see also [8]. $\qquad \square$

Lemma 2.5. *If* (7) *is satisfied, we have*

$$|K(U_i^m) - K(U_j^n)| = \mathcal{O}(1)\left(|(i-j)\Delta x| + \sqrt{|(m-n)\Delta t|}\right).$$

Proof. First, $|K(U_i^m) - K(U_j^n)| \le Q_1 + Q_2$, where $Q_1 = |K(U_i^m) - K(U_j^m)|$ and $Q_2 = |K(U_j^m) - K(U_j^n)|$. In view of (13), $\|D_+ K(U^m)\|_{L^\infty(\mathbb{Z})} = \mathcal{O}(1)$ and thus $Q_1 = \mathcal{O}(1)|(i-j)\Delta x|$. Kruzkov [10] has developed a technique for deriving a modulus of continuity in time from a known modulus of continuity in space of certain parabolic equations. To estimate Q_2 we apply a discrete version of this technique to the parabolic difference equation (16). To this end, let $\phi(x)$ be a test function, put $\phi_j = \phi(j\Delta x)$ and let $m < n$. Using the difference equation (16) and summation by parts (on the right-hand side of (16)), we easily find that

$$\left|\Delta x \sum_{j\in\mathbb{Z}} \phi_j \left(V_j^m - V_j^n\right)\right| = \mathcal{O}(1)\left(\|\phi\|_{L^\infty(\mathbb{R})} + \|\phi'\|_{L^\infty(\mathbb{R})}\right)\Delta t(m-n),$$

since, for all l, $a_{u,j}^l$, $a_{v,j}^l$ and $\left|V^{l+1}\right|_{BV(\mathbb{Z})}$ are uniformly bounded quantities. From this weak estimate and the BV regularity of V^n, it now follows that

$$\Delta x \sum_{j\in\mathbb{Z}} \left|V_j^m - V_j^n\right| = \mathcal{O}(1)\sqrt{|m-n|\Delta t},$$

see [8] for details. On the other hand, from (14) and Lemma 2.4, we also have

$$\Delta x \sum_{j\in\mathbb{Z}} \left|V_j^m - V_j^n\right| = \mathcal{O}(1)(m-n)\Delta t + \Delta x \sum_{j\in\mathbb{Z}} \left|D_+ K(U_j^m) - D_+ K(U_j^n)\right|.$$

We thus conclude that $\Delta x \sum_{j\in\mathbb{Z}} \left|D_+ K(U_j^m) - D_+ K(U_j^n)\right| = \mathcal{O}(1)\sqrt{(m-n)\Delta t}$. From this the desired Hölder estimate in time follows, since

$$Q_2 = \left|K(U_j^m) - K(U_j^n)\right| \le \Delta x \sum_{i\in\mathbb{Z}} \left|D_+ K(U_i^m) - D_+ K(U_i^n)\right| = \mathcal{O}(1)\sqrt{(m-n)\Delta t}.$$

This concludes the proof of the lemma. $\qquad\square$

Lemma 2.6. *If (7) is satisfied, then the following cell entropy inequality holds*

$$\frac{\left|U_j^{n+1} - c\right| - \left|U_j^n - c\right|}{\Delta t} + D_-\left(F(U_j^{n+1} \vee c, U_{j+1}^{n+1} \vee c) - F(U_j^{n+1} \wedge c, U_{j+1}^{n+1} \wedge c)\right)$$
$$- D_+\left|K(U_j^{n+1}) - K(c)\right|\right) \le 0.$$

Proof. The proof is similar to the one presented in [8], see also [5, 11]. $\qquad\square$

Let u_Δ (where $\Delta = (\Delta x, \Delta t)$) be the interpolate of degree one associated with the discrete data points $\{U_j^n\}$, see [8]. Note that u_Δ is continuous everywhere and differentiable almost everywhere. In view of Lemmas 2.2–2.4, we conclude that there is a constant $C = C(T) > 0$ such that

$$\|u_\Delta\|_{L^\infty(Q_T)} + |u_\Delta|_{BV(Q_T)} \le C,$$
$$|K(u_\Delta(y,\tau)) - K(u_\Delta(x,t))| \le C\left(|y-x| + \sqrt{|\tau-t|} + \Delta x + \sqrt{\Delta t}\right).$$

for all $(x,t),(y,\tau) \in \mathbb{R} \times [0,T]$. Consequently, since BV is compactly imbedded into L^1 on compacta, there is a subsequence of discretization parameters and a function $u \in L^\infty(Q_T) \cap BV(Q_T)$ such that $u_{\Delta_j} \to u$ a.e. in Q_T. Furthermore, via the Ascoli-Arszela theorem, $K(u_{\Delta_j}) \to K(u)$ uniformly on compact sets $\mathcal{K} \subset Q_T$, and $K(u) \in C^{1,\frac{1}{2}}(Q_T)$. Repeating the proof of the Lax-Wendroff theorem (with Lemma 2.6 in mind), it follows that u satisfies the entropy condition (3).

Summing up, we have proven the following main theorem:

Theorem 2.7. *The sequence $\{u_\Delta\}$, which is built from the implicit monotone difference schemes (6), converges a.e. to the BV entropy weak solution of (1).*

Acknowledgement. We thank Raimund Bürger, Bernardo Cockburn, Magne S. Espedal and Wolfgang L. Wendland for interesting discussions. This work has been supported by VISTA, a research cooperation between the Norwegian Academy of Science and Letters and Den norske stats oljeselskap a.s. (Statoil).

References

[1] H. Brezis and M. G. Crandall, *Uniqueness of solutions of the initial value problem for $u_t - \Delta\phi(u) = 0$*, J. Math. Pures et Appl., **58** (1979), 153–163.

[2] R. Bürger and W. L. Wendland, *Existence, uniqueness and stability of generalized solutions of an initial-boundary value problem for a degenerating parabolic equation*, J. Math. Anal. Appl., **218** (1998), 207–239.

[3] B. Cockburn and G. Gripenberg, *Continuous Dependence on the Nonlinearities of Solutions of Degenerate Parabolic Equations*, Preprint, IMA, University of Minnesota, Minneapolis, (1997).

[4] F. Concha, and R. Bürger, *Wave propagation phenomena in the theory of sedimentation. In: Toro, E.F. and Clarke, J.F. (eds.), Numerical Methods for Wave Propagation*, Kluwer Academic Publishers, Dordrecht, (May, 1998).

[5] M. G. Crandall and A. Majda, *Monotone difference approximations for scalar conservation laws*, Math. Comp., **34** (1980), 1–21.

[6] M. G. Crandall and T. M. Liggett, *Generation of semi-groups of nonlinear transformations on general Banach spaces*, Amer. J. Math., **93** (1971), 265–298.

[7] S. Evje and K. H. Karlsen, *Viscous splitting approximation of mixed hyperbolic-parabolic convection-diffusion equations.*, Submitted to Numer. Math., Preprint, Institut Mittag-Leffler, Stockholm, (1997).

[8] S. Evje and K. H. Karlsen, *Monotone difference approximations of BV solutions to degenerate convection-diffusion equations*, Submitted to Siam J. Num. Anal., Preprint, University of Bergen, (1998).

[9] S. N. Kruzkov, *First order quasilinear equations in several independent variables*, Math. USSR Sbornik, **10** (1970), 217–243.

[10] S. N. Kruzkov, *Results concerning the nature of the continuity of solutions of parabolic equations and some of their applications*, Mat. Zametki, **6** (1969), 97–108.

[11] B. J. Lucier, *On non-local monotone difference schemes for scalar conservation laws*, Math. Comp., **47** (1986), 19–36.

[12] A. I. Volpert and S. I. Hudjaev, *Cauchy's problem for degenerate second order quasilinear parabolic equations*, Math. USSR Sbornik, **7** (1969), 365–387.

[13] Z. Wu and J. Yin, *Some properties of functions in BV_x and their applications to the uniqueness of solutions for degenerate quasilinear parabolic equations*, Northeastern Math. J, **5** (1989), 395–422.

Department of Mathematics, University of Bergen,
Johs. Bruns. gt. 12, N-5008, Bergen, Norway
E-mail address: steinar.evje@mi.uib.no, kenneth.karlsen@mi.uib.no

International Series of Numerical Mathematics
Vol. 129, © 1999 Birkhäuser Verlag Basel/Switzerland

On the Numerical Solution of Multi-dimensional Non-linear Systems of Conservation Laws

Michael Fey

Abstract. The paper describes a general framework for the construction and analysis of numerical methods for systems of conservation laws with focus on multi-dimensional equations. The basic ideas are related to the Method of Transport originally derived for the Euler equations. The continuous extension of this approach to different systems and other methods results in a general description of genuinely multi-dimensional methods in terms of wave propagation. Many of the existing multi-dimensional schemes can be put into the resulting framework, which allows a comparison on a more analytic level. We first derive numerical methods in a very general setting. Next, we take the Euler equations as an example to demonstrate the use of this approach by means of the Method of Transport and the Steger-Warming flux-vector splitting. Finally we include fluctuation splitting schemes into this setting and show possible extensions.

1. Introduction

The existing finite volume approaches to solve conservation laws numerically can be grouped into three classes. We can distinguish between Godunov–type methods, which are related to the solution of a Riemann problem, flux–difference methods, which can be viewed as approximations of the first group, and flux-vector splitting methods, which are closely related to kinetic type schemes.

In one space dimension the latter two approaches are basically extensions and modifications of the first idea (computation of the flux across a cell boundary) with the goal to achieve better performance and higher accuracy. Thus, a separation is hardly necessary in this case.

The picture changes in several space dimensions. Here, upwinding is not a question of yes or no, i.e. of whether the flux is going to the left or to the right, but it is the question of selecting some directions out of the infinite set of possible directions.

For the first class of methods the choice of directions is dictated by the underlying grid since the solution of the Riemann problem is essentially a 1-D operator. Usually the interface normals are used as propagation directions. Some sophisticated methods that do use the solution of a Riemann problem but also allow for propagation directions other than the cell normals were derived by Collela [1], LeVeque [6, 7] and others.

The flux-difference splitting approach was also derived as a 1-D operator. The extension to multiple space dimensions was introduced by Roe [12, 11] and is called fluctuation splitting. The main implementations of this idea are based on a cell vertex discretization. The fluctuation on a given node is distributed to the adjacent vertices by some kind of wave model (see [10, 13, 14] for some examples). The different discretizations of the two groups complicate a comparison of the approaches while there is some relation.

Finally, the group of flux-vector splitting methods, if interpreted as kinetic schemes, are naturally multi-dimensional. Unfortunately, this property is often sacrificed for efficiency reasons so that most derived methods are basically related to 1-D operators.

The Method of Transport (MoT) is an exception (see [2, 3, 16]). Derived in the setting of flux-vector splitting, the multi-dimensional character is kept in the implementation. The first version uses infinitely many propagation directions at the cost of complicated formulas and large computation times, which restricts this approach to first order. Later versions overcome these problems and retain propagation directions independent of the grid which is only used to define the control volumes.

2. The general approach

There are different ways to motivate the approach outlined here. They range from more physical ones to pure mathematical formal manipulations. A very convenient interpretation views the method as a special linearization of a non-linear system. The general form of a conservation law reads

$$\mathbf{U}_t + \nabla \cdot \boldsymbol{F} = 0 \tag{1}$$

with $\mathbf{U}$ the vector of M conserved quantities and $\boldsymbol{F}$ the $M \times N$ flux matrix in N space dimensions.

Assume that we can pick a set of vector functions $\mathbf{S}_i$ and a corresponding set of space vectors $\mathbf{a}_i$ that both depend on $\mathbf{U}$ and fulfill the two conditions

$$\sum_{i=1}^{k} \mathbf{S}_i = \mathbf{U} \tag{2}$$

$$\sum_{i=1}^{k} \mathbf{S}_i \mathbf{a}_i^T = \boldsymbol{F}(\mathbf{U}) \tag{3}$$

for all $\mathbf{U}$ in a proper definition range. Then, equation (1) can be written as

$$\left(\sum_{i=1}^{k} \mathbf{S}_i(\mathbf{U}) \right)_t + \nabla \cdot \left(\sum_{i=1}^{k} \mathbf{S}_i(\mathbf{U}) \mathbf{a}_i^T(\mathbf{U}) \right) = 0. \tag{4}$$

For given $\mathbf{S}_i$ and $\mathbf{a}_i$ we call (4) an advection form of the non-linear system (1) because of the special structure. Equation (4) suggests that each part of $\mathbf{S}_i$ is propagated with velocity $\mathbf{a}_i$.

The linearization process exploits the form (4) and decomposes the system into independent linear advection equations. The non-linear system (4), which is equivalent to (1), can be approximated by k linear systems of the form

$$(\mathbf{S}_i(\mathbf{x}, \tau))_\tau + \nabla \cdot (\mathbf{S}_i(\mathbf{x}, \tau)\mathbf{a}_i^T(\mathbf{x})) = 0 \qquad i = 1, \ldots, k \tag{5}$$

with initial conditions

$$\mathbf{S}_i(\mathbf{x}, 0) = \mathbf{S}_i(\mathbf{U}(\mathbf{x}, t)); \quad \mathbf{a}_i = \mathbf{a}_i(\mathbf{U}(\mathbf{x}, t)). \tag{6}$$

An approximation of the solution at time Δt is given as

$$\mathbf{U}(\mathbf{x}, t + \Delta t) = \sum_{i=1}^k \mathbf{S}_i(\mathbf{x}, \Delta t). \tag{7}$$

It is simple to verify that this is in general a first order approximation. In a Taylor expansion, equation (2) gives the correct zeroth order term and (3) the first order one. Thus the error grows quadratically at most, which leads to first order.

Comparing the resulting method in (5)–(7) with the method of characteristics, there are some advantages. First, the resulting simpler equations are in conservation form. Second, the inverse of the decomposition is only an average, i.e. linear. Because of this the use of more than M vectors $\mathbf{S}_i$ and directions $\mathbf{a}_i$ is allowed. This is important for the construction of numerical methods since most relevant multidimensional conservation laws have infinitely many propagation directions. Even the simple linear wave equation for $N > 1$ shares this property which prohibits a diagonalization. An increasing number of directions can be used to account for this fact.

The decomposition (2) and the solution of (5) can be viewed as a general form of numerical schemes. Most of the existing methods can be cast in this framework. While this is quite obvious for all flux-vector splitting methods, we will show later that flux-difference splitting methods and even Riemann solver based methods fit in as well. The approach is general enough to include the existing 1-D methods but it also provides the platform for a proper multi-dimensional extension.

3. The Euler equations as an example

We will now explain the above idea for a given system of conservation laws, the compressible Euler equations. To demonstrate that existing methods fit in this context we derive a decomposition (4) for a flux-vector splitting method using a donor cell approach with 1-D operators to compute the flux. The approach is more commonly used for unstructured grids where a multiplicative operator splitting (of Strang type) is not possible. We can only include the additive version of operator splitting where all fluxes are computed simultaneously since the genuinely multidimensional method also evaluates the full operator in one step.

In the general form (1) the quantities and the flux matrix for the Euler equations are

$$\mathbf{U} = \begin{pmatrix} \rho \\ \rho\mathbf{u} \\ E \end{pmatrix}, \quad \mathbf{F} = \mathbf{U}\mathbf{u}^T + \begin{pmatrix} \mathbf{0}^T \\ \mathbf{I} \\ \mathbf{u}^T \end{pmatrix} p$$

with the usual notation for density ρ, momentum $\rho\mathbf{u}$, energy E and pressure p.

For simplicity we take the Steger-Warming flux-vector splitting method but other methods can be used as well with more complicated wave models. The flux at each cell interface on a cartesian equidistant mesh is computed by using the 1-D operator or the corresponding decomposition normal to the interface. In our example, the conserved quantities $\mathbf{U}$ can be decomposed into the eigenvectors of the Jacobian of the flux function normal to the cell interface. For two space dimensions x and y we denote the eigenvectors of the flux in x-direction by $\mathbf{r}_1, \mathbf{r}_2, \mathbf{r}_3, \mathbf{r}_4$ and in y-direction as $\mathbf{t}_1, \mathbf{t}_2, \mathbf{t}_3, \mathbf{t}_4$. Then, with $\mathbf{F} = (\mathbf{F}_1, \mathbf{F}_2)$ we get:

$$\mathbf{U} = \sum_{i=1}^{4} \alpha_i \mathbf{r}_i; \quad \mathbf{F}_1(\mathbf{U}) = \sum_{i=1}^{4} \alpha_i \mathbf{r}_i \lambda_i \tag{8}$$

$$\mathbf{U} = \sum_{i=1}^{4} \beta_i \mathbf{t}_i; \quad \mathbf{F}_2(\mathbf{U}) = \sum_{i=1}^{4} \beta_i \mathbf{t}_i \mu_i \tag{9}$$

We can generate the numerical scheme (8) and (9) with the general ansatz (2) and (3) using six waves, three in each coordinate direction (entropy wave and slip-line are combined since they have the same eigenvalue). The functions $\mathbf{S}_i$

$$\mathbf{S}_{1,3} = \frac{\rho}{4\gamma}\left(1, u_1 \pm c, 0, H \pm u_1 c\right)^T,$$

$$\mathbf{S}_2 = \frac{\gamma-1}{\gamma}\rho\left(1, u_1, u_2, |\mathbf{u}|^2/2\right)^T,$$

$$\mathbf{S}_{4,6} = \frac{\rho}{4\gamma}\left(1, 0, u_2 \pm c, H \pm u_2 c\right)^T,$$

$$\mathbf{S}_5 = \frac{\gamma-1}{\gamma}\rho\left(1, u_1, u_2, |\mathbf{u}|^2/2\right)^T,$$

with the corresponding speeds

$$\mathbf{a}_{1,3} = 2\begin{pmatrix} u_1 \pm c \\ 0 \end{pmatrix}, \mathbf{a}_2 = 2\begin{pmatrix} u_1 \\ 0 \end{pmatrix}, \mathbf{a}_{4,6} = 2\begin{pmatrix} 0 \\ u_2 \pm c \end{pmatrix}, \mathbf{a}_5 = 2\begin{pmatrix} 0 \\ u_2 \end{pmatrix}.$$

fulfill the consistency relations (2) and (3). Note that the factor 2 in front of the speeds is needed to satisfy equation (3). It appears naturally and does not change any number in the final computation. This reflects the fact that you can propagate

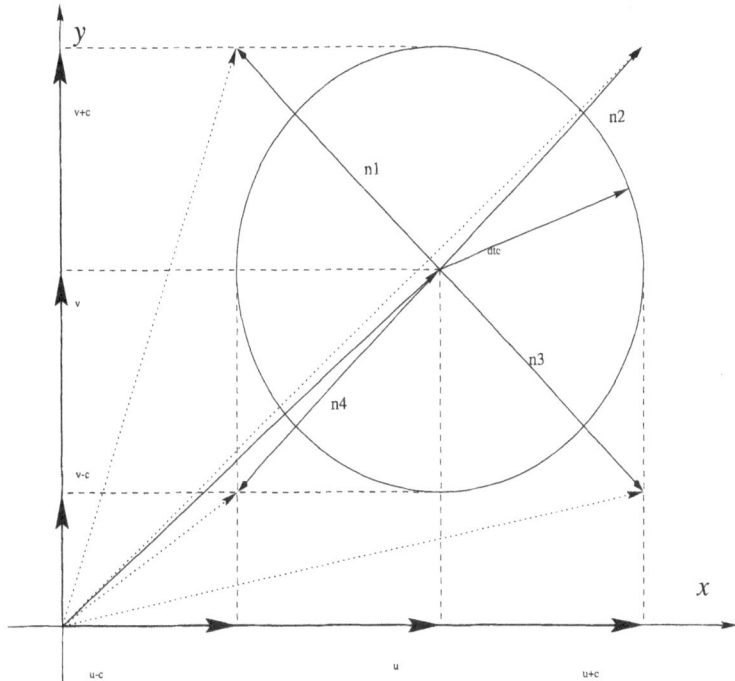

FIGURE 1. Relation of the Monge cone to the numerical propagation directions used for MoT (dotted lines) and Steger-Warming
splitting (full lines on axes).

more of the conserved quantity (e.g. density) out of a cell than you actually have,
which forces you to reduce the maximum time step for a stable computation.
Time-step restrictions of this kind are well known and are analysed in [6].

The approach in (2)–(6) is far more general than 1-D operator decompositions since genuinely multidimensional methods are possible. The idea is to use
propagation directions which are more closely related to the physical ones and
which are not necessarily normal to cell interfaces (see Figure 1).

In our approach we follow this idea. Similar to the splitting in (8) and (9)
we use a combination of eigenvectors, but now the velocities $\mathbf{a}_i$ are chosen to
approximate the truly multidimensional propagation directions. The vectors $\mathbf{S}_i$,
some generalized eigenvectors of the 1-D Jacobian matrices, are given as

$$\mathbf{S}_i = \frac{\rho}{4\gamma}\left(\begin{pmatrix} 1 \\ \mathbf{u} \\ H \end{pmatrix} + c \begin{pmatrix} \mathbf{0}^T \\ \mathbf{I} \\ \mathbf{u} \end{pmatrix} \mathbf{n}_i\right) \quad i = 1,\dots,4 \quad \text{and}$$

$$\mathbf{S}_5 = \frac{\gamma-1}{\gamma}\rho \begin{pmatrix} 1 \\ \mathbf{u} \\ |\mathbf{u}|^2/2 \end{pmatrix}.$$

(10)

The velocities are chosen to follow the entropy wave with the flow velocity and to cover, like quadrature support points, the Monge cone. It turns out that the velocities

$$\mathbf{a}_i = \mathbf{u} + \mathbf{n}_i\, c, \quad i = 1, \ldots, 4, \qquad \mathbf{a}_5 = \mathbf{u}$$

with the four vectors $\mathbf{n}_i$ pointing in the four corners of the unit square, i.e. $\mathbf{n}_i \in \{(1,1)^T, (1,-1)^T, (-1,1)^T, (-1,-1)^T\}$, have some advantages in terms of stability and maximal time-step size.

A second advantage of this approach is that it separates the decomposition and the choice of the propagation directions from the solution process which is basically the solution of a scalar linear equation. It allows a simple extension to other systems since only the decomposition needs to be changed.

4. Extensions related to other approaches

The MoT is strongly related to flux-vector splitting methods and the idea of kinetic schemes (see also [15]). In this section we will show that other methods like flux-difference splitting and Riemann solver based methods can be put into this framework as well.

Note that we have not used information on the final realization of the method. Especially the discretization and the numerical solution of the scalar equations do not enter at this point. All investigations are related to partial differential equations, linear or non-linear ones.

4.1. Fluctuation splitting

We exploit the fact that the numerical computations are hidden in the solution of the scalar equations. Thus, to use others than the conserved quantities $\mathbf{U}$ in the decomposition process (2), two things are needed: an evolution equation for the new dependent variables, and a proper decomposition of the quantities and the related fluxes.

The fluctuation splitting uses the residual or time change of the conserved quantities. We define the new variables as:

$$\mathbf{res} := \mathbf{U}_t = -\nabla \cdot \mathbf{F} \tag{11}$$

The evolution equation can be obtained by formal differentiation of equation (11). We get

$$\mathbf{res}_t = \frac{\partial}{\partial t}(-\nabla \cdot \mathbf{F}) = -\nabla \cdot \frac{\partial}{\partial t}\mathbf{F} = -\nabla \cdot \left(\frac{d\mathbf{F}}{d\mathbf{U}}\mathbf{U}_t\right) = -\nabla \cdot \left(\frac{d\mathbf{F}}{d\mathbf{U}}\mathbf{res}\right), \tag{12}$$

which leads to the conservation law

$$\mathbf{res}_t + \nabla \cdot \left(\frac{d\mathbf{F}}{d\mathbf{U}}\mathbf{res}\right) = 0. \tag{13}$$

To advance the solution $\mathbf{U}$, or better the cell averages $\mathbf{U}^n$, to the next time step we have the relation:

$$\mathbf{U}^{n+1} = \mathbf{U}^n + \int_{t_n}^{t_{n+1}} \mathbf{res}(\tau)d\tau.$$

Thus, what remains is solving the linearized equation (13), which fits perfectly into the concept of MoT. The advection form of (13) looks like (5) with $\mathbf{U}$ replaced by $\mathbf{res}$. The condition (2) is replaced by

$$\sum_{i=1}^{k} \tilde{\mathbf{S}}_i = \mathbf{res} \tag{14}$$

which is sufficient to get first order consistency. If in addition the analogue of (3) holds, i.e.

$$\sum_{i=1}^{k} \tilde{\mathbf{S}}_i \mathbf{a}_i^T = \frac{d\mathbf{F}}{d\mathbf{U}}\mathbf{res} \tag{15}$$

the method is even second order accurate without any correction terms (see [3]).

Since the vectors $\mathbf{S}_i$ in (10) are generalized eigenvectors they describe the linear part of the equations and are good candidates here. Thus, we take $\tilde{\mathbf{S}}_i$ as a multiple of $\mathbf{S}_i$ in equation (10), i.e. $\tilde{\mathbf{S}}_i = \mu_i \mathbf{S}_i$. To get a decomposition in (14) an inverse operation is convenient. Since we still want to use more directions than unknowns this 'inverse' is not a square matrix and underdetermined. We can add the constraint that all amplitudes, i.e. all factors μ_i, are almost equal. Then, an inverse exists and we get

$$(\mu_1, \ldots, \mu_k) = \mathbf{S}^{-1}\mathbf{res}$$

where $\mathbf{S} = (\mathbf{S}_1, \ldots, \mathbf{S}_k)$ and $\mathbf{S}\mathbf{S}^{-1} = \mathbf{I}$. Under these conditions the decomposition is unique, and in addition it satisfies equation (15) which leads to a simple way for second order.

There are many more wave models that try to include nonlinear waves in the decomposition like in Riemann solvers. The wave propagation package CLAW-PACK and the fluctuation splitting schemes differ only in the way to compute the residual. While the first one uses discontinuous reconstruction and evaluates the residual using a number of Riemann solvers, the second approach is based on a continuous reconstruction in a cell vertex discretization and computes the residual directly.

4.2. Other linearizations

There is no particular reason to use the residual in the linearization, unless a steady state solution is sought where the residual is zero. For general time dependent problems one may want to linearize at a different function, depending on the problem under investigation. The simplest choice is a constant if the solution has only small perturbations but nonlinear effects are still important, e.g. the von Neumann paradox. Another example is to use the solution of a Riemann problem.

Let us assume that $\hat{\mathbf{U}}$ is some approximate solution of the conservation law under consideration. An evolution equation for the difference

$$\mathbf{W} = \mathbf{U} - \hat{\mathbf{U}}$$

can be derived by formal differentiation:

$$
\begin{aligned}
\mathbf{W}_t = (\mathbf{U} - \hat{\mathbf{U}})_t &= \nabla \cdot (\boldsymbol{F}(\mathbf{U}) - \boldsymbol{F}(\hat{\mathbf{U}})) \\
&= \nabla \cdot \left(\frac{d\boldsymbol{F}}{d\mathbf{U}} (\mathbf{U} - \hat{\mathbf{U}}) \right) \\
&= \nabla \cdot \left(\frac{d\boldsymbol{F}}{d\mathbf{U}} \mathbf{W} \right).
\end{aligned}
$$

The final conservation law looks similar to (13), and the same technique can be used. Thus, we have the possibility to follow time-dependent structures.

The advantage of this approach is an incorporation of special features of the problem without ignoring the nonlinear effects. Even if the solution $\mathbf{U}$ does not behave as expected, i.e. $\mathbf{W}$ is large, $\mathbf{U}$ is still a solution to the nonlinear problem but possibly with bad resolution.

5. Conclusions

The concept of linearization as described here is a very powerful tool to generate numerical methods for special problems. Since the decomposition is independent of the numerical implementation an existing method can easily be adapted to the needs of various applications (see [9, 4, 5] for examples).

Since most existing methods fit into this framework it is possible to compare these schemes on a more analytical level and extract their advantages. We are able to compare MoT as a flux-difference method to CLAWPACK, a Riemann solver base method, or to fluctuation splitting, which is implemented with a different discretization.

References

[1] P. Collela, *Multidimensional upwind methods for hyperbolic conservation laws*, J. of Comp. Phys., **87** (1990), 171–200.

[2] M. Fey, *Multidimensional upwinding, Part I: The method of transport for solving the Euler equations*, J. Comp. Phys., **143** (1998), 159–180.

[3] M. Fey, *Multidimensional upwinding, Part II: Decomposition of the Euler Equations into advection equations*, J. Comp. Phys., **143** (1998), 181-199.

[4] G. Giese, *Decomposition of the elastic-plastic wave equation into advection equations*, Proceedings of Conference on Hyperbolic Problems, Zürich, **1998**.

[5] T. Gutzmer, *Adaptive Mesh Coarsening in CFD*, Proceedings of Conference on Hyperbolic Problems, Zürich, (1998).

[6] J. O. Langseth and R. J. LeVeque, *Three-dimensional Euler computations using CLAWPACK*, in: P. Arminjon, editor, Conf. on Numer. Meth. for Euler and Navier-Stokes Eq., **1995**.

[7] R. J. LeVeque, *High-resolution conservative algorithms for advection in incompressible flow*, SIAM J. Numer. Anal., to appear.

[8] J. Maurer, *The Method of Transport for Mixed Hyperbolic–Parabolic Systems*, Technical Report 97-13, Seminar for Applied Mathematics, ETH Zürich, (1997).

[9] J. Maurer, *The Method of Transport for Mixed Hyperbolic–Parabolic Systems*, Proceedings of Conference on Hyperbolic Problems, Zürich, (1998).

[10] H. Paillere, H. Deconinck, R. Struijs, P. Roe, L. Mesaros, and J. Müller, *Technical report*, AIAA Report No. 93-3301 CP, (1993).

[11] P. L. Roe, *Linear advection schemes on triangular meshes*, Technical Report CoA Rep. No. 8720, Cranfield, (1987).

[12] P.L. Roe, *A basis for upwind differencing of the two-dimensional unsteady Euler equation*, in: K.W. Morton and M.J. Baines, editors, Numerical Methods for Fluid Dynamics II, **1986**, 59–80.

[13] P.L. Roe, R. Struijs, and H. Deconinck, *A conservative linearisation of the multidimensional Euler equations*, (1991).

[14] M. Rudgyard, *A comparison of multidimensional upwinding for cell vertex schemes on triangular meshes*, in: Proceedings of Conference on Numerical Methods for Fluid Dynamics, Oxford University Press, Oxford, **1992**.

[15] S. A. Zimmermann, *The Method of Transport for the Euler Equations Written as a Kinetic Scheme*, Proceedings of Conference on Hyperbolic Problems, Zürich, (1998).

[16] Seminar of Applied Mathematics, ETH Zürich, Research Reports. http://www.sam.math.ethz.ch/.

Seminar of Applied Mathematics,
ETH Zürich,
CH-8092 Zürich, Switzerland
E-mail address: fey@sam.math.ethz.ch

International Series of Numerical Mathematics
Vol. 129, © 1999 Birkhäuser Verlag Basel/Switzerland

3D Radiative Transfer Under Conditions of Non-local Thermodynamic Equilibrium: A Contribution to the Numerical Solution

Doris Folini and Rolf Walder

Abstract. A new approach to the solution of the 3D NLTE optically thick radiative transfer problem for moving media is presented. The radiative transfer problem basically consists of determining consistent values for the radiation field and the state of the matter. A first task, therefore, is the solution of the radiative transfer equation, which describes the propagation of radiation in the presence of matter. The second task is the determination of the state of the matter in the presence of a radiation field. These two parts are then coupled iteratively. We present the first application of the generalized mean intensity approach, suggested by Turek [9], [10] for the solution of the radiative transfer equation, to NLTE problems. The resulting linear system is solved using BiCGStab. The corresponding code has successfully been applied to some first test problems.

1. Introduction

The problem of radiative transfer is crucial in astrophysics. Most information we have from objects in the universe reaches us in the form of radiation and the physics of many astrophysical objects is essentially influenced by strong radiation fields. For example, radiation fields play a significant role in the driving of most stellar winds. They are important for the energy transport in accretion disks and young stellar objects, and it has been suggested that they contribute to the complex shaping of planetary nebulae. The radiation we observe with our telescopes contains information about the physical conditions where it emerged as well as about the conditions on the way to us. To retrieve this information from an observed spectrum usually requires model simulations, in particular when, which is the rule, the distant object being observed appears as a point source only. It is this kind of astrophysical problems which to attack is our task. The complexity of the problem, however, also requires us to make use of the best numerical techniques we can access.

The problem of radiative transfer under conditions of non-local thermodynamic equilibrium (NLTE) basically consists of determining consistent values for the radiation field and the state of the matter. In the frame of this work, the state of the matter includes the ionization state of the matter and the atomic level populations, while the dynamics of the matter is assumed to be known. The problem can then be regarded as being divided into two parts: the radiative transfer equation, which describes the transport of radiation in the presence of matter, and the statistical equilibrium equations together with some conservation laws, which describe the state of the matter in the presence of a radiation field.

Our approach follows this partition. This means that, first, a solver for the radiative transfer equation is required, second, a solver for the statistical equilibrium part is needed, and, third, these two parts have to be coupled in a favorable way. Each of these three parts has its own difficulties. In addition, the storage requirements are enormous, at least for multi dimensions, as at each spatial grid point the radiation field for each frequency and the populations of all atomic levels and ionization stages have to be stored. While for one dimensional problems well established methods exist in the astrophysical community (see e.g. Hubeny [4]) the solution of the multidimensional problem is still in its infancy. Among the most recent multidimensional approaches to the problem are those of Fabiani Bendicho, Trujillo Bueno, and Auer [2], of Dykema, Klein, and Castor [1], of Stone, Mihalas, and Norman [8], and, for the transfer equation alone, of Kanschat [5] and Turek [9], [10]. The present approach takes up the idea of Turek for the solution of the radiative transfer equation. He suggested to solve for a generalized mean intensity instead of directed intensities. Although the entire radiative transfer problem is solved, including the statistical equilibrium equations, the strength of the present approach lies in the solution of the transfer equation itself.

Although the approach presented here aims at radiative transfer for astrophysical applications it seems worth mentioning the wide variety of other applications. The general problem of radiative transfer occurs in such different fields as room climate engineering, temperature measurements of melts, the optimization of combustion devices, atmospherical sciences, or in connection with laser technology.

The organization of this paper is as follows. In section 2 the solution of the radiative transfer equation is outlined and the convergence properties of BiCGStab, applied to the discretized transfer equation, are analyzed. Section 3 deals with the statistical equilibrium equations and section 4 addresses the question of parallelization. Finally, section 5 is devoted to discussion and conclusions.

2. Solution of the transfer equation

The transfer equation to be considered describes non-relativistic, stationary, frequency decoupled, and unpolarized radiative transfer. For a particular direction of propagation $n \in S^2$ it is given by

$$n \nabla_x I(x, n, \nu) + \chi(x, n, \nu) I(x, n, \nu) =$$
$$\lambda(x, \nu) \int_{S^2} P(n, n') I(x, n', \nu) d\omega' + f(x, n, \nu)$$
$$\text{in} \quad D_1 \times S^2 \times D_2. \tag{1}$$

Here $I(x, n, \nu)$ denotes the specific or directed intensity at point $x \in D_1$ and frequency $\nu \in D_2$ propagating in direction $n \in S^2$. It has to be emphasized that the integral term couples all directions $n \in S^2$ and that, therefore, such a transfer equation has to be solved for each direction $n \in S^2$. The coefficient f stands for atomic emission, λ is the scattering coefficient, and $\chi = \kappa + \lambda$ describes the total losses. κ is the absorption coefficient and $P(n, n')$ is the scattering redistribution function. It should be stressed that χ, λ, and f depend on the atomic level populations which, in turn, depend on the radiation field described by $I(x, n, \nu)$. The computation of these level populations and the coupling to the transfer equation are outlined in section 3. Further, $D_1 \subset R^3$, S^2 is the unit sphere in R^3, $D_2 \subset R^1$, n is the unit vector in the direction of propagation of I, $d\omega$ is the solid angle element associated with n. The boundary conditions are problem dependent and will not be considered here any further.

2.1. Discretization

FREQUENCY DISCRETIZATION: The transfer equation as given in equation 1 is decoupled in frequency. The choice of discrete frequency points is guided by the physics, for example by the location of ionization edges and spectral lines. Due to the decoupling of the transfer equation with respect to frequency, the frequency indices will be omitted in the following. One should bear in mind, however, that I, χ, λ, and f actually depend on frequency.

DISCRETIZATION OF S^2: A parameterization of S^2 over $[0, 2\pi] \times [0, \pi]$ is used, choosing the discrete ordinates n^m equidistantly in both intervals. For these directions $I(n^m)$ is computed. Using the same directions the trapezoidal rule is used for the integral over the sphere, with $K + 1$ support points on $[0, \pi]$ and $L + 1$ points on $[0, 2\pi]$:

$$\int_{S^2} P(n, n')I(n')d\omega' \approx \sum_{k=1}^{K-1} \sum_{l=0}^{L-1} c^{lk}(P)I(\phi^l, \theta^k) = \sum_{m=1}^{M} c^m(P)I^m, \qquad (2)$$

where $M \equiv (K-1)L$. For the special choice of $P(n, n') = 1/4\pi$ the quadrature weights become $c^{lk} = \frac{\pi}{2KL} \sin\theta^k$.

The resulting, physically unnecessary, concentration of discrete ordinate directions near the poles of the sphere at $\theta = 0$ and $\theta = \pi$ can be partly overcome by omitting some of the ordinates in ϕ-direction towards the poles, such that $\Delta\phi$ at latitude θ is at least $1/2\Delta\phi_{equator}$, a procedure which is physically well justified.

Using the discrete ordinate directions n^m defined above and writing I^m instead of $I(n^m)$ the transfer equation discretized over S^2 and for a particular direction n^m, $m \in \{1, \ldots, M\}$, becomes

$$n^m \nabla_x I^m(x) + \chi^m(x)I^m(x) = \lambda(x) \sum_{m=1}^{M} c^m(P)I^m(x) + f^m(x). \qquad (3)$$

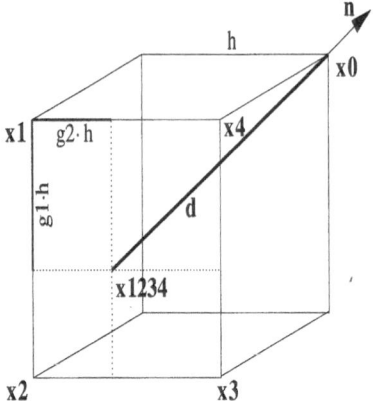

FIGURE 1. Spatial discretization of the transport term in 3D.

SPATIAL DISCRETIZATION: For the spatial discretization a Cartesian grid is used. The transport part is discretized using a first order finite difference scheme together with upwinding. Figure 1 illustrates the discretization of the transport term at point x_0 in 3D. Denoting by I_*^m the discrete approximation to I^m at point x_*, where $* = \{0, 1, 2, 3, 4\}$, the discretization of the transport term is given by

$$
\begin{aligned}
n^m \nabla_x I^m(x_0) &= \frac{I_0^m - I_{1234}^m}{d} + O(h) \\
&= d^{-1} \left[I_0^m - I_1^m (1 - g_1 - g_2 + g_1 g_2) - \right. \\
&\quad \left. I_2^m (g_1 - g_1 g_2) - I_3^m (g_1 g_2) - I_4^m (g_2 - g_1 g_2) \right] + O(h) \quad (4)
\end{aligned}
$$

Now let I_h^m denote the vector containing the specific intensity of direction n^m at each discrete grid point. The transfer equation for one frequency point can then be written as

$$
T_h^m I_h^m = L_h \sum_{m=1}^{M} c^m(P) I_h^m + f_h^m. \quad (5)
$$

Here T_h^m describes the discretized transport term as well as the discretized loss coefficient $\chi^m(x_0)$. L_h and f_h^m stand for the discretized scattering and emission coefficients at point x_0 on the right hand side of equation 3. It should be noted that this spatial discretization gives good results only for χh considerably smaller than 1. This may be illustrated by the special case of a transfer equation of the form $\partial I/\partial x + \chi I = 0$, with $\chi = const.$, whose analytical solution is given by $I_2 = I_1 \cdot e^{-hx}$ with $h = x_2 - x_1$, while the above discretization yields $I_2 = I_1/(1 + \chi h)$.

2.2. The generalized mean intensity approach

The generalized mean intensity is given by

$$
\tilde{J}(x) = \int_{S^2} P(n, n') I(x, n') d\omega' \quad (6)
$$

or, in its discrete version, by

$$\tilde{J}_h = \sum_{m=1}^{M} c^m(P) I_h^m, \tag{7}$$

where now $\tilde{J}_h$ denotes the vector containing the discrete generalized mean intensities at each spatial grid point. Remembering that $P(n, n')$ can depend on angle, it can be seen that $\tilde{J}_h$ deviates from the physical mean intensity

$$J(x) = \frac{1}{4\pi} \int_{S^2} I(x, n') d\omega' \tag{8}$$

by the term $P(n, n')$ in the integral if $P(n, n') \neq 1/4\pi$.

Equation 5 can now be rewritten in terms of the discretized generalized mean intensity $\tilde{J}_h$. For this purpose the quadrature sum in the equation is replaced by $\tilde{J}_h$ and the operator T_h^m is inverted and taken to the right hand side. Applying now a quadrature sum over all ordinate directions to the entire equation over all ordinate directions, using again the function $P(n, n')$, leads to

$$\begin{aligned}
\tilde{J}_h &= \sum_{m=1}^{M} c^m(P) (T_h^m)^{-1} L_h \tilde{J}_h + \sum_{m=1}^{M} c^m(P) (T_h^m)^{-1} f_h^m \\
&= T_h L_h \tilde{J}_h + F_h.
\end{aligned}$$

Taking all $\tilde{J}_h$ terms to the left one obtains

$$A_h \tilde{J}_h = F_h. \tag{9}$$

where $A_h = 1 - T_h L_h$. For the special choice of $P(n, n') = 1/4\pi$ this is an equation for the discrete counterpart of the physical mean intensity. If desired, the specific intensities I_h^m can be recovered from $\tilde{J}_h$ by solving equation 5 with the quadrature sum replaced by $\tilde{J}_h$.

Note that it is possible to invert T_h^m efficiently if it is brought to lower triangular form. This can be achieved due to the upwinding requirement and by an appropriate renumbering of the grid points, depending on the ordinate direction.

2.3. Iterative solution: analysis of the convergence behavior

For the solution of equation 9, the BiCGStab algorithm developed by Van der Vorst [11] is used. Its advantages are that it is applicable to non-symmetric problems as the one presented here, that its memory requirements are rather small, and that it usually shows a rather smooth and fast convergence. As it works reasonably well for the problems considered up to now no other solver has been tested. A solution is assumed to have converged if the l_2-norm of the BiCGStab residual is smaller than a prescribed tolerance ϵ.

In the following, the convergence properties of the diagonally preconditioning BiCGStab are analyzed. (See also the note below on preconditioning). Here only the numerical results are presented. An analytical analysis leading to the same qualitative results can be found in Folini [3].

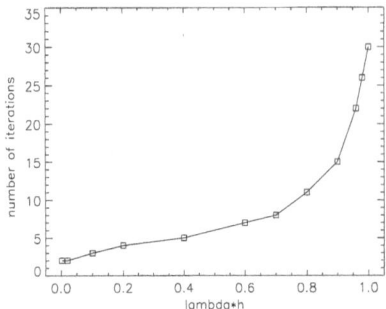 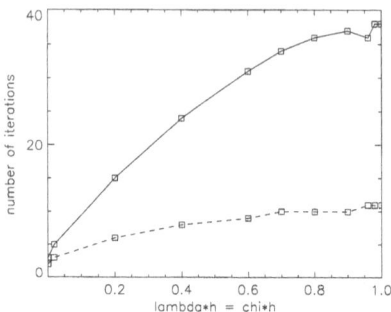

FIGURE 2. **Left:** Convergence of BiCGStab as a function of λh for fixed $\chi = 50$, $h = 0.02$, for test example 2. **Right:** Convergence of BiCGStab as a function of $\lambda h = \chi h$ for test example 1 and two different grid spacings. The solid line denotes $h = 0.02$, the dashed line $h = 0.1$. Note that the different convergence behaviour is not due to h but to χ. For details see text.

Test example 1: The computational domain is given by $[0, 1]^3$. A square shaped beam is considered, entering the computational domain at the boundary $x = 0$. It is directed along the discrete ordinate direction ($\phi = 0, \theta = \pi/2$) which is parallel to the positive x-direction. Its axis goes through the point $(x, y, z) = (0, 0.5, 0.5)$ and it has a cross-section of 0.41×0.41. The specific intensity I of the beam is set to 10^4 at the incoming boundary. $(n_\phi, n_\theta) = (24, 13)$ ordinates have been used. The error bound for BiCGStab was set to $\epsilon = 10^{-10}$. χ, λ, and h are variable and given along with the individual test cases.

Test example 2: The same as test example 1, but using $(n_\phi, n_\theta) = (16, 7)$ ordinates.

Convergence as a function of λ for $\lambda h \to \chi h$: Figure 2 (left) shows the convergence behavior of preconditioned BiCGStab as a function of λh for fixed $\chi = 50$ and $h = 0.02$ with $\chi h = 1$. The convergence properties depend strongly on the value of λh and start to deteriorate strongly above $\lambda h \approx 0.7$. In physical terms this means that the convergence properties are quite good as long as scattering contributes less than about 70% to the total losses.

Convergence as a function of $\chi h = \lambda h$: Considering the "worst case" (see Figure 2, left) $\chi h = \lambda h$, for increasing χh Figure 2 (right) illustrates that the number of iterations increases quite rapidly already for relatively small values of $\chi h = \lambda h$. Physically speaking, this means that the convergence properties deteriorate already for a small extinction term χ if this extinction is mostly due to scattering.

Convergence as a function of h: Figure 3 (left) shows that the convergence properties hardly depend on the grid spacing h, at least for the above test example and for the "worst case" $\chi = \lambda$. While h is reduced by a factor of 5, from $h = 0.1$ to $h = 0.02$, the number of iterations increases at most by 36%. For a more detailed discussion of this h dependence see Folini [3].

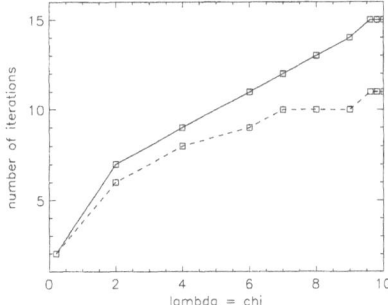 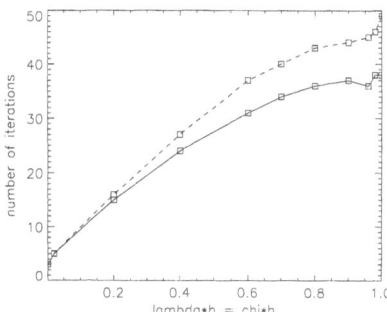

FIGURE 3. **Left:** Convergence behaviour of BiCGStab as a function of $\lambda = \chi$ for test example 1 and two different grid spacings. The solid line denotes $h = 0.02$, the dashed line $h = 0.1$. **Right:** Convergence behaviour of preconditioned (solid line) and unpreconditioned (dashed line) BiCGStab as a function of $\chi h = \lambda h$ with $h = 0.02$ for test example 1.

Convergence for non-constant χ and λ: One additional test was performed with test example 1 for the following, non-constant, choice of χ and λ: $\chi = \lambda = 5$ for $x < 0.5$ and $\chi = \lambda = 50$ for $x \geq 0.5$. In this case 26 iterations were needed. For comparison, the same example but with $\chi = \lambda = 5$ in the entire domain requires 10 iterations, for $\chi = \lambda = 50$ in the entire domain 38 iteration are needed.

Conclusions: The convergence properties of BiCGStab are strongly determined by the ratio λ/χ. If for constant χ and λ the ratio λ/χ increases convergence will slow down, even for relatively small values of χh. For small ratios the convergence will be fast, even for $\chi h = 1$. If the ratio varies over the computational domain the convergence will be better than for the maximum ratio alone and the number of iterations will lie somewhere between the number of iterations required for the maximum and minimum ratio alone. If $\lambda \approx \chi$ the number of iterations increases with increasing χ for constant χh. The grid spacing h has hardly any influence on the number of iterations.

A note on preconditioning: Preconditioning here is difficult for several reasons. As A is a full matrix it cannot be stored and it is necessary to implement the problem in a matrix free way. A itself is never computed explicitly. This makes it difficult to analyze A and to extract parts of A for use as a preconditioner. Also, many entries are of roughly the same size and A is neither necessarily diagonally dominant nor is it symmetric. As already mentioned, in a first attempt, diagonal preconditioning is used and BiCGStab is applied to the preconditioned system $D^{-1}AJ = D^{-1}F$ where $D = diag\{A\}$. However, Figure 3 (right) indicates that, at least for the considered example, this hardly reduces the number of iterations and that the overall computational costs may even increase. But under the above mentioned conditions, the search for a good preconditioner is a demanding task.

3. Statistical equilibrium equations and coupling

The coefficients χ, λ, and f of the radiative transfer equation at each spatial grid point are determined by the atomic level populations N_r at the corresponding grid point. The second big task in optically thick NLTE radiative transfer, therefore, is the determination of these atomic level populations N_r at each spatial grid point. The populations are determined by a non-linear system of equations, consisting of several conservation laws and the actual statistical equilibrium equations for the individual level populations,

$$\frac{dN_r}{dt} = \sum_{s \neq r}(R_{sr} + C_{sr})N_s - (R_{rs} + C_{rs})N_r. \tag{10}$$

The transition probabilities R_{sr} depend on the radiation field, the probabilities C_{sr} depend on the electron density. Both R_{sr} and C_{sr} depend on the temperature of the matter, which in turn is a function of N_r and the radiation field.

In general, the solution of this non-linear system can be rather delicate. First, the resulting population numbers easily differ by more than ten orders of magnitude. Second, the non-linearities not only require an iterative solution but, possibly more important, may allow for more than one stationary solution. For example, bistable solution exist if molecules are considered (see e.g. Bourlot et al. [6]).

In the present approach only stationary situations are considered. The appropriate electron density N_e to fulfill charge conservation is determined by fix-point iteration, computing N_e for given N_r, then computing new coefficients C_{sr} for the statistical equilibrium equations, and so on. The remaining linear system of algebraic equations presently consists of some ten equations, which is small enough to be solved by Householder transformations. It should be noted here that in 1D problems often several thousand equations are considered instead. These systems then require other solution techniques than just Householder transformations. Optically thick lines are treated in the frame of slightly extended Sobolev theory. Details on this point can be found in Folini [3].

Having solvers for the transfer equation and the statistical equilibrium equation part the two now have to be brought together. In the frame of this work only pure fix-point iteration between the two parts is applied. The simulation is assumed to have converged if between succeeding iterations the relative changes of all level populations and of the radiation field are smaller than a given global tolerance. However, from 1D simulations it is known that under extreme conditions (very high matter density, for example) mere iteration between the transfer equation and the rate equation part often fails. Several approaches have been suggested to overcome this problem. Among them are the so called ALI, MALI, and MUGA techniques (see e.g. Hubeny [4], Rybicki and Hummer [7], Fabiani Bendicho [2]), where the basic idea is to incorporate part of the transfer operator into the statistical equilibrium equations.

4. Aspects of parallelization

Due to the huge memory and CPU requirements multidimensional NLTE radiative transfer is a candidate for parallelization, preferably on distributed memory machines. Up to now we have parallelized the transfer part using MPI. In fact, the radiative transfer problem as considered here is very well suited for parallelization. Being decoupled in frequency, the transfer part can be split according to frequency. The rate equation part can be treated by domain decomposition. Load balancing, however, is difficult in real size problems. The coefficients of the transfer equation depend strongly on frequency and so does the number of iterations which can be hardly estimated a priori. The convergence properties of the rate equations can also vary considerably from point to point making an a priori estimate of the work to be done again difficult. Also the practical implementation on distributed memory machines is somewhat delicate. For the present problem the distribution of the memory among different nodes requires that most of the memory has to be communicated twice in each global iteration step, because the solution of the transfer equation for one particular frequency point needs data for this frequency point for all the spatial grid points, while the solution of the rate equation part at one particular spatial grid point needs data for this grid point and for all frequencies. Although the time required for this communication is not too important, as one global iteration step takes much longer anyhow, the 'logistics' of the memory exchange are quite difficult to program. From this point of view shared memory machines seem better suited, but they usually offer less memory.

5. Discussion and conclusions

The numerical solution of the multidimensional, NLTE radiative transfer problem is still in its infancy. Only very few groups have attack this problem up to now.

The approach presented here differs from other existing approaches in that it uses a generalized mean intensity approach for the solution of the transfer equation. It is the first time that such a generalized mean intensity approach is used in the context of the NLTE radiative transfer problem. At the moment, the code features an advanced solver for the transfer equation, a state of the art solver for the rate equation part, and an iterative coupling between the two. Extended Sobolev theory is used for the treatment of optically thick lines. Its main advantages lie in the transfer part, in its h-independence, the use of a modern, iterative solver, and the modest memory requirements of this part. The code has been successfully tested for some simple NLTE radiative transfer problems and in its present state is on the break of being applied to "real" problems. So far, we had no possibility to compare our code with other codes.

The problem of multidimensional NLTE radiative transfer is one of the most demanding problems to be solved in numerical astrophysics. The presented approach, which brings together astrophysical knowledge and modern and advanced numerical techniques, is a contribution to this project. Hopefully the mathematical community will continue to contribute to the solution of this problem.

References

[1] P.G. Dykema, R.I. Klein and J.I. Castor, *A new scheme for multidimensional line transfer III. A two-dimensional Lagrangian variable tensor method with discontinuous finite-element S_n transport*, The Astrophysical Journal, **457** (1996), 892–921.

[2] P. Fabiani Bendicho, J. Trujillo Bueno and L. Auer, *Multidimensional radiative transfer with multilevel atoms II. The non-linear multigrid method*, Astronomy and Astrophysics, **324** (1997), 161–176.

[3] D. Folini, *Computational approaches to multidimensional radiative transfer and the physics of radiative colliding flows*, PhD Thesis, Nr. 12606, ETH Zürich, Zürich, (1998).

[4] I. Hubeny, *Accelerated lambda iteration*, in: U. Heber C.S. Jeffery, Eds., The atmospheres of early-type stars, Springer Verlag, (1992), 377–392.

[5] G. Kanschat, *Parallel and adaptive Galerkin methods for radiative transfer problems*, PhD Thesis, Ruprecht-Karls-Universität, Heidelberg, (1996).

[6] J. Le Bourlot, G. Pineau des Forêts and E. Roueff, *Complex dynamical behavior in interstellar chemistry*, Astronomy and Astrophysics, **297** (1995), 251–260.

[7] G.B. Rybicki and D.G. Hummer, *An accelerated lambda iteration method for multilevel radiative transfer I. Non-overlapping lines with background continuum*, Astronomy and Astrophysics, **245** (1991), 171–181.

[8] J.M. Stone, D. Mihalas and M.L. Norman, *Zeus-2D: A radiation magnetohydrodynamics code for astrophysical flows in two space dimensions. III. The radiation hydrodynamic algorithms and tests*, The Astrophysical Journal Supplemental Series, **80** (1992), 819–845.

[9] S. Turek, *An efficient solution technique for the radiative transfer equation*, Impact of Computation in Science and Engineering, **5** (1993), 201–214.

[10] S. Turek, *A generalized mean intensity approach for the numerical solution of the radiative transfer equation*, Technischer Report, Interdisziplinäres Zentrum für wissenschaftliches Rechnen, Universität Heidelberg, (1994).

[11] H.A. Van der Vorst, *Bi-CGStab: A fast and smoothly converging variant of Bi-CG for the solution of nonsymmetric linear systems*, SIAM J. Sci. Stat. Comput., **13** (1992), 631–644.

Institut für Astronomie, HAA C15, ETH Zentrum, 8092 Zürich, Switzerland
and Observatoire de Strasbourg, Université Louis Pasteur, Strasbourg, France
(R. Walder)
E-mail address: folini@astro.phys.ethz.ch,
http://astro.phys.ethz.ch/staff/folini/folini.html

International Series of Numerical Mathematics
Vol. 129, © 1999 Birkhäuser Verlag Basel/Switzerland

Flow Simulations on Cartesian Grids Involving Complex Moving Geometries

Hans Forrer and Marsha Berger

Abstract. We describe a method to solve the compressible time-dependent Euler equations using Cartesian grids for domains involving fixed or moving geometries. We describe the concept of a mirror flow extrapolation of a given solution over a reflecting wall which may be curved or moving at a fixed or varying speed. We use this mirror flow to develop a Cartesian grid method to treat the cells along a reflecting boundary avoiding the "small-cell" problem. Numerical Results are presented.

1. Introduction

We are developing numerical methods based on Cartesian grids for compressible inviscid time-dependent flows involving complex fixed or moving geometries. Cartesian grids offer great speed, robustness, and flexibility in dealing with complex industrial applications. In addition, they are relatively automated. However, to be able to use Cartesian grids we need to develop a treatment of the irregular boundary cells along reflecting walls of moving or fixed objects.

In previous work by the authors ([3], [6]) other boundary treatments for time-dependent flows described by the Euler equations have been developed. These boundary treatments are stable (without limiting the time step due to the arbitrarily small cut cells), accurate (more than first order along the boundary) and flexible (applicable for any finite-volume method).

Here we describe a numerical method to treat objects moving at a prescribed motion or in interaction with the fluid. As in [1] and in [5] we are using a fixed Cartesian grid and let the object move through it. Many of the same difficulties of accuracy and stability of Cartesian boundary treatments are also common to front tracking algorithms (cf. [2]).

2. Problem description

The main idea of our approach is to use regular Cartesian grid cells as much as possible. To avoid the "small-cell" problem, we fill in the cut cells and a set of ghost cells, so that cell updates are performed on regular grid cells.

To obtain ghost cell values in our method, the flow is extrapolated beyond the boundary by a mirror flow reflection. The mirror flow is a smooth extrapolation of the flow variables beyond the boundary such that the extrapolated solution fulfills the governing equations.

The Euler equations in two dimensions are given as

$$\mathbf{U}_t + \mathbf{F}_x + \mathbf{G}_y = 0, \tag{1}$$

$$\mathbf{U} = \begin{pmatrix} \rho \\ \rho u \\ \rho v \\ \rho e \end{pmatrix}, \quad \mathbf{F} = \begin{pmatrix} \rho u \\ \rho u^2 + p \\ \rho u v \\ u(\rho e + p) \end{pmatrix}, \quad \mathbf{G} = \begin{pmatrix} \rho v \\ \rho u v \\ \rho v^2 + p \\ v(\rho e + p) \end{pmatrix},$$

$$p = (\gamma - 1)(\rho e - \frac{1}{2}\rho(u^2 + v^2)),$$

where ρ is the mass density, $\mathbf{u} = (u, v)^T$ is the velocity vector, ρe is the energy density, p is the pressure and $\gamma = 1.4$.

In the following the one-dimensional mirror flow extrapolation is described. The reflecting wall boundary condition at $x = x_w$ is

$$u(x_w, t) = 0. \tag{2}$$

To describe a mirror flow extrapolation at the wall boundary, we reflect the space coordinate x at x_w

$$\hat{x}(x) = 2x_w - x. \tag{3}$$

A mirror flow extrapolation is then given for $x \le x_w$ as follows:

$$\begin{pmatrix} \hat{\rho}(x, t) \\ \hat{u}(x, t) \\ \hat{p}(x, t) \end{pmatrix} := \begin{pmatrix} \rho(\hat{x}, t) \\ -u(\hat{x}, t) \\ p(\hat{x}, t) \end{pmatrix}. \tag{4}$$

This mirror flow defines a smooth extrapolation of the solution beyond the wall boundary fulfilling the governing equations.

If the reflecting wall is moving at constant speed, i.e.,

$$x_w = x_w(t), \quad \ddot{x}_w(t) = 0, \tag{5}$$

the boundary condition is

$$u(x_w(t), t) = \dot{x}_w. \tag{6}$$

The reflection of the space coordinate is time-dependent now:

$$\hat{x}(x, t) = 2x_w(t) - x. \tag{7}$$

A solution $\rho(x, t), u(x, t), p(x, t)$ for $x > x_w(t)$ fulfilling the boundary condition (6) can be smoothly extrapolated beyond the boundary by the following mirror flow:

$$\begin{pmatrix} \hat{\rho}(x, t) \\ \hat{u}(x, t) \\ \hat{p}(x, t) \end{pmatrix} := \begin{pmatrix} \rho(\hat{x}, t) \\ 2\dot{x}_w - u(\hat{x}, t) \\ p(\hat{x}, t) \end{pmatrix}. \tag{8}$$

Consider now the case of a wall moving at varying velocity:

$$x_w = x_w(t), \ddot{x}_w \neq 0. \tag{9}$$

The boundary condition is the same as for the wall moving at constant speed, i.e., condition (6). In general a mirror flow extrapolation is only possible in a small neighborhood of the reflecting wall in this case. The solution has a non-zero pressure gradient at the wall according to

$$p_x(x_w(t), t) = -\rho(x_w(t), t)\, \ddot{x}_w(t). \tag{10}$$

Thus particles moving along the wall change their speed due to this pressure gradient. Not only pressure but also density has a non-zero gradient at the wall.

FIGURE 1. Mirror flow for a wall moving with changing speed

In two dimensions the inviscid boundary condition is

$$\mathbf{u}(\mathbf{x}, t) \cdot \mathbf{n} = 0, \quad \mathbf{x} \in \Gamma(t), \tag{11}$$

where $\Gamma(t)$ is the reflecting wall and $\mathbf{n}$ is the normal vector at the point $\mathbf{x}$. For a two-dimensional mirror flow extrapolation the normal velocity component u^n is treated as u in one dimension. The other variables as the tangential velocity component u^t, p, and ρ are treated as p and ρ in one dimension. Note that u^t, p, and ρ may have non-zero normal derivatives even at a fixed reflecting wall, namely if the wall curvature is non-zero, i.e.,

$$\frac{\partial p}{\partial n} = -\frac{u^t u^t}{R}, \quad \frac{\partial u^t}{\partial n} = -\frac{u^t}{R}, \tag{12}$$

where n is the normal coordinate and R is the curvature radius of the reflecting wall. The pressure gradient lets the particles move along the wall, the gradient of the tangential velocity is required by zero vorticity along the inviscid wall.

3. Numerical method

Here we describe how we incorporate a boundary treatment into the discretization of the interior flow on a regular Cartesian grid. For the computational results, the boundary treatment is coupled to the Clawpack method of LeVeque [7], a multi-dimensional shock-capturing finite-volume method for describing inviscid flows.

Let h be the grid parameter of a Cartesian grid. Then we set $x_i = x_0 + ih$, $y_j = y_0 + jh$, $i, j \in \mathbb{Z}$. The regular Cartesian grid cell C_{ij} is then given by:

$$C_{ij} = [x_i, x_{i+1}] \times [y_j, y_{j+1}]. \tag{13}$$

The numerical solution at time t_n is given by approximations of the cell averages of the exact solution $\mathbf{U}(x, y, t_n)$ over the grid cells:

$$\mathbf{U}_{ij}^n \approx \frac{1}{h^2} \int_{C_{ij}} \mathbf{U}(x, y, t_n) dx\, dy. \tag{14}$$

This numerical solution can then be updated using the Euler equations in integral form

$$\mathbf{U}_{ij}^{n+1} = \mathbf{U}_{ij}^n - \frac{\Delta t}{h} (\mathbf{F}_{i+1,j}^n - \mathbf{F}_{ij}^n + \mathbf{G}_{i,j+1}^n - \mathbf{G}_{ij}^n), \tag{15}$$

where we calculate the fluxes $\mathbf{F}_{ij}, \mathbf{G}_{ij}$ using Clawpack.

If at time t_n there is a reflecting wall along $\Gamma(t_n)$ going through the Cartesian grid, we divide the cells into regular cells, boundary cells and empty cells. Figure 2 on the left shows the regular and the boundary cells – all other cells are empty cells. The exact solution at time t_n can be extrapolated over the reflecting wall

FIGURE 2. Different types of grid cells at time t_n (left), splitting of a boundary cell (right)

such that we can assign a numerical value also for boundary cells. If at time t_n the Cartesian grid cell C is a boundary cell, we split this cell into two parts, C^1 is the part lying in the fluid domain and C^2 is the part lying on the solid domain as sketched in Figure 2 on the right. The numerical solution for such a boundary cell C at time t_n is then given by $\mathbf{U}_C^n$ approximating the following cell average

$$\mathbf{U}_C^n \approx \frac{1}{h^2} \left(\int_{C^2} \hat{\mathbf{U}}(x, y, t_n)\, dx\, dy + \int_{C^1} \mathbf{U}(x, y, t_n)\, dx\, dy \right). \tag{16}$$

Here $\hat{\mathbf{U}}(x, y, t_n)$ denotes the exact mirror flow solution at time t_n. For the regular cells the numerical solution is given by (14).

To advance the numerical solution of the regular cells and the boundary cells using (15), we fill in a set of ghost cells next to the boundary cells. These ghost cells

C_{kl} are filled with flow variables $\hat{\mathbf{U}}^n_{kl}$ using a numerical mirror flow extrapolation of the numerical solution at time t_n. How many ghost cells are needed depends on the specific numerical method. Two ghost cells are needed for the second-order accurate Clawpack method, as shown in Figure 3 on the left.

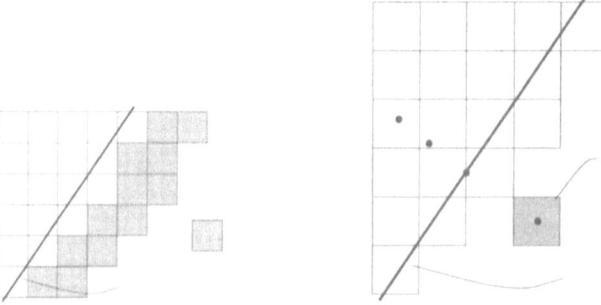

FIGURE 3. Two layers of ghost cells for a second-order accurate method (left), a ghost cell C lying beyond a reflecting wall (right)

In the following we describe how to calculate ghost cell values using a mirror flow extrapolation of the numerical solution. Suppose at time t_n there is a ghost cell C beyond a reflecting wall along $\Gamma(t_n)$ which may be moving (cf. Figure 3 on the right). The midpoint of cell C is denoted as $\mathbf{x}_C$. The point $\mathbf{x}_w$ is the point on $\Gamma(t_n)$ closest to $\mathbf{x}_C$. The point $\hat{\mathbf{x}}$ is the reflection of $\mathbf{x}_C$, i.e.,

$$\hat{\mathbf{x}} = 2\mathbf{x}_w - \mathbf{x}_C. \tag{17}$$

The normal vector $\mathbf{n}$ on $\Gamma(t_n)$ through $\mathbf{x}_w$ is given by

$$\mathbf{n} = \frac{\mathbf{x}_w - \mathbf{x}_C}{|\mathbf{x}_w - \mathbf{x}_C|}. \tag{18}$$

The flow variables of the ghost cell C are given by the pressure $\hat{p}$, the density $\hat{\rho}$, and the normal and tangential velocity components $\hat{u}^n$, $\hat{u}^t$ with respect to the $\mathbf{n}$ direction. The reflected point $\hat{\mathbf{x}}$ is lying between 4 cell centers of regular or boundary cells, such that $u^n(\hat{\mathbf{x}})$ can be obtained using a bilinear interpolation of the normal velocity at $\hat{\mathbf{x}}$. With $\dot{x}_w$ being the normal wall velocity at $\mathbf{x}_w$, the normal velocity of the ghost cell is obtained using

$$\hat{u}^n(\mathbf{x}_C) = 2\dot{x}_w - u^n(\hat{\mathbf{x}}). \tag{19}$$

We experimented with two different strategies to obtain the other ghost cell variables (as described only for pressure in the following). One is to use the bilinearly-interpolated values at the reflected point $\hat{\mathbf{x}}$, i.e.,

$$\hat{p}(\mathbf{x}_C) = p(\hat{\mathbf{x}}). \tag{20}$$

This is simpler than the volume weighted averaging used in the h-box method [3], and yields only a first-order boundary treatment for curved reflecting walls or walls moving with varying speed.

Another strategy is to extrapolate the corresponding values from the nearest boundary point $\mathbf{x}_w$. If $\mathbf{x}_w$ is lying between 4 cell centers of regular or boundary cells, a value $p(\mathbf{x}_w)$ is obtained using a bilinear interpolation, else $p(\mathbf{x}_w)$ is obtained by a linear interpolation from the nearest boundary cell center using finite-differences. By a bilinear interpolation we obtain also a value at the point $\mathbf{x}_w^h$ (cf. Figure 3, right), where

$$\mathbf{x}_w^h = \mathbf{x}_w + h\mathbf{n}. \tag{21}$$

Then the corresponding ghost cell values are obtained by

$$\hat{p}(\mathbf{x}_C) = p(\mathbf{x}_w) + |\mathbf{x}_w - \mathbf{x}_C| \frac{p(\mathbf{x}_w) - p(\mathbf{x}_w^h)}{h}. \tag{22}$$

Note that the boundary cells themselves can be updated using (15), so they do not need repeated application of (16); however the price of this is a lack of conservation (cf. [6] for more details).

4. Numerical results

First we look at a one-dimensional test case, namely a gas confined between two reflecting walls at $x_l = 0.5 + v_l t + \frac{a_l}{2} t^2$ and $x_r = 1.0$, with constants v_l and a_l. The initial conditions are

$$\rho(x,0) = 1.0 + 0.2 \cos(\pi \frac{x - 0.5}{0.5}), \tag{23}$$

$$v(x,0) = 2.0(1.0 - x)v_l, \tag{24}$$

$$p(x,0) = \rho(x,0)^\gamma, \tag{25}$$

such that entropy is constant, i.e., $s(x,0) = p(x,0)/rho(x,0)^\gamma = 1.0$. As long as the solution stays smooth, the entropy stays constant, such that we can use this variable for a numerical error analysis. In the numerical experiment we study the following quantities at the final time t_e:

$$err_{bry} = |s_{i_w}^{final} - 1.0|, \tag{26}$$

$$err_{tot} = \frac{\sum_i |s_i^{final} - 1| |C_i|}{\sum_i |C_i|}, \tag{27}$$

$$\Delta m = \left(\sum_i \rho_i^{initial} |C_i| - \sum_i \rho_i^{final} |C_i| \right) / \sum_i \rho_i^{initial} |C_i|, \tag{28}$$

where i_w is the index of the left boundary cell. The cell volume $|C_i|$ is h for regular cells, the length in the fluid domain for the boundary cells and zero for empty cells.

For the first test case we set: $v0 = -0.5, a0 = 0.0, t_e = 0.5$. The results in Table 1 for err_{bry} suggest that the boundary treatment is only first-order accurate

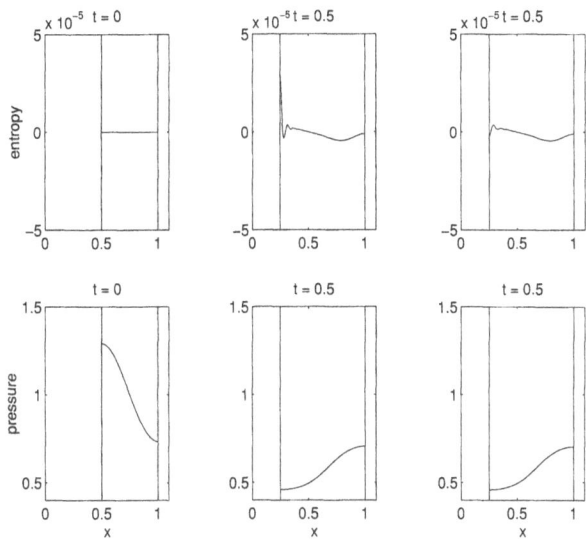

FIGURE 4. Wall moving with constant speed – ghost cell extra-polation using (20) (middle) or (22) (right), $h = 0.005$

	extrapolation using (20)			extrapolation using (22)		
$1/h$	Δm	err_{bry}	err_{tot}	Δm	err_{bry}	err_{tot}
200	4.27-6	2.93-5	2.00-6	3.19-5	3.35-6	1.67-6
400	9.71-7	1.16-5	5.31-7	8.08-6	7.31-7	4.14-7
800	2.39-7	5.57-6	1.41-7	2.04-6	1.70-7	1.03-7

TABLE 1.   Wall moving with constant speed – error analysis using the two different ghost cell extrapolations

if pressure and density are extrapolated using (20) in this case of a wall moving with constant speed, whereas using (22) the err_{bry} values suggest second-order accuracy also at the boundary in this case. But the lack of conservation is smaller using (20). Figure 4 shows the results for $h = 0.005$.

For the second test case we set: $v0 = 0.0, a0 = -2.0, t_e = 0.5$. The results in Table 2 suggest that for this second test case of a wall moving at varying speed, using (22) for the pressure and density extrapolation yields a second-order accurate boundary treatment.

For the following two-dimensional test cases we are using (22) for the ghost cell values but in the second test case we fix pressure and density to a small value (0.01), in case the ghost cell values drop below this small value.

The first test problem is taken from [4] for a numerical convergence study. It is a supersonic vortex in a channel formed by concentric circular arcs. The

$1/h$	Δm	err_{bry}	err_{tot}
400	6.65-5	1.90-5	3.22-6
800	2.29-5	4.55-6	8.71-7
1600	4.73-6	9.34-7	2.46-7

TABLE 2. Wall moving with varying speed – error analysis using (22) for the ghost cell extrapolations

boundaries of the channel form one quarter of a circle, with inner radius r_i and outer radius r_o. A smooth analytic solution exists for this problem, so the errors in the computation can be evaluated. The density is

$$\rho(r) = \rho_i \left[1 + \frac{\gamma - 1}{2} M_i^2 (1 - (\frac{r_i}{r})^2) \right]^{\frac{1}{\gamma - 1}}, \tag{29}$$

and the velocity varies inversely with the radius. We use the same geometry and test parameters as [4], $\rho_i = 1.0, r_i = 1.0, r_o = 1.384, M_i = 2.25, p_i = 1.0/\gamma$. We take the exact solution as initial condition and run the calculation until time $t = 5.0$, where the analysed values have converged to the 4-th digit. In the numerical study we look at the following relative errors using a discretization of the continuous L_1-norm (cf. [4])

$$err_{tot} = \frac{\sum_k |\rho_{exact} - \rho_k| |C_k|}{\sum_k \rho_{exact} |C_k|}, \tag{30}$$

$$err_{bry} = \frac{\sum_{k \in \partial} |\rho_{exact} - \rho_k| \sqrt{|C_k|}}{\sum_{k \in \partial} \rho_{exact} \sqrt{|C_k|}}, \tag{31}$$

$$\tag{32}$$

where $\sum_k$ is a summation over the regular cells and the boundary cells and $\sum_{k \in \partial}$ is a summation over the boundary cells only. $|C_k|$ is h^2 for regular cells and the area in the fluid domain for the boundary cells. Our boundary treatment is not strictly conservative. Therefore we look also at the difference of mass-inflow and mass-outflow in the final solution:

$$\Delta m = \sum_{k \in in} \rho_k |\mathbf{u}_k| l_k - \sum_{k \in out} \rho_k |\mathbf{u}_k| l_k, \tag{33}$$

where $\sum_{k \in in}$ is a summation over the inflow cells and $\sum_{k \in out}$ over the outflow cells. l_k is the length of the inflow/outflow interface. Table 3 (left) shows an error analysis of the above errors. The boundary treatment is of order $\log_2 \frac{16.3}{5.49} = 1.57$.

For a moving-boundary example, we show the cylinder lift-off by a strong shock wave, an example found in [5]. The movement of the cylinder is induced by the flow-field. To describe this motion within second-order accuracy, we use a staggered time grid. The center of the cylinder is given at full time-steps $\mathbf{X}_n$ and its velocity at staggered time-steps $\mathbf{V}_{n-\frac{1}{2}}$. At time t_n the force $\mathbf{F}_n$ on the cylinder

h	err_{tot}	err_{bry}	Δm
2.68-2	1.52-3	1.83-3	2.49-3
1.32-2	3.58-4	4.77-4	4.13-4
6.70-3	8.95-5	1.63-4	7.09-5
3.35-3	1.96-5	5.49-5	1.89-5

h	X	Y	Δm
3.33-3	6.89-1	1.429-1	1.79-2
2.50-3	7.00-1	1.392-1	1.05-2
2.00-3	7.06-1	1.379-1	9.50-3

TABLE 3. Error analysis for the supersonic vortex (left) – convergence history for the cylinder liftoff (right)

is calculated by a numerical integration of the flow field pressure times the normal vector along the surface of the cylinder; the velocity is updated by

$$\mathbf{V}_{n+\frac{1}{2}} = \mathbf{V}_{n-\frac{1}{2}} + \frac{\Delta t}{M}\mathbf{F}_n, \tag{34}$$

where M is the mass of the cylinder. The position of the cylinder is updated by

$$\mathbf{X}_{n+1} = \mathbf{X}_n + \Delta t\mathbf{V}_{n+\frac{1}{2}}. \tag{35}$$

We use the same parameters as in [5]. A cylinder with radius 0.05 is initially located at the lower wall at $x = 0.15$ of a channel with width 0.2. A Mach 3 shock wave starts at $x = 0.08$ moving towards the cylinder and lifting it off. The density and pressure of the resting gas are $\rho = 1.4$ and $p = 1.0$. The density of the cylinder is 10.77. The calculation is stopped at time 0.3282. In our calculation the cylinder hits the upper wall, opposed to the results of [5]. Table 3 (right) lists the final position of the center of the cylinder X, Y, and the final relative mass loss as a function of the grid parameter h. Figure 5 shows pressure contours of the initial condition, the solution at half-time and the final solution after the cylinder has hit the upper wall.

Acknowledgments

The authors were supported in part by DOE Grant De-FG02-92ER25139, and by AFOSR Grant F49620-97-1-0322.

References

[1] S.A. Bayyuk, K.G. Powell, and B. van Leer. *A Simulation Technique for 2-D Unsteady Inviscid Flows Around Arbitrarily Moving and Deforming Bodies of Arbitrary Geometry*. AIAA paper 93-3391, 1993.

[2] J.B. Bell, P. Colella, and M. Welcome. *Conservative Front Tracking for Inviscid Compressible Flow*. 10th AIAA Computational Fluid Dynamics Conference, Honolulu, pp. 814–822, 1991.

[3] M.J. Berger and R.J. LeVeque. *Stable Boundary Conditions for Cartesian Grid Calculations*. ICASE Report No. 90-37, May, 1990.

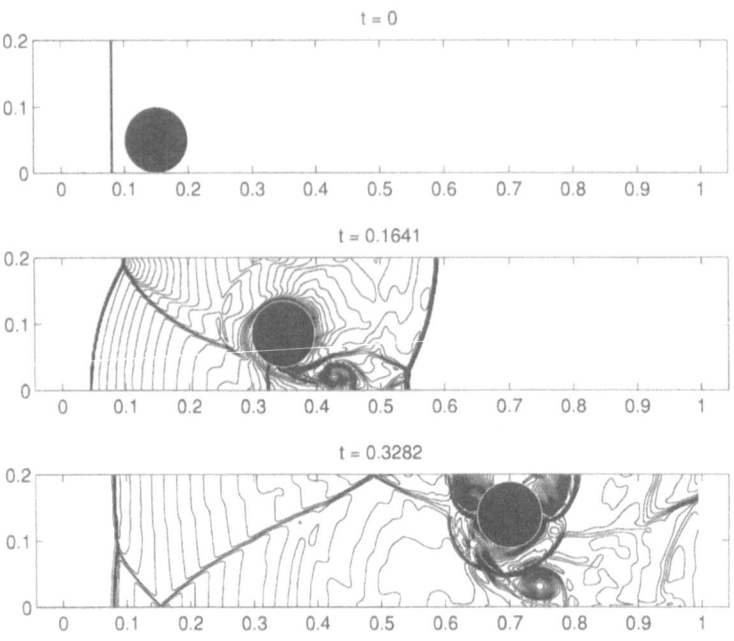

FIGURE 5. Cylinder lift-off, $h = 0.0025$

[4] M.J. Berger and J. Melton. *An Accuracy Test of a Cartesian Grid Method for Steady Flow in Complex Geometries.* Proceedings of 5th International Conference on Hyperbolic Problems, Stony brook, NY, 1994.

[5] J. Falcovitz, G. Alfandary, and G. Hanoch. *A Two-Dimensional Conservation Laws Scheme for Compressible Flows with Moving Boundaries.* Journal of Comp. Phys., 138, pp. 83–102, 1997.

[6] H. Forrer and R. Jeltsch. *A Higher-Order Boundary Treatment for Cartesian-Grid Methods.* To appear, J. Comp. Phys., 1998. Also, ETH Report No. 96-13, available via netscape in http://www.sam.math.ethz.ch/Reports/1996-13.html.

[7] R.J. LeVeque. *CLAWPACK software.* available from http://www.amath.washington.edu/~rjl/clawpack.html.

Courant Institute,
251, Mercer Street
New York, NY 10012
E-mail address: forrer@cims.nyu.edu

International Series of Numerical Mathematics
Vol. 129, © 1999 Birkhäuser Verlag Basel/Switzerland

Comparisons of Cell Centered and Cell Vertex Finite Volume Methods for Internal Flow Problems

J. Fořt, J. Fürst, J. Halama, M. Hrušová and K. Kozel

Abstract. The work deals with a numerical solution of laminar viscous and inviscid compressible 3D flows. Numerical methods with cell centered and cell vertex type of finite volumes are compared for some internal flow problems.

1. Governing equations

A laminar viscous flow or inviscid compressible 3D flow is described by the system of Navier-Stokes or Euler's equations written in conservative non-dimensional form:

$$W_t + F_x + G_y + H_z = \frac{1}{Re}\left(R_x + S_y + T_z\right) \tag{1}$$

with $W = (\rho, \rho u, \rho v, \rho w, e)^T$. The right hand side of (1) is considered zero for inviscid flow. Here ρ denotes density, (u, v, w) are the cartesian components of velocity vector, e is total energy per unit volume and p is pressure. Reference values are a proper characteristic length and reservoir values of density ρ_0 and speed of sound a_0. The system is closed by the equation of state for ideal gas:

$$p = (\kappa - 1)\left(e - \frac{1}{2}\rho(u^2 + v^2 + w^2)\right). \tag{2}$$

F, G, H denotes the cartesian components of convective fluxes and R, S, T are components of viscous fluxes.

2. Numerical solution

Numerical solution is computed on structured grid with hexahedral cells using following cell centered and cell vertex finite volume methods:

- simplified TVD Mac Cormack scheme published by Causon in [1]
- modification of Causon's TVD Mac Cormack scheme for quasi-stationary flows which uses less artificial dissipation (see [2])
- Runge-Kutta cell centered scheme
- cell vertex scheme based on the Ron-Ho-Ni's scheme (see [3], [4]).

We approximate an integral form of(1)

$$\int_t^{t+\Delta t} \int\int\int_D \mathbf{W}_t + \int_t^{t+\Delta t} \int\int_{\partial D} (\mathbf{F}, \mathbf{G}, \mathbf{H}_z) \mathbf{n} dS = 0 \qquad (3)$$

in case of 3D inviscid flow. Most of presented methods (expect method 3) are Lax-Wendroff type methods, i.e. they approximate in some proper way also the term corresponding formally to the second order term in Taylor expansion of $\mathbf{W}(t+\Delta t)$:

$$\frac{\Delta t}{2} \int\int_{\partial \tilde{D}} (\mathbf{F_W} \mathrm{div}(\mathbf{F}, \mathbf{G}, \mathbf{H}), \mathbf{G_W} \mathrm{div}(\mathbf{F}, \mathbf{G}, \mathbf{H}), \mathbf{H_W} \mathrm{div}(\mathbf{F}, \mathbf{G}, \mathbf{H})) \, \mathbf{n} dS \qquad (4)$$

For simplicity we describe the schemes in 2D. We denote vertices of cell centered quadrilateral finite volume with point i, j in center by indexes $1, \ldots, 4$ so, that cell face 1-2 is a boundary between volumes i, j and $i, j - 1$ and indexes are increased in direction of positive oriented curve. We use the same notation for vertices of cell vertex finite volume i, j with point 1 equal to i, j and point 2 equal to $i + 1, j$.

Mac Cormack cell centered scheme approximate (3), (4) in two stages:

1. predictor

$$\mathbf{W}_{i,j}^{n+1/2} = \mathbf{W}_{i,j}^n - \frac{\Delta t}{\mu_{i,j}} \left(\sum_{k=1}^4 \mathbf{F}_k^n \Delta y_k - \mathbf{G}_k^n \Delta x_k \right) \qquad (5)$$

$\Delta y_k = y_{k+1} - y_k, \ y_5 = y_1, \ \Delta x_k = x_{k+1} - x_k, \ x_5 = x_1, \ k = 1, \cdots, 4$
$\mathbf{F}_2 = \mathbf{F}_{i+1,j}, \mathbf{F}_3 = \mathbf{F}_{i,j+1}, \mathbf{F}_1 = \mathbf{F}_4 = \mathbf{F}_{i,j},$ and the similar is true for $\mathbf{G}$.

2. corrector

$$\widetilde{\mathbf{W}}_{i,j}^{n+1} = \frac{1}{2} \left(\mathbf{W}_{i,j}^n + \mathbf{W}_{i,j}^{n+1/2} \right) - \frac{1}{2} \frac{\Delta t}{\mu_{i,j}} \left(\sum_{k=1}^4 \mathbf{F}_k^{n+1/2)} \Delta y_k - \mathbf{G}_k^{n+1/2} \Delta x_k \right)$$

$\mathbf{F}_1 = \mathbf{F}_{i,j-1}, \mathbf{F}_4 = \mathbf{F}_{i-1,j}, \mathbf{F}_2 = \mathbf{F}_3 = \mathbf{F}_{i,j},$ and the similar is true for $\mathbf{G}$.

The third step of scheme involves damping terms in each grid direction:

$$\mathbf{W}_{i,j}^{n+1} = \widetilde{\mathbf{W}}_{i,j}^{n+1} + D_x(\mathbf{W}_{i,j}^n) + D_y(\mathbf{W}_{i,j}^n) \qquad (6)$$

The TVD version of scheme uses the terms proposed by Causon:

$$D_x(\mathbf{W}_{i,j}^n) = DW_{i+1/2,j}^n - DW_{i-1/2,j}^n$$
$$DW_{i+1/2,j}^n = (G^+i, j + G_{i+1,j}^-)\delta_{i+1/2,j}(\mathbf{W})$$
$$G_{i,j}^\pm = 0.5 \, C(\nu)[1 - \Phi(r^\pm)] \qquad (7)$$
$$C(\nu) = \begin{cases} \nu(1-\nu) & \nu \le \epsilon \\ \epsilon(1-\epsilon) & \nu > \epsilon, \ \epsilon \in (0, 1/2 > \end{cases}$$
$$\nu = \Delta t/\Delta x \, \rho(\mathbf{F_W}) = \Delta t/\Delta x(|u| + a)$$
$$r_{i,j}^\pm = \frac{(\delta_{i+1/2\pm1,j}(\mathbf{W}), \delta_{i+1/2,j}(\mathbf{W}))}{(\delta_{i+1/2,j}(\mathbf{W}), \delta_{i+1/2,j}(\mathbf{W}))}$$
$$\Phi(r) = \mathrm{minmod}(2r, 1),$$

where difference operator $\delta_{k+1/2,j}(\mathbf{W}) = \mathbf{W}_{k+1,j} - \mathbf{W}_{k,j}$ and $(\ ,\ )$ denotes a common scalar product of vectors. A less dissipative modification of this scheme uses a minimal eigenvalue instead of spectral radius of Jacobi matrices:

$$\nu = \frac{\Delta t}{\Delta x}\min(\Psi(u+a),\Psi(u),\Psi(u-a)) \tag{8}$$

$$\Psi(x) = \begin{cases} |x| & x > \epsilon \\ \frac{x^2+\epsilon^2}{2\epsilon} & |x| \le \epsilon \end{cases}$$

A next mentioned method is a Runge-Kutta multi-stage cell centered method. Values in new time level are computed in m steps:

$$\begin{aligned}
\mathbf{W}_{i,j}^{(0)} &= \mathbf{W}_{i,j}^n \\
\mathbf{W}_{i,j}^{(k+1)} &= \mathbf{W}_{i,j}^{(k)} - \alpha_{k+1}\Delta t L\mathbf{W}_{i,j}^{(k)} \quad k = 1,\cdots,m \\
\mathbf{W}_{i,j}^{n+1} &= \mathbf{W}_{i,j}^m.
\end{aligned} \tag{9}$$

The residual term $L\mathbf{W}$ consists of central approximation of fluxes and damping terms. It can be written as (see notation in (5)):

$$L\mathbf{W}_{i,j} = \frac{1}{\mu_{i,j}}\left[\sum_{k=1}^{4}(\mathbf{F}_k\Delta y_k - \mathbf{G}_k\Delta x_k) + DT_x(\mathbf{W}_{i,j}) + DT_y(\mathbf{W}_{i,j})\right]$$

$$\begin{aligned}
\mathbf{F}_1 &= (\mathbf{F}_{i,j-1}+\mathbf{F}_{i,j})/2 & \mathbf{F}_2 &= (\mathbf{F}_{i+1,j}+\mathbf{F}_{i,j})/2 \\
\mathbf{F}_3 &= (\mathbf{F}_{i,j+1}+\mathbf{F}_{i,j})/2 & \mathbf{F}_4 &= (\mathbf{F}_{i-1,j}+\mathbf{F}_{i,j})/2
\end{aligned} \tag{10}$$

and the similar is true for $\mathbf{G}$

$$DT_x(\mathbf{W}_{i,j}) = \epsilon_1\max(\gamma_i,\gamma_{i-1})\mathbf{W}_{xx} + \epsilon_2\mathbf{W}_{xxxx}$$

$$\gamma_i = \left|\frac{p_{i+1,j}-2p_{i,j}+p_{i+1,j}}{p_{i+1,j}+2p_{i,j}+p_{i+1,j}}\right|$$

The term (4) and terms of artificial viscosity are used as a redistribution terms in cell vertex scheme. Computation of unknowns in new time level consists of three steps:

1. computation of cell residuals $R\mathbf{W}$:

$$R\mathbf{W}_{i,j} = (\Delta t)/(2\mu_{i,j})\sum_{k=1}^{4}\left((\mathbf{F}_k^n + \mathbf{F}_{k+1}^n)\Delta y_k - (\mathbf{G}_k^n + \mathbf{G}_{k+1}^n)\Delta x_k\right) \tag{11}$$

2. distribution of cell residuals into vertices:

$$\begin{aligned}
\delta\mathbf{W}_1 &= -R\mathbf{W}_{i,j} - \mathbf{f} + \mathbf{g} + D_{L_1}, & \delta\mathbf{W}_2 &= -D\mathbf{W}_{i,j} - \mathbf{f} - \mathbf{g} + D_{L_2}, \\
\delta\mathbf{W}_3 &= -D\mathbf{W}_{i,j} + \mathbf{f} - \mathbf{g} + D_{L_3}, & \delta\mathbf{W}_4 &= -D\mathbf{W}_{i,j} + \mathbf{f} + \mathbf{g} + D_{L_4},
\end{aligned}$$

$$\mathbf{f} = (\Delta t/\mu_{i,j})\left(\mathbf{F_W}\Delta_y y - \mathbf{G_W}\Delta_y x + \epsilon_N|\Delta_x\mathbf{W}|\Delta_x\mathbf{W}\Delta t/(2\Delta_x x\mu_{i,j})\right)$$

$$\mathbf{g} = (\Delta t/\mu_{i,j})\left(\mathbf{G_W}\Delta_x x - \mathbf{F_W}\Delta_x y + \epsilon_N|\Delta_y\mathbf{W}|\Delta_y\mathbf{W}\Delta t/(2\Delta_y y\mu_{i,j})\right)$$

$$D_{L_l} = \epsilon_L\left(\frac{\Delta t}{\Delta_x x} + \frac{\Delta t}{\Delta_y y}\right)\left(\sum_{k=1}^{4}(\mathbf{W}_k - \mathbf{W}_l)\right) \tag{12}$$

where

$$\Delta_y a = a_2 + a_3 - a_4 - a_1 \qquad \Delta_x a = a_2 + a_1 - a_3 - a_4 \qquad (13)$$

3. computation of unknowns $\mathbf{W}^{n+1}$ in cell vertices:

$$\mathbf{W}_{i,j}^{n+1} = \mathbf{W}_{i,j}^{n} + (\mu_{i,j} + \mu_{i-1,j} + \mu_{i-1,j-1} + \mu_{i,j-1})^{-1} \qquad (14)$$

$$(\mu_{i,j}(\delta\mathbf{W}_4)_{i,j} + \mu_{i-1,j}(\delta\mathbf{W}_1)_{i-1,j} + \mu_{i-1,j-1}(\delta\mathbf{W}_2)_{i-1,j-1} + \mu_{i,j-1}(\delta\mathbf{W}_3)_{i,j-1})$$

All described methods can be used for computation of 3D inviscid internal flow problems.

3. Inviscid flow through an axial blade row

The first solved problem is an inviscid transonic flow through a 3D axial blade row. Solved geometry is a real stator geometry of a steam turbine developed by Škoda Turbines Pilsen, Czech Republic. A shape of hub and tip is a general axisymmetrical surface and blade is a little twisted.

Domain of solution is a one period of cascade. Its boundary consists of surfaces of blades, hub and tip (walls), surfaces with periodicity of unknowns in cylindrical coordinates and finally inlet and outlet surfaces.

A common non-permeability condition $(u, v, w)\mathbf{n} = 0$, where $\mathbf{n}$ is a vector of normal to the wall surface, is prescribed on the walls. Prescribed inlet and outlet conditions for subsonic inlet and outlet Mach number or normal velocity are constant in circumferential direction and depend on radius r. A distribution of stagnation pressure $p_0 = p_0(r)$, stagnation density $\rho_0 = \rho_0(r)$ and direction of velocity (two angles) $\alpha_1 = \alpha_1(r)$ and $\mu_1 = \mu_1(r)$ are prescribed on the inlet. Values of static pressure of Mach number are computed on the inlet from interior of domain of solution and then components of $\mathbf{W}$ are evaluated. One condition, distribution of the outlet pressure as a function of r $(p_2 = p_2(r))$ is given on the outlet. A missing outlet energy e_2 is evaluated from the equation of state (2) by using density and momentum computed from the interior of domain of solution. Necessary quantities on the inlet and outlet boundaries are obtained by extrapolation in cell centered methods or by regular scheme for modified one-sided control volumes in cell vertex scheme. Non-permeability condition is fulfilled by using auxiliary cells and mirroring principle in cell centered methods or by one-sided control volumes in cell vertex method. We present results achieved by modified TVD Mac Cormack scheme (cell centered) and Ron - Ho -Ni's scheme (cell vertex). Iso-Mach lines lines on the hub and tip surfaces achieved by both methods are on Fig 1-4. We can observe similar shape of iso-Mach lines for supersonic as well as subsonic outlet flow in corresponding cross-section. Shock waves are mapped more sharply by modified TVD Mac Cormack scheme. Relatively good agreement was achieved also for distribution of inlet Mach number (Fig. 5,6). Convergence history shows a better convergence of cell vertex method, but it depends on parameters in damping terms used in each method.

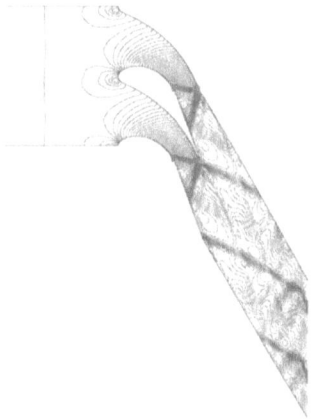

FIGURE 1. Iso-Mach
lines, surface of hub,
cell centered scheme

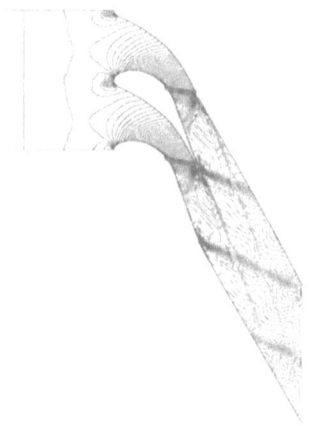

FIGURE 2. Iso-Mach
lines, surface of hub,
cell vertex scheme

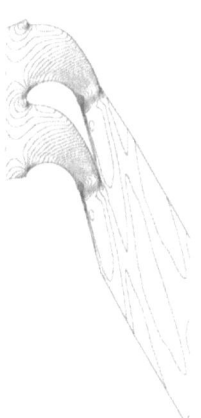

FIGURE 3. Iso-Mach
lines, surface of tip,
cell centered scheme

FIGURE 4. Iso-Mach
lines, surface of tip,
cell vertex scheme

4. Viscous laminar flow in a channel

Runge-Kutta cell centered (9) and cell vertex method (11),(12) were extended for computation of 3D laminar viscous flow. We use formally the same method as for inviscid flow, but we replace the convective fluxes $\mathbf{F},\mathbf{G},\mathbf{H}$ in (3) by a sum of convective and viscous fluxes, i.e. by $\mathbf{F} - \mathbf{R}/Re$, $\mathbf{G} - \mathbf{S}/Re$, $\mathbf{H} - \mathbf{T}/Re$ respectively.

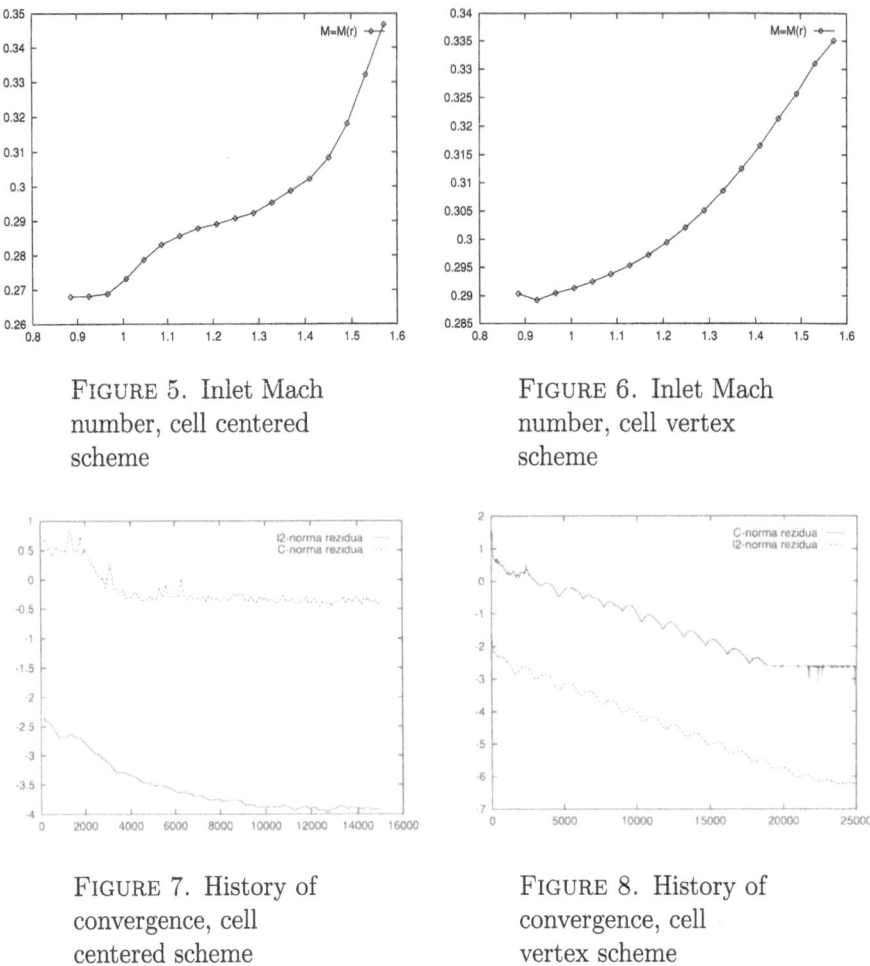

FIGURE 5. Inlet Mach number, cell centered scheme

FIGURE 6. Inlet Mach number, cell vertex scheme

FIGURE 7. History of convergence, cell centered scheme

FIGURE 8. History of convergence, cell vertex scheme

To evaluate viscous fluxes, we need to compute derivatives of velocity components. We use again finite volume approximation (mean value theorem). Derivatives of u, v are computed in the center of cell face for cell centered finite volume schemes. Auxiliary cells with vertices in centers of basic finite volumes and vertices of cell face are used. Derivatives of u, v in vertices are computed in two steps for cell vertex method. At first derivatives in the center of grid cell are computed and then we use averaging similar to (14). Influence of damping terms (artificial viscosity) is more significant in computation of viscous flow. The problem of interference of damping term and strong grid refinement near walls and small ratio of numerical and physical viscosity have to be solved, also time accuracy is more important, because the flow could be unsteady.

We compare results for subsonic flow in a curved channel with constant curvature and constant square cross-section. Important feature of this flow is a secondary flow, which structure becomes stable in sufficient distance from the inlet.

Typical structure (e.g. two or four vortices) depends on geometry of channel and Reynolds number. We compare results achieved by both methods for subsonic flow with inlet Mach number approximately 0.3. Computed cases and also formulation of boundary conditions are not exactly the same, but from theory is known, that a structure of secondary flow could be similar for both cases. We can observe this fact also on numerical results – Fig. 9–12.

5. Conclusion

Numerical methods with different types of finite volumes (cell centered, cell vertex) differs also in technique, how the boundary conditions are fulfilled or formulated (e.g. on the inlet). From this point of view the cell vertex method is more sensitive. It needs a special care also in some points lying on the wall, e.g. sharp trailing edge of profile or corners in 3D. Both types of method can yield a similar results for a relatively complicated 3D flows. Mutual comparisons of results is a suitable way for validation of numerical results.

Acknowledgement. This work was supported by grants 101/96/1696, 101/96/0193, 201/96/0313 of the Grant Agency of Czech Republic.

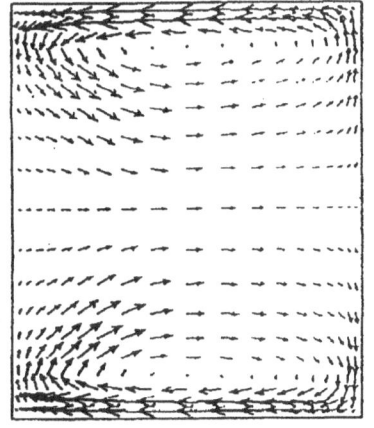

FIGURE 9. Re=5000, secondary flow, 63% of length, cell centered scheme

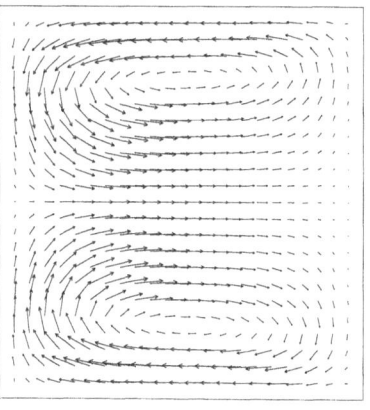

FIGURE 10. Re=1400, secondary flow, 50% of length, cell vertex scheme

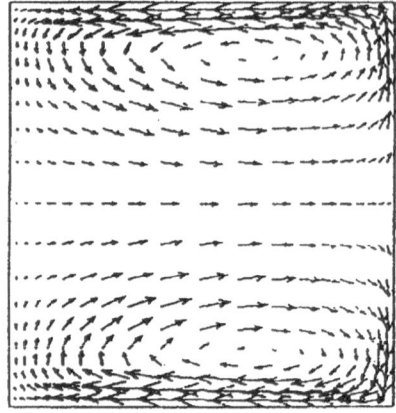

FIGURE 11. Re=5000, secondary flow, 88% of length, cell centered scheme

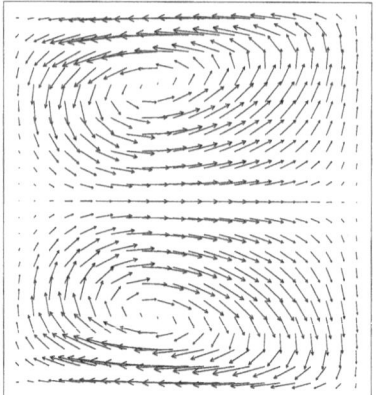

FIGURE 12. Re=1400, secondary flow, 88% of length, cell vertex scheme

References

[1] D. M. Causon, *High Resolution Finite Volume Schemes and Computational Aerodynamics,* Nonlinear Hyperbolic Equations, 1989, J. Ballmann, R. Jeltsch Eds., Notes on Numerical Fluid Mechanics **24** (1989), 63–74.

[2] P. Angot, J. Fürst, K. Kozel, *TVD and ENO Schemes for Multidimensional Steady and Unsteady Flow. A comparative Analysis,* Finite Volumes for Complex Applications, 1996, F. Benkhaldoun, R. Vilsmeier Eds., Hermes (1989), 283–290.

[3] R.H. Ni, *A Multiple Grid Scheme for Solving Euler Equations,* AIAA Journal **20 (1)** (1981).

[4] J. Fořt, M.Huněk, J. Lain, K. Kozel, M. Šejna, M. Vavřincová, *Numerical Simulation of Steady and Unsteady Flows Through Plane Cascades,* 14 ICNMFD, 1994, S. M. Desphane, S.S. Desai, R. Narasimha Eds, Lecture Notes in Physics (1995), 461–465.

Department of Technical Mathematics,
Faculty of Mechanical Eng. TU Prague,
Karlovo nám 13
121 35 Prague 2, Czech Republic
E-mail address: %fort:furst%@marian.fsik.cvut.cz, kozelk.fsik.cvut.cz

International Series of Numerical Mathematics
Vol. 129, © 1999 Birkhäuser Verlag Basel/Switzerland

Numerical Methods for Viscous Profiles of Non-classical Shock Waves

Heinrich Freistühler and Christian Rohde

Abstract. Numerical methods are proposed for the determination of viscous profiles of non-classical shock waves and for the detection of loci of corresponding heteroclinic bifurcations. The methods are applied to the case of magnetohydrodynamics.

1. Introduction

Consider a hyperbolic system

$$u_t + (f(u))_x = 0 \tag{1}$$

of conservation laws, together with a corresponding family

$$u_t + (f(u))_x = (D(u, \delta)u_x)_x, \quad \delta \in \Delta \tag{2}$$

of parabolic or hyperbolic-parabolic systems. A viscous profile for a shock wave, with boundary states u^-, u^+ and speed s, is a solution of the ODE boundary value problem

$$D(\phi, \delta)\phi' = f(\phi) - s\phi - q \quad (q = f(u^\pm) - su^\pm), \qquad \phi(\pm\infty) = u^\pm. \tag{3}$$

on the real line. Depending on the number of equations (= dimension of u) and the properties of f and D, problem (3) can be relatively subtle, and analytical methods alone are sometimes not sufficient for its solution. In magnetohydrodynamics (MHD)—our main example—, the state variable u ranges in (a part of) $\mathbf{R}^7$ while the range Δ of δ is $[0, \infty)^4$. The components of δ correspond to (two coefficients of) fluid viscosity, heat conductivity, and electrical resistivity. Depending on these viscosity parameters, the dynamical system (3) undergoes a global heteroclinic bifurcation: For fixed q, non-classical shock waves have viscous profiles for values of δ in a certain range, and no viscous profiles for values of δ in some other range. These results do not cover the full range Δ. Also, the proofs of the existence results are not constructive and do thus not provide any concrete representations of the profiles. This talk is concerned with numerical methods to approximate the solutions of problem (3), methods which are general enough to treat at least the example of MHD. For concreteness, we concentrate on this example.

Two problems are addressed. Problem I is the tracing of a critical heteroclinic orbit whose existence for certain values of (q, δ) characterizes the bifurcation locus $B \subset \mathbf{R}^7 \times \Delta$. Our first algorithm serves the purpose of determining B numerically. One goal behind this is to verify a certain conjecture about the shape of B. If this conjecture proves true, then the same algorithm is likely to provide a tool to decide, a priori, whether or not any profiles for non-classical shock waves exist for arbitrarily given values of q and δ. We present this algorithm for the full MHD system and show its convergence for a simpler subsystem. Problem II refers to situations in which profiles for non-classical shocks are already known to exist (e.g., after application of the algorithm described above). An interesting property of these situations is that the most of these profiles come in several-parameter families. To calculate them numerically, we develop an extension of an algorithm due to Beyn and Doedel/Friedman. This extension builds on a generalized phase condition which is motivated by the PDE stability theory for viscous profiles. We present this second algorithm and a convergence proof for this kind of algorithm in a rather general situation.

2. Short review on the MHD system

The equations governing plane magnetohydrodynamic waves are

$$\rho_t + (\rho v)_x = 0,$$

$$(\rho v)_t + (\rho v^2 + p + \frac{1}{2} \mid \mathbf{b} \mid^2)_x = \lambda v_{xx},$$

$$(\rho \mathbf{w})_t + (\rho v \mathbf{w} - \mathbf{b})_x = \mu \mathbf{w}_{xx},$$

$$\mathbf{b}_t + (v\mathbf{b} - \mathbf{w})_x = \nu \mathbf{b}_{xx},$$

$$E_t + (v(E + p + \frac{1}{2} \mid \mathbf{b} \mid^2) - \mathbf{w} \cdot \mathbf{b})_x = \lambda(v v_x)_x + \mu(\mathbf{w} \cdot \mathbf{w}_x)_x + \nu(\mathbf{b} \cdot \mathbf{b}_x)_x + \kappa \theta_{xx}.$$

Here $\rho, \theta, p = R\theta\rho > 0$ denote density, temperature, and pressure of the fluid, $v \in \mathbf{R}$ its longitudinal and $\mathbf{w} \in \mathbf{R}^2$ its transverse velocity, $\mathbf{b} \in \mathbf{R}^2$ the transverse magnetic field, and

$$E = \rho(\epsilon + \frac{1}{2}(v^2 + \mid \mathbf{w} \mid^2)) + \frac{1}{2} \mid \mathbf{b} \mid^2$$

the total energy with $\epsilon = c_v \theta$ the internal energy per mass. The "longitudinal" fluid viscosity λ and the "transverse" fluid viscosity μ are positive combinations of the two viscosity coefficients of the fluid, ν describes its electrical resistivity, and κ its heat conductivity. Abbreviating $(\rho, \rho v, \rho \mathbf{w}, \mathbf{b}, E) \in (0, \infty) \times \mathbf{R} \times \mathbf{R}^2 \times \mathbf{R}^2 \times (0, \infty)$ as u, and $(\lambda, \mu, \nu, \kappa) \in [0, \infty)^4$ as δ, we arrive at (2), with obvious meanings of f and D. After appropriate rescaling, renaming, and making use of Galilean invariance (in particular by setting $s = 0$), (3) reads

$$\nu \dot{\mathbf{b}} = -d\mathbf{w} + \tau \mathbf{b} - \mathbf{c}$$

$$\lambda \dot{\tau} = \tau + \frac{R\theta}{\tau} + \frac{1}{2} \mid \mathbf{b} \mid^2 - j$$

$$\mu \dot{\mathbf{w}} = \mathbf{w} - d\mathbf{b}$$

$$\kappa \dot{\theta} = c_v \theta - \frac{1}{2}(\mid \mathbf{w} \mid^2 - 2d\mathbf{w} \cdot \mathbf{b} + \tau \mid \mathbf{b} \mid^2) - \frac{\tau^2}{2} + j\tau + \mathbf{b} \cdot \mathbf{c} - e.$$

In this system of 6 equations, which we will refer to as Σ^6, $\bar{q} = (d, \mathbf{c}, \mathbf{j}, e) \in \mathbf{R} \times \mathbf{R}^2 \times \mathbf{R} \times \mathbf{R}$ plays the rôle of q. As long as $\mathbf{c} \neq \mathbf{0}$, Σ^6 has up to four rest points, traditionally numbered as u_0, u_1, u_2, u_3 (according to the dimension of their stable manifolds). Certain combinations (u_i, u_j) among these — namely those for which $(i, j) \in \{0, 1\} \times \{2, 3\}$ — give rise to non-classical or "intermediate" shock waves. Results in [8] (and in older papers referenced therein) support the following

Conjecture 2.1. *There exists a threshold $w^* = w^*(\bar{q}, \mu/\lambda, \kappa/\lambda) > 0$ such that the following holds for all $\bar{q}$ for which (3) has three or four rest points and for all $\delta = (\nu, \lambda, \mu, \kappa) \in (0, \infty)^4$: If $\nu/\lambda > w^*$, then all intermediate shocks (for the given $\bar{q}$) have viscous profiles (for the given δ). Conversely, if $\nu/\lambda < w^*$, then all intermediate shocks have no profiles.*

In other words, a global heteroclinic bifurcation takes place for the profiles of intermediate shocks; this bifurcation is the subject-matter of extended numerical investigations we carry out using the methods we present in this paper.

Instead of the six-dimensional system Σ^6, we also study, as a first step, the three-dimensional system Σ^3:

$$\nu \dot{\mathbf{b}} = (\tau - d^2)\mathbf{b} - \mathbf{c},$$

$$\lambda \dot{\tau} = \frac{1}{2}|\mathbf{b}|^2 + \tau - j + \frac{1}{K\tau}\left(-\frac{\tau^2}{2} - \frac{d^2}{2}|\mathbf{b}|^2 - \mathbf{b} \cdot \mathbf{c} + e\right),$$

which is obtained from Σ^6 by letting $\kappa = \mu = 0$ and substituting the variables θ and $\mathbf{w}$ in terms of τ and $\mathbf{b}$. The analogue of Conjecture 2.1 for Σ^3 was proved in [8].

3. Numerical detection of the bifurcation locus

To give numerical algorithms for the determination of the critical bifurcation ratio $w^* = w^*(\bar{q}, \mu/\lambda, \kappa/\lambda)$ in the case of Σ^6, resp. $w^* = w^*(\bar{q})$ in the case of Σ^3, we need the following observations. Assume $\mathbf{c} \neq \mathbf{0}$.

Let E^4 denote the fourdimensional submanifold of the state space of Σ^6 given by $\mathbf{b} \cdot \mathbf{c} = \mathbf{w} \cdot \mathbf{c} = 0$, and similarly E^2 the twodimensional manifold $\mathbf{b} \cdot \mathbf{c} = 0$ in the state space of Σ^3. E^4 and E^2 are invariant manifolds which contain all four rest points and many of their heteroclinic connections; the bifurcation can be studied by analyzing the subsystems $\Sigma^4 := \Sigma^6|_{E^4}$ resp. $\Sigma^2 := \Sigma^3|_{E^2}$. For the rest of Section 3 we restrict attention to Σ^4 resp. Σ^2.

Two ways to determine ω^* are proposed. Both rely on the fact that the stable manifold $M^+(u_2)$ is onedimensional for Σ^4 and Σ^2. Let C^ω continuously denote that one of the two connected components of $M^+(u_2)\backslash\{u_2\}$ which for certain values of ω, has the rest point u_0 as its α–limit $\alpha(C^\omega)$.

The first method shoots for ω^*, starting from initial guesses $\underline{\omega}^0, \overline{\omega}^0, 0 < \underline{\omega}^0 < \overline{\omega}^0$, made so that $\alpha(C^{\underline{\omega}}) \neq u_0$ and $\alpha(C^{\overline{\omega}}) = u_0$. The following procedure gives two converging sequences $\{\underline{\omega}^n\}$ and $\{\overline{\omega}^n\}$.

Algorithm:

For $n = 0, 1, 2, \ldots,$ do

    1.) Set $\tilde{\omega} = (\underline{\omega}^n + \overline{\omega}^n)/2$.

    2.) If $\alpha(C^{\tilde{\omega}}) \neq u_0$ $\begin{cases} \underline{\omega}^{n+1} &= \tilde{\omega} \\ \overline{\omega}^{n+1} &= \overline{\omega}^n. \end{cases}$

    If $\alpha(C^{\tilde{\omega}}) = u_0$ $\begin{cases} \underline{\omega}^{n+1} &= \underline{\omega}^n \\ \overline{\omega}^{n+1} &= \tilde{\omega}. \end{cases}$

If Conjecture 2.1 holds, then the said choice of $\underline{\omega}^0, \overline{\omega}^0$ is possible and also necessarily

$$\lim_{n\to\infty} \underline{\omega}^n = \lim_{n\to\infty} \overline{\omega}^n = \omega^*.$$

By the last sentence of Section 2, this is thus known for Σ^2 resp. Σ^3. It is not known at present for Σ^4 resp. Σ^6.

The second method finds ω^* through pathfollowing. Assume in addition to Conjecture 2.1, a critical orbit $u_1 \to u_2$ exists in E^4 resp. E^2 if and only if $\omega = \omega^*$. (This is known to be true for Σ^3; it is not known for Σ^6.) This orbit must be identical with C^{ω^*}. The pair (ω^*, C^{ω^*}) is now calculated by continuation from a known reference situation. Methods that are suitable for this task can be extracted from the work of Beyn [3] or Doedel&Friedman [6].

4. Numerical computation of heteroclinic manifolds

4.1. A parameterization of heteroclinic manifolds

Fix a vector field $F \in C^2(\mathbf{R}^m, \mathbf{R}^m)$ and consider a family Φ of trajectories which are heteroclinic to two given zeros u^-, u^+ of F, i. e., $\phi \in \Phi$ satisfy

$$\dot{\phi} = F(\phi), \quad \phi(\pm\infty) = u^\pm.$$

Assume that $u^\pm$ are hyperbolic rest points, (i. e., all eigenvalues of the Jacobians $DF(u^\pm)$ have non-vanishing real part), that with m_u^-, m_s^+ the numbers of unstable and stable eigenvalues at u^- resp. u^+,

$$d \equiv m_u^- + m_s^+ - m \geq 1, \tag{4}$$

and that the intersection M of the unstable manifold of u^- and the stable manifold of u^+ is a d-dimensional differentiable manifold. Coordinates on M, and thus on Φ, can then be chosen such that variations of one of their d components correspond to a phase shift along the fixed orbits constituting M while varying the remaining $d-1$ components relates to switching between the different orbits.

With $A : \mathbf{R} \times \mathbf{R}^m \to \mathbf{R}^{m \times m}$ define a mapping

$$\Pi : \Phi \to \mathbf{R}^m$$

through

$$\Pi(\Phi) \equiv \int_{\mathbf{R}} A(x, \phi(x))(\phi(x) - \phi_*(x)) dx,$$

where

$$\phi_*(x) = \begin{cases} u^- & : & x < 0 \\ u^+ & : & x > 0. \end{cases}$$

The idea of the method we present below for the purpose of numerical identification of Φ is to assume that

$$\Pi \text{ is injective, } S = \Pi(\Phi) \text{ is a } d\text{-dimensional manifold,} \tag{5}$$

and compute the individual orbits ϕ^τ, which are defined through $\mathbf{P}(\Pi(\phi^\tau)) = \tau$, $\tau \in \mathbf{R}^d$ lying in the range of a coordinate chart $\mathbf{P}$ of S.

Before turning to the numerical side, we show that for viscous profiles, the choice $A = Id$ (identity matrix of order m) allows for a natural interpretation of assumption (5). Let $w^\pm(., ., \kappa^\pm)$ denote the diffusion waves ([10]) with base states $u^\pm$ and masses

$$\int_{\mathbf{R}} w^\pm(x, t, \kappa^\pm) dx = \kappa^\pm \quad \in \quad R^\pm(u^\pm, s) \equiv \sum_{\pm(\lambda - s) > 0} \ker(Df(u) - sId).$$

$\phi^* \in \Phi$ is said to be asymptotically stable if for any function $\bar{u}_0$ in a interestingly large subclass of $L^1(\mathbf{R}) \cap L^\infty(\mathbf{R})$, there exist $\phi \in \Phi$ and $\kappa^-, \kappa^+ \in R^\pm(u^\pm, s)$ such that the solution u of (2) to the initial datum $\phi^* + \bar{u}_0$ satisfies

$$lim_{t \to \infty} \left\| u(., t) - (\phi(. - st) + w^-(., t, \kappa^-) + w^+(., t, \kappa^+)) \right\|_{L^1} = 0.$$

By conservation of mass, (4.1) implies

$$\int_{\mathbf{R}} \bar{u}_0 = (\int_{\mathbf{R}} \phi - \phi^*) + \kappa^- + \kappa^+.$$

Note now that in many situations it is either known or generally believed (e. g., for physical reasons) that for a stable profile ϕ^*, the corresponding asymptotic profile ϕ is determined by the mass $\int_{\mathbf{R}} \bar{u}_0$ of the initial perturbation alone. Obviously, any such situation is an example of injectivity of Π (with $A = Id$). To be more concrete, we mention that we conjecture that the overcompressive shocks occurring in MHD are all stable and have the property (5) with $A = Id$, i. e., one can distinguish the orbits which are heteroclinic to the same rest points by their relative masses. The numerical verification of this conjecture is part of our ongoing investigations.

Remark 4.1. Examples of systems with families of overcompressive shock profiles which do not share these properties (including perturbed versions of the cubic model $u_t + (|u|^2 u)_x = \nu u_{xx}$) are given in [9]. Our method, then with $A \neq Id$, remains useful in such cases.

4.2. The algorithm

Let now $\Psi \equiv P \circ \Pi$. Fix $\tau \in \Psi(\Phi)$. The goal is to compute ϕ^τ, the element of Φ with $\Psi(\phi^\tau) = \tau$, or, more concretely speaking, to solve

Problem 4.2. *Find* $\phi^\tau \in C^1(\mathbf{R})$ *with*

$$\dot{\phi}^\tau = F(\phi^\tau),$$

$$\phi^\tau(\pm\infty) = u^\pm,$$

$$\Psi(\phi^\tau) - \tau = 0.$$

To obtain numerical approximations for ϕ^τ, we restrict the ODE to a finite interval $I = [x_-, x_+]$ and use asymptotic boundary conditions in x_- and x_+. I. e., we consider

Problem 4.3. *Find* $\phi_I^\tau \in C^1(I)$ *with*

$$\dot{\phi}_I^\tau = F(\phi_I^\tau),$$

$$L_s^-(\phi_I^\tau - u^-) = 0, \quad L_u^+(\phi_I^\tau - u^+) = 0, \tag{6}$$

$$\Psi_I(\phi_I^\tau) - \tau = 0. \tag{7}$$

In (6), the matrices $L_s^- \in \mathbf{R}^{m_s^- \times m}$ resp. $L_u^+ \in \mathbf{R}^{m_u^+ \times m}$ are defined such that their rows form a basis of the maximal left stable invariant space of $DF(u^-)$ and the maximal left unstable invariant space of $DF(u^+)$, respectively. This means that the approximation ϕ_I^τ lies within the tangent space of the unstable manifold of u^- resp. of the stable manifold of u^+. The asymptotic boundary conditions (6) are called projection conditions and known to minimize the approximation error [3, 4]. Ψ_I is the analogue of Ψ for these approximations on the interval I, i. e., $\Psi_I(\phi) = \mathbf{P} \int_I A(., \phi)(\phi - \phi_*)$. Note that (7) and (6) give $m_u^+ + m_s^- + d$ relations, which sums up to m by (4). This matches with the m unknown components of ϕ.

Problem 4.3 is a two-point boundary value problem (which one readily sees upon introducing $\partial\phi/\partial x$ as a new unknown). There is a large variety of numerical methods to solve nonlinear problems of that kind [1]. In the numerical calculations presented in Section 5 we have used shooting methods as well as a spline collocation method.

We prove convergence of solutions ϕ_I^τ of Problem 4.3 to the corresponding solution ϕ^τ of Problem 4.2 on the whole line if x_- resp. x_+ tend to $-\infty$ resp.

$+\infty$. The approximation error turns out to decay exponentially with respect to the length of the interval I, due to the hyperbolicity of the fixed points.

We need two assumptions. The first assumption is a nondegeneracy property of the exact solution ϕ^τ under linearization.

Assumption 4.4. *If* $y \in C^1(\mathbf{R})$ *satisfies* $\dot{y} = DF(\phi^\tau)y$, $y(\pm\infty) = 0$ *then* $y \in$ $span\left\{\frac{\partial\phi^\tau}{\partial\tau_1}, \ldots, \frac{\partial\phi^\tau}{\partial\tau_d}\right\}$.

The next assumption corresponds to (5).

Assumption 4.5. *For* $I = [x_-, x_+]$ *sufficiently large we have*

$$span\left\{\mathbf{P}\int_I \frac{\partial\phi^\tau}{\partial\tau_1}, \ldots, \mathbf{P}\int_I \frac{\partial\phi^\tau}{\partial\tau_d}\right\} = \mathbf{R}^d.$$

Theorem 4.6. *Let* $\tau \in \mathbf{R}^d$ *and let* ϕ^τ *be a solution of Problem 4.2 such that Assumption 4.4 is satisfied. If additionally Assumption 4.5 holds then we have*

(i) *There are* $\rho, k > 0$ *such that for* $|x_-|, |x_+| > k$ *a unique solution* $\phi_I \in$ $C^1(I)$ *of Problem 4.3 in* $B_\rho(\phi^\tau)$ *exists.*

(ii) *There is a constant* $C > 0$ *such that for* $|x_-|, |x_+| > k$ *we have:*

$$\|\phi_I^\tau - \phi^\tau\|_{C^1(I)} \leq C(x_+ - x_-)\exp(-\min\{|\mu_u(u^-)x_-|, |\mu_s(u^+)x_+|\}). \quad (8)$$

Here $\mu_u(u)$ $(\mu_s(u))$ *denotes the smallest unstable (stable) eigenvalue of the matrix* $DF(u)$ *for* $u \in \mathbf{R}^m$ *with respect to absolute values. The constant* C *in (8) does not depend on* I.

Proof. We outline the main steps of the proof. For details we refer to [7]. The proof relies on the following perturbation argument which can be found for example in [12].

Let Y, Z be Banach spaces and $B_\rho(y_0) \subset Y$ be a ball of radius ρ. Consider a C^1-mapping $F : B_\rho \to Z$ such that $F'(y_0)$ is a homeomorphism. Assume that there are constants $\kappa, \sigma > 0$ such that

(a) $\||F'(y) - F'(y_0)\|| \leq \kappa < \sigma \leq \frac{1}{\||[F'(y_0)]^{-1}\||} \quad \forall y \in B_\rho(y_0),$ (9)

(b) $\|F(y_0)\| \leq (\sigma - \kappa)\rho.$ (10)

Here $\||.\||$ denotes the operator norm. Then there is a unique $\bar{y} \in B_\rho(y_0)$ such that $F(\bar{y}) = 0$ and

$$\|\bar{y} - y_0\| \leq \frac{\|F(y_0)\|}{\sigma - \kappa}. \quad (11)$$

In our case we choose $Y = C^1(I)$, $Z = C^0(I) \times \mathbf{R}^{m_s^-} \times \mathbf{R}^{m_u^+} \times \mathbf{R}^d$ and $y_0 = \phi^\tau$. For fixed interval I define $F = F_I : C^1(I) \to Z$, F_I given by

$$F_I(u) = (\dot{u} - f(u), L_s^-(u - u^-), L_u^+(u - u^+), \Psi_I(u)).$$

For the solution u of the inhomogeneous linear problem $F'_I(\phi^\tau)(u) = z$ —and this is the main step— it can be shown that there is a constant $\tilde{C} > 0$ such that for all $z \in Z$ the stability estimate

$$\|u\|_{C^1(I)} \leq \tilde{C}(x_+ - x_-)\|z\|_Z \tag{12}$$

holds. Using (12) and the differentiability properties of F_I, condition (9) follows for $2\kappa = \sigma = 1/(\tilde{C}(x_+ - x_-))$ and $\rho = O(1/(x_+ - x_-))$. Condition (10) can be obtained using the hyperbolicity of the rest points . Note that ϕ^τ is exponentially decaying with the rates $\mu_u(u^-)$ resp. $\mu_s(u+)$ which leads to (8). $\qquad\square$

Remark 4.7. The decay estimate in Theorem 4.6 seems not to be optimal. Numerical tests that we have done support the conjecture that on the right-hand side of (8) a purely exponential decay should hold, i. e., it should be possible to drop the algebraic factor $x_+ - x_-$.

5. Numerical results for Σ^3

We present numerical results that we have obtained for the MHD subsystem Σ^3. The results relate to the following specific set of parameters (cf. Section 2).

$$\bar{q} = (d, j, \mathbf{c}, e) = (1.0, 0.15, 0, 1.8, 1.0).$$

With the algorithm described in Section 3 we get $\omega^* \approx 0.019$. Let us at first consider the cases $\omega < \omega^*$ and $\omega = \omega^*$, for which all orbits lie in a two dimensional linear subspace E^2. The connections $u_0 \to u_1$ and $u_2 \to u_3$ correspond to classical Laxian shocks. For $\omega = \omega^*$ additionally the heteroclinic orbit $u_1 \to u_2$ arises (Figure 1). If we let $\omega > \omega^*$ we additionally get connections $u_0 \to u_2$, $u_1 \to u_3$.

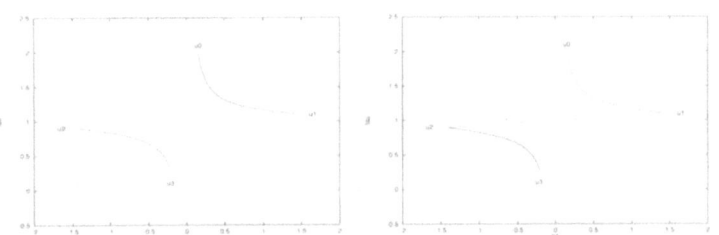

FIGURE 1. Orbits of Σ^3 for $\omega < \omega^*$ and $\omega = \omega^*$

Furthermore there is, for Σ^2, a one-parameter family of orbits connecting u_0 and u_3. Figure 2 shows the numerical results for two values of ω. For $\omega > \omega^*$ the heteroclinic structure of Σ^3 is not completely described by the restriction to E^2. In Figure 3, we present all orbits (except the orbits $u_0 \to u_3$) arising in Σ^3 for some $\omega > \omega^*$. Additionally to the viscous profiles of Laxian type (bold in Figure 3) there are one-parameter families of heteroclinic orbits connecting u_0 and u_2 respectively u_1 and u_3 (fine in Figure 3). Then there is a ring of two orbits from

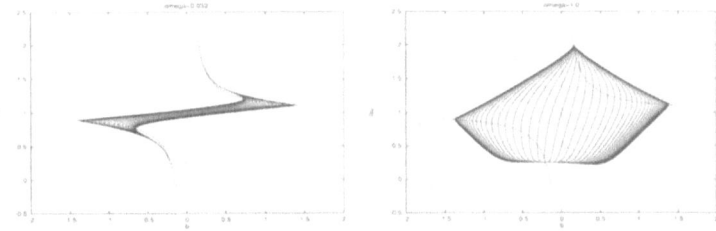

FIGURE 2. Orbits of Σ^2 for $\omega = 0.033$ and $\omega = 1.0$

u_1 to u_2 (again bold in Figure 3). All these orbits lie on the boundary of an open set of points which is filled by a two-parameter family of orbits $u_0 \to u_3$. Some

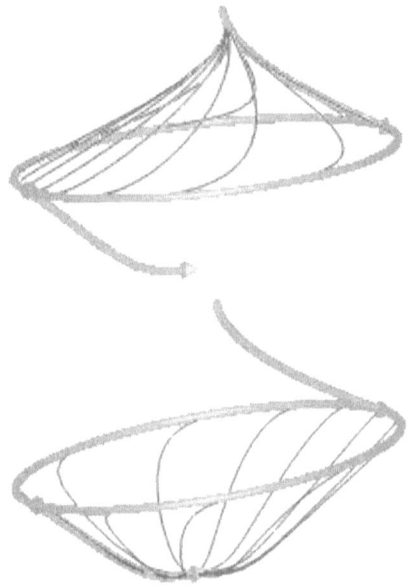

FIGURE 3. Orbits of Σ^3 (in two pieces)

of the computations presented above were performed using the code COLNEW of Bader&Kunkel.

References

[1] U. M. Ascher, R. M. M. Mattheij, and R. D. Russell, *Numerical solution of boundary value problems for ordinary differential equations*, Philadelphia, (1995).

[2] F. Bai, A. Spence and A. M. Stuart, *The numerical computation of heteroclinic connections in systems of gradient partial differential equations*, SIAM J. Num. Anal., **53 (3)** (1993), 743–769.

[3] W.-J. Beyn, *The numerical computation of connecting orbits in dynamical systems*, IMA J. Numer. Anal., **9** (1990), 379–405.

[4] F. R. DeHoog and R. Weiss, *An approximation theory for boundary value problems on infinite intervals*, Computing, **24** (1980), 227–239.

[5] E. J. Doedel, *AUTO: A program for the automatic bifurcation analysis of autonomous systems*, Cong. Num. 30, Proc. 10th Manitoba Conf. Num. Math. and Comp., Winnipeg, Canada, (1990), 265–284

[6] E. J. Doedel and M. J. Friedman, *Computation and continuation of invariant manifolds* SIAM J. Numer. Anal., **28 (3)** (1991), 789–808.

[7] H. Freistühler and C. Rohde, *Numerical computation of heteroclinic manifolds, with applications to shock waves*, in preparation.

[8] H. Freistühler and P. Szmolyan, *Existence and bifurcation of viscous profiles for all intermediate magnetohydrodynamic shock waves*, SIAM J. Appl. Math., **26** (1995), 112–128.

[9] H. Freistühler and K. Zumbrun, *Examples of unstable viscous shock waves*, preprint.

[10] T.-P. Liu, *Nonlinear stability of shock waves for viscous conservation laws*, Am. Math. Soc. Mem. Providence, AMS, **328** (1985).

[11] D. Serre, *Systèmes de lois de conservation I*, Paris, **1996**.

[12] G. Vainikko, *Funktionalanalysis der Diskretisierungsmethoden*, Leipzig, **1976**.

Institut für Mathematik,
Rheinisch-Westfälische Technische Hochschule Aachen,
55, Templergraben
D-52062 Aachen, Germany
E-mail address: hf@instmath.rwth-aachen.de

Institut für Angewandte Mathematik,
Albert–Ludwigs–Universität Freiburg,
10, Hermann–Herder–Straße
D-79104 Freiburg, Germany
E-mail address: chris@mathematik.uni-freiburg.de

International Series of Numerical Mathematics
Vol. 129, © 1999 Birkhäuser Verlag Basel/Switzerland

On the Asymptotic Behavior of Solutions of Certain Multi-D Viscous Systems of Conservation Laws

Hermano Frid

Abstract. We present a recent result establishing the asymptotic stability of planar Riemann solutions for certain systems of viscous conservation laws in several space variables. The systems to which our result applies are those whose flux fields are endowed with a common coordinate structure of Riemann invariants whose level sets are hyperplanes. No smallness restrictions are necessary, neither on the Riemann data nor on the perturbations.

1. Introduction

We consider solutions of initial-value problems for multi-D viscous systems of conservation laws given by

$$\partial_t u + \cdot \partial_{x_i} f^i(u) = \Delta u, \qquad (t, \mathbf{x}) \in (0, \infty) \times \mathbf{R}^n, \tag{1}$$

where $\mathbf{x} = (x_1, \ldots, x_n)$, $u \in \mathbf{R}^m$, $f^i(u) \in \mathbf{R}^m$, is a smooth vector field, $i = 1, \ldots, n$, $\Delta = \partial_{x_1}^2 + \cdots + \partial_{x_n}^2$ is the Laplacian operator, and we adopt the convention of summation for repeated indexes.

Here we are interested in systems (1) whose flux functions $f^1(u), \ldots, f^n(u)$ satisfy the hyperbolicity condition that, given any vector $\sigma = (\sigma_1, \ldots, \sigma_n) \in \mathbf{R}^n$, the matrix $\sigma_i \nabla f^i(u)$ is diagonalizable.

We form the initial value problem for (1) by prescribing an initial condition of the form

$$u(t, \mathbf{x})|_{t=0} = u_0(\mathbf{x}), \qquad \mathbf{x} \in \mathbb{R}^n, \tag{2}$$

where, here, we will assume $u_0 \in L^\infty(\mathbb{R}^n)$.

In this paper we are concerned with the question of the asymptotic behavior of the solution of (1), (2), $u(t, x)$, when the initial condition is a perturbation of a planar Riemann data. Namely, we assume that u_0 satisfies

$$u_0(\mathbf{x}) = R_0(x_1) + \psi(\mathbf{x}), \tag{3}$$

Research supported by CNPq, proc.352871/96-2.

where $R_0(x) = u_L$, $x < 0$, $R_0(x) = u_R$, $x > 0$, and $\psi \in L^\infty(\mathbb{R}^n)$ with

$$\lim_{T \to \infty} \psi(T\mathbf{x}) = 0 \quad \text{in } L^1_{\text{loc}}(\mathbb{R}^n). \tag{4}$$

It is natural to conjecture that the large-time behavior of $u(t, \mathbf{x})$ will be governed somehow by the solution $R(\xi)$, $\xi = \mathbf{x}/t$, of the Riemann problem given by (1) and

$$u(t, x)|_{t=0} = R_0(x_1), \tag{5}$$

which satisfies $R(\xi) = \bar{R}(\xi_1)$, $\xi_1 = x_1/t$, where $\bar{R}(x/t)$ is the solution of the one-dimensional Riemann problem

$$u_t + f^1(u)_x = 0, \tag{6}$$
$$u(t, x)|_{t=0} = R_0(x). \tag{7}$$

Nevertheless, if one intends to address this question in a broad context, which should include the trivial case when $f^i(u) = a^i u$, with $a^i \in \mathbb{R}$, $i = 1, \ldots, n$, one immediately realizes that convergence when t goes to infinity in, say, L^p_{loc} relatively to the $\mathbf{x}$ variables is not appropriate, for any $p \geq 1$, since this is false already in the one-dimensional case for the trivial equation $u_t = \partial_x^2 u$. On the other hand, the invariance of the Riemann solution under the scaling transformation $(t, \mathbf{x}) \mapsto (ct, c\mathbf{x})$, for any $c > 0$, indicates that the choice of the variables (t, ξ), $\xi = \mathbf{x}/t$, is a natural one for the study of the asymptotic behavior of the solution of the perturbed initial value problem. We are then led to the question whether the solution of the perturbed problem, u, satisfies

$$u(t, t\xi) \longrightarrow R(\xi), \quad \text{as } t \to \infty, \text{ in } L^1_{\text{loc}}(\mathbb{R}^n). \tag{8}$$

This question was addressed in [2], for a large number of one-dimensional systems with interest in applications, and for multi-D scalar conservation laws, both in the inviscid and in the viscous cases.

Here we answer positively the issue of the validity of (8) for a particular class of multi-D systems (1). Namely, we consider the systems (1) in which the flux functions f^i, $i = 1, \ldots, n$, are Temple fields (see Definition 3.1) whose Jacobian matrices ∇f^i, $i = 1, \ldots, n$, admit a common basis of eigenvectors. An example of Temple field is given by $f(u) = \frac{1}{D(u)}(k_1 u_1, \ldots, k_m u_m)^\top$, $0 < k_1 < \cdots < k_m$, $D(u) = 1 + u_1 + \cdots + u_m$, which is found in a model system in chromatography (see [1]). One-dimensional systems in which the flux functions are Temple fields have been studied by many authors after the first analysis carried out in [10], followed by [6, 7]. The class of systems for which our result applies clearly includes the case where $f^i(u) = a^i f(u)$, $a^i \in \mathbb{R}$, $i = 1, \ldots, n$, and f is any Temple field in $\mathbb{R}^m$. For $m = 2$, it was proved in [10] that the Temple fields whose Riemann invariants are $w_1(u)$, $w_2(u)$, are given by

$$f_1(u) = \frac{h_1(w_1) - h_2(w_2)}{w_1 - w_2}, \qquad f_2(u) = \frac{w_2 h_1(w_1) - w_1 h_2(w_2)}{w_2 - w_1}, \tag{9}$$

where h_1 and h_2 are arbitrary smooth functions of one variable. This shows that, for $m = 2$, it is possible to give examples of systems (1) with flux functions satisfying our assumptions, beyond the case when all of them are multiples of one only Temple field. We now state our main theorem.

Theorem 1.1. *Let f^i, $i = 1, \ldots, n$, be Temple fields possessing the same Riemann invariants. Let (6) be a hyperbolic, genuinely nonlinear system whose eigenvalues satisfy $c_1 \leq \lambda_1(u) \leq c_2 \leq \lambda_2(u) \leq c_3 \leq \cdots \leq c_m \leq \lambda_m(u) \leq c_{m+1}$, for all u in the domain of f_1 and certain constants $c_j \in \mathbb{R}$, $j = 1, \ldots, m+1$. Assume that $u_0 \in L^\infty(\mathbb{R}^n)$ satisfies (3), (4). Then, the only solution of (1), (2) verifies (8).*

Remark 1.2. This result extends to several space variables an earlier one, for the corresponding one-dimensional system, obtained in [2].

Remark 1.3. No smallness condition is needed, neither for the initial data nor for the perturbation. For the latter the only requirement is the one given by (5).

The method for obtaining (8) follows the general approach in [2]. The main point in the strategy is to prove the relation

$$\lim_{T \to \infty} \frac{1}{T} \int_0^T |u(t, t\xi) - R(\xi)| \, dt = 0, \quad \text{in } L^1_{\text{loc}}(\mathbb{R}^n). \tag{10}$$

An important aspect of (10) is its equivalence to the convergence in $L^1_{\text{loc}}(\mathbb{R}^{n+1}_+)$ of the scaling sequence $\{u^T\}$, given by $u^T(t, \mathbf{x}) = u(Tt, T\mathbf{x})$, to $R(\mathbf{x}/t)$, when $T \to +\infty$, if u and R belong to $L^\infty(\mathbb{R}^{n+1}_+)$. This fact is frequently useful when one is trying to prove (10). Once (10) is proved, a standard procedure established in [2] is then used to strengthen (10) into (8). This strengthening is similar to the ones encountered in [3, 5].

The remaining of this paper is organized in the following way. In Section 2 we state several results concerning more general systems (1). The goal of the section is to establish Lemma 2.5 which enables one to reduce the proof of (10) to a problem of analysing the support of certain probability measures and trying to show that it consists of just one point. The latter is similar in spirit to an usual procedure in the theory of compensated compactness, since the pioneering paper [9], although here the probability measures have nothing to do with the Young measures. In Section 3 we then specify our discussion to systems (1) in which all flux functions are Temple fields possessing the same Riemann invariants. We give an outline of the proof of (10) which requires to recall some results valid for Temple fields. We finally recall through a brief outline the procedure to pass from the weak time asymptotics given by (10) to the stronger one given by (8).

2. Results for general systems (1)

In this section we state some results which are valid for more general systems of the form (1). We start with a result concerning the regularity in the large of solutions of (1) which are uniformly bounded in the half-space $\mathbb{R}^{n+1}_+$.

Proposition 2.1. *Suppose that $u(t, \mathbf{x})$ is a solution of (1)), (2), uniformly bounded in $\mathbb{R}_+^{n+1}$, with*

$$u_0 \in W^{N-1,\infty}\left(\mathbb{R}^n; \mathbb{R}^m\right),$$

for some $N \in \mathbb{Z}^+$. Then,

$$u \in L^1\left(0, T; W^{N,\infty}\left(\mathbb{R}^n; \mathbb{R}^m\right)\right),$$

for any $T > 0$. More precisely, we have for $|\alpha| \leq N - 1$

$$\|\partial_{\mathbf{x}}^\alpha u\|_{L^\infty(\mathbf{R}_+^{n+1})} < +\infty, \tag{11}$$

and, if $|\alpha| = N$, for $t_0 > 0$ sufficiently small, there exists $C = C(t_0) > 0$ such that

$$\|\partial_{\mathbf{x}}^\alpha u(t)\|_\infty \leq C, \qquad \text{for } t > t_0, \tag{12}$$

and

$$\|\partial_{\mathbf{x}}^\alpha u(t)\|_\infty \leq \frac{C}{\sqrt{t}}, \qquad \text{for } 0 < t \leq t_0, \tag{13}$$

where we adopt the multi-indice notation $|\alpha| = \alpha_1 + \cdots + \alpha_n$, $\partial_{\mathbf{x}}^\alpha = \partial_{x_1}^{\alpha_1} \ldots \partial_{x_n}^{\alpha_n}$.

For the next results we shall need to assume (1) is endowed with a strictly convex entropy η with associated entropy flux $\mathbf{q} = (q_1, \ldots, q_n)$. We recall that a pair $(\eta(u), \mathbf{q}(u))$ is an entropy-entropy flux pair for (1) if one has

$$\nabla \mathbf{q}(u) = \nabla \eta(u) \nabla \mathbf{f}(u), \tag{14}$$

where we denote $\mathbf{f} = (f^1, \ldots, f^n)$.

For $0 < t \leq T$, $K, M > 0$, $\xi \in \mathbb{R}$, $|\xi| < M$, we define the hypersurfaces of $\mathbb{R}^n$

$$S_1^t(T) \equiv \{\mathbf{x} \in \mathbb{R}^n : x_1 = \xi t, \ |(x_2, \ldots, x_n)|_\infty \leq MT + K(T - t)\},$$

$$S_{j-}^t(T) \equiv \{\mathbf{x} \in \mathbb{R}^n : -Mt \leq x_1 \leq \xi t, \ |x_j| = MT + K(T - t),$$
$$|(x_2, \ldots, \hat{x}_j, \ldots, x_n)|_\infty \leq MT + K(T - t)\},$$

$$S_{j+}^t(T) \equiv \{\mathbf{x} \in \mathbb{R}^n : \ \xi t \leq x_1 \leq Mt, \ |x_j| = MT + K(T - t),$$
$$|(x_2, \ldots, \hat{x}_j, \ldots, x_n)|_\infty \leq MT + K(T - t)\},$$

$j \geq 2$, where, for $\mathbf{z} \in \mathbb{R}^s$, we denote $|\mathbf{z}|_\infty = \max\{|z_j| \mid j = 1, \ldots, s\}$, and the hat over one of the variables means that it should be omitted.

An important feature of our method for obtaining (10), for the systems (1) satisfying the hypotheses of Theorem 1.1, is the use of the Gauss-Green formula after an integration of the entropy inequality

$$\partial_t \eta(u) + \nabla \cdot \mathbf{q}(u) \leq \Delta \eta(u), \tag{15}$$

over domains laterally bounded by surfaces whose intersections with the hyperplanes $t = const.$ are unions of hypersurfaces in $\mathbb{R}^n$ of the form $S_1^t(T)$, $S_{j\sigma}^t(T)$, $j = 1, \ldots, n$, $\sigma = \pm$. The entropy inequality (15) is obtained by multiplying (1) by

$\nabla\eta(u)$, using (14) and the convexity of η. The integration procedure is intended to produce boundary terms, all of which should be discarded, except one involving the integration of the normal trace of the entropy vector $(\eta, \mathbf{q})$ over a hypersurface whose sections by $t = const.$ is of the type $S_1^t(T)$. The following propositions are used to show that we may discard the uninteresting boundary terms coming out from the integration.

Proposition 2.2. *Suppose that there is a strictly convex entropy associated with* (1) *and* $u \in L^\infty(\mathbb{R}_+^{n+1})$ *is solution of* (1), (2). *Then, for a.e.* $\mathbf{b} \in \mathbb{R}^n$, *we have*

$$\lim_{T\to\infty} \frac{1}{T} \int_0^T |\nabla_{\mathbf{x}} u(t, \mathbf{b}t)|\, dt = 0. \tag{16}$$

In particular, for a.e. $\xi \in \mathbb{R}$, $M > 0$,

$$\lim_{T\to\infty} \frac{1}{T^n} \int_0^T dt \int_{S_1^t(T)} |\nabla_{\mathbf{x}} u(t, \mathbf{x})|\, dS_1^t(T) = 0, \tag{17}$$

where S_1^t *is defined above.*

Proposition 2.3. *Suppose the hypotheses of Proposition 2.2 hold. Then* $\frac{1}{T}\nabla_{\mathbf{x}} u^T \to$ 0 *as* $T \to \infty$ *in* $L_{loc}^1(\mathbb{R}_+^{n+1})$. *As a consequence, for some subsequence* $T_k \to \infty$, *a.e.* $\xi \in \mathbb{R}$, $K, M > 0$, *one also has*

$$\lim_{T_k\to\infty} \frac{1}{T_k^n} \int_0^{T_k} dt \int_{S_{j\sigma}^t(T_k)} |\nabla_{\mathbf{x}} u(t, \mathbf{x})|\, dS_{j\sigma}^t(T_k) = 0, \tag{18}$$

where $\sigma = \pm$ *and* $j = 2, \ldots, n$.

Proposition 2.4. *Assume again that the hypotheses of Proposition 2.2 are valid, and that* u_0 *satisfies* (3), (4). *Then, there exists* $M_0 > 0$, *such that* $u^T \to u_L$ *in* $L_{loc}^1(\Omega_{M_0}^-)$, *and* $u^T \to u_R$ *in* $L_{loc}^1(\Omega_{M_0}^+)$, *as* $T \to \infty$, *where*

$$\Omega_{M_0}^- = \{ (t, \mathbf{x}) \mid x_1 < -M_0 t \},$$
$$\Omega_{M_0}^+ = \{ (t, \mathbf{x}) \mid x_1 > M_0 t \}.$$

The following lemma is an important tool which allows one to reduce the study of the asymptotic behavior of the solution to a question of proving that the support of certain probability measures is concentrated in the state corresponding to the exact value of the Riemann solution.

Lemma 2.5. *Suppose* η *is a nonnegative convex entropy for* (1), *with associated entropy flux* $\mathbf{q}$, *and assume that the hypotheses of Proposition 2.4 hold. Let* T_l *be a sequence, obtained from Proposition 2.4, such that* $u^{T_l} \to u_L$ *a.e. in* $\Omega_{M_0}^-$, *and* $u^{T_l} \to u_R$ *a.e. in* $\Omega_{M_0}^+$. *Then, for a.e.* $\xi \in \mathbb{R}$, $K > 0$ *such that* $|\mathbf{q}(u)|_\infty \leq K\eta(u)$, *and any* $M > 0$ *we have:*
(a) *if* $\eta(u_L) = q(u_L) = 0$,

$$\limsup_{T_l\to\infty} \frac{1}{T_l^n} \int_0^{T_l} \int_{S_1^t(T_l)} (-\xi\eta + q_1)\, (u(t,\mathbf{x}))\, dS_1^t(T_l)dt \leq 0; \tag{19}$$

(b) *if $\eta(u_R) = q(u_R) = 0$,*

$$\limsup_{T_l \to \infty} \frac{1}{T_l^n} \int_0^{T_l} \int_{S_1^t(T_l)} (\xi\eta - q_1)\,(u(t,\mathbf{x}))\,dS_1^t(T_l)dt \leq 0. \tag{20}$$

3. Temple systems

In this section we specify our discussion to systems (1) whose flux functions f^i satisfy the hypotheses of Theorem 1.1. We recall the definition of Temple fields.

Definition 3.1. *(cf. [8]) A hyperbolic system (1) is said a Temple system in a domain $V \subset \mathbb{R}^m$ if, for any $i \in \{1, \cdots, m\}$, it satisfies the following:*

1. *there exists an i-Riemann invariant $w_i(u)$, i.e.,*
   $$\nabla w_i(u) \nabla f(u) = \lambda_i(u) \nabla w_i(u);$$
2. *the level sets $\{u \in V \mid w_i(u) = const.\}$ are intersections of affine hyperplanes with V.*

We now outline the proof of the Theorem 1.1. For ξ out of the set of measure zero where (a) and (b) of Lemma 2.5 do not hold, we consider the probability measures defined for all $h \in C(\mathbb{R}^m)$ by

$$\langle \mu_l^\xi, h \rangle \equiv \frac{1/\sqrt{1+\xi^2}}{(2M + K)^{n-1}T_l^n} \int_0^{T_l} \int_{S_1^t(T_l)} h(u(t,\mathbf{x}))\,dS_1^t(T_l)dt. \tag{21}$$

The proof is then reduced to showing that $\mu_l^\xi \rightharpoonup \delta_{\bar{R}(\xi)}$, as $l \to \infty$, for a.e. $\xi \in \mathbb{R}$ as in [2]. We briefly recall some important steps of the analysis in [2].

We need the following three lemmas concerning fields in the Temple class. The first is due to D. Serre [7], the second to A. Heibig [4] and the third to Chen-Frid [2]; we refer to the corresponding cited papers for the proofs.

Lemma 3.2. *(D. Serre [7]) For a Temple field $f \colon \mathcal{O} \subset \mathbb{R}^m \to \mathbb{R}^m$, defined in a in a convex domain $\mathcal{O} \subset \mathbb{R}^m$ convex.*

$$l_i(v) \cdot (u - v) = 0 \Longrightarrow l_i(v) \cdot (f(u) - f(v)) = 0,$$

$(u,v) \in \mathcal{O} \times \mathcal{O}$. In particular, if (1) satisfies the hypotheses of Theorem 1.1 for $u \in \mathcal{O}$, then, for any $v \in \mathcal{O}$, the following are entropy-entropy flux pairs for (1):

$$\left(\eta_i^{0+}, \mathbf{q}_i^{0+}\right) \equiv ((l_i(v) \cdot (u - v))_+ ,$$
$$H(l_i(v) \cdot (u - v))l_i(v) \cdot (\mathbf{f}(u) - \mathbf{f}(v))),$$
$$\left(\eta_i^{+0}, \mathbf{q}_i^{+0}\right) \equiv ((l_i(v) \cdot (v - u))_+ ,$$
$$H(l_i(v) \cdot (v - u))l_i(v) \cdot (\mathbf{f}(v) - \mathbf{f}(u))),$$
$$\left(\eta_i^{++}, \mathbf{q}_i^{++}\right) \equiv (|l_i(v) \cdot (u - v)| ,$$
$$sgn(l_i(v) \cdot (u - v))l_i(v) \cdot (\mathbf{f}(u) - \mathbf{f}(v))).$$

Here, $H(s)$ is the well-known Heaviside function and we use the notation $(s)_+ = sH(s)$.

Lemma 3.3. (A. Heibig [4]) *If $f: \mathcal{O} \subset \mathbb{R}^m \to \mathbb{R}^m$ is a Temple field, defined in a $\mathcal{O} \subset \mathbb{R}^m$ convex, then there exists an unique map $\bar{A}: \mathcal{O} \times \mathcal{O} \to M_{m \times m}(\mathbb{R})$ such that*

(i) *For all $(u, v) \in \mathcal{O} \times \mathcal{O}$,*

$$f(u) - f(v) = \bar{A}(u, v)(u - v),$$

$\bar{A}(u, u) = A(u)$, where $A(u) = \nabla f(u)$; is the Jacobian matrix of f;

(ii) *for all $(u, v) \in \mathcal{O} \times \mathcal{O}$, $A(v)$ and $\bar{A}(u, v)$ have the same (left and right) eigenvectors;*

(iii) *$\bar{A}$ is a smooth function.*

Lemma 3.4. (Chen-Frid [2]) *Let $\bar{\lambda}_i(u, v)$ be the i-th eigenvalue of the matrix $\bar{A}(u, v)$, given by Lemma 3.3 and suppose $\mathcal{O}$ has the form*

$$\mathcal{O} = \bigcap_{i=1}^{m} \{ u \in \mathbb{R}^m : |w_i(u) - w_i(\bar{u})| \leq M_i \}, \tag{22}$$

for certain $M_i > 0$, $i = 1, \ldots, m$. Then,

$$\min_{u \in \mathcal{O}} \lambda_i(u) \leq \bar{\lambda}_i(u, v) \leq \max_{u \in \mathcal{O}} \lambda_i(u), \qquad (u, v) \in \mathcal{O} \times \mathcal{O}. \tag{23}$$

Assume $c_j \leq \lambda_j(u) \leq c_{j+1}$, and so $c_j \leq \bar{\lambda}_j(u, v) \leq c_{j+1}$, $j = 1, \ldots, m$. where $\bar{\lambda}_j(u, v)$ is the j-th eigenvalue of $\bar{A}(u, v)$, with $\bar{A}$ given by the Lemma 3.3, for $f = f^1$.

We give an idea of how to prove (10), using Lemma 2.5 to analyze the support of certain probability measures. For instance, let us consider first the simplest cases when $\xi < c_1$ or $\xi > c_{m+1}$.

For $\xi < c_1$ we use (19) with $(\eta, \mathbf{q})$ replaced by $(\eta_j^{++}(u, u_L), \mathbf{q}_j^{++}(u, u_L))$, $j = 1, \ldots, m$. The sequence of probability measures μ_l^ξ is weakly compact and, so, we may find a subsequence $\mu_{l_s}^\xi$ and a probability measure μ^ξ such that $\mu_{l_s}^\xi \rightharpoonup \mu^\xi$. From (19) with

$$(\eta, \mathbf{q}) = (\eta_j^{++}(u, u_L), \mathbf{q}_j^{++}(u, u_L)),$$

$j = 1, \ldots, m$, we get

$$\langle \mu^\xi, -\xi |l_j(u_L) \cdot (u - u_L)| + \mathrm{sgn}(l_j(u_L) \cdot (u - u_L))(l_j(u_L) \cdot (f^1(u) - f^1(u_L))) \rangle \leq 0,$$

$j = 1, \ldots, m$. Applying the By Lemma 3.3, we find

$$\langle \mu^\xi, |l_j(u_L) \cdot (u - u_L)|(-\xi + \bar{\lambda}_j(u, u_L)) \rangle \leq 0,$$

$j = 1, \ldots, m$. Since $\bar{\lambda}_j(u, u_L) \geq c_1 > \xi$ (Lemma 3.4), we readily conclude that $\mu^\xi = \delta_{u_L}$, the latter being the Dirac measure concentrated at u_L. Since this holds for any weakly convergent subsequence of μ_l^ξ, we conclude that $\mu_l^\xi \rightharpoonup \delta_{u_L}$, for $\xi < a_1$. X-Mozilla-Status: 0000

For $\xi > c_{m+1}$, using (20) with (η, q) replaced by $(\eta_j^{++}(u, u_R), \mathbf{q}_j^{++}(u, u_R))$, $j = 1, \ldots, m$, analogously we obtain $\mu_l^\xi \rightharpoonup \delta_{u_R}$.

Now, assume

$$\xi \in (c_{j_0}, c_{j_0+1}),$$

for some $j_0 \in \{1, \ldots, m\}$. Again, we consider a probability measure μ^ξ, obtained as weak limit of some subsequence of μ_l^ξ. Let $u_1 = u_L$, $u_{m+1} = u_R$, $u_j = R(c_j)$, $j = 2, \ldots, m$, be the constant states appearing in of the Riemann solution $R(x/t)$. Let $r_j(u)$ be the j-th right eigenvector of $A(u) = \nabla f^1(u)$, with l_j and r_j normalized so that $l_j(u) \cdot r_j(u) = 1$ and $\nabla \lambda_j(u) \cdot r_j(u) = 1$, $j = 1, \ldots, m$. We show that

$$\text{supp}\,\mu^\xi \subset L_{j_0} \equiv \{u \in \mathcal{O} \mid u = u_{j_0} + s r_{j_0}(u_{j_0}),\ s \in \mathbb{R}\}. \tag{24}$$

We have

$$L_{j_0} = \frac{\cap_{i=1}^{j_0-1}\{u \in \mathcal{O} \mid l_i(u_R) \cdot (u - u_R) = 0\}}{\cap_{j=j_0+1}^{m}\{u \in \mathcal{O} \mid l_j(u_L) \cdot (u - u_L) = 0\}}.$$

In (20), we choose

$$(\eta, q) = \left(\eta_i^{++}(u, u_R), \mathbf{q}_i^{++}(u, u_R)\right),$$

$i = 1, \ldots, j_0 - 1$. Using Lemma 3.3, we obtain

$$\langle \mu^\xi, |l_i(u_R) \cdot (u - u_R)|(\xi - \bar{\lambda}_i(u, u_R))\rangle \leq 0, \tag{25}$$

for $i = 1, \ldots, j_0 - 1$. Similarly, in (19) we choose

$$(\eta, q) = \left(\eta_j^{++}(u, u_L), \mathbf{q}_j^{++}(u, u_L)\right),$$

$j = j_0 + 1, \ldots, m$. We apply Lemma 3.3, to get,

$$\langle \mu^\xi, |l_j(u_L) \cdot (u - u_L)|(-\xi + \bar{\lambda}_j(u, u_L))\rangle \leq 0, \tag{26}$$

for $j = j_0 + 1, \ldots, m$. Now, $\bar{\lambda}_1(u, v) \leq \cdots \leq \bar{\lambda}_{j_0-1}(u, v) \leq c_{j_0} < \xi < c_{j_0+1} \leq \bar{\lambda}_{j_0+1}(u, v) \leq \cdots \leq \bar{\lambda}_m(u, v)$, for $u, v \in \mathcal{O}$ (Lemma 3.4). Hence, (25) and (26) imply (24). The remaining follows by similar procedures and (24).

We finally give a brief outline of how (10) can be strengthened into (8). This goal is achieved using a method established in [2], motivated by [5]. The strategy is similar to the one for obtaining the decay of periodic solutions in L^p, $1 \leq p < \infty$, from the decay in time-average along rays (see [3]): we come back to the entropy inequalities and use the convergence in time-average to get the usual convergence in time. Since the Riemann solution depends only on x_1, the situation here is similar to the one-dimensional case, so its enough to show the arguments for the latter. So, assume u is a solution of of the 1-D parabolic system

$$u_t + f(u)_x = u_{xx}, \tag{27}$$

and $R(x/t)$ is a self-similar solution of a Riemann problem for the inviscid system corresponding to (27), which is piecewise Lipschitz relatively to $\xi = x/t$. We shall assume that (10) holds and then show how (8) can be obtained. Let $(\eta(u), q(u))$ denote a strictly convex entropy pair for (1), and $(\alpha(u, v), \beta(u, v))$ be given by

$$\alpha(u, v) = \eta(u) - \eta(v) - \nabla\eta(v)(u - v),$$
$$\beta(u, v) = q(u) - q(v) - \nabla\eta(v)(f(u) - f(v)).$$

After some manipulations with (27) and the equation for R, using the change of coordinates $(t, x) \mapsto (t, \xi)$, $\xi = x/t$, one obtains

$$\partial_t\alpha(u, R) - \frac{\xi}{t}\partial_\xi\alpha(u, R) + \frac{1}{t}\partial_\xi\beta(u, R) \le \frac{1}{t}\partial_\xi(\partial_x\eta(u)) + \frac{1}{t}\partial_\xi(\nabla\eta(R)\partial_x u)$$

$$- \frac{1}{t}\partial_\xi(\nabla\eta(R)\partial_x u) - \frac{1}{t}\nabla^2\eta(R)(\partial_\xi R, \partial_x u) - \frac{1}{t}\nabla^2\eta(R)(\partial_\xi R, Qf(u, R)),$$

where $Qf(u, v) = f(u) - f(v) - \nabla f(v)(u - v)$. Let I be an interval in the ξ-axis such that R is Lipschitz in I. Integrating the above inequality in the variable ξ over I, and using Proposition 2.1 which guarantees the uniform boundedness of u_x for $t \ge t_0 > 0$, one obtains

$$\frac{d}{dt}Y(t) \le \frac{C}{t}, \tag{28}$$

for some constant $C > 0$, where

$$Y(t) = \int_I \alpha(u(t, t\xi), R(\xi)) \, d\xi.$$

Now, it is easy to prove that (28) plus the fact, given by (10), that

$$\lim_{T \to \infty} \frac{1}{T} \int_0^T Y(t) \, dt = 0, \tag{29}$$

together imply

$$\lim_{t \to \infty} Y(t) = 0, \tag{30}$$

with a rate $O((\frac{1}{T} \int_0^T Y(t) \, dt)^{1/2})$. To extend (30) to the case where I is any bounded interval, possibly containing points of jump discontinuity of R, we observe that I is the union of a finite number of open intervals where R is Lipschitz continuous plus a finite number of points. Then, the integral of $|u(t, t\xi) - R(\xi)|$ over I is equal to the sum of the integrals of this function over a finite number of intervals, each of which, as has been proven, goes to zero when $t \to +\infty$, and the convergence then follows for a general I.

Acknowledgement. The author gratefully acknowledges the partial support received from CNPq, Brazil, proc. 352871/96-2.

References

[1] R. Aris, N. Amundson, *Mathematical Methods in Chemical Engineering*, Vol.2, Prentice-Hall, Englewood Cliffs, N. J., 1966.

[2] G.-Q. Chen and H. Frid, *Large-time behavior of solutions of conservation laws*, submitted (1997).

[3] G.-Q. Chen and H. Frid, *Asymptotic decay of solution to conservation laws*, C. R. Acad. Sci. Paris, Serie I (1996), 257–262.

[4] A. Heibig, *Existence and uniqueness of solutions for some hyperbolic systems of conservation laws*, Arch. Rational Mech. Anal. **26** (1994), 79–101.

[5] L. Hsiao and D. Serre, *Asymptotic behavior of large weak entropy solutions of the damped p-system*, J. Partial Diff. Eqs., **10** (1997), 355–368.

[6] R.J. Leveque and B. Temple, *Stability of Godunov's method for a class of* 2×2 *systems of conservation laws*, Trans. Amer. Math. Soc. **288** (1985), 115–123.

[7] D. Serre, *Solutions à variations bornées pour certains systèmes hyperboliques de lois de conservations*, J. Diff. Eqs. **68** (1987), 137–169.

[8] D. Serre, *Richness and the classification of quasilinear hyperbolic systems*, In: Multidimensional Hyperbolic Problems and Computations, ed. J. Glimm and A. Majda, IMA Vol. **29**, Springer-Verlag, New York, (1991), 315–333.

[9] L. Tartar, *Compensated compactness and applications to partial differential equations*, In: Research Notes in Mathematics, Nonlinear Analysis and Mechanics ed. R. J. Knops, **4** (1979), Pitman Press, New York, 136–211.

[10] B. Temple, *Systems of conservation laws with invariant submanifolds*, Trans. Amer. Math. Soc. **280** (1983), 781–795.

Instituto de Matemática,
Universidade Federal do Rio de Janeiro,
Caixa Postal 68530, CEP,
21945-970, RJ, Brazil
E-mail address: hermano@lpim.ufrj.br

International Series of Numerical Mathematics
Vol. 129, © 1999 Birkhäuser Verlag Basel/Switzerland

Stability of General Shock profiles – a Novel Weight Function for the Non-convex Case

Christian Fries

Abstract. Consider a system

$$u_t + f(u)_x = \mu u_{xx} \tag{1}$$

of viscous conservation laws. Let ϕ be a profile for a shock wave associated with a simple eigenvalue λ of f', with end states $\phi(\pm\infty) = u_\pm$.

Then there are numbers ϵ_0, $\beta_0(u_-, u_+) > 0$ such that whenever $|u_+ - u_-| < \epsilon_0$ and $\bar{u}_0 \in L^1$ satisfies $U_0 := \int_{-\infty}^\infty \bar{u}_0(x)\, dx \in H^{2,2}$, $\|U_0\|_{H^{2,2}} < \beta_0$, then the solution $u(x,t)$ to (1) with data $u(\cdot,0) = \phi + \bar{u}_0$ exists for all times $t > 0$ and has

$$\lim_{t\to\infty} \sup_x |u(x,t) - \phi(x - st)| = 0$$

(i.e. ϕ is asymptotically stable).

The novelty consists in the absence of any convexity assumption on λ; thus the theorem generalizes a previous result by J. Goodman. In this note we only sketch the proof of this theorem and focus on motivating the choice of a new weight function that enables overcoming the technical difficulties caused by the absence of convexity.

1. Introduction

Consider a system of viscous conservation laws

$$u_t + f(u)_x = \mu u_{xx} \tag{1}$$

whose inviscid part $u_t + f(u)_x$ is hyperbolic, i.e. the Jacobian $f'(u)$ of the flux function $f : \mathbb{R}^n \to \mathbb{R}^n$ can be diagonalized over $\mathbb{R}$ at any state u. Assume that f' has a simple eigenvalue λ in a neighborhood of some reference state u_* and let ϕ denote the profile of a Laxian shock wave, i.e., $\phi : \mathbb{R} \to \mathbb{R}^n$ solves

$$\mu\phi' = h(\phi) := f(\phi) - s\phi - (f(u_-) - su_-) \quad, \quad \phi(\pm\infty) = u_\pm \ ,$$

where u_-, u_+, s satisfy $f(u_-) - su_- = f(u_+) - su_+$, $\lambda(u_-) > s > \lambda(u_+)$, and u_-, u_+ are close to u_*. The following is proved in [1]:

Theorem 1.1. *There are numbers ϵ_0, $\beta_0(u_-, u_+) > 0$ such that whenever $|u_\pm - u_*| < \epsilon_0$ and $\bar{u}_0 \in L^1$ satisfies $U_0 := \int_{-\infty}^{\cdot} \bar{u}_0(x) \, dx \in H^{2,2}$, $\|U_0\|_{H^{2,2}} < \beta_0$, then the solution $u(x, t)$ to (1) with data $u(\cdot, 0) = \phi + \bar{u}_0$ exists for all times $t > 0$ and has*

$$\lim_{t \to \infty} \sup_x |u(x, t) - \phi(x - st)| = 0 .$$

The same result was previously known by Goodman [3] under the additional assumption that the eigenvalue λ be convex, i.e. $r \cdot \nabla \lambda \neq 0$ where $\mathbb{R}r = \ker(f' - \lambda I)$. In other words the novelty of Theorem 1.1 consists in showing the asymptotic stability of traveling-wave profiles for small-amplitude shocks associated with possibly non-convex modes.

In this note we only sketch the proof of Theorem 1.1 and focus on motivating the choice of a new weight function for the primary field that enables overcoming the technical difficulties caused by the absence of convexity.

The theorem is restricted to zero total mass, because $U_0 \in H^{2,2}$ implies $\int_{-\infty}^{\infty} \bar{u}_0(x) \, dx = 0$. A similar result can be obtained for non-zero mass perturbation, see [2]. In its proof – using the approach of Szepessy and Xin, [6] – the same weight function is essential in different ways.

2. Basic steps

The proof of Theorem 1.1 consists of a short time existence result and an a priori estimate from which global existence and the desired stability result follow. Focusing on the latter, we consider the difference between the solution $u^*(x, t) = \phi(x - st)$ and a solution u (corresponding to the perturbed initial data $\phi + \bar{u}_0$). Let $\bar{u}(x, t) := u(x, t) - \phi(x - st)$, then from (1)

$$\bar{u}_t + (f(u^* + \bar{u}) - f(u^*))_x = \mu \bar{u}_{xx} . \tag{2}$$

Multiplying (2) and (2)$_x$ by $\bar{u}$ and integrating $\int_{-\infty}^{\infty} dx$, $\int_{t_0}^{T} dt$ we easily arrive at

$$\|\bar{u}(\cdot, T)\|_{H^{1,2}} + \int_{t_0}^{T} \|\bar{u}_x(\cdot, t)\|_{H^{1,2}} \, dt \leq C \big(\|\bar{u}(\cdot, t_0)\|_{H^{1,2}} + \int_{t_0}^{T} \|\bar{u}(\cdot, t)\|_{L^2} \, dt \big). \tag{3}$$

Without the last term on the right-hand side we would have an a priori estimate for $\|\bar{u}(\cdot, T)\|_{H^{1,2}}$ and the decay of $\|\bar{u}_x(\cdot, t)\|_{H^{1,2}}$. To absorb the term on the r.h.s. and to get the decay of $\|\bar{u}(\cdot, t)\|_{H^{1,2}}$ (and hence of $\sup_x |\bar{u}|$) we thus consider the anti-derivative (or "integrated perturbation")

$$U(x - st, t) = \int_{-\infty}^{x} \bar{u}(\xi, t) \, d\xi = \int_{-\infty}^{x} u(\xi, t) - \phi(\xi - st) \, d\xi .$$

We integrate equation (2) and pass to the moving coordinates $(x - st, t)$:

$$-sU_x + U_t + f(\phi + U_x) - f(\phi) = \mu U_{xx}$$

$$U(x, 0) = U_0(x) = \int_{-\infty}^{x} u(\xi, 0) - \phi(\xi) \, d\xi .$$

By using Taylor expansion

$$f(\phi + U_x) - f(\phi) = f'(\phi)U_x - F(\phi, U_x)$$

we thus arrive at the so-called *integrated equation*

$$U_t + h'(\phi)U_x - \mu U_{xx} = F(\phi, U_x) \tag{4}$$

$$U(\cdot, 0) = U_0 = \int_{-\infty}^{\cdot} u(x, 0) - \phi(x) \, dx \ ,$$

where $h' = f' - sI$. Together with (3) it is thus sufficient to show

$$\|U(\cdot, T)\|_{L^2}^2 + \int_{t_0}^{T} \|U_x(\cdot, t)\|_{L^2}^2 \, dt \le C\|U(\cdot, t_0)\|_{L^2}^2 \ , \tag{5}$$

which leads to

$$\|U(\cdot, T)\|_{H^{2,2}}^2 + \int_{t_0}^{T} \|U_x(\cdot, t)\|_{H^{2,2}}^2 \, dt \le C\|U(\cdot, t_0)\|_{H^{2,2}}^2 \ . \tag{6}$$

From the above, global existence can be obtained and thus, using (6) with $T = \infty$, we get the existence of a sequence (t_n) with $t_n \to \infty$ and

$$\|U_x(\cdot, t_n)\|_{L^2}^2 \to 0 \quad , \quad \int_{t_n}^{\infty} \|U_x(\cdot, t)\|_{L_2}^2 \, dt \to 0 \ . \tag{7}$$

From (3) (with $t_0 = t_n$) and (7) we get

$$\lim_{t \to \infty} \sup_x |\phi(x - st) - u(x, t)|^2 = \lim_{t \to \infty} \sup_x |U_x(x, t)|^2$$

$$\le \lim_{t \to \infty} \|U_{xx}(\cdot, t)\|_{L^2}^2 \cdot \|U_x(\cdot, t)\|_{L^2}^2 \le \lim_{t \to \infty} C\|U_x(\cdot, t)\|_{L^2}^2$$

$$\overset{(3)}{\le} C \lim_{t \to \infty} \left(\|U_x(\cdot, t_n)\|_{L^2}^2 + \int_{t_n}^{t} \|U_x(\cdot, \tau)\|_{L^2}^2 \right) \to 0 \quad (t_n \to \infty) \ .$$

Thus we concentrate on the proof of (5) where a weight function is essential in the non-convex case.

3. Motivation of the weight function

Our choice of the weight function is inspired by the choice of Matsumura and Nishihara for the non-convex scalar case [5]. Here we present the proof of (5) for the convex and non-convex scalar case to motivate the choice of our weight and to discuss some difficulties that arise in the case of a non-convex system.

For simplicity we assume $F \equiv 0$ in (4):

$$U_t + h'(\phi)U_x - \mu U_{xx} = 0 \ . \tag{8}$$

3.1. The scalar case with convex flux

Multiplying (8) by U and integrating by parts $\int_{-\infty}^{\infty} dx$ we obtain

$$\frac{1}{2}\frac{\partial}{\partial t}\int_{-\infty}^{\infty}|U|^2\,dx + \int_{-\infty}^{\infty}Uh'(\phi)U_x + \mu(U_x)^2\,dx = 0$$

and by integrating $\int_{t_0}^{T} dt$

$$\frac{1}{2}\|U(\cdot,T)\|_{L^2}^2 \;+\; \int_{t_0}^{T}\int_{-\infty}^{\infty}-\frac{1}{2}(h'(\phi))_x U^2\,dx\,dt$$

$$+\;\int_{t_0}^{T}\|U_x\|_{L^2}^2\,dt \;\leq\; \frac{1}{2}\|U(\cdot,t_0)\|_{L^2}^2 \;.$$

Together with the assumption of convexity $-(h'(\phi))_x > 0$, the inequality above implies (5).

3.2. The scalar case with non-convex flux

Multiplying (8) by $U \cdot w$, where $w = w(x)$, and integrating by parts $\int_{-\infty}^{\infty} dx$ we obtain

$$\frac{1}{2}\frac{\partial}{\partial t}\int_{-\infty}^{\infty}(w|U|^2)\,dx + \int_{-\infty}^{\infty}Uwh'(\phi)U_x + \mu U w_x U_x + \mu w(U_x)^2\,dx = 0$$

and by integrating $\int_{t_0}^{T} dt$

$$\frac{1}{2}\|\sqrt{w}U(\cdot,T)\|_{L^2}^2 \;+\; \int_{t_0}^{T}\int_{-\infty}^{\infty}-\frac{1}{2}(wh'(\phi)+w_x)_x U^2\,dx\,dt$$

$$+\;\int_{t_0}^{T}\|\sqrt{w}U_x\|_{L^2}^2\,dt \;\leq\; \frac{1}{2}\|\sqrt{w}U(\cdot,t_0)\|_{L^2}^2 \;.$$

Due to the presence of the weight w we get $-\frac{1}{2}(wh'(\phi)+w_x)_x$ in place of $-\frac{1}{2}(h'(\phi))_x$. Therefore the task is to find a positive w such that:

$$-\frac{1}{2}(wh'(\phi)+\mu w_x)_x > 0 \;.$$

Using the ansatz $w(x) = \tilde{w}(\phi(x))$ we get

$$-\frac{1}{2}(wh'(\phi)+\mu w_x)_x \;=\; -\frac{1}{2}(\tilde{w}(\phi)h'(\phi)+\mu\tilde{w}'(\phi)\phi_x)_x \tag{9}$$

$$=\; -\frac{1}{2}((\tilde{w}h)'(\phi))_x = -\frac{1}{2}(\tilde{w}h)''(\phi)\phi_x$$

and by choosing

$$\tilde{w}(u) = -\frac{(u-u_+)(u-u_-)}{h(u)}\cdot\operatorname{sign}\phi_x > 0 \tag{10}$$

we obtain

$$-\frac{1}{2}(wh'(\phi)+\mu w_x)_x = |\phi_x| \;.$$

Obviously, in the case of a system (9) is not valid and (10) has no meaning, because then $h(u)$ is a vector.

4. Weight for the system case with non-convex mode

Like Goodman [3] we diagonalize $h'(\phi)$. As f' (and hence h') is $\mathbb{R}$-diagonalizable in a neighborhood of the reference state u_* comprising the values of ϕ, we find smooth matrix valued functions

$$L(x) = \tilde{L}(\phi(x)), \quad R(x) = \tilde{R}(\phi(x))$$

such that $LR \equiv I$ and

$$L(h'(\phi))R = \operatorname{diag}(\lambda_1, \ldots, \lambda_n) ,$$

where

$$\lambda_p = \lambda - s, \quad \lambda_i < 0 < \lambda_j \quad (i < p < j)$$

We substitute $U =: RV$ in (4), multiply by $V^T W L$ and integrate $\int_{-\infty}^{\infty} dx$, where $W = \operatorname{diag}(1, \ldots, 1, w, 1, \ldots, 1)$:

$$\int_{-\infty}^{\infty} \frac{1}{2} \frac{\partial}{\partial t} (V^T W V) + V^T W \Lambda V_x + V^T W \Lambda L R_x V$$

$$+ \mu (V^T W L)_x (RV)_x - V^T W L F(\phi, (RV)_x) \, dx = 0 .$$

We group the terms above as

(A1) $\frac{1}{2} \frac{\partial}{\partial t} (V^T W V)$

(A2) $(w\lambda_p + \mu w_x) V_p (V_p)_x$

(A3) $\sum_{k \neq p} (-\frac{1}{2} (\lambda_k)_x + l_k (r_k)_x \lambda_k)(V_k)^2$

(A4) $\mu w ((V_p)_x)^2 + \mu \sum_{k \neq p} ((V_k)_x)^2$

(B1) $\sum_{j \neq p} (w\lambda_p + \mu w_x) l_p (r_j)_x V_p V_j$

(B2) $\sum_{i \neq p, i \neq j} \lambda_i l_i (r_j)_x V_i V_j$

(B3) $\mu V^T W L_x R V_x + \mu V_x^T W L R_x V$

(B4) $\mu V^T W L_x R_x V$

(B5) $-V^T W L F(\phi, (RV)_x)$

where $L = (l_1, \ldots, l_n)^T$ and $R = (r_1, \ldots, r_n)$.

The terms (B1)–(B5) have to be estimated; they mainly consist of coupling terms. (A1)–(A4) are what we expect from a decoupled system. In particular (A2) – the term referring to the primary field – corresponds to the scalar model case of Section 3.2. Our first task is thus to "find" a positive weight w such that

$$-\frac{1}{2} (w\lambda_p + \mu w_x) = |\phi_x| .$$

For this we can prove

Lemma 4.1. *For $\epsilon := |u_- - u_+|$ sufficiently small there exists $w : \mathbb{R} \to \mathbb{R}$ with $\inf_x w(x)$, $\inf_x (1/w(x)) > 0$ and*

$$-\frac{1}{2}(w\lambda_p + \mu w_x)_x = |\phi_x| \tag{11}$$

Proof. For fixed x_0, w_0 let w be the solution of

$$\mu w_x + w\lambda_p = -2 \int_{x_0}^{x} |\phi_x| \, dx \tag{12}$$

$$w(0) = w_0,$$

i. e.,

$$w(x) = e^{-\int_0^x \frac{\lambda_p(\xi)}{\mu} \, d\xi} \left(w_0 - \int_0^x e^{\int_0^y \frac{\lambda_p(\xi)}{\mu} \, d\xi} \frac{2}{\mu} \left[\int_{x_0}^y |\phi_x| \, d\xi \right] dy \right). \tag{13}$$

We now choose x_0 such that

$$I(x_0) \equiv \int_{-\infty}^{\infty} e^{\int_0^y \frac{\lambda_p(\xi)}{\mu} \, d\xi} \frac{2}{\mu} \left[\int_{x_0}^y |\phi_x| \, d\xi \right] dy = 0 \, ,$$

and correspondingly

$$w_0 = \int_0^{\pm\infty} e^{\int_0^y \frac{\lambda_p(\xi)}{\mu} \, d\xi} \frac{2}{\mu} \left[\int_{x_0}^y |\phi_x| \, d\xi \right] dy, \, .$$

By $\lambda_p(-\infty) > 0 > \lambda_p(\infty)$ the function I is well defined and $I(-\infty) > 0 > I(\infty)$ implies the existence of x_0. This choice is equivalent to the boundedness of w. By (13)

$$\lim_{x \to \pm\infty} e^{\int_0^x \frac{\lambda_p}{\mu} \, d\xi} \cdot w(x) = 0 \, ,$$

by (12)

$$\left(e^{\int_0^x \frac{\lambda_p}{\mu} \, d\xi} \cdot w \right)' = \frac{-2}{\mu} e^{\int_0^x \frac{\lambda_p}{\mu} \, d\xi} \int_{x_0}^{x} |\phi_x| \, dx \begin{cases} > 0 & \text{for } x < x_0 \\ = 0 & \text{for } x = x_0 \\ < 0 & \text{for } x > x_0 \, . \end{cases}$$

Consequently $e^{\int_0^x \frac{\lambda_p}{\mu} \, d\xi} \cdot w > 0$ and thus $w > 0$.
As $w(\pm\infty) = -2 \int_{x_0}^{\pm\infty} |\phi_x| dx / \lambda_p(\pm\infty) > 0$ by (12), both $\inf w$ and $\inf(1/w)$ are positive. $\qquad\square$

We now turn to the discussion of the coupling terms. Guided by the explicit choice (10) which was used for the non-convex scalar case we see that if the flux $h(\phi)$ is of the order ϵ^{-3} (a natural thing in the non-convex case) then the weight w is of the order ϵ^{-1}. On the other hand – for a system ϵ has to be chosen small to control the coupling between the different fields. Although this seems to cause difficulties, taking a closer look to our grouping of the coupling terms, we see that either w appears as $w\lambda_p + \mu w_x$ or as $w\phi_x$ where both can be estimated to be of order ϵ or ϵ^2. In the scalar case of Section 3.2 this can be explicitly verified. For (11) we can prove

Lemma 4.2. *For the weight function w of Lemma 4.1 the following holds:*

$$|w\lambda_p + \mu w_x| \leq O(1)|u_+ - u_-| \tag{14}$$
$$|\mu(w\phi_x)_x| = |(wh(\phi))_x| \leq O(1)|u_+ - u_-| \cdot |\phi_x| \tag{15}$$
$$|\mu w\phi_x| = |wh(\phi)| \leq O(1)|u_+ - u_-|^2 \tag{16}$$

For the proof we refer to [1].

By Lemma 4.2 the proof of (5) can be completed. E.g. (11) gives positivity of (A2) and (14), (15) and (16) enable estimating (B1), (B3) and (B4). With (A3) and (B2) we deal as suggested in [4] for the case of a convex system.

References

[1] Fries, C., *Nonlinear asymptotic stability of general small-amplitude viscous Laxian shock waves*, J. Differ. Equations, to appear.

[2] Fries, C., Doctoral Thesis, in preparation.

[3] Goodman, J., *Nonlinear Asymptotic Stability of Viscous Shock Profiles for Conservation Laws*, Arch. Rational Mech. Anal., **95** (1986), 325–344.

[4] Goodman, J., *Remarks on the Stability of Viscous Shock Waves*, Viscous Profiles and Numerical Methods for Shock Waves, Ed. M. Shearer, SIAM, Philadelphia, 1991, 66–72.

[5] Matsumura, A., Nishihara K., *Asymptotic Stability of Traveling Waves for Scalar Viscous Conservation Laws with Non-convex Nonlinearity*, Commun. Math. Phys., **165** (1994), 83–96.

[6] Szepessy, A., Xin, Z., *Nonlinear Stability of Viscous Shock Waves*, Arch. Rational Mech. Anal., **122** (1993), 53–103.

Institut für Mathematik
RWTH Aachen
D 52072 Aachen, Germany
E-mail address: `fries@instmath.rwth-aaachen.de`

International Series of Numerical Mathematics
Vol. 129, © 1999 Birkhäuser Verlag Basel/Switzerland

Global Correctness of Cauchy Problem for Nonlinear Conservation Laws Systems and one Example for the Gas Dynamics

V.A. Galkin

Abstract. The paper is devoted to the background of correctness for systems of nonlinear equations possessing applied significance in mathematical physics particularly in gas and fluid dynamics, Boltzmann and Smoluchovskii equations in physical kinetics. There are discussed general mathematical structures, connected with approximate methods convergence. The existence and uniqueness theorems for global solutions of Cauchy problem for quasilinear and semilinear systems are proved. The problems of computations in above models are discussed too.

1. Introduction

The mathematical models of physical systems, consisting of statistically plenty of particles (rare gases, dispersible systems, plasma) and continuous mechanics models are based on fundamental relations of the balance which general name is *conservation laws*. The significant quantity of modern researches on the conservation laws theory is connected with questions of correctness problem for nonlinear systems of differential and integrodifferential equations

$$\partial_t u^{(\omega)}(x,t) + \sum_{j=1}^{n} \partial_{x_j} f_j^{(\omega)}(u,x,t) = S^{(\omega)}(u,x,t),$$

$$x \in \mathbb{R}_n, \quad t > 0, \quad \omega \in \Omega, \tag{1}$$

where $u = \{u^{(\omega)}\}$ is unknown vector-function, the kind of flows f_j and source (collision operator) S are considered as given by character of simulated physical process, $x \in \mathbb{R}_n$ are space coordinates, t is the time, Ω are parameters, numbering equations. Below we shall name such systems of equations as *conservation laws systems*. Their applications are well-known, in particular, in connection with gas dynamics equations and hydrodynamics, physical kinetics equations by Boltzmann and Smoluchovskii, plasma theory etc.

Along with correctness in a circle of problems for the conservation laws (1) traditionally the special role are played by such problems of nonlinear mathematical physics as the background of passage to the limit and asymptotics for small parameters of approximate methods, used during search of unknown solution. Until recently the difficulties connected with passage to the limit in nonlinear (quasilinear and semilinear) systems of the conservation laws (1) for many methods seemed insuperable.

The extention of the concept of a solution (functional solutions) [1–5] makes it possible to justify the convergence of approximate methods in presence of an a priori estimate of an approximations in $L_1^{loc}(Q, \nu)$, which is uniform in the parameter.

The general idea is to obtain reasonable background for computations of solutions for equation (1) which is describing the movement of great number interacting particles whose behavior is very complicated. From this point of view the local characteristics based on vector of variables u (gas density, impulse density etc.) are usually irregular functions especially in presence of turbulence. In the last case the main role are playing mean values of physical variables related to the state-space-time volumes V in $\Omega \times \mathbb{R}_n \times \mathbb{R}_+$, namely

$$\int_{\Omega \times \mathbb{R}_n \times \mathbb{R}_+} I_V u^{(\omega)}(x, t) \mu \otimes d_x \otimes d_t,$$

where μ is the measure on particles states Ω and d_x, d_t are the Lebesgue measures on space-time variables and I_V is indicator function for volume V. The values of above integrals (functionals) for different indicator functions I_V make up the *functional solution* which exact definition is below. So the functional solutions concept is devoted to background of global correctness of Cauchy problem for equation (1) and global convergence of approximate methods to unique solution is proved provided weak approximation and weak stability take place for given approximate method $\mathcal{AM}$.

2. General definitions of functional solutions theory

Let Ω is locally compact separable metric space, being σ-finite concerning dense Borel measure μ, which is finite on compact sets, strictly positive on the open sets in Ω. (The Borel measure is called dense, if $\mu(E) = \sup_{E \supset K \in \mathcal{K}} \mu(K)$, where $\mathcal{K}$ is class of compact subsets from the space Ω). We shall denote by the symbols d_x and d_t the Lebesgue measures on $\mathbb{R}_n = \{x = (x_1, \ldots, x_n)\}$ and $\mathbb{R}_1 = \{t\}$ respectively; $\dot{B}$ is set of compactly supported bounded Borel functions on the topological product $Q = \Omega \times \mathbb{R}_n \times \mathbb{R}_1^+$; $Q_0 = \Omega \times \mathbb{R}_n$; $\dot{B}^\infty$ is set of infinitely differentiable on variable $(x, t) \in \mathbb{R}_n \times \mathbb{R}_1$ functions from the space $\dot{B}$, which derivatives lie in $\dot{e}$; $L_1^{loc}(A, \nu)$ is set of locally summable functions relatively to the Borel measure ν on the set $\mathcal{C}$.

Let on the set $\mathcal{D} \subset L_1^{loc}(\Omega, \nu)$ the following mappings are defined

$$f_j : \mathcal{D} \times \mathbb{R}_n \times \mathbb{R}_1^+ \to L_1^{loc}(\Omega, \nu), \quad 0 \le j \le n,$$

where $\nu = \mu \otimes d_x \otimes d_t$. On the functions family

$$M = \{u \in L_1^{loc}(Q, \nu) : u|_{(x,t) \in \mathbb{R}_n \times \mathbb{R}_1^+} \in \mathcal{D}, (f_j \circ u) \in L_1^{loc}(Q, \nu), 0 \le j \le n+1\},$$

a system of integral equations concerning unknown variable $u \in M$ is considered

$$\int_Q [u \partial_t g + \sum_{j=1}^n (f_j \circ u) \partial_{x_j} g + (f_{n+1} \circ u) g] \nu(dQ) +$$

$$\int_{Q_0} g|_{t=0} (f_0 \circ u_0)|_{t=0} \nu_0(dQ_0) = 0, \quad \forall g \in \dot{B}^\infty, \tag{2}$$

where the measure $\nu_0 = \mu \otimes d_x$, u_0 is given function from the set $L_1^{loc}(Q_0, \nu_0)$ such, that the superposition $(f_0 \circ u_0)|_{t=0}$ belongs $L_1^{loc}(Q_0, \nu_0)$. The system of integral equations (2) serves for definition of generalized solution $u \in M$ of the following Cauchy problem

$$J(u) \overset{\text{def}}{=} \partial_t f_0^{(\omega)}(u, x, t) + \sum_{j=1}^n \partial_{x_j} f_j^{(\omega)}(u, x, t) - f_{n+1}^{(\omega)}(u, x, t) = 0, \tag{3}$$

$$(\omega, x) \in Q_0, \quad t > 0,$$

$$u|_{t=0} = u_0, \quad (\omega, x) \in Q_0. \tag{4}$$

The most complete description of the correctness class for this problem was obtained in [6] when $card\ \Omega = 1$. For the case $card\ \Omega > 1$ the concept of the generalized solution determined by system (2) is restrictive from the point of view of the correctness justification for the Cauchy problem and computing methods. The expansion offered below for the solutions concept (functional solutions) permits to justify convergence of approximate methods at availability uniform on parameter a priori estimation of approximations in the space L_1^{loc}.

We shall denote by $\mathcal{E}$ the vector space consisting of the linear combinations

$$F_{g,g_1} = \sum_{j=0}^n f_j \partial_{x_j} g + f_{n+1} g + u g_1, \quad u \in \mathcal{D},$$

$$x \overset{\text{def}}{=} t,$$

with any $g \in \dot{B}^\infty, g_1 \in \dot{B}$. For each vector $F \in \mathcal{E}$ the operators π, π_0, π_1 are defined by the following relations

$$\pi(F) \overset{\text{def}}{=} F_{g,0}, \quad \pi_1(F) \overset{\text{def}}{=} g|_{t=0} f_0|_{t=0} + g_1|_{t=0} u, \quad \pi_0(F) = \pi_1(F_{g,0}).$$

Let $\mathcal{E}^+ = \{l\}$ is algebraic dual space to $\mathcal{E}(\mathcal{E}^+$ on definition consists of the finite linear functionals on the linear space $\mathcal{E})$. On $\mathcal{E}^+$ we shall set a topology $\sigma(\mathcal{E}^+, \mathcal{E})$ by means of a system of the seminorms $\{p_F\}_{F \in \mathcal{E}}$, where $p_F(l) = |l(F)|, \forall l \in \mathcal{E}^+, F \in \mathcal{E}$. The topological space $\mathcal{E}^+, \sigma(\mathcal{E}^+, \mathcal{E})$ is locally convex Hausdorff one.

In a specified topology each mapping $l \mapsto l(F), l \in \mathcal{E}^+, (\forall F \in \mathcal{E})$, is continuous.

We shall consider embedding of set $\hat{\imath}$ in $\mathcal{E}^+$, which we shall define by the formula

$$\forall u \in M : u \mapsto l_u \in \mathcal{E}^+, \qquad l_u(F) \overset{\text{def}}{=} \int_Q (F \circ u)\nu(dQ), F \in \mathcal{E}. \qquad (5)$$

It should note that integral in the right part of this formula for the functions $u \in M$ without fail finite as by virtue of finiteness of the test functions g and g_1 included in F integration is distributed only on a compact support of these functions.

Thus, the integrals for the terms which make integrand expression in (5) are finite because of their local summability, stipulated by requirements for the class $\hat{\imath}$ and by boundness of the test functions g, g_1. Thus, each functional l_u set by the formula (5) belongs to the space $\mathcal{E}^+$.

The embedding (5) is monomorphic on equivalence classes of functions from M coinciding almost everywhere on the set Q concerning the measure ν.

We denote $[M]$ the closure of an image of the embedding (5) for set $\hat{\imath}$ into the space $\mathcal{E}^+, \sigma(\mathcal{E}^+, \mathcal{E})$. On the set $[M]$ we shall consider the induced topology $\sigma(\mathcal{E}^+, \mathcal{E})$. By a similar way we shall enclose set of initial data M_0 consisting of the functions u_0 which together with superpositions $(f_0 \circ u_0)|_{t=0}$ belong to the space $L_1^{loc}(Q_0, \nu_0)$, into the space $\mathcal{E}_0^+$, being algebraic dual to the vector space $\mathcal{E}_0 \overset{\text{def}}{=} \pi_1(\mathcal{E})$. The topology $\sigma(\mathcal{E}_0^+, \mathcal{E}_0)$ which introduced on the set $\mathcal{E}_0^+$ is defined by family of the seminorms $\{p_F^{(0)}\}_{F \in \mathcal{E}}$, where $p_F^{(0)} = |l^{(0)}(\pi_1(F))|, l^{(0)} \in \mathcal{E}_0^+$. The appropriate embedding of the set $\hat{\imath}_0$ in $\mathcal{E}_0^+$ is executed by mapping

$$\forall u_0 \in M_0 : u_0 \mapsto l_{u_0}^{(0)} \in \mathcal{E}_0^+, \qquad l_u^{(0)}(F) \overset{\text{def}}{=} \int_{Q_0} (\pi_1(F) \circ u_0)\nu_0(dQ_0), \qquad F \in \mathcal{E}.$$

The topology $\sigma(\mathcal{E}_0^+, \mathcal{E}_0)$ is induced on the image of the set $\hat{\imath}_0$ in $\mathcal{E}_0^+$.

If the function $u \in M$ is the solution of the problem (2) (i.e. generalized solution of Cauchy problem (3), (4)) then the equalities (2) are equivalent to the following relations

$$l_u(\pi(F)) + l_{u_0}^{(0)}(\pi_0(F)) = 0, \quad \forall F \in \mathcal{E}. \qquad (6)$$

Definition 2.1. *The element $l \in [M]$ we shall name as the functional solution of equation (2) with the initial data $u_0 \in M_0$ if for each element $F \in \mathcal{E}$ the following equality holds*

$$l(\pi(F)) + l_{u_0}^{(0)}(\pi_0(F)) = 0. \qquad (7)$$

Hereinafter we shall consider the problem (7) concerning unknown $l \in [M]$ instead of relations (2), (3), (4), (6).

We shall speak that an *approximate method* (denoted as $\mathcal{AM}$) for solving of the problem (7) is given if for some directed parameters set A it is specified the choice of parametric elements family in the set $\hat{\imath}$ as generalized sequence

$$\alpha \mapsto u_\alpha \in \hat{\imath}, \qquad \alpha \in A.$$

Definition 2.2. *We shall name the functional solution $l \in [M]$ as regular if it is limit of an approximations set by regular method [4,5].*

Definition 2.3. *Method $\mathcal{AM}$ converges if it defines the converging generalized subsequence $\{l_{u_{\alpha_n}}\}$ in the space $[M], \sigma(\mathcal{E}^+, \mathcal{E})$, which limit is the functional solution of the problem (7).*

Definition 2.4. *We shall name a method $\mathcal{AM}$ as uniformly weakly stable if the following uniform estimation for approximations u_α takes place*

$$\sup_{\alpha \in A} \int_K |u_\alpha| \nu(dQ) < \infty, \tag{8}$$

on each compact set $K \subset Q$.

2. For the methods based on the solving of homogeneous linear difference schemes, the necessary condition (8) equivalent to the spectral stability criterion on Neumann which leads to the known Courant-Friedrichs-Levi conditions.

3. Convergence of approximate methods to the functional solution

The efficiency of functional solution concept is stipulated by its verifying and opportunity to prove the existence of the global functional solutions being by limit points of method with properties of weak approximation and weak stability.

Theorem 3.1. *Suppose the method $\mathcal{AM}$ is stable and there is property of weak approximation. Then there exists uniqueness class of global functional solutions which contain limit points of given method. So this method converges to unique functional solution of Cauchy problem for system (3).*

Theorem 3.2. *Every regular functional solution is defined by unique element of space $L_1^{loc}(Q, \nu)$.*

Remark The regular functional solutions are identified as locally summable functions $u \in L_1^{loc}$.

The proof of above theorems is based on consideration of approximations in Tikhonov topology constructed on the vector space for the system (1). The closure of locally summable functions in this topology produced by Young embedding contains global functional solutions as limit points [1–5]. There exists the directed sequence of values for regularization parameters when approximations are tending to the unique limit for all initial data under consideration in the Cauchy problem connected with the system (1). Thus the uniqueness classes for given approximated method $\mathcal{AM}$ are described.

According to the definition the functional solutions belong to closure of locally summable functions in the space of linear functionals equipped by Tikhonov topology. The functionals are defined on the vector space produced by weak integral form of conservation laws system and mean values of unknown which are related to the time-spatial sets. But the set of these limit points is not a convex closed set in this space and well known idea by R.DiPerna on application of Milman-Krein theorem to obtain uniqueness for measure-valued solutions can not be used in this case. The correctness classes are described for given approximate method $\mathcal{AM}$ due to existence of generalized sequence of approximations which is *the same for given class of initial data*. So we have obtained description for uniqueness classes of functional solutions which are limit points of approximations for given approximate method provided weak uniform stability and weak uniform approximation properties are established. The stability property is merely uniform boundness of approximations in L_1^{loc}.

More complicated is weak approximation property. But the most of reasonable methods have it.

We consider the exact global solutions of gas dynamics system

$$\partial_t \rho + \partial_x(\rho v) = 0,$$

$$\partial_t(\rho v) + \partial_x(\rho v^2 + R\rho T) = \partial_x(\mu \partial_x v),$$

$$\partial_t(\rho(v^2 + \frac{3}{2}RT) + \partial_x(\rho v(v^2 + \frac{3}{2}RT) + R\rho Tv - \frac{\mu}{2}\partial_x v^2) = \partial_x(\kappa \partial_x T),$$

$$t > 0, \qquad x \in \mathbb{R}, \qquad R, \mu, \kappa > 0,$$

consisting of regular function and singular Dirac measure for density ρ caused by initial shock of velocity v. Path to the limit $\mu \to 0, \kappa \to 0$ for this solution gives us the limit solution of functional type.

We consider the following exact solution for above problem

$$v(x,t) = -\operatorname{sgn}(x), \quad \rho(x,t) = 1 + 2t\delta_0, \quad t > 0, \quad x \in \mathbb{R},$$

$$T(x,t) = \begin{cases} \dfrac{\mu}{Rt}, & x = 0, \quad t > 0, \\ T_0, & x \neq 0, \quad t > 0. \end{cases}$$

References

[1] V.A. Galkin, *The functional solutions of conservation laws*, Doklady Acad. Nauk SSSR, 1990, v.310, N4, 834–839; English transl. in Sov. Phys. Dokl., **35**(2) (1990), 133–135.

[2] V.A. Galkin, *The methods of solving for the physical kinetics problems*, IATE Ed., (1995), 171,

[3] V.A. Galkin and V.V. Russkikh, *On the background of limit path for the Korteweg de Vries Equation as the Dispersion Vanishes*, Acta Applicande Matematicae, Kluwer Academic Publishers. Netherlands, **39** (1995), 307–314.

[4] V.A. Galkin, *The background of approximate methods for conservation laws systems*, Vestnik Mosk. Univ., (1995), N6, 55–59.

[5] V.A. Galkin, *The functional solutions theory of quasilinear conservation laws systems and its applications*, Proc. I.G.Petrowskii seminar, Moscow Univ. Ed., Vol.20 (1997), 81–120.

[6] S.N. Kruzkov, *First order quasilinear equations for many-dimensional independing variables*, Math. Sb., **81** (1970), 228.

Dept. of Applied Mathematics,
Institute of Atomic Energetics,
249020, RUSSIA
E-mail address: iate@storm.iasnet.com

International Series of Numerical Mathematics
Vol. 129, © 1999 Birkhäuser Verlag Basel/Switzerland

Asymptotics for Hyperbolic Equations with a Skew-symmetric Perturbation

Isabelle Gallagher

We are interested in symmetric hyperbolic systems of the type

$$(\mathcal{H}^\varepsilon) \quad \begin{cases} \partial_t v^\varepsilon + q(v^\varepsilon, v^\varepsilon) + \dfrac{L(v^\varepsilon)}{\varepsilon} &= 0 \quad \text{in} \quad \mathbf{T}^d \\[2mm] v^\varepsilon_{|t=0} &= v_0, \end{cases}$$

where v^ε and v_0 are d'-dimensional vector fields, defined respectively on $\mathbf{R} \times \mathbf{T}^d$ and $\mathbf{T}^d$, where $\mathbf{T}^d$ is a periodic box: functions defined on $\mathbf{T}^d$ are $2\pi b_j$-periodic in each of the d directions of $\mathbf{R}^d$, where the b_j's are strictly positive real numbers. The linear operator L is a skew-symmetric matrix of Fourier multipliers, defined, for all functions v, by $L(v) = \mathcal{F}^{-1} \sum_{a=1}^{d'} i\omega^a(n) \, (\mathcal{F}v(n), e^a(n)) \, e^a(n)$, where $(\cdot, \cdot)$ denotes the usual scalar product. We have written $\mathcal{F}v(n)$ for the discrete Fourier transform of v, the $\omega^a(n)$ are real numbers, and $(e^a(n))_{1 \leq a \leq d'}$ is a set of orthonormal vectors for all n. Finally $q(u, v)$ is a bilinear form, of the type

$$q(u, v) = (A(u) \cdot \nabla v + A(v) \cdot \nabla u) / 2, \tag{1}$$

where $A(u) = \left(A^j u \right)_{1 \leq j \leq d'}$, and for all j, $A^j u$ is a smooth, symmetric matrix. It can also be a form of the type $q(u, v) = P\left(A(u) \cdot \nabla v + A(v) \cdot \nabla u \right) / 2$, where P is the L^2-orthogonal projector onto the space of divergence free vector fields. In that case, we are not strictly speaking in the setting of hyperbolic equations; however, the energy estimates used are true for (1) as well as for that bilinear form.

Typical examples of such systems occur in fluid mechanics. For instance, the primitive equations of the quasigeostrophic system can be put in the form $(\mathcal{H}^\varepsilon)$ (see [2], [7], [11] for example). The three-component, divergence free, velocity field v^ε and the scalar temperature field T^ε satisfy, writing $V^\varepsilon = (v^\varepsilon, T^\varepsilon)$,

$$(PE^\varepsilon) \quad \partial_t V^\varepsilon + P(v^\varepsilon \cdot \nabla V^\varepsilon) + \varepsilon^{-1} P(-v_2^\varepsilon, v_1^\varepsilon, F^{-1}T^\varepsilon, -F^{-1}v_3^\varepsilon) = 0,$$

where F is a physical constant. Another example we shall be considering here is the slightly compressible Euler system. Let v^ε be the velocity field, and ρ^ε the density. If the fluid is isentropic, then the pressure field and the density are related by $p(\rho^\varepsilon) = \rho^{\varepsilon^\gamma}$, where $\gamma > 1$ is a constant. The system we shall consider gives the evolution of v^ε and of the sound speed c^ε, defined as $c^\varepsilon = \frac{2}{\gamma-1}\sqrt{\partial_{\rho^\varepsilon} p(\rho^\varepsilon)}$. Before writing the system, we linearize the problem by writing $c^\varepsilon = \bar{c}_0 + \varepsilon\, \tilde{c}^\varepsilon$, where $\bar{c}_0$ is a constant, and we get finally the following system for $V^\varepsilon = (v^\varepsilon, \tilde{c}^\varepsilon)$ (see [9], [10], [13]),

$$(CE^\varepsilon)\quad \begin{cases} \partial_t v^\varepsilon + v^\varepsilon \cdot \nabla v^\varepsilon + \bar{\gamma}\tilde{c}^\varepsilon\nabla\tilde{c}^\varepsilon + \bar{\gamma}\bar{c}_0\dfrac{\nabla\tilde{c}^\varepsilon}{\varepsilon} & = & 0 \quad \text{in} \quad \mathbf{T}^d \\[2mm] \partial_t\tilde{c}^\varepsilon + v^\varepsilon \cdot \nabla\tilde{c}^\varepsilon + \bar{\gamma}\tilde{c}^\varepsilon\mathrm{div}\, v^\varepsilon + \bar{\gamma}\bar{c}_0\dfrac{\mathrm{div}\, v^\varepsilon}{\varepsilon} & = & 0 \quad \text{in} \quad \mathbf{T}^d \\[2mm] V^\varepsilon_{|t=0} & = & V_0, \end{cases}$$

where $\bar{\gamma} = \frac{\gamma-1}{2}$, which is a system of the type $(\mathcal{H}^\varepsilon)$.

Equations of the type $(\mathcal{H}^\varepsilon)$ have been studied by a number of authors (see [6], [8], [12], [13]). It is convenient, for the rest of the study, to introduce the following operator (see [6], [13]): $\mathcal{L}(t)v = \mathcal{F}^{-1}\sum_{a=1}^d e^{-i\omega^a(n)t}\,(\mathcal{F}v(n), e^a(n))\, e^a(n)$, as well as to define the "filtered" associate of v^ε, $u^\varepsilon = \mathcal{L}(-\frac{t}{\varepsilon})v^\varepsilon$, which satisfies, defining $\mathcal{Q}^\varepsilon(u,v) = \mathcal{L}(-\frac{t}{\varepsilon})q\left(\mathcal{L}(\frac{t}{\varepsilon})u, \mathcal{L}(\frac{t}{\varepsilon})v\right)$,

$$(\tilde{\mathcal{H}}^\varepsilon)\quad \begin{cases} \partial_t u^\varepsilon + \mathcal{Q}^\varepsilon(u^\varepsilon, u^\varepsilon) & = & 0 \quad \text{in} \quad \mathbf{T}^d \\[2mm] u^\varepsilon_{|t=0} & = & v_0. \end{cases}$$

The following theorem sums up known results on $(\mathcal{H}^\varepsilon)$, and is proved in [13].

Theorem 1.1. *If $v_0 \in H^s(\mathbf{T}^d)$, with $s > \frac{d}{2}+3$, and if T_0 is the life span of the limit system*

$$(\mathcal{H}_0)\quad \begin{cases} \partial_t u + \mathcal{Q}(u,u) & = & 0 \quad \text{in} \quad \mathbf{T}^d \\[2mm] u_{|t=0} & = & v_0, \end{cases}$$

where $\mathcal{Q}(u,u)$ is the limit in $\mathcal{D}'$ of $\mathcal{Q}^\varepsilon(u^\varepsilon, u^\varepsilon)$, then for all $T < T_0$, and for ε small enough, the solution v^ε of the system $(\mathcal{H}^\varepsilon)$ is defined on $[0,T]$, and we have moreover $v^\varepsilon - \mathcal{L}(\frac{t}{\varepsilon})u = o(1)$ in $C^0([0,T], H^{s-1}(\mathbf{T}^d))$.

In this paper, we are interested in studying more precisely that $o(1)$ term. In order to do so, we shall need to control the three-wave interactions introduced by $\mathcal{L}$, given by the following definition.

Definition 1.2. *We will say that L satisfies a second order small divisor estimate of degree α if there exists a constant C_α such that for all (k,m) in $\mathbf{Z}^{2d}$, for all (a,b,c) in $\{1,\ldots,d'\}^3$, we have, if $\omega^a(k)+\omega^b(m) \neq \omega^c(k+m)$,*

$$|\omega^a(k)+\omega^b(m) - \omega^c(k+m)|^{-1} \leq C_\alpha\,(1+|k|)^\alpha\,(1+|m|)^\alpha.$$

The aim of this paper is to prove the following result.

Theorem 1.3. *Suppose the perturbation L satisfies a second order small divisor estimate of degree α. Then if $v_0 \in H^{s+\alpha}$, where $s > \frac{d}{2}+4$, and if T_0 is the life* *span of $(\mathcal{H}_0)$, then for all $T < T_0$ and for ε small enough, the filtered solution u^ε satisfies*

$$u^\varepsilon - u = \varepsilon\, h - \varepsilon\, R^\varepsilon_{osc}(u) + o(\varepsilon) \quad in \quad C^0([0,T], H^{s-3}(\mathbf{T}^d)),$$

where h is the solution in $C^0([0,T], H^{s-2}(\mathbf{T}^d))$ of a linearized limit equation

$$(\mathcal{H}_1) \qquad \partial_t h + 2\mathcal{Q}(u,h) = R(u)$$

where $R^\varepsilon_{osc}(u)$ is a highly oscillating term in time, and where both $R(u)$ and $R^\varepsilon_{osc}(u)$ can be computed explicitly in terms of u.

Remark 1.4. That theorem means that the first order correction between u^ε and u can be measured by a "slow" function h, independent of ε, and a purely oscillating function R^ε_{osc}, of the type $R^\varepsilon_{osc}(u) = \mathcal{F}^{-1}(\mathbf{1}_{\{n\,|\,\beta(n)\neq 0\}} e^{i\frac{t}{\varepsilon}\beta(n)} r(n,u))$, where β and r can be computed explicitly. The forcing term in the linearized limit equation $(\mathcal{H}_1)$ is due to the fact that when two highly oscillatory waves interact, the oscillations can cancel out, and that contributes to the slow part h in the first order correction.

Definition 1.5. *A function $R^\varepsilon_{osc}(t)$ will be said to be (p,σ)-oscillating if it can be written as $R^\varepsilon_{osc}(t) = \sum_{q\in\{1,\dots,p\}} \sum_{\vec{k}_q\in K^n_q} e^{-i\frac{t}{\varepsilon}\beta_q(n,\vec{k}_q)} r_0(n,\vec{k}_q) f_1(t,k_1)\dots f_q(t,k_q)$, where we have written $K^n_q = \{\vec{k}_q \in \mathbf{Z}^{dq} \mid \sum_{i=1}^q k_i = n \text{ and } \beta_q(n,\vec{k}_q) \neq 0\}$, and where we suppose that*

$$\exists\, (\alpha_i)_{1\leq i\leq q}, \ \alpha_i \geq 0, \quad such \ that \quad |r_0(n,\vec{k}_q)| \leq C\,(1+|k_1|)^{\alpha_1}\dots(1+|k_q|)^{\alpha_q};$$
$$\forall\, i \in [1,\dots,q], \quad \mathcal{F}^{-1}(f_i(t,\cdot)) \quad is \ an \ element \ of \quad C^0([0,T], H^{\sigma+\alpha_i}), \quad and$$
$$\exists\, \sigma_i > -\sigma \quad such \ that \quad \mathcal{F}^{-1}(\partial_t f_i(t,\cdot)) \quad is \ an \ element \ of \quad C^0([0,T], H^{\sigma_i}).$$

Remark 1.6. If $R^\varepsilon_{osc}(t)$ is a (p,σ) oscillating function, then in particular $\partial_t R^\varepsilon_{osc}$ is not bounded uniformly in ε, whereas $\varepsilon\,\partial_t R^\varepsilon_{osc}$ is.

Lemma 1.7. *Let $T > 0$ and $\sigma > \frac{d}{2} + 2$ be two real numbers, let b^ε be a family of functions bounded in the space $C^0([0,T], H^\sigma(\mathbf{T}^d))$, and let a^ε_0 be a function going to zero with ε in $H^{\sigma-1}(\mathbf{T}^d)$. Finally let R^ε_{osc} be a $(p,\sigma-1)$-oscillating function, with $p \geq 1$, and F^ε be a function going to zero with ε in $C^0([0,T], H^{\sigma-1}(\mathbf{T}^d))$. Then the function a^ε, solution of*

$$\begin{cases} \partial_t a^\varepsilon + \mathcal{Q}^\varepsilon(a^\varepsilon, b^\varepsilon) &= R^\varepsilon_{osc} + F^\varepsilon \quad in \quad \mathbf{T}^d \\ a^\varepsilon_{|t=0} &= a^\varepsilon_0, \end{cases} \qquad (2)$$

is an $o(1)$ in the space $C^0([0,T], H^{\sigma-1}(\mathbf{T}^d))$.

The proof of that lemma is based on a method introduced by S. Schochet in [13], enabling one to transform (2) into an equation where the oscillating part has disappeared, up to an ε, by a change of function. The idea behind that remark is

the same as the one transforming the model equation $\partial_t \varphi - F(\varphi) = \cos \frac{t}{\varepsilon}$ into the
equation $\partial_t \psi = F\left(\psi + \varepsilon \sin \frac{t}{\varepsilon}\right)$, where $\psi = \varphi - \varepsilon \sin \frac{t}{\varepsilon}$.

Then to prove Theorem 1.3., one notices that the function $w^\varepsilon = u^\varepsilon - u$
satisfies

$$\partial_t w^\varepsilon + \mathcal{Q}^\varepsilon(w^\varepsilon, w^\varepsilon + 2u) = \mathcal{Q}(u, u) - \mathcal{Q}^\varepsilon(u, u). \tag{3}$$

The non stationary phase theorem implies that $\mathcal{F}(\mathcal{Q}^\varepsilon(u,u) - \mathcal{Q}(u,u))(n)$ is equal
to

$$\sum_{\substack{k+m=n \\ \omega^a(k)+\omega^b(m) \neq \omega^c(n)}} e^{-i\frac{t}{\varepsilon}(\omega^a(k)+\omega^b(m)-\omega^c(n))} \left(P(n)A^j u^a(k)m_j u^b(m), e^c(n)\right) e^c(n),$$

where $u^a(n) = (\mathcal{F}u(n), e^a(n)) e^a(n)$, so the right hand side of (3) is an oscillating
function, in the sense of Definition 1.5. Then the repeated action of Lemma 2,
associated with transformations similar to the ones of the model problem above,
enable one to construct the expansion of Theorem 1.3.

Remark 1.8. Theorem 1.3. can be applied to the primitive equations (PE^ε), as well
as the compressible equations (CE^ε). One can prove, in both those cases, that the
perturbation operator satisfies a second order small divisor estimate , at some
order, for almost all periodic boxes $\mathbf{T}^d$; then one computes the limit system, and
Theorem 1.3. is applied directly to get the first order asymptotic expansion. In the
case of the compressible equations (CE^ε), the limit system can be separated into
two equations: if $U = (u, c)$ is the limit of $U^\varepsilon = \mathcal{L}(-\frac{t}{\varepsilon})V^\varepsilon$, then we decompose U
into $U = (\bar{u}, 0) + U_{osc}$, where $\bar{u}$ is the divergence free part of u, and we prove
that $\bar{u}$ satisfies the incompressible Euler equation, whereas U_{osc} satisfies a coupled
equation.

In the case of (PE^ε), one also can separate the limit equation into two parts:
one is an equation on U_{QG}, the projection of U onto the kernel of PA, and the other
is, for almost all periodic boxes,a linear equation on the remainder $U_{osc} = U - U_{QG}$.
That enables one to prove that the limit system of (PE^ε) is generically globally
well posed, which implies, by Theorem 1.3., that the life span of (PE^ε) goes to
infinity as ε goes to zero.

Let us note that precise proofs of the results given here can be found in [4].
Moreover, similar results in the parabolic case (typically when $-\Delta v^\varepsilon$, or any other
elliptic operator, is added to the system $(\mathcal{H}^\varepsilon)$) have been obtained in [5], and
announced in [3].

References

[1] A. Babin, A. Mahalov, and B. Nicolaenko, *Global Splitting, Integrability and Regularity of 3D Euler and Navier-Stokes Equations for Uniformly Rotating Fluids*, European Journal of Mechanics, **15 (3)** (1996), 291–300.

[2] J.-Y. Chemin, *A propos d'un problème de pénalisation de type antisymétrique*, Journal de Mathématiques Pures et Appliquées, **76** (1997), 739–755.

[3] I. Gallagher, *Existence globale pour des équations des fluides géostrophiques*, Notes aux Comptes-Rendus de l'Académie des Sciences de Paris, **325** (1997), Série I, 623–626.

[4] I. Gallagher, *Asymptotics of the Solutions of Hyperbolic Equations With a Skew-Symmetric Perturbation*, accepted for publication in Journal of Differential Equations, and Preprint of the Laboratoire d'Analyse Numérique, Paris 6, **1997**.

[5] I. Gallagher, *Applications of Schochet's Methods to Parabolic Equations*, accepted for publication in Journal de Mathématiques Pures et Appliquées, and Preprint of the Laboratoire d'Analyse Numérique, Paris 6, **1998**.

[6] E. Grenier, *Oscillatory Perturbations of the Navier-Stokes Equations*, Journal de Mathématiques Pures et Appliquées, **76** (1997), 477–498.

[7] D. Iftimie, *La résolution des équations de Navier-Stokes dans des domaines minces et la limite quasigéostrophique*, Thèse de l'Université Paris 6, **1997**.

[8] J.-L. Joly, G. Métivier and J. Rauch, *Generic Rigorous Asymptotic Expansions for Weakly Nonlinear Multidimensional Oscillatory Waves*, Duke Math. J., **70** (1993), 373–404.

[9] S. Klainerman and A. Majda, *Singular Limits of Quasilinear Hyperbolic Systems with Large Parameters, and the Incompressible Limit of Compressible Fluids*, Comm. Pure Appl. Math., **34** (1981), 481–524.

[10] H.-O. Kreiss, J. Lorenz, and M.J. Naughton, *Convergence of the Solutions of the Compressible to the Solutions of the Incompressible Navier-Stokes Equations*, Adv. Appl. Math., **12** (1991), 187–214.

[11] J.-L. Lions, R. Temam, and S. Wang, *Geostrophic Asymptotics of the Primitive Equations of the Atmosphere*, Top. Meth. Non Lin. Anal., **4** (1994), 1–35.

[12] A. Majda, *Compressible Fluid Flow and Systems of Conservation Laws in Several Space Variables*, Applied Mathematical Sciences, **53**, Springer Verlag, New York, **1984**.

[13] S. Schochet, *Fast Singular Limits of Hyperbolic PDEs*, Journal of Differential Equations, **114** (1994), 476–512.

Laboratoire d'Analyse Numérique, Paris 6
E-mail address: gallagher@ann.jussieu.fr

International Series of Numerical Mathematics
Vol. 129, © 1999 Birkhäuser Verlag Basel/Switzerland

Decomposition of the Elastic-plastic Wave Equation into Advection Equations

Guido Giese

Abstract. In this paper we present a numerical method of high order for solving the multidimensional elastic-plastic wave equation. The basic idea is to decompose the conservation law into advection equations which can be solved numerically. Furthermore, the occurrence of hysteresis makes it necessary to compute numerical approximations of the stress-strain relationship.

1. Introduction

In the last years, a large variety of numerical schemes have been developed for the simulation of hyperbolic conservation laws (s. e.g. [5]). Most of the methods were constructed for solving partial differential equations in fluid dynamics, e.g. the Euler equations. However, similar equations can be found in many other fields, e.g. in elasto-dynamics. The wave equation of an elastic material represents a hyperbolic conservation law for the momentum and the strain variables. For solids which undergo plastic deformation, only the momentum is conserved.

In this work, we follow the ansatz of Fey [1], [2] who developed a high order scheme called Method of Transport for solving the multidimensional Euler equations. The basic idea is to decompose the partial differential equation into scalar advection equations which can be solved numerically. Each of these advection equations represents a wave transporting some physical information.

Although the elasto-plastic wave equation is not a pure conservation law, we were able to use the ansatz of Fey to construct a genuinely multidimensional numerical scheme of high order for the "plastic" wave equation. In addition to the flux calculation which allows the update of strain and velocity variables in each time step, we have to construct an appropriate update for the stress variables.

In this paper, we first give a brief introduction into the governing equations of elasto-plasticity. In the second section, the basic idea of our scheme is explained with the help of a very simple example, the so called anti-plane shear waves. Furthermore, we shall show that in the general case where compression and shear waves occur, more physical knowledge has to be invested in order to construct a suitable decomposition.

2. Governing equations

We use a formulation of the elastic-plastic wave equation as a first order system, which means we have to use three kinds of variables:

- Velocities v_i $i = 1, \ldots, 3$
- stress tensor σ_{ij} $i, j = 1, \ldots, 3$
- strain tensor ϵ_{ij} $i, j = 1, \ldots, 3$

With these variables, we get three equations describing the Conservation of momentum (ρ = density):

$$\rho \frac{\partial}{\partial t} v_i = \sum_{j=1}^{3} \frac{\partial}{\partial x_j} \sigma_{ij} \qquad i = 1, \ldots, 3$$

and six Compatibility relations between velocity and strain variables

$$\frac{\partial}{\partial t} \epsilon_{ij} = \frac{1}{2} \left(\frac{\partial}{\partial x_j} v_i + \frac{\partial}{\partial x_i} v_j \right)$$

which are due to the fact that both strain and velocity variables are derivatives of the displacement vector u.

Altogether we obtain nine equations describing wave phenomena in solids. However, we still need an equation of state connecting stress and strain variables. We will discuss two cases:

- Linear relationship (Hooks' Law)

$$\epsilon_{ij} = \frac{1+\nu}{E} \sigma_{ij} - \frac{\nu}{E} \delta_{ij} \sigma_{kk} \tag{1}$$

  with ν = Poisson's ratio, E = Young's modulus.
- Plastic deformation: Due to Hysteresis, only a relationship between infinitesimal changes of stress and strain can be found:

$$\frac{d}{dt} \epsilon_{ij} = \frac{1+\nu}{E} \frac{d}{dt} \sigma_{ij} - \frac{\nu}{E} \delta_{ij} \frac{d}{dt} \sigma_{kk} + \sigma_{ij} \frac{d}{dt} \chi \tag{2}$$

  In order to distinguish between elastic and plastic deformation we use the so called von Mises yield condition.

For anti-plane shear, the von Mises yield function is

$$f(\sigma_{13}, \sigma_{23}) = \sigma_{13}^2 + \sigma_{23}^2 =: \kappa^2 \tag{3}$$

Furthermore, for this example the function $\frac{d}{dt}\chi$ can be written in the form

$$\dot{\chi} = \frac{1}{2\kappa^2} \left(\frac{1}{\mu_p(\kappa)} - \frac{1}{\mu} \right) (\sigma_{13} \dot{\sigma}_{13} + \sigma_{23} \dot{\sigma}_{23}) \tag{4}$$

with a function $\mu_p(\kappa) \leq \mu$ and μ the shear modulus. Hence, for anti-plane shear we have

$$\begin{pmatrix} \dot{\epsilon}_{13} \\ \dot{\epsilon}_{23} \end{pmatrix} = \begin{pmatrix} \frac{1}{2\mu} + \frac{1}{2\kappa^2}\left(\frac{1}{\mu_p(\kappa)} - \frac{1}{\mu}\right)\sigma_{13}^2 & \frac{1}{2\kappa^2}\left(\frac{1}{\mu_p(\kappa)} - \frac{1}{\mu}\right)\sigma_{13}\sigma_{23} \\ \frac{1}{2\kappa^2}\left(\frac{1}{\mu_p(\kappa)} - \frac{1}{\mu}\right)\sigma_{13}\sigma_{23} & \frac{1}{2\mu} + \frac{1}{2\kappa^2}\left(\frac{1}{\mu_p(\kappa)} - \frac{1}{\mu}\right)\sigma_{23}^2 \end{pmatrix} \begin{pmatrix} \dot{\sigma}_{13} \\ \dot{\sigma}_{23} \end{pmatrix} \tag{5}$$

For a numerical scheme it is necessary to integrate this stress-strain relationship.

3. First order decomposition for anti-plane shear

Under the condition for anti-plane shear, i.e.

$$\sigma_{11} = \sigma_{22} = \sigma_{33} = \sigma_{12} = \epsilon_{11} = \epsilon_{22} = \epsilon_{33} = \epsilon_{12} = v_1 = v_2 \equiv 0 \tag{6}$$

the conservation law reads

$$\begin{pmatrix} v_3 \\ \epsilon_{13} \\ \epsilon_{23} \end{pmatrix}_t = \begin{pmatrix} \sigma_{13}/\rho \\ \frac{1}{2}v_3 \\ 0 \end{pmatrix}_x + \begin{pmatrix} \sigma_{23}/\rho \\ 0 \\ \frac{1}{2}v_3 \end{pmatrix}_y \tag{7}$$

Using Hook's Law for linearly elastic deformation

$$\epsilon_{13} = \frac{1}{2\mu}\sigma_{13} \qquad \epsilon_{23} = \frac{1}{2\mu}\sigma_{23} \tag{8}$$

we obtain the linear hyperbolic equation

$$\begin{pmatrix} v_3 \\ \sigma_{13} \\ \sigma_{23} \end{pmatrix}_t = \begin{pmatrix} \sigma_{13}/\rho \\ \mu v_3 \\ 0 \end{pmatrix}_x + \begin{pmatrix} \sigma_{23}/\rho \\ 0 \\ \mu v_3 \end{pmatrix}_y \tag{9}$$

or simply

$$U_t + \nabla c L(U) = 0$$

where $c = \sqrt{\dfrac{\mu}{\rho}}$ is the wave speed.

If we choose a set of direction vectors $n_i \in \mathbb{R}^2$, $i = 1, \ldots, k$ we can rewrite equation (9) in the equivalent form:

$$\frac{1}{k}\sum_{i=1}^{k}\{\frac{\partial}{\partial t}(U + Ln_i) + \nabla c(U + Ln_i)n_i^T\} = 0 \tag{10}$$

provided the vectors n_i satisfy

$$\sum_{i=1}^{k} n_i = 0 \qquad \frac{1}{k}\sum_{i=1}^{k} n_i n_i^T = I$$

Equation (10) allows us to interpret the wave equation as a set advection equations transporting the quantity $R_i := U + Ln_i$ at speed c into direction n_i. Our numerical scheme consists of decoupling the system, i.e. solving each of the advection equations

$$(R_i)_t + \nabla(R_i n_i^T c) = 0 \qquad i = 1, \ldots, k \tag{11}$$

separately which is a first order approximation to the wave equation in 2D. In 1D, the decomposition into advection equations is exact!

For the elastic-plastic wave equation, we use a slightly different approach: Instead of inserting the stress-strain relationship (5) into the conservation law and decomposing the full wave equation, we decompose the conservation law (7) directly, which can be written in the form

$$V_t + \nabla c L(U) = 0 \tag{12}$$

with $V = (\rho w, \epsilon_{13}, \epsilon_{23})^T$ and $U = (\rho w, \sigma_{13}, \sigma_{23})^T$. The equivalent form of equation (12) is

$$\frac{1}{k} \sum_{i=1}^{k} \{ \frac{\partial}{\partial t}(V + Ln_i) + \nabla c(V + Ln_i)n_i^T \} = 0 \tag{13}$$

Again, we solve the advection equations (11) for the quantities $R_i := V + L(U)n_i$. However, solving these advection equations will only yield an update for V, i.e. the velocity and strain variables. Since we haven't used the stress-strain relationship in the decomposition, we have to integrate equation (2) after each time step in each cell in order to obtain the stress update. Solving equation (5) for anti-plane shear for $\dot{\sigma}_{13}$ and $\dot{\sigma}_{23}$ yields an ODE of the form

$$\begin{pmatrix} \dot{\sigma}_{13} \\ \dot{\sigma}_{23} \end{pmatrix} = \left(\begin{matrix} \frac{1}{2\mu} + \frac{1}{2\kappa^2}\left(\frac{1}{\mu_p(\kappa)} - \frac{1}{\mu}\right)\sigma_{13}^2 & \frac{1}{2\kappa^2}\left(\frac{1}{\mu_p(\kappa)} - \frac{1}{\mu}\right)\sigma_{13}\sigma_{23} \\ \frac{1}{2\kappa^2}\left(\frac{1}{\mu_p(\kappa)} - \frac{1}{\mu}\right)\sigma_{13}\sigma_{23} & \frac{1}{2\mu} + \frac{1}{2\kappa^2}\left(\frac{1}{\mu_p(\kappa)} - \frac{1}{\mu}\right)\sigma_{23}^2 \end{matrix} \right)^{-1} \begin{pmatrix} \dot{\epsilon}_{13} \\ \dot{\epsilon}_{23} \end{pmatrix} \tag{14}$$

which we solve with a Runge-Kutta method, starting at the point $(\sigma_{13}^n, \sigma_{23}^n)$ to the unknown point $(\sigma_{13}^{n+1}, \sigma_{23}^{n+1})$.

A problem arises if a cell changes from elastic to plastic deformation during a time step. Then, we have to use the elastic update according to equation (8) until we reach the yield surface and afterwards solve the ODE in (14) (s. Figure 1) up to the new point. Alternatively, one can use an integration along the approximate stress path $A \rightarrow B \rightarrow C$ according to Figure 1, which was first proposed by Lin in [6]. It can be proved that Lin's stress update reduces to a first order ODE-solver for equation (14) for plastic deformation (s. [4]).

It is interesting to note that for pure elastic waves, the decomposition of the form (12) together with the stress update according to Hook's law is equivalent to the decomposition for elastic waves of the form (10), so our scheme for plastic waves is a generalization to the scheme for elastic waves.

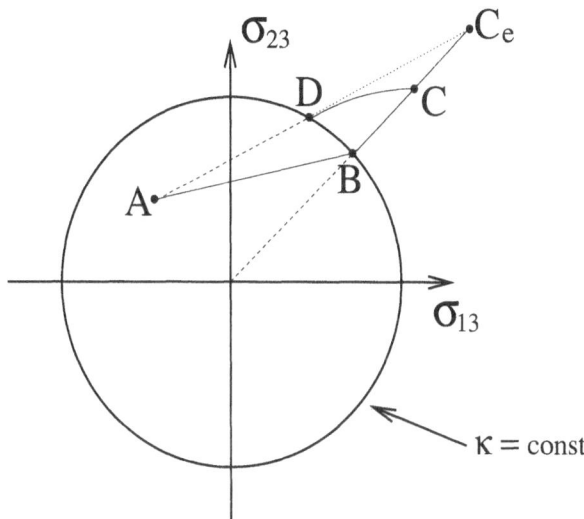

FIGURE 1. Stress update: At time level t^n point A lies in the elastic region. However, during the following time step, plasticity occurs which means that the elastic update i.e. point C_e lies outside the yield surface. In this case, an elastic update is used until point D and then an ODE has to be solved from D to the new stress point C. Alternatively, one can follow the idea of Lin and choose a stress path of the form $A \to B \to C$. For pure plastic deformation, i.e. $A = D$, Lin's update can be proved to be a first order ODE-solver for the stress-strain relationship.

4. Higher order approximation

The same approach can be extended to a higher order approximation. Therefore, we modify the transported quantities by adding some correcting matrix K to the flux L, so that the transported quantities become

$$R_i := V + (L + K)n_i \tag{15}$$

The matrix K can be found by comparing the Taylor expansion for V in time for the exact and the approximate solution:

$$V(x, t_0 + \Delta t) = V(x, t_0) + \Delta t V_t(x, t_0) + \frac{\Delta t^2}{2} V_{tt}(x, t_0) + \dots$$

$$+ \frac{\Delta t^N}{N!} V^{(N)}(x, t_0) + \mathcal{O}(\Delta t^{N+1})$$

For the exact solution, the time derivatives V_t^e, V_{tt}^e etc. can be found by expressing time derivatives by spatial derivatives via the exact partial differential equation.

Therefore, we can write equation (7) in the form

$$V_t = AU_x + BU_y$$

with constant matrices A and B. Furthermore, we write the stress-strain relationship in the form

$$U_t = \mathcal{A}(\sigma_{13}, \sigma_{23})V_t$$

Hence, the time derivatives are

$$
\begin{aligned}
V_t^e &= AU_x + BU_y = \nabla\{[A, B]U\} \\
V_{tt}^e &= A(U_t)_x + B(U_t)_y = \nabla\{[A, B]U_t\} = \nabla\{[A, B](AV_t)\} \\
V_{ttt}^e &= \nabla\{[A, B](\dot{A}V_t + AV_{tt})\}
\end{aligned}
$$

where $[A, B]$ is a matrix vector and $[A, B]U$ is defined as the matrix $[AU, BU]$. For our approximation V^a, the corresponding time derivatives V_t^a, V_{tt}^a can be found by our scheme to be:

$$V_t^a = \frac{1}{k}\sum_{i=1}^{k}(R_i)_t, \qquad V_{tt}^a = \frac{1}{k}\sum_{i=1}^{k}(R_i)_{tt}$$

and the derivatives of the quantities R_i can be found using the advection equations:

$$
\begin{aligned}
(R_i)_t &= -\nabla(cR_in_i^T) \\
(R_i)_{tt} &= -\nabla(c(R_i)_tn_i^T) = \nabla(\nabla(cR_in_i^T)cn_i^T)
\end{aligned}
$$

Comparing the Taylor expansions of our scheme to the expansion of the exact solution allows us to choose the correction matrix K in such a way that our approximation is of fourth order for elastic waves and third order for plastic waves. For example, order $\mathcal{O}(\Delta t^3)$ i.e. second order can be achieved by choosing the correction matrix K as

$$\nabla K = \frac{\Delta t}{2}(V_{tt}^e - V_{tt}^a) \tag{16}$$

5. Plane strain condition

The Plane strain condition can be characterized as follows:

$$\sigma_{13} = \sigma_{23} = \epsilon_{13} = \epsilon_{23} = \epsilon_{33} = v_3 \equiv 0$$

We discuss the elastic wave equation for plane strain only, since the generalization for plastic waves can be done analogously to the case of anti-plane shear. With the bulk modulus K the wave equation reads

$$
\begin{pmatrix}
v_1 \\
v_2 \\
\sigma_{11} \\
\sigma_{22} \\
\sigma_{33} \\
\sigma_{12}
\end{pmatrix}_t
= \nabla
\begin{pmatrix}
\frac{1}{\rho}\sigma_{11} & \frac{1}{\rho}\sigma_{12} \\
\frac{1}{\rho}\sigma_{12} & \frac{1}{\rho}\sigma_{22} \\
(K+4/3\mu)v_1 & (K-2/3\mu)v_2 \\
(K-2/3\mu)v_1 & (K+4/3\mu)v_2 \\
(K-2/3\mu)v_1 & (K-2/3\mu)v_2 \\
\mu v_2 & \mu v_1
\end{pmatrix}
\tag{17}
$$

or simply

$$U_t + \nabla c L(U) = 0$$

The basic difference to anti-plane shear waves is the existence of two different wave modes:

- Compression waves with speed $c_1 = \sqrt{\dfrac{K+4/3\mu}{\rho}}$
- Shear waves with speed $c_2 = \sqrt{\dfrac{\mu}{\rho}}$

Of course, the same decomposition as above

$$\frac{1}{k} \sum_{i=1}^{k} \left\{ \frac{\partial}{\partial t}(U + Ln_i) + \nabla c(U + Ln_i)n_i^T \right\} = 0 \tag{18}$$

with $c \geq c_1 > c_2$ would work out. However, it turns out that this ansatz leads to poor numerical results in first and second order. The reason for this is obvious: Our decomposition (18) does not distinguish between information transported at the two different speeds c_1 and c_2. Our goal is to transport the compression and shear waves at the corresponding wave-speeds c_1 and c_2. In order to achieve such a decomposition, it is necessary to use a different decomposition for each component of equation (17).

Let us analyse for instant the first component of equation (17):

$$(v_1)_t = \nabla \left(\frac{\sigma_{11}}{\rho}, \frac{\sigma_{12}}{\rho} \right) \tag{19}$$

A one dimensional wave propagating along the x-axis would transport the quantity $v_1 + \frac{\sigma_{11}}{c_1 \rho}$ in the right eigenvector at the speed c_1, whereas a one dimensional wave along the y-axis transports $v_1 + \frac{\sigma_{12}}{c_2 \rho}$ with c_2. Our genuinely two-dimensional decomposition uses four wave directions n_i, $i = 1, \ldots, 4$ with

$$n_i = \begin{pmatrix} \pm 1 \\ \pm \frac{c_2}{c_1} \end{pmatrix} \tag{20}$$

and the decomposition of equation (19) is

$$\left(v_1 + \left(\frac{1}{c_1} \frac{\sigma_{11}}{\rho}, \frac{c_1}{c_2^2} \frac{\sigma_{12}}{\rho} \right) n_i \right)_t - \nabla c_1 \left(v_1 + \left(\frac{1}{c_1} \frac{\sigma_{11}}{\rho}, \frac{c_1}{c_2^2} \frac{\sigma_{12}}{\rho} \right) n_i \right) n_i^T = 0 \qquad \forall i \tag{21}$$

which means that we transport the quantity

$$v_1 + \left(\frac{1}{c_1} \frac{\sigma_{11}}{\rho}, \frac{c_1}{c_2^2} \frac{\sigma_{12}}{\rho} \right) n_i$$

at the speed vectors

$$c_1 n_i = \begin{pmatrix} \pm c_1 \\ \pm c_2 \end{pmatrix}$$

Decomposition for Plain Strain		
abbreviation: $a = K + 4/3\mu$; $b = K - 2/3\mu$		
Equation	Transported quantity	Wave model
$\dfrac{\partial v_1}{\partial t} = \nabla\left(\dfrac{\sigma_{11}}{\rho}, \dfrac{\sigma_{12}}{\rho}\right)$	$v_1 + \left(\dfrac{1}{c_1}\dfrac{\sigma_{11}}{\rho}, \dfrac{c_1}{c_2^2}\dfrac{\sigma_{12}}{\rho}\right)n_i$	$c_1 n_i^T = (\pm c_1, \pm c_2)$
$\dfrac{\partial v_2}{\partial t} = \nabla\left(\dfrac{\sigma_{12}}{\rho}, \dfrac{\sigma_{22}}{\rho}\right)$	$v_2 + \left(\dfrac{c_1}{c_2^2}\dfrac{\sigma_{12}}{\rho}, \dfrac{1}{c_1}\dfrac{\sigma_{22}}{\rho}\right)n_i$	$c_1 n_i = \left(\begin{smallmatrix}\pm c_2\\ \pm c_1\end{smallmatrix}\right)$
$\dfrac{\partial \sigma_{11}}{\partial t} = \nabla(av_1, bv_2)$ $\dfrac{\partial \sigma_{22}}{\partial t} = \nabla(bv_1, av_2)$ $\dfrac{\partial \sigma_{33}}{\partial t} = \nabla(bv_1, bv_2)$	$\sigma_{11} + \dfrac{1}{c_1}(av_1, bv_2)n_i$ $\sigma_{22} + \dfrac{1}{c_1}(bv_1, av_2)n_i$ $\sigma_{33} + \dfrac{1}{c_1}(bv_1, bv_2)n_i$	$c_1 n_i^T = (\pm c_1, \pm c_1)$
$\dfrac{\partial \sigma_{12}}{\partial t} = \nabla(\mu v_2, \mu v_1)$	$\sigma_{12} + \dfrac{1}{c_2}(\mu v_2, \mu v_1)n_i$	$_1 n_i^T = (\pm c_2, \pm c_2)$

TABLE 1. Decomposition of the elastic wave equation.

Our decomposition is consistent with the exact equation (19) since we have

$$\frac{1}{4}\sum_{i=1}^{4}\left(v_1+\left(\frac{1}{c_1}\frac{\sigma_{11}}{\rho},\frac{c_1}{c_2^2}\frac{\sigma_{12}}{\rho}\right)n_i\right)_t = (v_1)_t$$

$$\frac{1}{4}\sum_{i=1}^{4}\nabla c_1\left(v_1+\left(\frac{1}{c_1}\frac{\sigma_{11}}{\rho},\frac{c_1}{c_2^2}\frac{\sigma_{12}}{\rho}\right)n_i\right)n_i^T = \nabla\left(\frac{\sigma_{11}}{\rho},\frac{\sigma_{12}}{\rho}\right)$$

A similar analysis can be done for the other components. Table 1 on page 382 shows the wave models we use for all components.

6. Numerical example

As an example, we consider a cracked plate under the plain strain condition (s. Figure (2)). When the two compression waves arrive simultaneously at the crack tip, a singularity in the stress component σ_{22} will occur.
Figure (2) shows a third order solution of the stress distribution of $\sigma_{22}(x,y,t = 0.57)$ with $c_1 = 1$ and $c_2 = \frac{1}{\sqrt{3}}$ and a resolution of 200×100 points:
   The singularity in the stress is very well resolved.

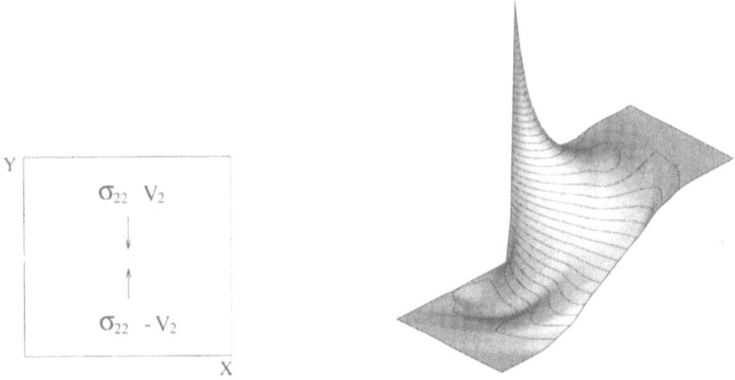

FIGURE 2.
Left Figure: Two compression waves approaching a crack.
Right Figure: Stress distribution σ_{22} in half of the plate.

References

[1] M. Fey. *Ein echt mehrdimensionales Verfahren zur Lösung der Eulergleichungen*. PhD thesis, ETH Zürich, 1993

[2] M. Fey. *Decomposition of the multidimensional Euler equations into advection equations*. Seminar for Applied Mathematics, ETH Zurich. Research Report No. 95–14

[3] M. Fey, R. Jeltsch, A. -T. Morel. *Multidimensional Schemes for Non-Linear Systems of Hyperbolic Conservation Laws*. In D. Griffiths and G.A. Watson, editors, *Numerical Analysis 1995*. Longman, 1996

[4] G. Giese *Numerical Solution of anti-plane shear waves*.

[5] R. J. LeVeque *Numerical Methods for Conservation Laws. Lectures in Mathematics*. Birkhäuser, 1992.

[6] X. Lin. *Numerical Computation of stress waves in solids*. Akademie Verlag, 1996

Seminar for Applied Mathematics,
ETH Zürich,
CH-8092, Switzerland
E-mail address: ggiese@sam.math.ethz.ch

International Series of Numerical Mathematics
Vol. 129, © 1999 Birkhäuser Verlag Basel/Switzerland

Adaptive Finite Volume Schemes for Conservation Laws Based on Local Multiresolution Techniques

Birgit Gottschlich-Müller and Siegfried Müller

Abstract. A new adaptive strategy for finite volume schemes is presented based on local multiscale decompositions. By this approach, the complexity of the scheme can be restricted to the number of significant coefficients. For one–dimensional, scalar conservation laws, the efficiency of the concept is verified for several test configurations.

1. Introduction

We consider partial differential equations of the form

$$\frac{\partial\, u(t,x)}{\partial\, t} + \frac{\partial\, f(u(t,x))}{\partial\, x} = 0, \qquad 0 < t < T,\ a < x < b \tag{1}$$

with $T > 0$ and $a, b \in \mathbb{R}$ fixed, the so-called conservation laws. Herein, $u : [a, b] \to \mathbb{R}$ is the conservative quantity and $f : \mathbb{R} \to \mathbb{R}$ the flux. In order to ensure a unique solution, an initial condition $u(0, x) = u_0(x)$, $a < x < b$, has to be imposed. In practical applications, boundary conditions of the form $g_a(u(t,a)) = 0$, $g_b(u(t,b)) = 0$, $0 \le t \le T$ are also needed. The one-dimensional scalar equation (1) is a simple model for more complex systems arising from fluid dynamics, e.g. Euler equations.

During the last decade, finite volume schemes have been proven to be very robust and reliable approximation schemes for conservation laws as has been verified by numerous applications. However, the construction of robust and efficient schemes is not yet finished. More advanced approaches are based on the heterogeneity of the flow field, i.e., several physical effects proceeding on different spatial and temporal scales have to be appropriately resolved. In order to resolve these effects by a numerical scheme, different discretization lengths are required where the highest resolution is determined by the finest scale. In the case of multidimensional flow fields this will, in general, exceed presently available memory capacities or computational time will become prohibitively long. Therefore, algorithms have to be developed with computational effort and memory capacity proportional to the complexity of the underlying problem, i.e., only in regions where small-scale effects are present the discretization has to be chosen sufficiently fine, and elsewhere a much coarser discretization will be sufficient. In view of realistic simulations of multidimensional flow fields around complex geometries uniform discretizations are no longer adequate since they exceed presently accessible resources of modern computer systems. In principle, new computer generations will not overcome this problem. Therefore, discretizations have to be adaptive rather than uniform, in order to utilize efficiently the available resources. Nowadays, numerous software packages especially applied in engineering are not in general based on adaptive strategies, although adaptive concepts are developed and investigated for some time in basic research. Herein, adaptivity is controlled by local gradients since error estimates are, in general, not available for problems of interest in practice. The quantity by which adaptivity is performed depends, in general, on the problem configuration at hand. Considering another problem, this quantity can turn out to be useless. Hence, concepts which are independent of the underlying problem have to be developed, i.e., concepts which can be universally applied and are capable to resolve the different scales. To this end, techniques are required by which information can be decomposed into the different scales. A well-known example is the Fourier expansion of a signal. More advanced techniques are based on wavelets. The incorporation of wavelets into the discretization of partial differential equations is subject of current research, yet it is still in the beginning, at least as far as applications in engineering are concerned.

In the present paper, we introduce an new concept how to incorporate multiscale techniques into already existing finite volume schemes in order to achieve a genuinely adaptive scheme. This approach represents a significant extension of recent investigations in this field based on hybrid techniques (cf. [5, 7, 1, 4]). Firstly, we describe a multiscale decomposition applied to cell averages. This decomposition is incorporated into an arbitrary finite volume scheme. Several test calculations have been performed verifying the efficiency of our adaptive strategy. In the end, we give a short outlook on future work and possible improvements.

2. Multiscale decomposition

In the following, we summarize the derivation of a multiscale decomposition for a sequence of discrete averages corresponding to an integrable function $u : \Omega \to \mathbb{R}$ defined on Ω representing the computational domain. In our applications, we choose $\Omega = [a, b] \subset \mathbb{R}$. However, the presented concept also holds for $\Omega \subset \mathbb{R}^d$, $d > 1$. The starting point is a sequence of $L + 1$ decompositions of Ω corresponding to increasingly higher resolution, i.e., $\Omega = \bigcup_{j \in I_k} V_{k,j}$, $k = 0, \ldots, L$, where k indicates the level of resolution with $k = 0$ denoting the coarsest discretization and $k = L$ the finest, and j represents the position of the cell corresponding to this level. The only assumption we have to impose on this decomposition is the nestedness of the sequence, i.e., $V_{k,j} = \bigcup_{l \in D_{k,j}} V_{k+1,l}$, $D_{k,j} \subset I_{k+1}$, $j \in I_k$, which means that a coarse-grid cell is composed of $\#D_{k,j}$ fine-grid cells. To any of these cells we now define the cell average

$$u_{k,j} := \frac{1}{|V_{k,j}|} \int_{V_{k,j}} u(x)\, dV, \quad j \in I_k, \ k = 0, \dots, L. \tag{2}$$

The main task is to determine a completion of $\mathrm{I\!R}^{\# I_k}$ to $\mathrm{I\!R}^{\# I_{k+1}}$ such that

$$u_{k+1,j} = \sum_{r \in I_k} m_{j,r}^{k,0} u_{k,r} + \sum_{r \in J_k} m_{j,r}^{k,1} d_{k,r}, \quad j \in I_{k+1}, \tag{3}$$

holds and there exist coefficients $g_{j,r}^{k,0}$ and $g_{j,r}^{k,1}$ such that

$$u_{k,j} = \sum_{r \in I_{k+1}} g_{j,r}^{k,0} u_{k+1,r}, \quad j \in I_k, \tag{4}$$

$$d_{k,j} = \sum_{r \in I_{k+1}} g_{j,r}^{k,1} u_{k+1,r}, \quad j \in J_k, \tag{5}$$

where J_k is the index set corresponding to the coefficients $d_{k,j}$ such that $\#J_k = \#I_{k+1} - \#I_k$. The coefficients $d_{k,j}$ may be seen as difference information between the data of level k and $k+1$, which can be interpreted as prediction error.

From a numerical point of view, we have to impose four reasonable conditions to the coefficients of $\mathbf{M}_{k,i} = \{m_{j,r}^{k,i}\}_{j,r}$ and $\mathbf{G}_{k,i} = \{g_{j,r}^{k,i}\}_{j,r}$, $i = 0, 1$:

(C1) the representation (3) and (4, 5) have to be equivalent, i.e.,

$$\mathbf{G}_{k,r} \mathbf{M}_{k,s} = \delta_{r,r} \mathbf{I}, \ r, s = 0, 1, \quad \text{and} \quad \mathbf{M}_{k,0}\, \mathbf{G}_{k,0} + \mathbf{M}_{k,1}\, \mathbf{G}_{k,1} = \mathbf{I};$$

(C2) the matrices $\mathbf{M}_{k,i}$, $\mathbf{G}_{k,i}$, $i = 0, 1$ have to be uniformly sparse;
(C3) the matrices $\mathbf{M}_k = (\mathbf{M}_{k,0}, \mathbf{M}_{k,1})$, $\mathbf{G}_k^T = (\mathbf{G}_{k,0}^T, \mathbf{G}_{k,1}^T)$ have to be spectrally bounded uniformly in k;
(C4) the identity $\mathbf{G}_{k,1}\, \mathbf{1} = \mathbf{0}$ holds.

The conditions (C1), (C2) and (C3) guarantee an efficient and stable transformation of the single-scale representation $\mathbf{u}_L = \{u_{L,j}\}_{j \in I_L}$ into the multiscale representation composed of the coarse-scale data $\mathbf{u}_0 = \{u_{0,j}\}_{j \in I_0}$ and a sequence of difference data $\mathbf{d}_k = \{d_{k,j}\}_{j \in J_k}$, $k = 0, \dots, L-1$, obtained by successive application of (4,5). This transformation can be conversed by (3). The most important feature is condition (C4), which means that the difference coefficients $d_{k,j}$ become small or even vanish in regions where the data vary moderately. Due to this property data compression can be performed. We will see later on how to incorporate this into an adaptive strategy. For the construction of appropriate coefficients, we refer to those known from literature (see e.g. [5, 4, 2, 3]). We want to emphasize that, by the definition of the average (2) and the nestedness of the grids, the coefficients $g_{j,r}^{k,0}$ are uniquely determined by the grid hierarchy.

3. Multiscale transformation and adaptivity

The single-scale representation $\mathbf{u}_L$ can be transformed into a multiscale representation $\mathbf{u}_{L,MR} = (\mathbf{u}_0, \mathbf{d}_0, \dots, \mathbf{d}_{L-1})$ applying (4,5) and this operation is conversed using (3). The complexity of this transformation, i.e., the number of operations, is proportional to the number of cells $\# I_L$ of the finest grid, which is optimal, if you are interested in an *exact* transformation. From a numerical point of view, this is no longer desirable, since the finest resolution is only needed where strong variations locally occur in the data and elsewhere the changes are moderate. Hence, we are not in general interested in all data of the finest level, but in single-scale coefficients $u_{k,j}$ corresponding to different levels of resolution. In terms of the multiscale coefficients $d_{k,j}$ this means that some of them vanish or become negligibly small. In the following, we will explain how to reduce the complexity of the transformation to the number of non-vanishing multiscale coefficients. To this end, we introduce the sequence of *truncated* multiscale coefficients $\tilde{\mathbf{u}}_{L,MR} = (\mathbf{u}_0, \tilde{\mathbf{d}}_0, \ldots, \tilde{\mathbf{d}}_{L-1})$ where the vectors $\tilde{\mathbf{d}}_k = \{\tilde{d}_{k,j}\}_{j \in \tilde{J}_k}$ are composed of the significant coefficients, i.e., $\tilde{d}_{k,j} = d_{k,j}$, $j \in \tilde{J}_k \subset J_k$, iff $|d_{k,j}| > \varepsilon_k$, where the ε_k are fixed, level-depending tolerances. The number of significant coefficients is given by $\# I_0 + \sum_{k=0}^{L-1} \# \tilde{J}_k$ which can be much smaller than the number of all coefficients $\# I_0 + \sum_{k=0}^{L-1} \# J_k$. Corresponding to this truncated multiscale decomposition $\tilde{\mathbf{u}}_{L,MR}$ we define an *adaptive* single-scale representation $\tilde{\mathbf{u}}_{L,SR} = (\tilde{\mathbf{u}}_0, \ldots, \tilde{\mathbf{u}}_L)$ with $\tilde{\mathbf{u}}_k = \{\tilde{u}_{k,j}\}_{j \in \tilde{I}_k}$ composed of the significant coefficients $\tilde{u}_{k,j} = u_{k,j}$, $j \in \tilde{I}_k \subset I_k$. Obviously, the number of adaptive single-scale coefficients $\sum_{k=0}^{L} \# \tilde{I}_k$ can be much larger than $\# I_L$ in the worst case if $\tilde{I}_k = I_k$. Therefore, no one-to-one transformation between $\tilde{\mathbf{u}}_{L,SR}$ and $\tilde{\mathbf{u}}_{L,MR}$ necessarily exists as there is between $\mathbf{u}_{L,MR}$ and $\mathbf{u}_L$. For this purpose, we have to characterize the representation $\tilde{\mathbf{u}}_{L,SR}$. This is based on the fact that we are dealing with averages. Therefore, we demand that the representation is

(A1) *adaptive*, i.e.,

$$\Omega = \bigcup_{k=0}^{L} \bigcup_{j \in \tilde{I}_k} V_{k,j} \text{ with meas}(V_{k,j} \cap V_{k',j'}) = 0, \ (k,j) \neq (k',j') \text{ and}$$

(A2) *compressed*, i.e., $\left(\forall r \in J_{k-1} : m_{j,r}^{k-1,1} \neq 0 \Rightarrow r \notin \tilde{J}_{k-1} \right) \Rightarrow j \notin \tilde{I}_k$.

The basic idea is to define an *associated* grid from the cells $V_{k,j}$ corresponding to the averages $\tilde{u}_{j,k}$, $j \in \tilde{I}_k$, $k = 0, \ldots, L$. By condition (A1), this grid is a decomposition of the computational domain Ω. Since the cells belong to different levels, this grid is adaptive. Which cells are chosen for the associated grid, mainly depends on condition (A2). By this condition, the adaptive single-scale and multiscale representations are connected, since $r \notin \tilde{J}_{k-1}$ is equivalent to $|d_{k-1,r}| < \varepsilon_{k-1}$. Demanding condition (A2) without (A1), may result in an associated grid with overlapping cells and (or) gaps. These effects are excluded by (A1). Furthermore, we conclude that $u_{k,j}$, $j \in \tilde{I}_k$, can be reconstructed by the coarse-grid data, i.e., $u_{k,j} = \sum_{r \in I_{k-1}} m_{j,r}^{k-1,0} u_{k-1,r}$, where the error is proportional to ε_{k-1}.

In our application to finite volume schemes, we have to switch between adaptive single-scale and multiscale representation. The transformations are summarized in the following. The encoding, i.e., $\tilde{\mathbf{u}}_{L,SR} \to \tilde{\mathbf{u}}_{L,MR}$, consists of three steps for each level k where we initialize $\hat{I}_L := \tilde{I}_L$ and proceed in descending order:

(E1) $U_k^0 := \bigcup_{j \in \hat{I}_{k+1}} \left\{ r \in I_k : g_{r,j}^{k,0} \neq 0 \right\}$, $U_k^1 := \bigcup_{j \in \hat{I}_{k+1}} \left\{ r \in J_k : g_{r,j}^{k,1} \neq 0 \right\}$,

(E2) $u_{k,j} = \sum_{r \in E_{k,j}^0} g_{j,r}^{k,0} u_{k+1,r}, \quad j \in U_k^0 \backslash \tilde{I}_k,$

$d_{k,j} = \sum_{r \in E_{k,j}^1} g_{j,r}^{k,1} u_{k+1,r}, \quad j \in U_k^1,$

(E3) $\tilde{J}_k := \{j \in U_k^1 \; : \; |d_{k,j}| > \varepsilon_k\}, \; \hat{I}_k := \tilde{I}_k \cup U_k^0.$

In step (E1), we collect all indices corresponding to single-scale and difference coefficients, respectively, on level k, which depend on a significant single-scale coefficient on level $k + 1$. Next, we decompose the coefficient $u_{k+1,j}$ according to (4,5),
where the summation is only performed for the non-vanishing mask coefficients
$\{g_{j,r}^{k,i}\}_{r \in I_{k+1}}$ collected in the sets $E_{k,j}^i$, $i = 0, 1$. If we access to $u_{k+1,j}$, $j \notin \tilde{I}_{k+1}$,
then $u_{k+1,j}$ has to be calculated by recursively applying $u_{k+1,j} = \sum_{r \in I_k} m_{j,r}^{k,0} u_{k,r}$.
According to (A2), only coarse-grid coefficients are required on level k, whereas
the difference coefficients are not significant, i.e., they are smaller than the threshold value ε_k. Due to assumption (A1) and the special structure of $\mathbf{G}_{k,0}$, one has
$\Omega = \bigcup_{l=0}^{k-1} \bigcup_{j \in \tilde{I}_l} V_{l,j} \cup \bigcup_{j \in \hat{I}_k} V_{k,j}$ after each sweep in k and, thus, $\hat{I}_0 = I_0$.

The decoding, i.e., $\tilde{\mathbf{u}}_{L,MR} \to \tilde{\mathbf{u}}_{L,SR}$, is more complicated:

(D1) $\tilde{I}_0^p := I_0, \; \tilde{I}_{k+1}^p := \bigcup_{j \in \tilde{J}_k} \left\{r \in I_{k+1} \; : \; m_{r,j}^{k,1} \neq 0\right\}, k = 0, \dots, L - 1$

$u_{k+1,j} = \sum_{r \in D_{k,j}^0} m_{j,r}^{k,0} u_{k,r} + \sum_{r \in D_{k,j}^1} m_{j,r}^{k,1} d_{k,r}, \; j \in \tilde{I}_{k+1}^p,$

$k = 0, \dots, L - 1$

(D2)   1) $P_L := \emptyset, P_k := \bigcup_{j \in \tilde{I}_{k+1}^p \cup P_{k+1}} \left\{r \in I_k \; : \; g_{r,j}^{k,0} \neq 0\right\}, k = L - 1, \dots, 0$

2) $R_k := P_k \cap \tilde{I}_k^p, \; D_{k+1} := \bigcup_{j \in P_k} \left\{r \in I_{k+1} \; : \; g_{j,r}^{k,0} \neq 0\right\}$

$S_0 = \emptyset, \; S_{k+1} := D_{k+1} \backslash \left(D_{k+1} \cap \left(\tilde{I}_{k+1}^p \cup P_{k+1}\right)\right), \; k = 0, \dots, L - 1$

3) $u_{k+1,j} = \sum_{r \in D_{k,j}^0} m_{j,r}^{k,0} u_{k,r} + \sum_{r \in D_{k,j}^1} m_{j,r}^{k,1} d_{k,r}, \; j \in S_{k+1},$

$k = 0, \dots, L - 1$

4) $\hat{I}_L := \tilde{I}_L^p \cup S_L, \; \tilde{I}_k := \tilde{I}_k^p \backslash R_k \cup S_k, \; k = 0, \dots, L - 1$

The transformation is performed only locally according to (3) for all single-
scale coefficients $u_{k+1,j}$, $j \in \tilde{I}_{k+1}^p$, where a significant contribution of a difference
coefficient $d_{k,r}$ has to be taken into account. For efficiency reason, the summation in
(D1) is only done for the non-vanishing mask coefficients $\{m_{j,r}^{k,0}\}_{r \in I_k}$, $\{m_{j,r}^{k,1}\}_{r \in J_k}$
collected in the sets $D_{k,j}^i$, $i = 0, 1$. Proceeding this way, the associated grid may
contain overlappings or gaps. Hence, assumption (A1) is not satisfied. In a second
step, these have to be removed. To this end, we first have to determine all possible
overlappings, i.e., for any single-scale coefficient $u_{k+1,j}$, $j \in \tilde{I}_{k+1}^p$, we determine
all coarse-grid volumes $V_{k',j'}$ with $V_{k+1,j} \subset V_{k',j'}$. This step (D2.1) is successively
performed on each level proceeding from fine to coarse. These indices are collected
in the sets P_k for all levels $k = 0, \dots L$. Obviously, P_L is empty, since there is no
data on a finer level. The sets R_k, $k = 0, \dots, L$, contain the indices of P_k that
have actually to be removed from $\tilde{I}_k^p$. If a coarse-grid value $u_{k,j}$ is taken away, the

resulting gap associated to the volume $V_{k,j}$ in the computational domain has to be refilled by information on finer levels. These indices are collected in the sets S_k, $k = 1, \ldots, L$, according to (D2.2). Finally, the overlappings are removed from the predictive sets $\tilde{I}_k^p$, (D2.4), and the gaps are refilled by applying the two-scale equation (3), (D2.3). If we access to a not yet calculated coarse grid information $u_{k,j}$ in the summation, we recursively apply the same two-scale equation.

4. Adaptive finite volume schemes

We now incorporate the multiscale concept presented in Sec. 3 into a finite volume scheme in one spatial dimension. Proceeding as in [4, 2], the extension to the multidimensional case for structured as well as unstructured grids is straight forward.

In the following, we assume the computational domain $\Omega = [a, b] \subset \mathbb{R}$ to be uniformly decomposed on each level $k = 0, \ldots, L$, i.e., $V_{k,j} = [x_{k,j-1}, x_{k,j}]$, $j = 1, \ldots, N_k$, determined by the equidistant grid points $x_{k,j} = a + h_k j$, $j = 0, \ldots, N_k$, with spatial discretization length $h_k = (b - a)/N_k$. The nestedness of the grid hierarchy is guaranteed by demanding $N_{k+1} = 2N_k$ or $h_{k+1} = h_k/2$, respectively. Hence, the sets of indices I_k, J_k introduced in Sec. 2 are given by $I_k = J_k = \{1, \ldots, N_k\}$. Furthermore, the time interval $[0, T]$ is also uniformly discretized by $0 = t_0 < \ldots < t_{\overline{N}} = T$, $\overline{N} \in \mathbb{N}$, with $\tau = t_n - t_{n-1} = const$. The ratio of the spatial and temporal discretization length is defined by $\lambda_k := \tau/h_k$.

The starting point is a finite volume scheme on the finest grid

$$v_{L,j}^{n+1} = v_{L,j}^n - \lambda_L \left[F_{L,j}^n - F_{L,j-1}^n \right], \qquad j \in I_L, \tag{6}$$

by which the average $u_{L,j}^n := h_L^{-1} \int_{V_{L,j}} u(t_n, x)\, dx$ of the solution u of (1) is approximated by $v_{L,j}^n$. Here, we assume that the numerical flux can be written as

$$F_{L,j}^n := F(v_{L,i}^n, i \in S_{L,j}), \tag{7}$$

where $S_{L,j} \subset I_L$ is the stencil on which the numerical flux $F_{L,j}^n$ is calculated. Except for periodic boundary conditions, one-sided stencils have to be applied near the boundary according to Dirichlet or von Neumann boundary conditions.

First, we apply the multiscale transformation (4,5) to the quantities $v_{L,j}^n$ where we recursively introduce the single-scale coefficients $v_{k,j}^n$, $j \in I_k$, and the difference coefficients $d_{k,j}^n$, $j \in J_k$, of the levels $k = 0, \ldots, L - 1$. Together with (6), we obtain the multiscale coefficients of the new time level

$$v_{0,j}^{n+1} = v_{0,j}^n - \lambda_0\, \delta_{0,j}^n, \quad j \in I_0 \tag{8}$$

$$d_{k,j}^{n+1} = d_{k,j}^n - \lambda_{k+1}\, d_{k,j}^n(\delta), \quad j \in J_k, \ k = 0, \ldots, L - 1, \tag{9}$$

with the flux balances $\delta_{k,j}^n$, $j \in I_k$, and the corresponding multiscale coefficients $d_{k,j}^n(\delta)$, $j \in J_k$, recursively defined by

$$\delta_{k,j}^n := \frac{\lambda_{k+1}}{\lambda_k} \sum_{r \in I_{k+1}} g_{j,r}^{k,0}\, \delta_{k+1,r}^n, \quad d_{k,j}^n(\delta) := \sum_{r \in I_{k+1}} g_{j,r}^{k,1}\, \delta_{k+1,r}^n, \tag{10}$$

initialized on level L according to (7) and $\delta_{L,j}^n := F_{L,j}^n - F_{L,j-1}^n$. Since we consider a one-dimensional dyadic grid hierarchy, the flux balances of the coarser levels can be rewritten in terms of the finest level, i.e.,

$$\delta_{k,j}^n = F_{k,j}^n - F_{k,j-1}^n \quad \text{with} \quad F_{k,j}^n := F_{L,2^{L-k}j}^n. \tag{11}$$

Up to now, the system of evolution equations (6) for the averages $v_{L,j}^n$, $j \in I_L$, of the finest level has been *equivalently* transformed into a system of evolution equations (8, 9) for the multiscale coefficients $v_{0,j}^n$, $j \in I_0$, and $d_{k,j}^n$, $j \in J_k$, $k = 0, \ldots, L-1$. The efficiency of the scheme can now be improved by applying a threshold process to the difference coefficients $d_{k,j}^{n+1}$, i.e., the evolution equation (9) is only performed for $j \in \tilde{J}_k^{n+1} \subset J_k$, $k = 0, \ldots, L-1$. Following Harten's original concept [5], the truncation of the difference coefficients, i.e., $d_{k,j}^n := 0$, $j \in J_k \backslash \tilde{J}_k^{n+1}$, can be used to *replace* the calculation of certain coarse-grid fluxes $F_{k,j}^n$ according to (10) by a linear combination of already calculated numerical fluxes of the coarser level $k-1$ (see [5, 2, 3]). This leads to a *hybrid* scheme where only a few terms are computed according to (7) and (11) and the remaining fluxes are calculated by difference formulae. In general, this accelerates the computation, if the calculation of (7) is sufficiently expensive. For more details see [5, 4, 2]. However, the complexity of the scheme still remains that of the *finest* level (see e.g. [2]). This is essentially based on the fact that the numerical fluxes $F_{k,j}^n$ which have to be computed by (11) are calculated from the data $v_{L,j}^n$ of the finest level. In order to reduce the complexity of the computation, we have to proceed differently. The main idea is to compute the numerical fluxes not necessarily from the single-scale coefficients $\mathbf{v}_L^n$ of the finest level but from the adaptive single-scale representation $\tilde{\mathbf{v}}_{L,SR}^n$ defined according to $\tilde{\mathbf{u}}_{L,SR}$ introduced in Sec. 3. This results in an *adaptive* instead of an hybrid scheme:

(S1) Decoding: $\tilde{\mathbf{v}}_{L,MR}^n \to \tilde{\mathbf{v}}_{L,SR}^n = (\tilde{\mathbf{v}}_0^n, \ldots, \tilde{\mathbf{v}}_L^n)$ with $\tilde{\mathbf{v}}_k^n = \{v_{k,j}^n\}_{j \in \tilde{I}_k^n}$

(S2) Prediction of $\tilde{J}_k^{n+1}$, $k = 0, \ldots, L-1$, according to [5]

(S3) $\bar{v}_{L,i}^n := v_{k,j}^n$ with $j \in \tilde{I}_k^n$, $V_{L,i} \subset V_{k,j}$, $\forall i \in \bigcup_{j \in F_n} S_{L,j}$,

$$F_n := \left\{ j \, 2^L \, : \, j \in I_0 \right\} \cup \bigcup_{k=0}^{L-1} \bigcup_{j \in \tilde{J}_k} \left\{ (r-1) \, 2^{L-k-1} \, : \, r \in E_{k,j}^1 \right\}$$

(S4) $F_{L,j}^n = F(\bar{v}_{L,i}^n \, ; \, i \in S_{L,j})$, $\quad j \in F_n$

(S5) $v_{0,j}^{n+1} = v_{0,j}^n - \lambda_0 \, \delta_{0,j}^n$, $\quad j \in I_0$

$$d_{k,j}^{n+1} = d_{k,j}^n - \lambda_{k+1} \sum_{r \in E_{k,j}^1} g_{j,r}^{k,1} \, \delta_{k+1,r}^n, \quad j \in \tilde{J}_k^{n+1}, \, 0 \leq k \leq L-1$$

Each time step is composed of five steps where a sequence of truncated multiscale coefficients $\tilde{\mathbf{v}}_{L,MR}^n = (\tilde{\mathbf{v}}_0^n, \tilde{\mathbf{d}}_0^n, \ldots, \tilde{\mathbf{d}}_{L-1}^n)$ with $\tilde{\mathbf{d}}_k^n = \{d_{k,j}^n\}_{j \in \tilde{J}_k^n}$ is supposed to be known at time level t_n. Herein, $\tilde{\mathbf{v}}_{L,MR}^n$ is defined according to $\tilde{\mathbf{u}}_{L,MR}$ introduced in Sec. 3. In a first step (S1) we decode the sequence of truncated multiscale coefficients into a sequence of adaptive single-scale coefficients according to (D1, D2) such that (A1, A2) hold. Since $\tilde{\mathbf{v}}_{L,MR}^{n+1}$ is not yet known, we next determine a prediction of $\tilde{J}_k^{n+1}$, $k = 0, \ldots, L-1$, from $\tilde{\mathbf{v}}_{L,MR}^n$. Here, we follow Harten's original concept [5]. This is based on the finite speed of propagation which is characteristic for hyperbolic problems. For more details see [2, 4]. The most important step of our algorithm is the projection of the single-scale coefficients of the coarser levels to the finest level (S3). This projection is unique, since the representation $\tilde{\mathbf{v}}_{L,SR}^n$ is supposed to satisfy (A1). Of course, we do not calculate all coefficients $\bar{v}_{L,j}^n$ of the finest level, but only those which are needed in the calculation of the numerical fluxes according to (11). To this end, we introduce the set F_n which consists of all indices corresponding to the required coefficients. From the projection $\hat{v}_L$, we now calculate the numerical fluxes (S4) according to definition (7). Finally, we can calculate the multiscale coefficients of the the new time level, (S5). Once again, the summation is only performed for those coefficients which give a non-vanishing contribution, i.e., $g_{j,r}^{k,1} \neq 0$.

5. Numerical calculations

In the following, we want to present preliminary results of the adaptive algorithm applied to Burgers' equation neglecting viscosity, i.e., $f(u) = 0.5\,u^2$, and periodic initial data $u_0(x) = \sin(\pi x)$, $x \in \Omega := [0, 2]$. Hence, a stationary shock develops at time $t_s = 1/\pi$ in position $x_s = 1$. We always choose $T = 0.4$. The exact solution is plotted in Fig. 1. The underlying finite volume scheme is a second order ENO scheme using polynomial reconstruction of degree 1 and an exact Riemann solver (see [6]). For all computations, we have chosen $N_0 = 20$ intervals on the coarsest level such that there are $N_L = N_0\,2^L$ intervals on the finest level according to the dyadic grid hierarchy. Correspondingly, the time discretization length is chosen such that the CFL number is always equal to 0.8 on the finest level. This corresponds to $\overline{N}_L = 5 \times 2^L$ time steps depending on L. Furthermore, the coefficients $m_{j,r}^{k,i}$ and $g_{j,r}^{k,i}$, respectively, are determined by a polynomial reconstruction of degree 2 (see [4]). According to Sec. 3, the truncation of the multiscale decomposition is applied with the threshold values $\varepsilon_{L-1} = 0.01$, $\varepsilon_k = 0.7\,\varepsilon_{k+1}$, $k = L-2, \ldots, 0$. All computations have been performed on an SGI O2 workstation.

As an example, the approximation is presented for $L = 14$ in Fig. 1. It is in good agreement with the exact solution, although the finest level is only involved close to the shock as can be seen from the level distribution $\tilde{I}_k$, $k = 0, \ldots, L$, of the adaptive single-scale coefficients (Fig. 2). In addition, the distribution $\tilde{J}_k$, $k =$

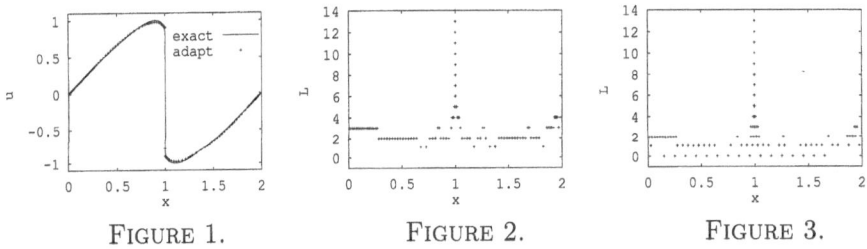

FIGURE 1.   FIGURE 2.   FIGURE 3.

L	N_L	$D_{\max}$	$F_{\max}$	C_{MR}	C_{FV}	S_{MR}
10	20480	186	73	233.3	419.9	1.80
11	40960	232	77	704.4	1975.0	2.80
12	81920	261	81	1746.3	8943.3	4.77
13	163840	295	81	4362.9	38152.5	8.74
14	327680	362	89	12388.0	153936.8	12.43

TABLE 1.

$0, \ldots, L-1$, of the corresponding multiscale coefficients is shown in Fig. 3. Here, the indices $j \in \tilde{I}_k$ and $j \in \tilde{J}_k$, respectively, are plotted with respect to the cell centers $\hat{x}_{k,j}$ of the corresponding intervals $V_{k,j}$. We notice that the smooth regions of the solution are, in fact, resolved by coarse levels, whereas the higher levels are only required close to the shock position x_s. This implies that the scheme is capable to resolve appropriately effects proceeding on different scales. Similar results are reported by other authors following Harten's concept (see [5, 7, 2]).

Furthermore, we are interested in the complexity of our scheme. To this end, we have performed several calculations refining the finest level, i.e., $L = 10, \ldots, 14$. The results are presented in Tab. 1. As a measure of the complexity, we introduce the maximum number $D_{\max}$ and $F_{\max}$ of significant multiscale coefficients, i.e., $\# I_0 + \sum_{k=0}^{L-1} \# \tilde{J}_k^n$, and of the calculated numerical fluxes, i.e., $\# F_n$, with respect to all time steps $n = 0, \ldots, \overline{N}_L$. We notice that the numbers $D_{\max}$ and $F_{\max}$, respectively, increase with L. This corresponds to the additional highest-scale coefficients close to the shock and a larger threshold value ε_k on the coarser levels. Moreover, these numbers are correlated. This is different to Harten's concept [5] where the number $D_{\max}$ corresponds to the number of original fluxes, but not to the number of all fluxes including the interpolated fluxes. Hence, following Harten's approach the computation can only be accelerated by a factor about 2–3 (see [2]) depending on the underlying finite volume scheme. Opposite to this, we achieve much higher speed-up factors as recorded in Tab. 1. For this, we have additionally performed the computation with the original finite volume scheme. By this the speed-up factor S_{MR} can be introduced as the ratio of the computational times corresponding to the original and the adaptive scheme, C_{FV} and C_{MR}. Here, the computational time is measured in CPU seconds. We notice that the speed-up factor increases with higher levels. This is based on the fact that C_{FV} increases by a factor of almost 4 corresponding to the refining of the spatial and temporal discretization lengthes by a factor of 2. However, C_{MR} only increases by an averaged factor of about 2.5 which is almost optimal, since the number of time steps is doubled. Thus, we conjecture that, theoretically, an arbitrary speed-up factor can be achieved for $L \to \infty$, since the complexity of the algorithm corresponds to that of the significant multiscale coefficients. This is not true for Harten's concept, where it has been proved that S_{MR} tends to a finite number for $L \to \infty$ (see [4, 2]).

6. Conclusion

In the present paper, we have introduced a new adaptive concept for finite volume schemes. Here, adaptivity is controlled by local multiscale decompositions. First test calculations have shown that the complexity of the resulting algorithm corresponds to the number of non-vanishing multiscale coefficients. This is an essential improvement of Harten's original idea of an hybrid scheme. Additionally, further investigations have already shown that this concept can also be applied to inhomogeneous conservation laws.

Our future work is devoted to time adaptivity. Presently, the temporal discretization length is significantly restricted by the CFL number because of a very small number of significant multiscale coefficients of the finest level. This constraint becomes even more serious if more levels are involved, i.e., $L \gg 1$. In order to overcome this obstruction, we want to combine our present concept with a local time-stepping strategy.

Acknowledgement. The authors are supported in parts by the DFG in the Collaborative Research Center SFB 401 at the University of Technology Aachen.

References

[1] B. Bihari and A. Harten, *Multiresolution schemes for the numerical solution of 2-D conservation laws I*, SIAM J. Sci. Comput., **18 (2)** (1997), 315–354.

[2] B. Gottschlich-Müller, *Multiscale Schemes for Multidimensional Conservation Laws*, PhD thesis, RWTH Aachen, 1998, submitted.

[3] B. Gottschlich-Müller and S. Müller, *Application of multiscale techniques to hyperbolic conservation laws,* Numerical Treatment of Multi-scale Problems, Proceedings of the 13th GAMM-Seminar held in Kiel, January 24–26, 1997, submitted.

[4] B. Gottschlich-Müller and S. Müller, *On multi-scale concepts for multi-dimensional conservation laws,* Proceedings of Guangzhou International Symposium on Computational Mathematics held in Guangzhou, August 11–15, 1997, to appear.

[5] A. Harten, *Multiresolution algorithms for the numerical solution of hyperbolic conservation laws*, Comm. Pure Appl. Math., **48 (12)** (1995), 1305–1342.

[6] A. Harten, B. Engquist, S. Osher and S. R. Chakravarthy, *Uniformly high order accurate essentially non-oscillatory schemes III*, J. Comp. Phys., **71** (1987), 231–303.

[7] B. Sjögreen, *Numerical experiments with the multiresolution scheme for the compressible Euler equations*, J. Comp. Phys., **117** (1995), 251–261.

Institut für Geometrie und Praktische Mathematik,
RWTH Aachen,
D-52056 Aachen, Germany
E-mail address: `mueller@igpm.rwth-aachen.de`

International Series of Numerical Mathematics
Vol. 129, © 1999 Birkhäuser Verlag Basel/Switzerland

Existence of Global Smooth Solutions to Euler Equations for an Isentropic Perfect Gas

M. Grassin

Abstract. One considers Euler equations for an isentropic perfect gas in $\mathbb{R}^d$ where $d \geq 1$. One shows that there exist global smooth solutions, provided that the initial data satisfy:
- $D^2 u_0 \in H^{m-1}(\mathbb{R}^d)$ with $m > 1 + d/2$ and $Du_0 \in L^\infty(\mathbb{R}^d)$,
- $\exists \delta > 0$ such that $\forall x \in \mathbb{R}^d$, $\quad \mathrm{d}(\mathrm{Sp}(Du_0(x)), \mathbb{R}_-) \geq \delta$,
- ρ_0 has a compact support and $\rho_0^{\frac{\gamma-1}{2}}$ is small enough in $H^m(\mathbb{R}^d)$,

where u_0 stands for initial velocity, ρ_0 for initial density and γ for the adiabatic constant of the gas. One denotes by $\mathrm{Sp}(Du_0(x))$ the spectrum of the jacobian matrix of u_0. One introduces an approximate problem on the velocity:

$$\begin{cases} \partial_t \bar{u} + (\bar{u} \cdot \nabla)\bar{u} = 0 & \text{on} \quad \mathbb{R}^d \times \mathbb{R}_+, \\ \bar{u}(x,0) = u_0(x) & \text{on} \quad \mathbb{R}^d. \end{cases}$$

Thanks to the hypotheses, this problem has a global solution $\bar{u}$. Using this approximate solution and a symmetrisation of the system, one obtains a local solution, such that $(\rho^{\frac{\gamma-1}{2}}, u - \bar{u}) \in \mathcal{C}^j([0, \infty[; H^{m-j}(\mathbb{R}^d))$ for $j \in \{0, 1\}$. With some accurate energy estimates, one shows that this solution is global in time.

1. Introduction

We consider Euler equations for an isentropic perfect gas:

$$\begin{cases} \partial_t \rho & + & \mathrm{div}(\rho u) & & & = & 0 \\ \rho(\partial_t u & + & (u \cdot \nabla)u) & + & \nabla p & = & 0 \end{cases} \tag{1}$$

where $t \in \mathbb{R}_+$, $x \in \mathbb{R}^d$ and $u : \mathbb{R}^d \times \mathbb{R}_+ \to \mathbb{R}^d$ stands for the velocity, $\rho : \mathbb{R}^d \times \mathbb{R}_+ \to \mathbb{R}_+$ for the density, $p = (\gamma - 1)\rho^\gamma$ for the pressure. The adiabatic constant of the gas is denoted by $\gamma > 1$ and $d \geq 1$ is the dimension of the space. The vector of all the spatial derivatives of order k is denoted by D^k. The differential with respect to x of $u(.,t)$ will be denoted by Du. The notation V^T stands for the transpose of V. We denote by $|.|_\infty$ the norm of $L^\infty(\mathbb{R}^d)$, by $\|.\|_0$ the one of $L^2(\mathbb{R}^d)$ and by $\|.\|_m$ the one of $H^m(\mathbb{R}^d)$. We also denote by $\|.\|_X$ the norm of the space $X = \{z : \mathbb{R}^d \times \mathbb{R}_+ \to \mathbb{R}^d / Dz \in L^\infty, D^2 z \in H^{m-1}(\mathbb{R}^d)\}$.

We are interested in the existence of global smooth solutions to the Cauchy problem for (1) with (ρ_0, u_0) as initial data. There exist few results concerning this problem, in which the choice of initial data depends on whether one wants to prove or to disprove global existence. In [4], T. Sideris has shown a non global existence result: the initial density is close to a constant at infinity – the constant should be different from 0 – and some global quantities have to be large. For $d = 1$, we have a 2×2 system. In this case, some results can be proved using P.D. Lax's works [3]. In the same case with less restrictive conditions, J.Y. Chemin has also proved a non global existence result: the initial velocity has to be smaller than the initial speed sound in each point – this quantity depends mostly on ρ_0. There are few results of global existence, in particular for $d > 1$, for gas dynamics. D. Serre has proved one in the non isentropic case with $\gamma \le 1 + 2/d$ in [1]. One restriction to this result is that the initial velocity must be close to a linear field. We give here a global existence result in the isentropic case without this particular hypothesis.

2. Existence of global smooth solutions

We state the following result:

Theorem 2.1. *Let $m > 1 + d/2$ and assume that*

(H1) $\rho_0^{\frac{\gamma-1}{2}}$ *is small enough in $H^m(\mathbb{R}^d)$,*

(H2) $u_0 \in X$,

(H3) $\exists \delta > 0$ *such that $\forall x \in \mathbb{R}^d, \quad d(Sp(Du_0(x)), \mathbb{R}_-) \ge \delta$.*

Let $\bar{u}$ be the global solution to

$$\begin{cases} \partial_t \bar{u} + (\bar{u} \cdot \nabla)\bar{u} = 0 & on \quad \mathbb{R}^d \times \mathbb{R}_+, \\ \bar{u}(x,0) = u_0(x) & on \quad \mathbb{R}^d. \end{cases} \tag{2}$$

Then there exists a global smooth solution to (1) with (ρ_0, u_0) as initial data, i.e. (ρ, u) such that $(\rho^{\frac{\gamma-1}{2}}, u - \bar{u}) \in \mathcal{C}^j([0, \infty[; H^{m-j}(\mathbb{R}^d))$ for $j \in \{0, 1\}$.

Note that the case $\rho = 0$ can occur, that is to say that there can be vacuum in some area. This requires some attention as we will notice later.

We shall give a general and naive interpretation of the hypotheses: since we do not want any shock to happen, we choose a small initial density and an initial velocity which makes the particles to spread out. That's why these hypotheses seem to be the good ones to expect some result. To be more precise, we can explain where our hypotheses come up. The exponent $\frac{\gamma-1}{2}$ which appears in the hypothesis (H1) comes from the first step of the proof: it is introduced by the symmetrisation of the system when we prove local existence. (H1) is not equivalent to $\rho_0 \in H^m(\mathbb{R}^d)$. But the two assumptions can sometimes be linked according to the value of γ. The hypothesis (H2) about u_0 is the usefull one to obtain a global solution to the simplified problem (2).

We prove first the theorem with the hypothesis (H4): ρ_0 has a compact support. Then, by truncation of ρ_0 and with a property of local – in space and time – uniqueness, we can show the theorem without this assumption.

The proof is based on local existence of a smooth solution and on energy estimates. But the classical method does not work here. Thus we consider a simplified problem which gives us a global approximate solution thanks to our hypotheses (H2)–(H3). Then we compare the two problems and we use the properties of the simpler one to improve the classical energy estimates. Our demonstration is split in three steps. First, we use a local existence theorem – cf. J.Y. Chemin [2] – and we construct a local solution with $u_0 \in X$. Then we study $\bar{u}$ the global smooth solution to the approximate problem (2). To conclude, we obtain good energy estimates on a certain spatial norm of the difference between the local solution and the approximate one.

2.1. Local existence

The first step in the proof of local existence consists in the symmetrisation of the system. The symmetrisation must be cautiously chosen since the case $\rho = 0$ can occur. Following T. Makino, S. Ukai, S. Kawashima [5], we take $\pi = \rho^{\frac{\gamma-1}{2}}$ to obtain:

$$\begin{cases} C_1(\partial_t + u \cdot \nabla)\pi & + & C_2 \pi \mathrm{div}(u) & = & 0, \\ (\partial_t + (u \cdot \nabla))u & + & C_2 \pi \nabla \pi & = & 0, \end{cases} \tag{3}$$

where $C_1 = \frac{4\gamma}{(\gamma-1)}$ and $C_2 = 2\gamma$. We set $\pi_0 = \rho_0^{\frac{\gamma-1}{2}}$.

This system is not equivalent to (1) because of the case $\rho = 0$, but we can pass from (3) to (1) multiplying by ρ. Thus if we find a global smooth solution to this problem, we obtain one for (1). We only loose the property of uniqueness of the solution.

Note that $u_0 \notin H^m$ – that would contradict (H3). Hence, we can not apply directly the theorem of local existence for symmetric hyperbolic systems – see [2]. But we apply it with $(\pi_0, u_0\varphi) \in H^m$ as initial data, where:

$\varphi \in C_0^\infty(\mathbb{R}^d)$ and $\varphi \equiv 1$ on $B(0, R + 2\eta)$,
R is a positive real number such that $\mathrm{supp}\rho_0 \subset B(0, R)$,
η is a positive real number.

Thus we obtain $(\pi^\varphi, u^\varphi) \in C^0([0, T_{ex}[; H^m(\mathbb{R}^d)) \cap C^1([0, T_{ex}[; H^{m-1}(\mathbb{R}^d))$, a local solution to (3) with $(\pi_0, u_0\varphi)$ as initial data. Note that $(0, \bar{u})$ is a solution of (3) with $(0, u_0)$ as initial data. Now we consider $K = \{(x, t)/0 \le t \le T, x \in B(0, R + \eta + Mt)\}$ with:

$$M = \sup_{0 \le t \le T_{ex} - \epsilon} \left(\frac{C_2}{\sqrt{C_1}} |\pi^\varphi|_{L^\infty} + |u^\varphi|_{L^\infty} \right), \text{ and}$$

$T = \min(T_{ex} - \epsilon, \eta/(2M) - \epsilon)$ for $\epsilon > 0$ given. We set

$$(\pi, u) = \begin{cases} (\pi^\varphi, u^\varphi) \text{ in } K, \\ (0, \bar{u}) \text{ outside } K. \end{cases}$$

This is a solution of (3) in and outside K. By a property of local uniqueness for the solution of (3), we show that $(\pi^\varphi, u^\varphi) = (0, \bar{u})$ in a neighbourhood of ∂K, the boundary of K. Thus (π, u) is smooth across ∂K and is a solution of

(3) with (π_0, u_0) as initial data. Moreover $(\pi, u - \bar{u})$ is in $C^0([0, T[; H^m(\mathbb{R}^d)) \cap C^1([0, T[; H^{m-1}(\mathbb{R}^d))$ and has a compact support.

2.2. Approximate problem

We have to suppose that the density is small to expect a global smooth solution. By neglecting ρ and $\nabla(\rho^{\gamma-1})$, we obtain the approximate problem:

$$\begin{cases} \partial_t \bar{u} + (\bar{u} \cdot \nabla)\bar{u} = 0 & \text{on} \quad \mathbb{R}^d \times \mathbb{R}_+, \\ \bar{u}(x, 0) = u_0(x) & \text{on} \quad \mathbb{R}^d. \end{cases}$$

The solution is $\bar{u}(X(x_0, t), t) = u_0(x_0)$ with $X(x_0, t) = x_0 + t u_0(x_0)$. This is well-defined thanks to (H2)–(H3). Note that this problem does not take in account any forces. That is why the choice of u_0 is decisive in the global existence of the smooth solution $\bar{u}$. Our hypothesis (H3) requires some uniformity: the aim is to obtain estimates on the spatial norm of the derivatives of $\bar{u}$ and to use them in energy estimates. We show:

Proposition 2.2. *Assume (H2),(H3). Let $\bar{u}$ be the global solution of (2). Then:*

- $D\bar{u}(x, t) = \dfrac{1}{(1+t)} I + \dfrac{1}{(1+t)^2} K(x, t), \qquad \forall x \in \mathbb{R}^d \quad and \quad \forall t \in \mathbb{R}_+,$
- $\|D^l \bar{u}(., t)\|_0 \leq K_l (1+t)^{d/2-(l+1)}, \text{ for } 2 \leq l \leq m+1,$
- $|D^2 \bar{u}(., t)|_\infty \leq C(1+t)^{-3},$

with $I = Id_{\mathbb{R}^d}$ and $K : \mathbb{R}^d \times \mathbb{R}_+ \to M_d(\mathbb{R})$, $|K|_{L^\infty(\mathbb{R}^d \times \mathbb{R}_+)} \leq M$ where M, C and K_l, for $2 \leq l \leq m+1$, are some positive numbers which depend only on m, d, δ, $\|u_0\|_X$.

We remark that the decays in t of the derivatives of $\bar{u}$ improve themselves with the order of the derivatives. We will show that it is the same for $V = (\pi, u - \bar{u})^T$.

2.3. Energy estimates

Then, we compare V and $\bar{U} = (0, \bar{u})^T$. We set $U = (\pi, u - \bar{u})^T$ and we have the following system:

$$A^0 \partial_t U + \sum_{\alpha=1}^{d} A^\alpha(U) \partial_\alpha U = -B(D\bar{U})U - \sum_{\alpha=1}^{d} C^\alpha(\bar{U}) \partial_\alpha U, \qquad (4)$$

where $A^0 \in M_{d+1}(\mathbb{R})$ is positive definite and symmetric, $\forall 1 \leq \alpha \leq d$, $A^\alpha(U) \in M_{d+1}(\mathbb{R})$ and $C^\alpha(\bar{U}) \in M_{d+1}(\mathbb{R})$ are symmetric.

Before we perform the energy estimates, we have to choose the spatial norm we want to estimate. We consider the semi-norm, equivalent to $\|D^k U\|_0(t)$, which appears naturally in the calculus:

$$Y_k(t) = \left(\int_{\mathbb{R}^d} D^k U(x, t) \cdot A^0 D^k U(x, t) dx \right)^{1/2}.$$

We expect Y_k for $k = 0, \ldots, m$ to behave more or less like $\|D^k \bar{u}\|_0$, that is to say to have different decays in t according to k. Thus instead of using the classical norm in $H^m(\mathbb{R}^d)$, we introduce

$$Z(t) = \sum_{k=0}^{m} (1+t)^{\gamma_k} Y_k(t),$$

where γ_k is chosen such that each term of the sum has the same decay in t. Thus we will obtain a good estimate on Z.

We apply D^k on (4) and we take the inner product with $D^k U$. Then we integrate on $\mathbb{R}^d$. We obtain:

$$\frac{1}{2} \frac{d}{dt} Y_k^2 + \frac{k+r}{(1+t)} Y_k^2 \leq C |DU|_\infty Y_k^2 + C' Y_k Z (1+t)^{-\gamma_k - 2}. \tag{5}$$

The first term comes from the time derivative in (4) thanks to the fact that A^0 is symmetric. The second term on the left-hand side of the inequality is obtained by an exact computation of the terms with one derivative of order one of $\bar{U}$. The first term on the right-hand side comes from the terms which involve only U. And the second term on the left comes from the term with a derivative of order at least two of $\bar{U}$. To obtain (5), we used Proposition 2.2.

This leads to:

$$\frac{dZ}{dt}(t) + \frac{a}{(1+t)} Z(t) \leq C(1+t)^\beta (Z(t))^2 + \frac{C'}{(1+t)^2} Z(t). \tag{6}$$

with $\gamma_k = k + r - a$, $\qquad r = \begin{cases} 1 - d/2 & \text{if} \quad \gamma \geq \gamma_c = 1 + 2/d, \\ \frac{\gamma-1}{2} d - d/2 & \text{if} \quad \gamma < \gamma_c, \end{cases}$

and $\beta = -\gamma_1 - d/2$.

The constants C, C' are positive, C depends only on γ, m, d and C' depends only on $\gamma, m, d, \delta, \|u_0\|_X$. We have to choose a in order to obtain a good estimate. The choice $a = 1 + r + d/2 > 1$ leads to:

$$\frac{dZ}{dt}(t) + \frac{a}{(1+t)} Z(t) \leq C(Z(t))^2 + \frac{C'}{(1+t)^2} Z(t). \tag{7}$$

2.4. Conclusion

We use the following result:

Proposition 2.3. *Since $a > 1$ then there exists $Z_{a,C,C'} > 0$ such that the Cauchy problem:*

$$\begin{cases} \dfrac{d\hat{Z}}{dt} + \dfrac{a}{(1+t)} \hat{Z} & = C\hat{Z}^2 + \dfrac{C'}{(1+t)^2} \hat{Z}, \\ \hat{Z}(0) & = Z_{a,C,C'}, \end{cases} \tag{8}$$

has a global solution for $t \geq 0$.

To prove this proposition, we solve the differential equation:

$$\hat{Z}(t) = \frac{(1+t)^{-a} \exp(C'(1 - \frac{1}{1+t}))}{\left(1/\hat{Z}(0) - \int_0^t C(1+\tau)^{-a} e^{C'} e^{-C'\frac{1}{1+\tau}}\right)},$$

and we note that the integral converges only if $a > 1$.

Finally, if $Z(0)$ is small enough, $Z(t)$ is less than the global solution $\hat{Z}$ given by Proposition 2.3. The condition on $Z(0)$ is: $Z(0) \leq Z_{a,C,C'}$. This corresponds to hypothesis (H1) since we have:

$$Z(0) = \sqrt{C_1} \sum_{k=0}^{m} \|D^k(\rho_0^{\frac{\gamma-1}{2}})\|_0 = \sqrt{C_1} \|\rho_0^{\frac{\gamma-1}{2}}\|_m.$$

We know $\hat{Z}$ and we can obtain easily the estimate:

$$Z(t) \leq K(1+t)^{-a}, \quad \text{where } K \text{ depends only on } \gamma, m, d, \delta, \|u_0\|_X \text{ and } \|\rho_0^{\frac{\gamma-1}{2}}\|_m.$$

Then we deduce that $Y_k(t) = O((1+t)^{-(k+r)})$ when t tends to $+\infty$. We conclude that the solution is global since its norm is bounded by a function of t which does not blow up.

References

[1] D. Serre, *Solutions classiques globales des équations d'Euler pour un fluide parfait compressible*, Annales de l'Institut Fourier, **47** (1997), 139–153.

[2] J.Y. Chemin, *Dynamique des gaz à masse totale finie*, Asymptotic Analysis, **3** (1990), 215–220.

[3] P.D. Lax, *Development of singularities of solutions of nonlinear hyperbolic partial differential equations*, J. Math. Phys., **5** (1964), 611–613.

[4] T. Sideris, *Formation of singularities in three-dimensional compressible fluids*, Commun. Math. Phys., **101** (1985), 475–485.

[5] T. Makino, S. Ukai and S. Kawashima, *Sur la solution à support compact de l'équation d'Euler compressible*, Japan J. Appl. Math., **3** (1986), 249–257.

[6] M. Grassin and D. Serre, *Existence de solutions globales et régulières aux équations d'Euler pour un gaz parfait isentropique*, C. R. Acad. Sci. Paris, **325** (1997), 721–726.

UMPA, ENS Lyon,
69364 LYON Cedex 07,
France
E-mail address: mgrassin@umpa.ens-lyon.fr

International Series of Numerical Mathematics
Vol. 129, © 1999 Birkhäuser Verlag Basel/Switzerland

Adaptive Mesh Refinement for Singular Structures in Incompressible Hydro- and Magnetohydrodynamic Flows

Rainer Grauer

Abstract. The question whether finite time singularities develop in incompressible hydro- and magnetohydrodynamic systems starting from smooth initial conditions is still an open problem. Here we present numerical simulations using the technique of adaptive mesh refinement which show evidence that in the 3D incompressible Euler equations a finite time blow-up in the vorticity occurs whereas in the 2D incompressible magnetohydrodynamic equations only exponential growth of vorticity and current density is observed.

1. Introduction

The formation of singular structures in incompressible hydrodynamic and related systems is still a controversial issue. Since mathematically only very little is known, numerical simulations may help to guide further analytical work. In this paper, the ideal incompressible magnetohydrodynamic equations and the incompressible Euler equations are studied in two, respectively three space dimensions.

Mathematically, only few theorems discussing criteria for finite time blow-up are available. The theorem of Beale, Kato and Majda [1] states that if the vorticity in the 3d incompressible Euler equations blows up in finite time T, then also the time integral of the maximum norm $\|\omega\|_{L^\infty}$ of the vorticity field ω diverges. This gives a bound to the speed of blow-up such that asymptotically the growth of vorticity must be faster than $\|\omega\|_{L^\infty} \geq \frac{C}{T-t}$. Any slower increase of vorticity would indicate severe problems in the numerical scheme.

Although there exist high resolution numerical experiments (see Bell and Marcus [4], Brachet *et al* [8], Boratav *et al* [7], Kerr [13]) the question whether the three dimensional incompressible Euler equations develop a finite time singularity in the vorticity is still under discussion. A main problem with non-adaptive three-dimensional simulations is the insufficient spatial resolution due to limited computing resources.

The situation is even worse for the ideal incompressible magnetohydrodynamic equations. Not even the question whether global solutions do exist in two dimensions is solved analytically, although numerical simulations indicate that the growth of vorticity and current density is only exponential in time.

Here we investigate such problems using adaptive mesh refinement introduced by Berger and Colella [5] and Bell et al. [2] for hyperbolic conservation laws. Special care has to be taken to handle the incompressible nature of the flow.

The results presented here [9, 11] were obtained commonly with C. Marliani (Courant Institute) and H. Friedel and K. Germaschewski (Düsseldorf).

The outline of the paper is as follows. In section 2 we comment on the numerical procedure of adaptive mesh refinement and on the integrator used in each grid. In section 3 and 4 we present results for the 2d MHD and the 3d Euler equations, respectively.

2. Adaptive mesh refinement

The idea of adaptive mesh refinement is near at hand. Starting with one grid of given resolution (in most two-dimensional cases we chose 256×256 mesh points and in the three-dimensional ones $64 \times 64 \times 64$) the partial differential equation is solved with some appropriate scheme which will be described below. After a certain number of time steps it is checked whether the local numerical resolution is still sufficient on the whole grid. If one detects that locally finer grids are needed, a first refinement is carried out. For that, critical points, i.e. those locations where the error of the discretization exceeds a given value, are marked on the grid. In addition to these grid points adjacent ones are included. The size of this neighborhood is determined by the local velocities and the time interval to the next regridding. This enables the grid hierarchy to follow moving structures. Now, the marked points of insufficient numerical resolution have to be covered with rectangular grids of finer resolution as efficiently as possible. Our algorithm for this purpose is very similar to the one by Berger and Colella [5] and was recently described in Friedel et al. [9]. On the grids of the newly built level the spatial discretization length and the time step are reduced by a factor r which is called refinement factor. The new grids are filled with data obtained by interpolation from the preceding level. On both levels integration advances until the resolution again becomes locally insufficient. The rebuilding of the grid hierarchy starting with the actual level and proceeding on all subsequent levels begins when the above mentioned threshold for the error is locally exceeded, e.g. if the critical regions of high vorticity (or current density) have moved out of the region covered with finer grids or if local gradients have developed such that the prescribed accuracy is not guaranteed. Next, the critical points are collected on all grids of each level. On the basis of the resulting lists of critical points new grids are generated. Before they can be filled with data it has to be checked whether they are correctly embedded in grids of the preceding level. If data existed on grids of the same level before the regridding, these have to be taken instead of the interpolated data.

In order to deal with the incompressible nature of the flow linear interpolation is performed for obtaining data of newly generated grids on the fields with the highest derivative, e.g. vorticity for the Euler equations and vorticity and current density for the MHD equations. Afterwards, the Poisson equations for the vector potentials are solved using additive Schwarz iteration on neighboring grids [12]. This approach is three times more expensive than interpolating the velocity fields and solving only one Poisson equation for the pressure, but no violations of the incompressibility condition are observed at the boundaries of different grids. Two questions remain. First, we have to comment on the integration scheme used on each grid. The single-grid integration scheme for the MHD equation is a projection method with second order upwinding which has already been used in former non-adaptive simulations [10]. The method is an adaption of work for the Navier–Stokes equations [3] to magnetohydrodynamics. The scheme for the 3d Euler equations is identical to the one used by Bell and Marcus [4]. In the current context it only is important to note that the upwind scheme is stable even in the presence of discontinuities and locally introduces considerable numerical dissipation when underresolved steep gradients occur.

The implementation of the adaptive mesh refinement strategy is done in C++. The handling of the data structures is separated from the actual problem under consideration. Therefore it is relatively easy to use the program for other types of problems. Computations were mostly performed on a Sun Ultra Sparc two-processor machine for the 2D simulations and on a SGI Origin 2000 distributed shared memory computer for the 3D problems. Since on each grid the upwind procedure and the Poisson equation can be solved independently and the number of grids (greater than 200) supersedes the number of processors available to us, parallelization is highly efficient. We implemented the parallelization using Posix threads so that the code is portable.

3. Current sheet formation in the MHD equations

We investigate the formation of singular current sheets described by the ideal incompressible magnetohydrodynamic equations (MHD equations) in two dimensions for the time evolution of the velocity field $\mathbf{u}$ and magnetic field $\mathbf{B}$. Using Elsässer variables $\mathbf{z}^\pm = \mathbf{u} \pm \mathbf{B}$, the MHD equations take the symmetric form

$$\partial_t \mathbf{z}^\pm + \mathbf{z}^\mp \cdot \nabla \mathbf{z}^\pm + \nabla p = 0, \quad \mathrm{div}\, \mathbf{z}^\pm = 0 \;. \tag{1}$$

The equations (1) are integrated in a periodic quadratic box of length 2π with Orszag–Tang or Biskamp–Welter initial conditions, the latter being made less symmetric by introducing arbitrary phase shifts [6]. In either case current sheets form and after a transitional phase the absolute maximum of current density grows exponentially while the width of the sheets decreases exponentially in time. Figure 1 shows vorticity and current density at the end of the calculation in the Orszag–Tang case.

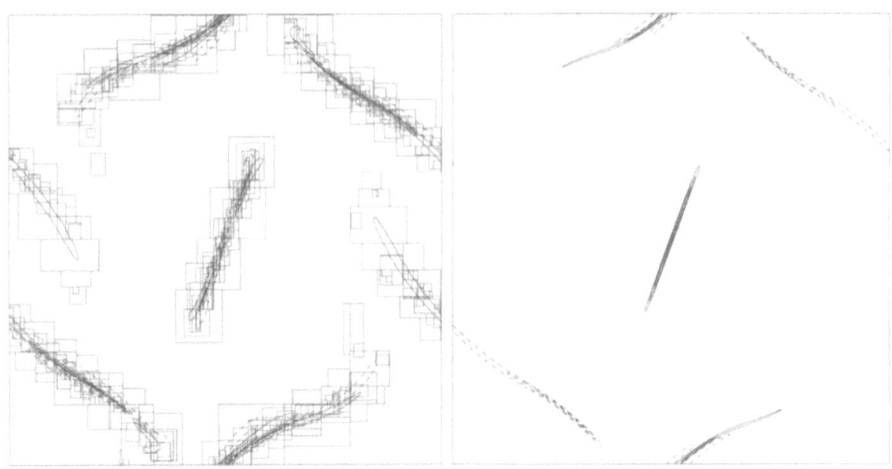

FIGURE 1. Contour plots of vorticity with grid hierarchy (left)
and current density (right).

The simulations start on an initial grid with 256^2 mesh points and whenever
the resolution becomes insufficient refinements are done. The resulting hierarchy
of finer and finer grids dynamically adjusts to the flow and ensures sufficient resolution over the whole integrational domain. Here, as in the following simulations
of the 3d incompressible Euler equations, we tested different criteria for defining
numerically underresolved points. We used criteria based on the local vorticity and
current density, on the deformation tensor and on comparing the nonlinear terms
calculated using the actual resolution of the grid and calculated using twice the
spacing of the grid. The results presented are not very sensitive to the choice of
the refinement criteria which indicates that the technique is very robust.

Whereas former simulations of the ideal equations [15] with classical integration schemes soon ran out of resolution, we can avoid any dissipation. Of course,
as the current density continues to increase the simulations have to end at some
point either due to the limited amount of memory available on the machine or because of becoming to expensive in computational time. In the present simulations
typically up to 7 refinements by a factor of two could be carried out leading to a
final resolution that compares to 32768^2 grid points with non-adaptive treatment.

The ratio of the number of grid points used in the adaptive simulations to the
number needed in a comparable non-adaptive treatment may serve as a measure
of the efficiency of the code. In the case of ideal 2D MHD it is typically about 2%.
The efficiency of an adaptive treatment is obviously related to the codimension of
the structures to be resolved. In addition, it increases with the number of refined
levels.

Figure 2 shows a comparison of the adaptive treatment to non-adaptive simulations of fixed resolution corresponding to 128^2, 256^2 and 512^2 grid points. The

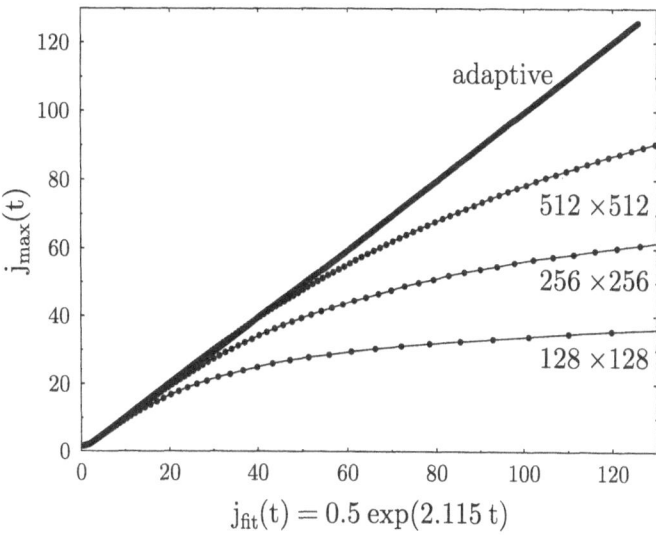

$$j_{fit}(t) = 0.5 \exp(2.115\, t)$$

FIGURE 2. Growth of vorticity versus time.

upwind-scheme produces a substantial amount of dissipation as soon as the current sheet gets underresolved which results in a flattening of the growth of vorticity. In contrast to the non-adaptive simulations, the adaptive treatment shows an exponential increase of current density in time to very high accuracy.

The exponential growth of the current density can be explained by the depletion of nonlinearity through the geometry of the current sheet. Taking the z-component of the curl of the ideal MHD equations (1) yields

$$\partial_t \omega^\pm + \mathbf{z}^\mp \cdot \nabla \omega^\pm = \mathbf{e}_z \cdot \sum_i \nabla z_i^\pm \times \nabla z_i^\mp \tag{2}$$

with $\omega^\pm = [\nabla \times \mathbf{z}^\pm]_z$ meaning the sum and difference of vorticity ω and current density j and with $z_i^\pm$ the i-th component of the vector $\mathbf{z}^\pm$ (see Pouquet [14]). Now it becomes clear that if one-dimensional current sheets develop the derivative in one direction remains smooth and only the derivative transverse to the sheet contributes to the vorticity and current density production. This means that the production term acts like a linear term which results in an only exponential growth of vorticity and current density.

4. Hairpin structure in the Euler equations

The simulations for the 3d incompressible Euler equations are the adaptive version of the simulations performed by Bell and Marcus [4]. The initial velocity is given

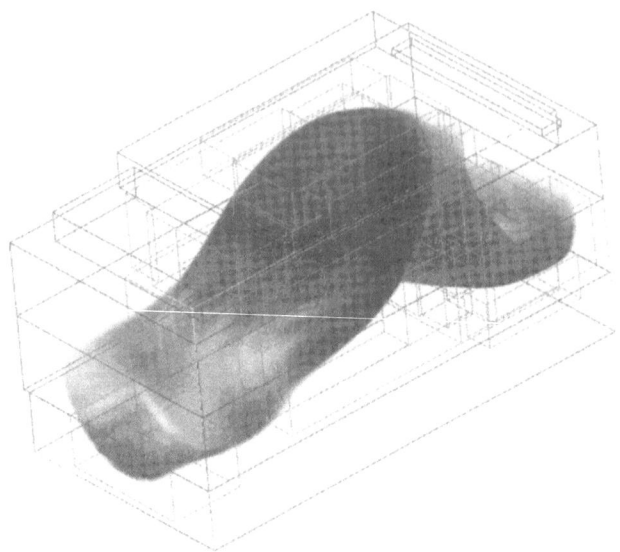

FIGURE 3. Volume rendering of $|\omega|$ at time 0.99.

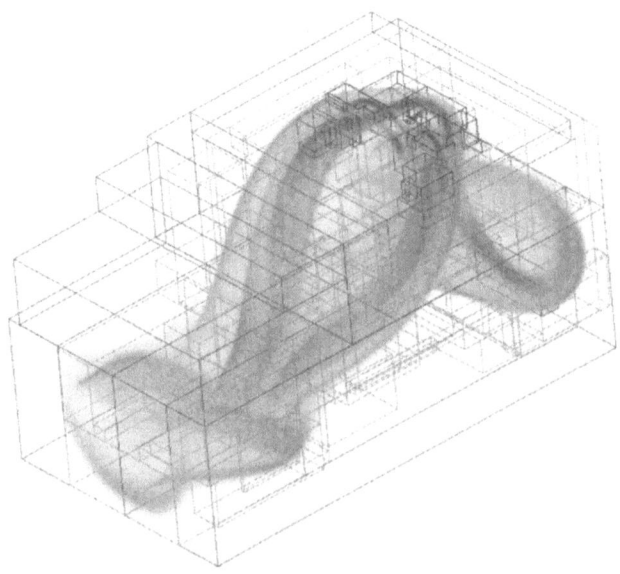

FIGURE 4. Volume rendering of $|\omega|$ at time 1.18.

FIGURE 5. Volume rendering of $|\omega|$ at time 1.32.

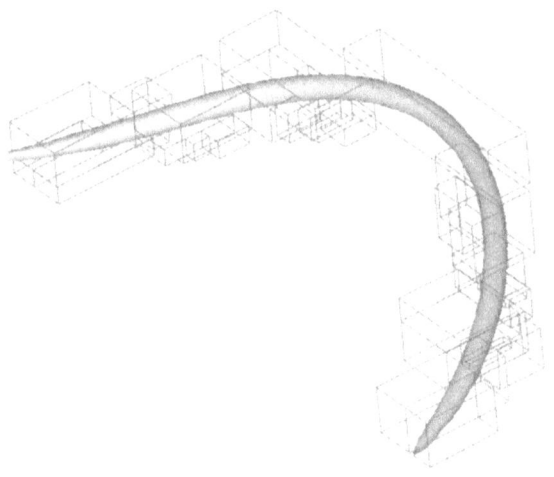

FIGURE 6. Hairpin structure at time 1.32.

by

$$\mathbf{u}_0 = \left(\tanh \left[\frac{(\rho - \sqrt{y^2 + z^2})}{\delta} \right], \ 0, \ \epsilon e^{-\beta(x^2 + y^2)} \right)$$

with $\rho = 0.15$, $\delta = 0.0333$, $\epsilon = 0.05$, $\beta = 15$. Initially, vorticity is concentrated only in the region of shear resulting in a tubular structure. The resolution of the coarsest grid is given by 64^3 mesh points. In Figure 3 the development of a Kelvin-Helmholtz instability is shown which forms from the slight perturbation in the initial condition. At the same time the resolution is automatically enlarged in those regions where steep gradients develop. Figure 3 shows two levels of refinement corresponding to 256^3 mesh points in a non-adaptive treatment. Figures 4–5 show the formation of a hairpin like structure. This is the region where the finite time singularity occurs. Figure 4 and Figure 5 contain three respectively five levels of refinement corresponding to 512^3 and 2048^3 mesh points. Since the grid structure would hide the hairpin structure in figure 5, the boxes indicating the grid

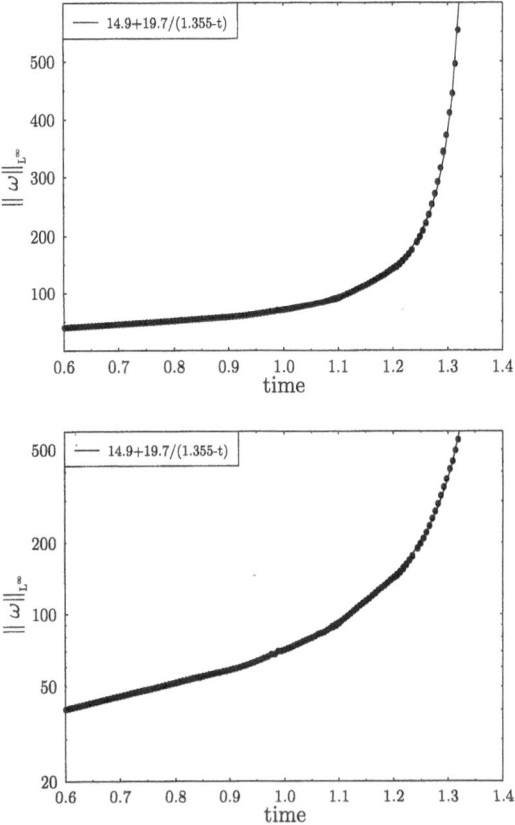

FIGURE 7. Growth of vorticity versus time.

hierarchy are omitted. The hairpin vortex visible in the upper right part is magnified in Figure 6. It shows the isosurface of 60% maximum vorticity, in addition the covering with the finest level is depicted. Using a second order upwind scheme has the advantage that one can detect whether the flow is resolved by monitoring the energy conservation. Upwind schemes produce substantial energy dissipation if steep gradients are not sufficiently resolved. Therefore, the conservation of energy is a measure whether the flow is correctly resolved. During the whole calculation the energy is conserved within less than 1%.

It is important to comment on the alignment properties of the flow. In our calculations we found perfect alignment between the eigenvalue corresponding to the middle eigenvalue of the deformation matrix $(\nabla \mathbf{u} +^T\nabla \mathbf{u})/2$ and the vorticity vector ω.

In Figure 7 we show the temporal evolution of the L_∞-norm of the vorticity. The semilogarithmic plot rules out an only exponential growth of vorticity. For comparison we include a fit of the form $a + b/(T^* - t)$. The constant $a = 14.9$ corresponds to the global background vorticity introduced by the shear flow initial condition. This type of blow-up is consistent with the result of Beale *et al.* [1] which is a necessary (not sufficient) check whether the simulations produce numerical artifacts.

References

[1] J. T. Beale, T. Kato, and A. Majda, *Remarks on the breakdown of smooth solutions for the 3d euler equations*, Comm. Math. Phys. **94** (1984), 61–64.

[2] J. Bell, M. Berger, J. Saltzman, and M. Welcome, *Three-dimensional adaptive mesh refinement for hyperbolic conservation laws*, SIAM J. Sci. Comput. **15 (1)** (1994), 127–138.

[3] J. B. Bell, P. Colella, and H. M. Glaz, *A second-order projection method for the incompressible navier–stokes equation*, J. Comput. Phys. **85** (1989), 257–283.

[4] J. B. Bell and D. L. Marcus, *Vorticity intensification and transition to turbulence in the three-dimensional euler equation*, Comm. Math. Phys. **147** (1992), 371–394.

[5] M. J. Berger and P. Colella, *Local adaptive mesh refinement for shock hydodynamics*, J. Comput. Phys. **82** (1989), 64–84.

[6] D. Biskamp and H. Welter, *Dynamics of decaying two-dimensional magnetohydrodynamic turbulence*, Phys. Fluids B **1** (1989), 1964–1979.

[7] O. N. Boratav and R. B. Pelz, *Locally isotropic pressure hessian in a high-symmetry flow*, Phys. Fluids **7 (5)** (1995), 895–897.

[8] M. E. Brachet, M. Meneguzzi, A. Vincent, H. Politano, and P. L. Sulem, *Numerical evidence of smooth self-similar dynamics and possibility of subsequent collapse for three-dimensional ideal flows*, Phys. Fluids A **4** (1992), 2845–2854.

[9] H. Friedel, R. Grauer, and C. Marliani, *Adaptive mesh refinement for singular current sheets in incompressible magnetohydrodynamic flows*, J. Comput. Phys. **134** (1997), 190–198.

[10] R. Grauer and C. Marliani, *Numerical and analytical estimates for the structure functions in two-dimensional magnetohydrodynamic flows*, Phys. Plasmas **2** (1) (1995), 41–47.

[11] R. Grauer, C. Marliani, and K. Germaschewski, *Adaptive mesh refinement for singular solutions of the incompressible euler equations*, submitted to Phys. Rev. Lett. (1997).

[12] W. Hackbusch, *Iterative solution of large sparse systems of equations*, Applied mathematical sciences, Springer, New York, 1994.

[13] R. M. Kerr, *Evidence for a singularity of the three-dimensional, incompressible euler equations*, Phys. Fluids A **5** (7) (1993), 1725–1746.

[14] A. Pouquet, *Turbulence, statistics and structures: An introduction*, Plasma Astrophysics (Berlin, Heidelberg) (C. Chiuderi and G. Einaudi, eds.), Lecture Notes in Physics, **468**, Springer-Verlag, 1996, 163–212.

[15] P. L. Sulem, U. Frisch, A. Pouquet, and M. Meneguzzi, *On the exponential flattening of current sheets near neutral x-points in two-dimensional ideal mhd flow*, J. Plasma Phys. **33** (1985), 191–198.

Institut für Theoretische Physik I,
Heinrich-Heine-Universität Düsseldorf,
D-40225 Düsseldorf, Germany
E-mail address: grauer@thphy.uni-duesseldorf.de

International Series of Numerical Mathematics
Vol. 129, © 1999 Birkhäuser Verlag Basel/Switzerland

Adaptive Sparse Grids
for Hyperbolic Conservation Laws

Michael Griebel and Gerhard Zumbusch

Abstract. We report on numerical experiments using adaptive sparse grid discretization techniques for the numerical solution of scalar hyperbolic conservation laws. Sparse grids are an efficient approximation method for functions. Compared to regular, uniform grids of a mesh parameter h contain h^{-d} points in d dimensions, sparse grids require only $h^{-1}|logh|^{d-1}$ points due to a truncated, tensor-product multi-scale basis representation.

For the treatment of conservation laws two different approaches are taken: First an explicit time-stepping scheme based on central differences is introduced. Sparse grids provide the representation of the solution at each time step and reduce the number of unknowns. Further reductions can be achieved with adaptive grid refinement and coarsening in space. Second, an upwind type sparse grid discretization in $d + 1$ dimensional space-time is constructed. The problem is discretized both in space and in time, storing the solution at all time steps at once, which would be too expensive with regular grids. In order to deal with local features of the solution, adaptivity in space-time is employed. This leads to local grid refinement and local time-steps in a natural way.

1. Sparse grids

Sparse grids were introduced for the solution of elliptic partial differential equations, see [4] and references in [2]. They provide an efficient approximation method of smooth functions, especially in higher dimensions. So far, Galerkin methods [2] and finite difference schemes [3, 9] for elliptic problems on sparse grids have been investigated. There are also attempts to solve parabolic problems [1] and Navier-Stokes equations [9].

The multi-dimensional approximation scheme of sparse grids can be constructed as a subspace of the tensor-product of one-dimensional spaces represented by a hierarchical multi-resolution scheme [5]. Consider piecewise linear interpolants on a d-dimensional unit hyper-cube. We start with the one-dimensional hierarchical basis [11]. The space of functions on the regular grid of dyadic level l and mesh parameter $h = 2^{-l}$ can be represented by the space of all tensor-products of one-dimensional basis functions of support larger than 2^{-l-1}. The corresponding sparse grid space consists of all products of hierarchical basis functions with support larger than a d-dimensional volume of size 2^{-l-1}, see Figure 1. This is

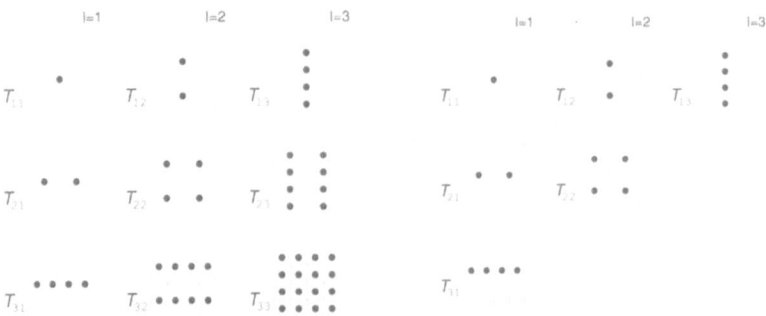

FIGURE 1. Tableau of supports of hierarchical basis functions for
a regular (left) and a sparse grid.

a subset of the regular grid space. A regular grid has about $2^{d \cdot l}$ nodes, which is
substantially more than the $2^l \cdot l^{d-1}$ nodes of the sparse grid.

Now it is straightforward to apply a Galerkin scheme to the spaces defined [2].
Another way to discretize equations on sparse grids is the combination technique,
which is an extrapolation scheme applied to solutions obtained on several regular
grids of about 2^l and 2^{l-1} nodes [4]. There exists also a Fourier-collocation method
on sparse grids [6].

A different way of a sparse grid discretization is a finite difference scheme [3],
which we will employ throughout this paper. We define the hierarchical transformation $\mathbf{H}$ as the hierarchical basis transformation on the regular grid from nodal
values to hierarchical values which is restricted to the sparse grid nodes. Based on
$\mathbf{H}$, we define the action of a one-dimensional finite difference operator for the discretization of a differential operator: We apply the associated standard difference
stencil $\mathbf{D}_i$ along the x_i-axis to values located on the sparse grid nodes in a specific
basis representation. To this end the values are given in nodal basis in direction
i and in hierarchical basis representation in all other directions $I \setminus \{i\}$. The associated transformation is denoted by $\mathbf{H}_{I\setminus\{i\}}$. The stencil $\mathbf{D}_i$ for each node itself
is chosen as the narrowest finite difference stencil available on the sparse grid. It
is equivalent to the corresponding stencil on a regular (non-equidistant) grid, e.g.
a first order upwind stencil for $\frac{\partial}{\partial x_i}$. In nodal values the finite difference operator
reads

$$\frac{\partial}{\partial x_i} u \approx \mathbf{H}_{I\setminus\{i\}}^{-1} \circ \mathbf{D}_i \circ \mathbf{H}_{I\setminus\{i\}} u$$

A general difference operator is then obtained by dimensional splitting. The linear
advection term, for example, can be discretized in nodal basis representation as

$$\nabla \cdot u \approx \sum_{i=1}^{d} \mathbf{H}_{I\setminus\{i\}}^{-1} \circ \mathbf{D}_i \circ \mathbf{H}_{I\setminus\{i\}} u \tag{1}$$

	regular grid	sparse grid
# nodes	$\mathcal{O}(h^{-d})$	$\mathcal{O}(h^{-1} \cdot |\log_2 h|^{d-1})$

TABLE 1. Number of nodes.

	$d = 1$	$d = 2$	$d = 3$	$d = 1$	$d = 2$	$d = 3$
order $p = 1$	$\mathbf{N^{-1}}$	$N^{-1/2}$	$N^{-1/3}$	$\mathbf{N^{-1}}$	$N^{-1+\gamma}$	$N^{-1+\gamma}$
order $p = 2$	N^{-2}	$\mathbf{N^{-1}}$	$N^{-2/3}$	N^{-2}	$N^{-2+\gamma}$	$N^{-2+\gamma}$
order $p = 3$	N^{-3}	$N^{-3/2}$	$\mathbf{N^{-1}}$	N^{-3}	$N^{-3+\gamma}$	$N^{-3+\gamma}$

TABLE 2. Storage-complexity of a regular (left) and a sparse grid discretization.

where the one dimensional difference operators $\mathbf{D}_i$ may be chosen as a two-point upwind stencil $\frac{1}{x_i - x_{i-1}} \cdot [-1\ 1\ .]$. On adaptively refined grids, the nearest neighbor nodes are chosen, which may lead to unsymmetric stencils.

The major advantage of sparse grids compared to regular grids is their smaller number of nodes for the same level l and resolution 2^{-l}. This is especially true in higher dimensions $d > 1$, see Table 1.

However, the question whether sparse grids have an advantage compared to regular grids does also depend on the accuracy of a solution obtained on a grid. We are interested in a comparison of accuracy versus number of nodes for both types of grids. We define the storage ε-complexity of an approximation method by the accuracy ε which can be achieved with a storage of N nodes. The accuracy depends on the smallest mesh parameter h and an approximation order p like $\varepsilon = \mathcal{O}(h^p)$ for smooth data. For regular grids the number of nodes depends on the space dimension d as $N_{storage} = h^{-d}$ which results in $\varepsilon = \mathcal{O}(N^{-p/d})$. In the case of sparse grids, the dependence on the dimension d is much weaker and we denote $\varepsilon = \mathcal{O}(N^{-p+\gamma})$ for every $\gamma > 0$ and an approximation order p. The sparse grid approximation breaks the curse of dimensionality.

Let us assume that a sparse grid approximation is of first order $p = 1$, which of course depends on the discretization order, the error norm, the smoothness of the solution and the sparse grid approximation itself. Then the sparse grid is competitive to a second order method in two dimensions and to a third order method in three dimensions, which is usually hard to construct, see Table 2.

Furthermore the number-of-operations complexity is of interest, because it is an estimate for the computing time a specific algorithm needs. In some cases the work count is proportional to the number of nodes. This is true for a single time-step of a standard explicit finite difference code. It is also true for the corresponding sparse grid code, because it is true for each operator $\mathbf{H}$ and $\mathbf{D}$. However, the work count usually is higher than the number of nodes for implicit discretizations and for stationary problems involving the solution of (non-) linear equation systems, and for time-dependent problems in general. The number of time-steps for an evolution problem of a fixed time interval is proportional to h due to the CFL condition, which leads to a higher work count complexity. On regular grids we obtain $N_{work} = \mathcal{O}(h^{-d-1})$, which is equivalent to the storage complexity in $d+1$ space dimensions. This means that any reduction in storage $N_{storage}$, e.g. through sparse grids, may reduce the number of operations even further.

We still have to check the assumption on the approximation order $p = 1$ (or some other constant) of finite difference sparse grids discretizations. Up to now such orders had been verified for the Poisson equation under strong regularity assumptions [9]. Furthermore similar estimates are available for the interpolation error under suitable smoothness assumptions [2]. Numerical experiments indicate that adaptive sparse grids can weaken the smoothness requirements.

It is the goal of this article to construct suitable finite difference discretizations for conservation laws and to investigate their numerical order of convergence.

2. Space-time schemes

We want to solve the scalar conservation law

$$
\begin{aligned}
u_t + \nabla \cdot f(u) &= 0 \quad \text{for } u(x,t), \\
x &\in \Omega \subset \mathbb{R}^d \\
t &\in [0, t_0]
\end{aligned}
\tag{2}
$$

written as an initial-boundary value problem. The standard procedure for the numerical solution of (2) is to discretize the space Ω and the initial value $u^0 = u(x, 0)$ on Ω and to step forward in time. The solution u^{t+1} at time step $t+1$ is computed from u^t and the boundary conditions. This 'time-stepping' scheme is iterated until $t = t_0$ is reached. This will be discussed in chapter 3.

An alternative solution algorithm uses a discretization of (2) in the 'space-time' domain $\Omega \times [0, t_0]$. The conservation law can be re-written as a boundary value problem

$$
\begin{aligned}
\nabla \cdot F(u) &= 0 \quad \text{for } u(x), \\
x &\in \Omega \times [0, t_0] \subset \mathbb{R}^{d+1}
\end{aligned}
\tag{3}
$$

with F given by $F_0(u) = u$ and $F_{1,\ldots,d}(u) = f(u)$. The algorithm requires the numerical solution of a single large (non-) linear equation system and returns an approximation of u at all time steps at once. This approach is related to waveform relaxation [10]. It has been used with finite elements for periodic parabolic equations on sparse grids in [1].

The storage requirements of the space-time formulation are often considered as too high. However, with sparse grid technique the additional dimension in storage is affordable. Furthermore there is the advantage of easy adaptive grid refinement in space-time. In any stage of the computation it is possible to introduce a finer grid, which gives better resolution in space and local time steps. This is often difficult or even impossible for time-stepping algorithms.

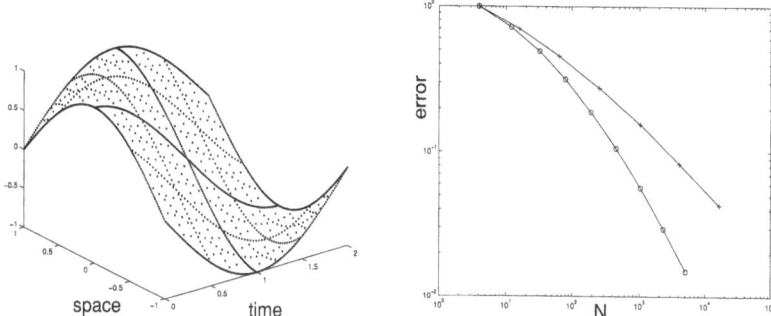

space   time   N

FIGURE 2. Sine wave $u = cos(\pi(x-t))$ (left); convergence history: L_∞ error vs. $N_{storage}$ (space-time), '+' regular grid and 'o' sparse grid.

2.1. Numerical examples

In this section we will consider the linear, oblique advection equation $f(u) := u \cdot 1\!\!1$ as a simple prototype for a conservation law. In space-time formulation it reads

$$\nabla \cdot u = 0 \quad \text{in } \Omega \subset \mathbb{R}^{d+1} \tag{4}$$

We use a first order upwind discretization in space-time, which is equivalent to an implicit Euler discretization in time.

For example on a two dimensional regular grid (one space and one time dimension), this is

$$\frac{1}{h}(u_{i,j} - u_{i-1,j}) + \frac{1}{h}(u_{i,j} - u_{i,j-1}) = 0$$

The corresponding finite difference sparse grid discretization is

$$\mathbf{H}^{-1}_{\{2\}} \circ \frac{u_{i,j} - u_{i-1,j}}{h_1} \circ \mathbf{H}_{\{2\}} + \mathbf{H}^{-1}_{\{1\}} \circ \frac{u_{i,j} - u_{i,j-1}}{h_2} \circ \mathbf{H}_{\{1\}} = 0 \tag{5}$$

First we test the rate of convergence for smooth data. We choose a sine wave as initial data and choose the Dirichlet data on the inflow boundary such that the solution in space-time is $u = cos(\pi(x - t))$ on the square $[-1, 1] \times [0, 2]$, see Figure 2. The linear equation system is solved with an iterative Krylov method.

Here, the convergence rates of the error in discrete L_1, L_2 and L_∞ norms, that is the ratio of $\varepsilon_{2h}/\varepsilon_h$ converges to the factor two for the regular gird and the sparse grid discretization, see Table 3. Hence both methods have the numerical order of convergence $p = 1$, which we expected for a first order scheme. Sparse grid Galerkin schemes often show a slightly slower convergence of $\mathcal{O}(h|logh|)$, which is not the case for finite differences, see also [9].

Convergence histories are given in Figure 2, demonstrating the rapid convergence of the sparse grid solution. In the ε-N diagram the complexity of the sparse

$2/h$	L_1	L_2	L_∞	L_1	L_2	L_∞
32	1.8973	1.8713	1.7746	1.4347	1.6189	1.6752
64	1.9348	1.9194	1.8696	1.5539	1.7186	1.7834
128	1.9610	1.9534	1.9297	1.6518	1.7985	1.8652
256				1.7247	1.8558	1.9235
512				1.7746	1.8932	1.9606

TABLE 3. Convergence rates on regular (left) and on sparse grids, sine wave.

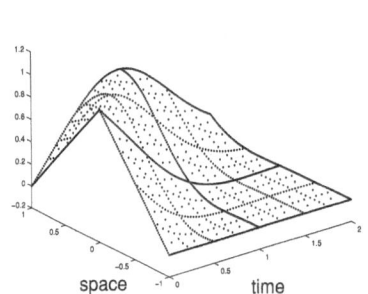
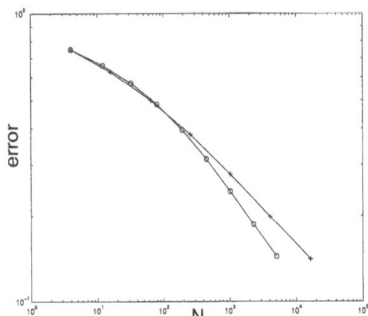

FIGURE 3. Hat function (left); convergence history: L_∞ error vs. $N_{storage}$, '+' regular grid and 'o' sparse grid.

grid can be seen. The steeper slope of the sparse grid approach is comparable to the performance of a second order method. Note that $N_{storage}$ accounts for the nodes in space-time.

The performance of sparse grids depends on the smoothness of the function to be approximated. The sine-wave example was analytic. In order to test the sparse grid method for data which does not match the smoothness requirements, we consider a C^0 hat function as initial data $u|_{t=0}$. The inflow data is set to zero. The hat function is advected towards the outflow boundary, see Figure 3. The solution is continuous and contains a jump in the first derivative. The jump is not resolved at each time step on a sparse grid, because there are not enough nodes on each time-slice. The interpolation procedure inherent to the sparse grid discretization introduces artificial viscosity in these cases.

The convergence rates in Table 4 indicate the numerical order of convergence $p = 1/2$ for the regular grid and an even lower order for the sparse grid solution. However, the complexity of the sparse grid algorithm still is better than of the regular grid in Figure 3, due to the lower number of nodes.

$2/h$	32	64	128	256	512
regular	1.3675	1.4021	1.4112		
sparse	1.2226	1.2606	1.2909	1.3030	1.2960

TABLE 4. L_∞ convergence rates on regular and on sparse grids, hat function.

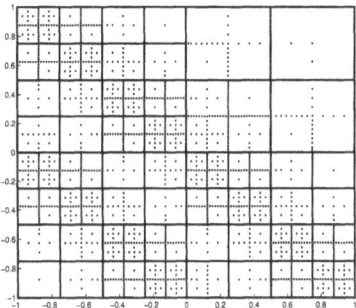
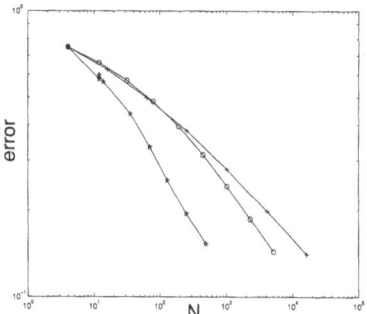

FIGURE 4. Adapted grid (left); convergence history: L_∞ error vs. $N_{storage}$, '+' regular grid, 'o' sparse grid and '*' adaptive strategy.

2.2. Adaptive space-time schemes

The performance of the sparse grid for the hat-function example degraded because of the jumps in the first derivative of the solution. Similar effects can be observed for many other discretizations. A common method to fix this is adaptivity.

The sparse grid is refined locally, to resolve such jumps. We employ an error indicator on the space-time mesh which indicates nodes with large errors. It is based on the absolute value of the associated hierarchical basis coefficient, see [3], and is related to residual based error indicators. The grid is refined in the neighborhood of nodes with large errors, inserting the hierarchical sons of a node. Locally a sparse grid of level $l + 1$ is obtained. A new solution is computed on the refined grid. The cycle of error indication, grid refinement and solution is iterated until a final error tolerance is matched, for further details see [2, 3].

A solution algorithm based on adaptive grid refinement is applied to the previous non-smooth test example. An adapted grid and the corresponding convergence history are depicted in Figure 4.

The complexity of the adaptive sparse grid in this case is similar to the second order complexity of the standard sparse grid in Figure 2. This means that, with the help of the adaptive grid refinement process, the original sparse grid performance is regained. The adaptive refinement does work as expected. However, the solve-refine cycle adds some overhead to the run time of the solution process.

Adaptive grid refinement added some features to the solution algorithm for time dependent problems: There is only one error indicator operating in the space-time domain, instead of several indicators, which operate separately in space and time and are coupled through CFL type of conditions. Grid refinement now enhances local space-resolution and at the same time introduces local time-stepping. However, due to the discretization implicit in time, there is no CFL condition.

3. Time-stepping schemes

Now we briefly look at the construction of time-stepping schemes based on sparse grids in space. We consider central difference schemes which are explicit in time in contrast to Godunov schemes. This means we are able to avoid Riemann solvers and can concentrate on finite difference stencils. We start with the first order prototype of all central scheme, the Lax-Friedrichs scheme [7]. We choose the non-staggered version

$$u_i^{t+1} = \frac{1}{2}(u_{i-1}^t + u_{i+1}^t) - \lambda \left(f(u_{i+1}^t) - f(u_{i-1}^t) \right) \tag{6}$$

with flux f, which can be extended to systems of equations easily. The construction of a sparse grid version of (6) requires a formulation that is invariant with respect to the hierarchical basis transformations $\mathbf{H}_{I \setminus \{i\}}$, e.g. a basis free formulation. In order to achieve this, we define the field

$$f_i^t := f(u_i^t)$$

for a given solution u^t in nodal basis. Equation (6) can be re-written as

$$u_i^{t+1} = \frac{1}{2}(u_{i-1}^t + u_{i+1}^t) - \lambda_i(f_{i+1}^t - f_{i-1}^t) \tag{7}$$

which is linear in u^t and in f^t. Hence equation (7) holds in any basis, especially in the hierarchical basis. As in the linear case, see equation (1), we apply (first order) dimensional splitting. We obtain the following algorithm for a single time step:

$$
\begin{aligned}
\hat{u}_{i,j}^t &= \mathbf{H}_{\{2\}} \, u_{i,j}^t \\
\hat{f}_{i,j}^t &= \mathbf{H}_{\{2\}} \, f(u_{i,j}^t) \\
\hat{u}_{i,j}^{t+1/2} &= \frac{1}{2}(\hat{u}_{i-1,j}^t + \hat{u}_{i+1,j}^t) - \frac{\Delta t}{\Delta 2 x_{i,j}}(\hat{f}_{i+1,j}^t - \hat{f}_{i-1,j}^t) \\
u_{i,j}^{t+1/2} &= \mathbf{H}_{\{2\}}^{-1} \, \hat{u}_{i,j}^{t+1/2} \\
\tilde{u}_{i,j}^{t+1/2} &= \mathbf{H}_{\{1\}} \, u_{i,j}^{t+1/2} \\
\tilde{g}_{i,j}^{t+1/2} &= \mathbf{H}_{\{1\}} \, g(u_{i,j}^{t+1/2}) \\
\tilde{u}_{i,j}^{t+1} &= \frac{1}{2}(\tilde{u}_{i,j-1}^{t+1/2} + \tilde{u}_{i,j+1}^{t+1/2}) - \frac{\Delta t}{\Delta 2 y_{i,j}}(\tilde{g}_{i,j+1}^{t+1/2} - \tilde{g}_{i,j-1}^{t+1/2}) \\
u_{i,j}^{t+1} &= \mathbf{H}_{\{1\}}^{-1} \, \tilde{u}_{i,j}^{t+1}
\end{aligned}
\tag{8}
$$

$1/h$	8	16	32	64	128	256	512
regular	1.4559	1.6623	1.8927				
sparse	1.2622	1.2575	1.5461	1.5009	1.4357	1.4422	1.4662

TABLE 5. L_∞ convergence rates on regular and sparse grids for CFL number $= 1/2$, sine example.

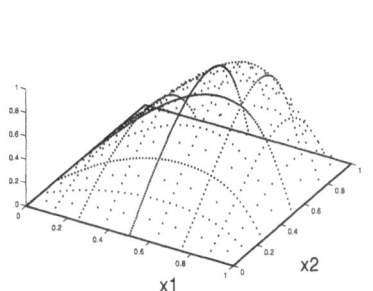
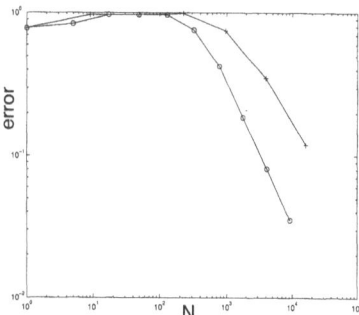

FIGURE 5. Solution for fixed Δt, sine example (left); convergence history: L_∞ error vs. $N_{storage}$ (space), '+' regular grid and 'o' sparse grid for fixed Δt.

Note that the hierarchical basis transform has to be applied to both the fields u^t and f^t and to $u^{t+1/2}$ and $g^{t+1/2}$.

3.1. Numerical experiments

We consider Burgers' equation in two space dimensions.

$$u_t + \frac{1}{2}\nabla(u^2 \cdot \mathbb{1}) = 0 \tag{9}$$

with smooth initial data $u|_{t=0} = \sin\pi x \, \sin\pi y$. We compute until $t_0 = .2$ which is before the shock at $t = 1/\pi$, with a CFL number $= 1/2$. This means that the solution remains smooth, see Figure 5 (left). We use the usual sparse grid without adaptive refinement. The numerical convergence rates are depicted in Table 5.

We obtain a numerical order of convergence of $p = 1$ for the regular grid case and a rate of about $p = 1/2$ for the sparse grid. This means that both methods seem to have comparable complexity. However, for the same number of nodes N, the sparse grid solution uses a finer h, which implies a smaller time step Δt and a better resolution of the non-linearity.

In order to compare both algorithms, we fix the time step Δt instead of the CFL number in the following. Choosing $\Delta t = 1/1024$, we obtain the convergence history and the solution of Figure 5.

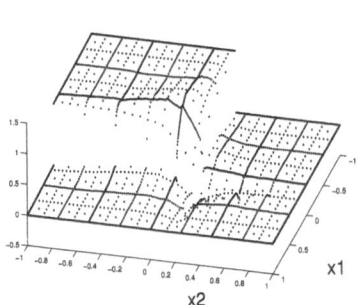
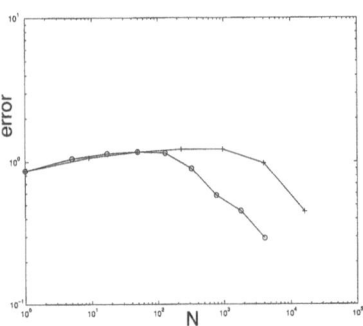

FIGURE 6. Solution for fixed Δt, jump example (left); convergence history: L_∞ error vs. $N_{storage}$, '+' regular grid and 'o' sparse grid for fixed Δt.

The sparse grid algorithm shows a better complexity and a better complexity order (slope) than the regular grid algorithm. This is due to the smooth solution, which can be resolved on the standard sparse grid. We expect an adaptive algorithm, which refines and coarsens the sparse grid in space and resolves the boundary layer at $x_i = 1$ to perform even better.

We again consider Burgers' equation (9) to test less smooth data, now with non-smooth initial data $u|_{t=0}$ chosen as a jump $t_0 = .5$. The jump is advected with constant velocity towards the corner $(1,1)$. We fix the time step $\Delta t = 1/256$. The solution and the convergence rates are depicted in Figure 6.

Both regular grid and sparse grid show slow convergence. The sparse grid demonstrates a better convergence than regular grid algorithm. The sparse grid is able to resolve the jump at specific locations such as short binary fractions $x_i = \mathbb{Z}$, $1/2\mathbb{Z}$, $1/4\mathbb{Z}$, ... It cannot resolve the jump at locations in between and uses linear interpolation through the hierarchical basis. Hence the time evolution algorithm creates slight oscillations. In our further experiments, even a piecewise constant basis, as proposed in [5] did not improve the performance significantly.

The initial data is no longer C^0, but discontinuous. However, we still use the linear hierarchical basis and C^0 approximations. We expect an adaptive grid in space, which resolves the jumps, to improve the convergence substantially.

3.2. Higher order central differences

As a brief outlook we propose a second order sparse grid central differencing scheme based on the scheme by Nessyahu-Tadmor [8]. We define the field f and a field u'^t_i of reconstructed slopes u', which can be obtained with a slope limiter in nodal basis.

$$f^t_i := f\left(u^t_i - \frac{\lambda_i}{2} f'(u^t_i)\right)$$

The scheme can be written as

$$u_i^{t+1} = \frac{1}{2}(u_{i-1}^t + u_{i+1}^t) - \lambda_i(f_{i+1}^t - f_{i-1}^t) - \frac{1}{8}(u'^t_{i+1} - u'^t_{i-1}) \qquad (10)$$

in any basis, because it is linear in u^t, in u'^t and in f^t. A second order sparse grid version can be obtained easily with second order Strang-splitting. In addition adaptive grid refinement will improve the performance of the scheme for non-smooth data.

4. Conclusion

We have constructed several numerical schemes based on sparse grids: An implicit discretization in space-time, an adaptive algorithm in space-time and an explicit central differencing scheme were proposed. Sparse grids have the advantage of a lower complexity than regular grids, especially in higher dimensions. Sparse grids can be interpreted as a global higher order method. Such methods usually require regularity, which either is given by smooth data or can be substituted by suitable adaptive grid refinement. Grid refinement has been demonstrated for the sparse grid space-time discretization, where a single error indicator controlled both refinement in space and in time.

References

[1] R. Balder, *Adaptive Verfahren für elliptische und parabolische Differentialgleichungen auf dünnen Gittern*, PhD thesis, TU München, Inst. für Informatik, **1994**.

[2] H.-J. Bungartz, T. Dornseifer and C. Zenger, *Tensor product approximation spaces for the efficient numerical solution of partial differential equations*, In *Proc. Int. Workshop on Scientific Computations*, Konya, **1996**. Nova Science Publishers, Inc. to appear.

[3] M. Griebel, *Adaptive sparse grid multilevel methods for elliptic PDEs based on finite differences*, In *Proc. Large Scale Scientific Computations*, Varna, Bulgaria. Vieweg, **1998**.

[4] M. Griebel, M. Schneider and C. Zenger, *A combination technique for the solution of sparse grid problems*, in P. de Groen and R. Beauwens, editors, Iterative Methods in Linear Algebra, IMACS, Elsevier, **1992**, 263–281.

[5] A. Harten, *Multi-resolution representation of data: A general framework*, SIAM J. Numer. Anal., **33** (1995), 1205–1256.

[6] F. Kupka, *Sparse Grid Spectral Methods for the Numerical Solution of Partial Differential Equations with Periodic Boundary Conditions*, PhD thesis, Universität Wien, Inst. für Math., **1997**.

[7] R. J. LeVeque, *Numerical Methods for Conservation Laws*, Birkhäuser, Basel, **1992**.

[8] H. Nessyahu and E. Tadmor, *Non-oscillatory central differencing for hyperbolic conservation laws*, J. Comput. Phys., **87** (1990), 408–448.

[9] Th. Schiekofer, *Die Methode der Finiten Differenzen auf Dünnen Gittern zur adaptiven Multilevel-Lösung partieller Differentialgleichungen*, PhD thesis, Universität Bonn, Inst. für Angew. Math., **1998**. to appear.

[10] S. Vandewalle, *Parallel Multigrid Waveform Relaxation for Parabolic Problems*, Teubner, Stuttgart, **1992**.

[11] H. Yserentant, *On the multilevel splitting of finite element spaces*, Numer. Math., **49** (1986), 379–412.

Institute for Applied Mathematics,
University Bonn,
Wegelerstr. 6
D-53115 Bonn, Germany
E-mail address: griebel@iam.uni-bonn.de
E-mail address: zumbusch@iam.uni-bonn.de

International Series of Numerical Mathematics
Vol. 129, © 1999 Birkhäuser Verlag Basel/Switzerland

Numerical Results for the Flux Identification in a System of Conservation Laws

G. Guiochon, F. James and M. Sepúlveda

Abstract. A numerical solution of the inverse problem of chromatography is described. This method consists in solving a constrained minimization problem for some cost function J. A Lagrangian formulation leads to a formal computation of the gradient of J. We then perform an exact computation for the gradient of the discrete cost function. This method allows the determination of numerical estimates of the coefficients for an isotherm model on a set of real experimental data.

1. Introduction

Chromatography is a very efficient separation process, used for instance to obtain highly purified compounds in pharmaceutical industry. For obvious economical reasons, the separation must be carried out at high concentrations, so that nonlinear phenomena dominate. In this setting, modelling the chromatographic process by a system of quasilinear conservation laws is reasonable (see [10, 3]). Indeed the separation relies on the interaction between a mobile phase and a stationary one. The nonlinear function involved in the hyperbolic system, which is called an *isotherm*, represents a thermodynamical equilibrium state between the two phases. It is a vector-valued function, $H(C)$, which gives the quantities in one phase, in chromatography the stationary solid phase, as a function of the quantities C in the other phase, here the mobile fluid phase. A more detailed description of the model is given in Section 2, followed in Section 3 by a discussion of the algorithm allowing the calculation of numerical solutions. Numerical results are presented in Section 4.

The knowledge of isotherms is of great importance in separation science. Indeed they allow numerical simulations of the processes, prediction of production rates, comparison between different separation schemes. Several experimental procedures exist for the determination of isotherms. Two of them will be mentioned here: the "frontal analysis" (FA) and the "elution by characteristic points" (ECP). We refer to [5] for other techniques, and for more precise references on this experimental aspect. We just emphasize that ECP is efficient only for a single compound (a real-valued H), and that FA allows the determination of "competitive" isotherms (vector-valued H), but is very slow, and requires significant amounts of expensive pure chemical products. Thus experimental determination of isotherms is a very tough problem.

Another possibility is to solve numerically the inverse problem corresponding to the above mentioned modelling of chromatography. We are thus led to the problem of identifying the flux in a nonlinear system of conservation laws. In theory, this approach is more interesting than the previous one, since it requires a smaller amount of experiments, which are moreover easier to perform. In practice, it turns out that the problem is mathematically ill-posed, and this leads to difficulties for the continuous equations as well as for the numerical resolution. The aim of this paper is to show an application of this method to a set of experimental data originating from the separation of a binary mixture. We evidence the fact that a careful choice of a model for the function H allows to recover a good agreement with the experimentally determined isotherms.

2. The equilibrium model for ideal chromatography

We consider a one-dimensional diphasic medium (a chromatographic or a distillation column) in which phase 1 moves with a velocity $u > 0$, and phase 2 with a velocity $v \leq 0$. The case in which $v = 0$ would correspond to a conventional chromatographic column, as operated in batch separations. The cases in which $v < 0$ are those of distillation and continuous counter-current chromatography, e.g., simulated moving bed chromatography (SMB). In the following, we assume that u is constant and $v = 0$. Let L be the length of the column. Although a more general theory could easily be presented, we consider for the sake of simplicity a mixture of only two components. We also assume that the temperature is constant and that the process is quasi-static. Because the mobile phase velocity and the temperature are constant, no momentum nor energy conservation equations are needed. The assumption of quasi-static equilibrium means that, at any time, the phase system is at thermodynamical equilibrium [3] and is equivalent to assume the ideal model. Therefore, the conservation of matter equations can be written

$$(C + H(C))_t + (uC)_z = 0, \qquad (z,t) \in \Omega = (0,L) \times (0,T), \qquad (1)$$

where $C = (C_1, C_2)$ is the vector of concentrations of the components in the fluid phase and H is a valued-vector function. Because experimental data do exhibit a certain amount of axial dispersion, the model does not give an exact band profiles. However, with the high efficiency columns used in many applications, the differences between actual band profiles and those predicted by the ideal model are often small [3].

We associate to Eq. (1) the following initial and boundary conditions

$$C(z,0) = C_0(z), \quad z \in (0,L), \qquad C(0,t) = C^{\text{inj}}(t), \qquad t > 0, \qquad (2)$$

where $C_0(z)$ and $C^{\text{inj}}(t)$ are given functions of the time and position in the column. Namely, $C_0(z)$ is the initial state of the column, which is often a constant, $C^{\text{inj}}(t)$

is the concentration profile of the injection performed. We shall assume that the injection has a compact support, that is $C^{inj}(t) = 0$ for $t > T_0 > 0$. From its thermodynamical origin, the function $H(C)$ has the property that the eigenvalues of its Jacobian matrix $H'(C)$ are real and positive. We can therefore perform the variable change $w = C + H(C)$, with $w = (w_1, w_2)$, where w_i is the total amount of component i in the column. We denote the reciprocal variable change by $C = g(w)$, and set $F(w) = ug(w)$. Thus, the system (1) can be rewritten in a more classical form:

$$w_t + F(w)_z = 0 \qquad (z, t) \in \Omega = (0, L) \times (0, T), \qquad (3)$$

where the initial and boundary conditions become

$$w(z, 0) = w_0(z), \quad z \in (0, L); \qquad w(0, t) = w^{inj}(t) \quad t > 0. \qquad (4)$$

The vector-valued function H will be the unknown of our identification problem. As we show later, the problem of identifying H as a function, say of class $C^2(\mathbb{R}^2)$, is particularly ill-posed. However, because of its thermodynamic origin, H has several properties which can help to restrict the domain of admissible functions. From a practical point of view, we can even compute explicit models for the function H, such that its shape is totally determined by a finite number of parameters. Any of the numerous models developed in liquid-solid or gas-solid adsorption studies can be used for this purpose [11]. Here we shall focus first on the bi-Langmuir isotherm which has been found to be an accurate model for the considered experiment [2]:

$$H(C_1, C_2) = \begin{pmatrix} \dfrac{\alpha_1 C_1}{1 + a_1 C_1 + a_2 C_2} + \dfrac{\beta_1 C_1}{1 + b_1(C_1 + C_2)} \\[4mm] \dfrac{\alpha_2 C_2}{1 + a_1 C_1 + a_2 C_2} + \dfrac{\beta_1 C_2}{1 + b_1(C_1 + C_2)} \end{pmatrix}. \qquad (5)$$

Next we consider a slightly different model, the Moreau model of degree 2 [8, 6], which writes

$$H_i(C_1, C_2) = \frac{C_i \dfrac{\partial P}{\partial C_i}}{2P} \qquad (6)$$

$$P(C_1, C_2) = 1 + k_1 C_1 + k_2 C_2 + e^{-\beta E_{2,0}} C_1^2 + e^{-\beta E_{1,1}} C_1 C_2 + e^{-\beta E_{0,2}} C_2^2$$

3. Identification

In the following, we emphasize the dependance of the solution to Eq. (3) with respect to F by denoting it w_F. The principle of identification is as follows. Being given a set of experimental data measured at the output of the column, $w^{obs}(t)$, we search for a function F such that the solution $w_F(L, t)$ is "as close as possible" to $w^{obs}(t)$. To give a precise meaning to this sentence, a cost function J is chosen, which measures, in a way described below, the distance between the experimental and the calculated functions. Then we set $\tilde{J}(F) = J(w_F)$, and we search to minimize $\tilde{J}$ over the set of all possible F-s.

As indicated earlier, the general problem, as stated above, is totally unrealistic. So we restrict the set of possible F-s by choosing a family of analytical isotherms depending on a finite set of parameters. Thus, we are led to a minimization problem on $\mathbb{R}^N$, for some given (possibly large) N. Even then, the minimization problem remains delicate, since the function $\tilde{J}$ is not globally convex, which may cause local minima to occur.

We give here only a brief presentation of the method. Further details and the mathematical justifications of the optimization process have been published elsewhere [4]. First, let us consider the choice of J. A natural idea is to consider a least-square distance

$$J_0(w) = \frac{1}{2} \int_0^T \|w(L, t) - w^{\mathrm{obs}}(t)\|^2 \, dt. \tag{7}$$

However, since the propagation velocity is finite and since the injection has a compact support, both $w(L, \cdot)$ and w^{obs} have a compact support. If

$$\mathrm{supp}(w(L, \cdot)) \cap \mathrm{supp}(w^{\mathrm{obs}}) = \emptyset, \tag{8}$$

one can easily see that J_0 remains constant if we move $w(L,.)$ along the t-axis. Actually, the function J_0 does not take into account the position of the profiles.

Therefore we complement J_0 with another cost function, J_1, which is defined as follows. For a given function $X(t) \in \mathbb{R}^2$, we define the first moment by

$$\mu_1(X) = \int_{-\infty}^{+\infty} tX(t) \, dt \quad \in \mathbb{R}^2. \tag{9}$$

This definition is meaningful because the functions that we consider have compact support. If X happens to be a concentration profile, then the two moments, $\mu_1(X)_1$ and $\mu_1(X)_2$, are related to the retention times of the two components [3]. Thus we have some information on the position of the profiles along the t-axis. By analogy, we define

$$J_1(w) = \frac{1}{2}\|\mu_1(w(L, .)) - \mu_1(w^{\mathrm{obs}})\|^2, \tag{10}$$

and the cost function that we will use is

$$J(w) = J_0(w) + \rho J_1(w), \tag{11}$$

where $\rho > 0$ is a parameter which has to be adjusted carefully.

The minimization of $\tilde{J}$ will be carried out using a classical optimization method, the so-called conjugate gradient method (see e.g. [7] and [9]). The first step of the method is the computation of the gradient of $\tilde{J}$. A formal computation is given below, assuming that all the functions involved are smooth. Actually, in practice, this is done on a discretized version of the cost function and of Eq. (3)

(see [4]), but the presentation is clearer this way. We will rewrite the minimization problem on $\tilde{J}$ as a constrained minimization problem on J, namely

$$\inf\{J(w) \quad ; \quad w \text{ is a solution of Eq. (3) for some } F\}. \tag{12}$$

It is well known that to any such constraint is associated a Lagrange multiplier [7]. Here this multiplier will appear as the solution of another partial differential equation, which we call the *adjoint equation*.

The constraint on w, that is "w is the solution of Eq. (3)", has to be rewritten in a weak form. First, it is possible to fix $T > 0$ such that $w(z, T) = 0$, because the injection is compactly supported. Then, let p be a smooth test function, and define $E(w, p; F)$ by

$$E(w, p; F) \quad = \quad -\int_0^T \int_0^L (w \cdot p_t + F(w) \cdot p_z) \tag{13}$$

$$+ \int_0^L w(z, T) \cdot p(z, T) \, dz + \int_0^T F(w(L, t)) \cdot p(L, t) \, dt$$

$$- \int_0^T w^{\text{inj}}(z) \cdot p(z, 0) \, dt - \int_0^L F(w_0(z)) \cdot p(z, 0) \, dz,$$

where the dot $\cdot$ denotes the scalar product in $\mathbb{R}^2$. The constraints are satisfied, that is $w = w_F$, if and only if $E(w, p; F) = 0$ for all p, which is the actual definition of a weak solution. We define now a Lagrangian for the constrained minimization problem by

$$L(w, p; F) = J(w) - E(w, p; F). \tag{14}$$

The computation of the gradient of $\tilde{J}$ is based on the remark that

$$L(w_F, p; F) = J(w_F) = \tilde{J}(F), \tag{15}$$

so that, when differentiating in the δF direction, we obtain

$$\tilde{J}'(F)\delta F = \left\langle \frac{\partial L}{\partial w}(w_F, p; F), \frac{\partial w}{\partial F}\delta F \right\rangle + \frac{\partial L}{\partial F}(w_F, p; F)\delta F \tag{16}$$

for every test function p.

Now, the principle of the method is that, since we cannot compute $\frac{\partial w}{\partial F}\delta F$, we will choose a particular p which cancels this factor. From a theoretical viewpoint, this can also be considered as a consequence of the inf-sup formulation for Eq. (7), which leads, at least formally, to cancel the gradient in w of L (see [9]). Let δw be a smooth function, and perform a formal differentiation of L in the δw direction. We obtain, taking Eq. (13) into account, and since w^{inj} and w_0 are given fixed functions,

$$\left\langle \frac{\partial L}{\partial w}, \delta w \right\rangle \quad = \quad \int_0^T \int_0^L \delta w \cdot (p_z + \partial_w F(w)^\top p_t)$$

$$+ \int_{z=L} \delta w \cdot \left(\frac{\partial J}{\partial w} - \partial_w F(w)^\top p \right) - \int_{t=T} \delta w \cdot p. \tag{17}$$

This quantity is equal to zero when p is a solution of the following backward linear problem, which is called the *adjoint problem*:

$$p_t + \partial_w F(w)^\top p_z = 0, \tag{18}$$

and which uses the following boundary conditions

$$\partial_w F(w(L,t))^\top p(L,t) = \left\langle \frac{\partial J}{\partial w}, \delta_t \right\rangle, \quad t \geq 0, \tag{19}$$

$$p(z,T) = 0, \quad z \in [0,L],$$

where δ_t is the Dirac mass at t. Note that Eq. $(19)_1$ is well defined since $\partial_w F(w)^\top$ is an invertible matrix – a consequence of the thermodynamical properties of H. To be more specific, if we consider the cost function J_ρ defined by Eqs. (7), (10) and (11), then the boundary condition in Eq. $(19)_1$ can be rewritten as

$$\partial_w F(w(L,t))^\top p(L,t) = w(L,t) - w^{\text{obs}}(t)$$

$$+ \rho \left[\mu_1(w(L,t)) - \mu_1(w^{\text{obs}}(t)) \right] t, \quad t \geq 0. \tag{20}$$

Now the gradient of $\widetilde{J}$ is given by the following expression

$$\widetilde{J}'(F)\delta F = \frac{\partial L}{\partial F}\delta F = -\frac{\partial E}{\partial F}\delta F$$

$$= \int_0^T \int_0^L \delta F(w) \cdot p_t + \int_0^L \delta F(w_0) \cdot p(0,z) \tag{21}$$

where p is any solution of Eqs. (18) and (19).

Provided the gradient of $\widetilde{J}$ is defined, we can apply the conjugate gradient method which belongs to the class of so-called steepest descent methods and is described elsewhere, for instance in [7]. Actually, this method has obviously to be applied to a discretized version of the equations. The long and tedious details of the discretization can be found in [4], as well as a discussion of some specific difficulties which arise here. Let us just mention that, once a numerical method is chosen to solve Eq. (3), one can perform the very same kind of computations at the discrete level, obtaining an "adjoint scheme" to solve Eq. (18), and a "discrete gradient", analogous to the one defined in Eq. (21).

4. Numerical results

Figures 1 to 3 compare the experimental band profiles obtained with two different samples, under different sets of experimental conditions (dotted lines), and those calculated from several identified isotherms (solid lines). Isotherms were identified at first experimentally (by the FA method), and the data were fitted to the bi-Langmuir isotherm (Eq. (5)), which is particularly relevant for this type of compounds ([3, 2]). The coefficients are reported in Table I, and the comparison between the profiles in Figure 1. As reported in [3], the agreement is excellent.

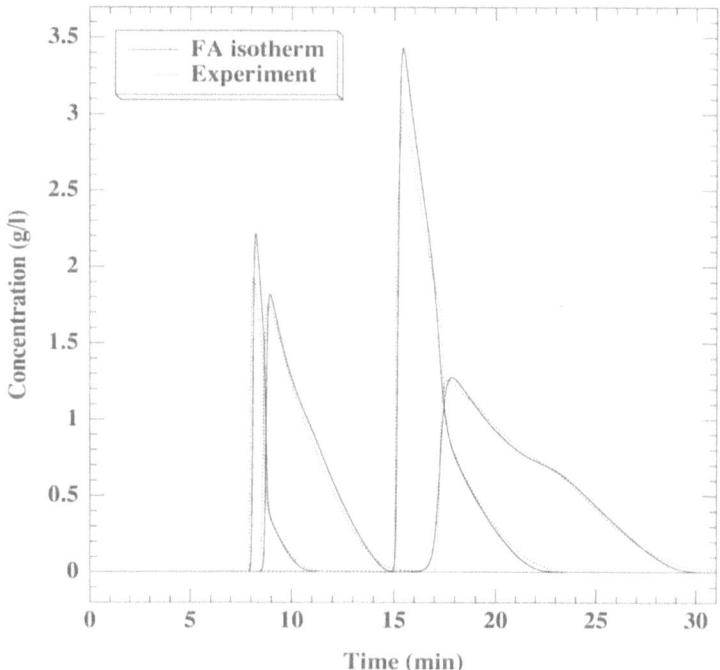

FIGURE 1. FA identification fitted on Bilangmuir

Coefficient	Estimate from FA data	Estimate from Band Profiles
α_1	4.40	4.62
β_1	2.89	2.89
a_1	0.202	0.205
a_2	0.312	0.305
α_2	6.79	6.92
b_1	0.0186	0.0313

TABLE 1. Best Estimates of the Numerical Coefficients
of the Bilangmuir Isotherm (Eq. (5))

Next, the coefficient of the bi-Langmuir isotherm were identified by the above described method, using as an observation only the couple of larger peaks in Figure 1. The best values of the numerical coefficients obtained are listed also in Table I. There is an excellent agreement between the two sets of coefficients, up to the order of a few percents. Figure 2 shows the comparison between the experimental (dotted lines) and the calculated profiles (solid lines). When comparing the large experimental profiles, we notice a better agreement with the calculated profiles in Figure 2 than in Figure 1 for the front of the first peak. This was expected since this peak was used as observation.

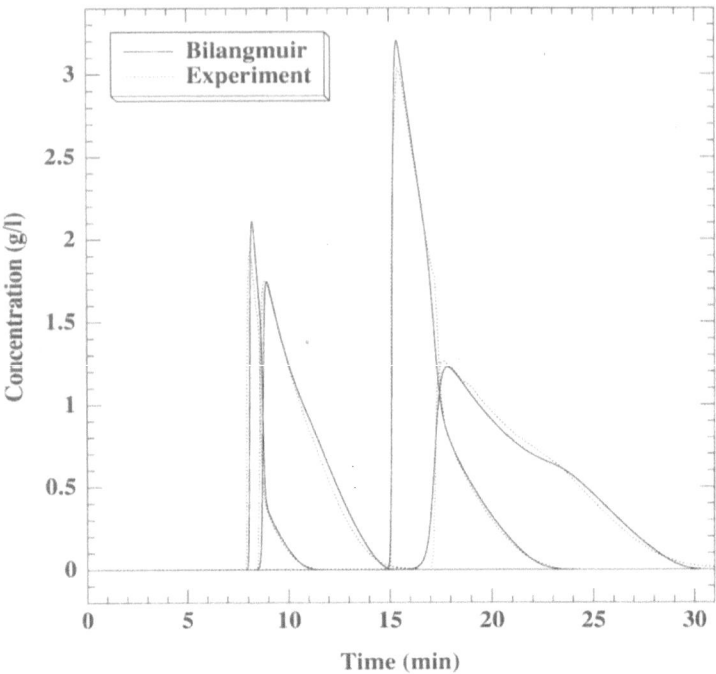

FIGURE 2. Bilangmuir identified by inverse problem

Finally, a third set of coefficients were computed using the Moreau isotherm (6), again with the right-hand couple of peaks as observation. The best estimates of the numerical values of the coefficients obtained with the calculation method described here are reported in Table II, together with those derived from an identification of the parameters of the same equilibrium isotherm model by nonlinear regression of the ECP data. The agreement between the values of k_i is good. The numerical values of the exponential coefficients are poor. This originates from the extreme sensitivity of the calculation method and from the ill-posed character of the problem.

In Figure 3, the band profiles calculated from the isotherms just obtained (Eq. (6) and Table II) are compared with the experimental data. The agreement between experimental and calculated profiles is quite good for the pair of large peaks used for the identification, as expected. More important is the poor agreement between the experimental (dotted lines) and calculated (solid lines) profiles obtained for the smaller pair of peaks (with a different relative composition). The important differences observed illustrate the marked deviations arising between the actual experimental isotherm data and those interpolated using Eq. (6).

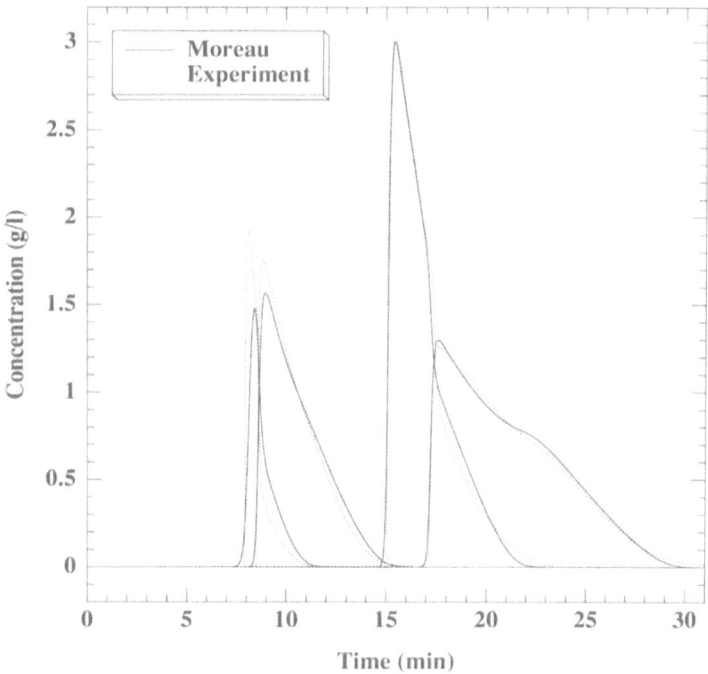

FIGURE 3. Moreau identified by inverse problem

Coefficient	Estimate from FA data	Estimate from Band Profiles
k_1	0.1103	0.1070
k_2	0.1563	0.1555
E_{20}	−95.59	−195.1
E_{11}	263.68	−196.8
E_{02}	121.08	304.1

TABLE 2. Best Estimates of the Numerical Coefficients
of the Moreau Isotherm (Eq. (6))

In summary, these results evidence that the particularly ill-posed problem of identifying the flux in a nonlinear system of conservation laws can be solved numerically, and fit to realistic experimental data. The success requires a careful modelling of the process, and in particular the introduction of specific analytic models for H was of great importance.

References

[1] F. Charton, M. Bailly and G. Guiochon *Recycling in preparative liquid chromatography*, J. Chromatogr., **687** (1994), 13–31.

[2] G. Guiochon, P. Sajonz and T. Fornstedt *A closer study of chiral retention mechanisms*, Chirality, In Press (1998).

[3] G. Guiochon, S. G. Shirazi and A. M. Katti *Fundamentals of Preparative and Nonlinear Chromatography*, Academic Press, Boston, MA. (1994).

[4] F. James, and M. Sepúlveda *Parameter Identification for a Model of Chromatographic Column*, Inverse Problems, **10**, (1994) 367–385.

[5] F. James, M. Sepúlveda, F. Charton and G. Guiochon *Determination of Competitive Equilibrium Isotherms from the Individual Chromatographic Band Profiles*. Technical Report 97-16. Departamento de Ingeniería Matemática, Universidad de Concepción (1997).

[6] F. James, M. Sepúlveda and P. Valentin *Statistical thermodynamic models for multicomponent diphasic isothermal equilibria*, Math. Models and Methods in Applied Science, **7 (1)**, (1997) 1–29.

[7] D. Luenberger *Linear and Nonlinear Programming*, Addison-Wesley, London, UK (1984).

[8] M. Moreau, P. Valentin, C. Vidal-Madjar, B. Lin and G. Guiochon *Adsorption Isotherm Model for Multicomponent Adsorbate-Adsorbate Interactions*, J. Coll. Interf. Sci., **141** (1991), 127–136.

[9] R.T. Rockafellar *Convex Analysis*, Princeton University Press, Princeton, USA (1970).

[10] P. Rouchon, M. Schœnauer, P. Valentin and G. Guiochon *Numerical Simulation of Band Propagation in Nonlinear Chromatography*, Separat. Sci. Technol., **22** (1987), 1793–1833.

[11] D. M. Ruthven *Principles of Adsorption and Adsorption Processes*, Wiley, New York, NY (1984).

Department of Chemistry,
The University of Tennessee,
Knoxville, TN, 37996-1600, USA
E-mail address: guiochon@novell.chem.utk.edu

Mathématiques, Applications et Physique Mathématique d'Orléans,
UMR CNRS 6628, Université d'Orléans,
BP 6759, F-45067 Orléans, France
E-mail address: james@cmapx.polytechnique.fr

Departamento de Ingenería Matemática,
Universidad de Concepción,
Casilla 4009, Concepción, Chile
E-mail address: mauricio@ing-mat.udec.cl

International Series of Numerical Mathematics
Vol. 129, © 1999 Birkhäuser Verlag Basel/Switzerland

Adaptive Mesh Coarsening in CFD

Tim Gutzmer

Abstract. In this paper we extend the Method of Transport to adaptivity in space and time. The algorithm is described in detail for the linear advection equation in conservation form. A numerical example is presented for two-dimensional Euler equations.

1. Introduction

In physics, flows are modelled by various systems of partial differential equations, e.g. the Euler equations, which describe inviscid compressible flow, or the Navier-Stokes equations, which include viscosity. In [2], Fey developed a special linearization strategy called Method of Transport (MoT) to solve these systems numerically. We review the MoT in Section 2.

Computationally the method is somewhat more expensive than more standard solvers like splitting methods or finite difference methods. The aim of the present paper is to reduce the computational costs by using adaptive methods. Based on a finite volume discretization on structured meshes Harten [9] suggested multiresolution algorithms depending on the local smoothness of the solution. Following these ideas we developed an adaptive method called Adaptive Mesh Coarsening using Interpolation Technique (AMCIT) for the calculation of unsteady solutions of the system of Euler equations in two space dimensions. The three basic phases of AMCIT are described in Section 3.

In this paper we emphasize the second phase of AMCIT, namely the calculation of the time step for the adapted grid by an extension of the MoT. While in principle this explicit method can handle hanging nodes, the following difficulty has to be addressed. To get the same numerical viscosity on all grids, it is necessary to perform different time steps on different grids. Basically this requires a flux computation with CFL number two on a fine grid at the transition to a coarser level. We can handle this grid transition without loss of order which is described in detail in Section 3.

Results are obtained for the two-dimensional Euler equations, cf. Section 4.

The efficient use of massively parallel computers for adaptive codes requires special efforts in the development of a dynamic load balancing. In the applications for which we propose AMCIT, highly adapted grids are necessary for accurately resolving various phenomena of the solution. Furthermore the problems are non stationary which results in a frequent modification of the grid. For an efficient implementation of AMCIT on massively parallel machines we refer to [8].

2. The method of transport

The Method of Transport (MoT) was originally developed in [1] for solving multi-dimensional Euler equations. It is a truly multidimensional explicit finite volume method that is based on the decomposition of the nonlinear system and the flux function into elementary waves. For the MoT, fluxes are not only considered between cells with common edges but also between cells which have a single point in common. The infinitely many propagation directions of the equations are approximated with some finite number of directions.

The MoT was extended in a series of papers to second order, cf. [2] and [5], and to other nonlinear system such as shallow-water equations [12], Navier-Stokes equations [11] and Magneto-Hydro-Dynamic equations [5].

2.1. Decomposition in linear advection equations

A multidimensional hyperbolic system can be written in the form

$$\boldsymbol{U}_t + \nabla \cdot \mathcal{F}(\boldsymbol{U}) = \boldsymbol{0}, \tag{1}$$

where $\boldsymbol{U}$ contains the conserved quantities and $\mathcal{F}$ represents the multidimensional flux. The divergence $\nabla \cdot$ acts on the rows of $\mathcal{F}$.

Assume we can decompose $\boldsymbol{U}$ and $\mathcal{F}$ in some waves $\boldsymbol{S}_i$ with velocities $\boldsymbol{a}_i$

$$\boldsymbol{U} = \sum_{i=1}^{k} \boldsymbol{S}_i, \quad \mathcal{F}(\boldsymbol{U}) = \sum_{i=1}^{k} \boldsymbol{S}_i \boldsymbol{a}_i^T.$$

Then we can rewrite (1) in

$$\sum_{i=1}^{k} \left[(\boldsymbol{S}_i)_t + \nabla \cdot (\boldsymbol{S}_i \boldsymbol{a}_i^T) \right] = \boldsymbol{0}.$$

The idea is to replace the nonlinear system by a number of linear advection equations in conservation form, namely

$$(\boldsymbol{S}_i)_t + \nabla \cdot (\boldsymbol{S}_i \boldsymbol{a}_i^T) = \boldsymbol{0}, \ i = 1, \dots, k. \tag{2}$$

This linear equation describes the transport of the quantity $\boldsymbol{S}_i$ in the direction $\boldsymbol{a}_i$, where $\boldsymbol{a}_i$ generally depends on the space vector $\boldsymbol{x}$.

In the case of two-dimensional Euler equations we get 5 advections equations of the kind (3) where the waves $\boldsymbol{S}_i$ are natural extensions of the right eigenvectors of the Jacobian of the flux function in one space dimension. The transport directions $\boldsymbol{a}_i$ are chosen in a way that the true domain of influence is covered by the numerical domain of influence. In the exact decomposition the $\boldsymbol{a}_i$ depend on the state vector $\boldsymbol{U}$. In order to discretise the equations in time the transport directions are frozen at time t_0.

In the separation of the transport equations in (2) an approximation error of second order is introduced. Due to the error analysis in [2] we end up with a second order scheme if the decomposition is modified by some correction terms.

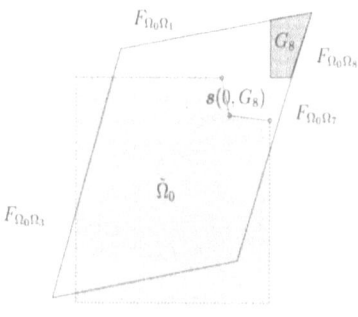

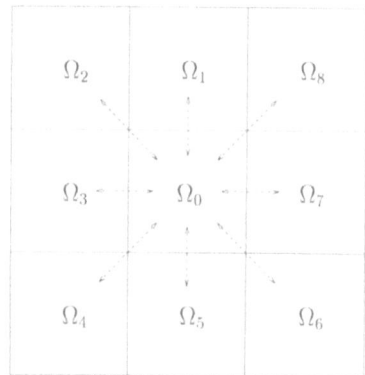

FIGURE 1. Sketch of transformation $z(t)$. The solid line represents the forward transformation of the original cell.

FIGURE 2. Interaction between the cell Ω_0 and its neighbouring cells for a two-dimensional cartesian grid.

2.2. Linear advection equation

The MoT for the linear advection equation in conservative form

$$w_t + \nabla \cdot (w \boldsymbol{a}(\boldsymbol{x})^T) = 0 \tag{3}$$

is an explicit finite volume method that is formulated for a cartesian grid. In the next section we will extend the idea to the transition between cartesian grids of different cell size. For the advection equation (3), the Method of Transport reads

$$w_{\Omega_i}^{n+1} = w_{\Omega_i}^n - \frac{1}{|\Omega_i|} \sum_{j \in V} (F_{\Omega_i \Omega_j} - F_{\Omega_j \Omega_i}), \tag{4}$$

where $|\Omega_i|$ is the area of the cell. The index of the central cell is denoted by i and V describes the set of indices of all neighbouring cells which have at least one point in common with Ω_i. For a cartesian grid in two space dimensions, the fluxes of the eight nearest neighbours have to be taken into account.

The flux $F_{\Omega_i \Omega_j}$ represents the quantity of information which is flowing from domain Ω_i into domain Ω_j. Let us assume that the characteristic curve $z(\tau)$ which is defined as the solution of

$$\dot{z}(\tau) = \boldsymbol{a}(z(\tau)), \quad z(t_0) = \boldsymbol{\xi} \tag{5}$$

exists and that the map $\boldsymbol{\xi} \mapsto z(t_0 + \Delta t)$ is bijective. Then for fixed $\boldsymbol{x}$ the variable transformation $\boldsymbol{v}(\boldsymbol{\xi}, \boldsymbol{x}) := z(t_0 + \Delta t, \boldsymbol{\xi}) - \boldsymbol{x}$ has an inverse $\boldsymbol{s}(\boldsymbol{\xi}, \boldsymbol{x})$, i.e. $\boldsymbol{v} \circ \boldsymbol{s} = \boldsymbol{\xi}$.

We can formulate the fluxes as an integral

$$F_{\Omega_i \Omega_j} = \int_{\boldsymbol{s}(0, G_j)} w(\boldsymbol{x}, t_0) \, d\boldsymbol{x}, \tag{6}$$

where $\boldsymbol{s}(0, G_j) \subset \Omega_i$ describes the inverse of the domain $G_j = z(t_0 + \Delta t, \Omega_i) \cap \Omega_j$.

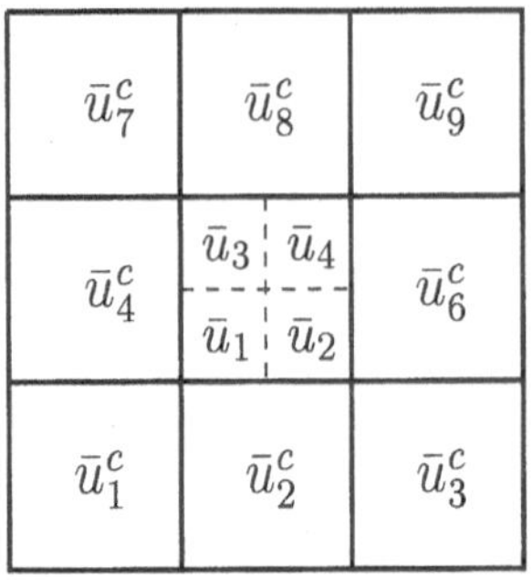

FIGURE 3.   Coarsening decision: $\left| \bar{u}_k - \overline{(s_{\bar{u}^c})}_k \right| \stackrel{!}{<} \epsilon, \; k = 1, \ldots, 4$
$s_{\bar{u}^c}$ polynomial interpolant of $\bar{u}_i^c, \; i = 1, \ldots, 9$

With a linear reconstruction of $\boldsymbol{a}(\boldsymbol{z})$ we get second order accuracy in time. Further the domains of integration become polygons which is illustrated in Figure 1. To get second order in space, we reconstruct $w(\boldsymbol{x}, t_0)$ linearly and integrate over this polygon to get the fluxes.

If the set of neighbours which have at least one point in common with Ω_i is considered, the CFL number

$$\mathrm{CFL} \; = \sup_{\boldsymbol{x}} \|\boldsymbol{a}\|_{\infty} \frac{\Delta t}{h}$$

is restricted to one. If enough neighbours are taken into account the method is unconditional stable for an arbitrary CFL number but the accuracy decreases for large CFL numbers.

3. AMCIT

Adaptive mesh coarsening using interpolation technique (AMCIT) was developed in [7] for the calculation of highly unsteady problems. The idea is to calculate the solution at any time on an overall fine grid. The computational costs are reduced by distinguishing regions near discontinuities, which have to be evaluated with a given complex upwinding scheme, from those in smooth regions which can be interpolated.

AMCIT consists of three basic phases. We start with the solution given on the finest grid which is assumed to be cartesian. A hierarchy of nested coarser grids is built up by combining together 2×2 cells to a coarser cell of double side length. An indicator based on polynomial interpolants determines locally which grid has to be used, cf. Figure 3.

In the second phase of AMCIT the solution is advanced in time with the MoT by calculating the fluxes between neighbouring cells. As described in Section 2 it is sufficient to solve the advection equation in conservation form with adaptation. To get adaptivity in time, it is necessary to perform different time steps on different cells depending on the size of the cells. We give the algorithm for the adaptive calculation of fluxes in the next section.

In the third phase, the solution advanced with one time step on the coarsest level is reconstructed from coarser to finer levels using thin plate spline interpolants.

3.1. Adaptation of MoT

In this section we extend the idea of the MoT for the linear advection equation in conservation form to the case of an adapted cartesian grid.

It is assumed that a cell Ω_j of side length $h/2$ is neighboured to a cell Ω_i with double side length h, cf. Figure 5. Further, we assume that a cell Ω_i has not both smaller and larger cells, which have a point in common with Ω_i. To advance both cells in time with the same numerical viscosity, it is necessary to calculate different time steps $\Delta t/2$ and Δt according to the magnitude of the cells. Two time steps on Ω_j correspond to one time step on Ω_i. This leads to adaptivity in space and time.

Let $V_{\Delta t/2}$ denote the set of all indices of the neighbours Ω_j of Ω_i which have at least one point in common with Ω_i. The subset $\tilde{V}_{\Delta t/2} \subset V_{\Delta t/2}$ contains the indices of cells with the same magnitude as Ω_i. The superset $V_{\Delta t} \supset V_{\Delta t/2}$ includes the indices of smaller cells which have a distance $h/2$ to Ω_i and are connected with Ω_i by a cell of the same size as Ω_i. Similarly $\tilde{V}_{\Delta t} \supset V_{\Delta t/2}$ includes the indices of all cells which have a distance h to Ω_i and are connected with Ω_i by a cell which is larger than Ω_i.

In the example of Figure 5 we have $V_{\Delta t/2} = \{1, \dots, 8\}$, $\tilde{V}_{\Delta t/2} = \{1, \dots, 7\}$, $V_{\Delta t} = V_{\Delta t/2} \cup \{9, 10\}$ and $\tilde{V}_{\Delta t} = V_{\Delta t/2}$.

With this definition we modify the MoT in (4) to

$$w_{\Omega_i}^{n+1/2} = w_{\Omega_i}^n - \frac{1}{|\Omega_i|} \sum_{j \in V_{\Delta t/2}} (\alpha_{ij} F_{\Omega_i \Omega_j}^{1/2} - \alpha_{ji} F_{\Omega_j \Omega_i}^{1/2}),$$

$$w_{\Omega_i}^{n+1} = w_{\Omega_i}^{n+1/2} - \frac{1}{|\Omega_i|} \sum_{j \in V_{\Delta t}} (\alpha_{ij} F_{\Omega_i \Omega_j}^1 - \alpha_{ji} F_{\Omega_j \Omega_i}^1),$$

where α_{ij} is the scaling factor

$$\alpha_{ij} = \begin{cases} 1, & \text{if } \Omega_i \text{ has the same side length as } \Omega_j, \\ \frac{1}{4}, & \text{if } \Omega_i \text{ has half the side length of } \Omega_j, \\ 4, & \text{if } \Omega_i \text{ has double the side length of } \Omega_j. \end{cases}$$

The algorithm for an update of a cell Ω_i with time step Δt is given by:

1. If all Ω_j, $j \in V_{\Delta t/2}$ have equal size or larger than Ω_i :
   (a) Calculate the fluxes $F_{\Omega_i \Omega_j}^{1/2} = F_{\Omega_i \Omega_j}$, $j \in V_{\Delta t/2}$ with time step Δt as in (6).

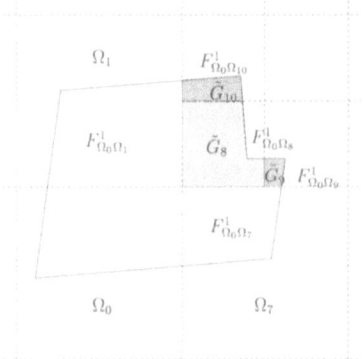

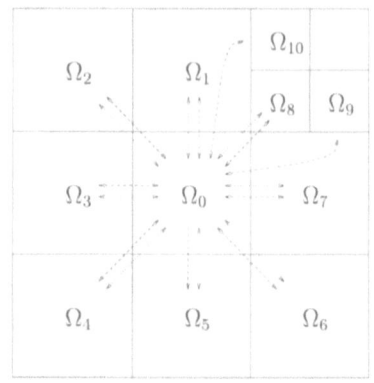

FIGURE 4. Sketch of transformation $z(t)$. The solid line represents the forward transformation of the modified cell $\tilde{\Omega}_0$ with Δt.

FIGURE 5. Interaction between the cell Ω_0 and its neighbouring cells for an adapted grid. The dotted lines show the fluxes at the first half of the time step, the dashed at the second half.

(b)
$$\text{Let}\quad \tilde{\Omega}_i := \Omega_i - \left(\bigcup_{j \in (V_{\Delta t/2} - \tilde{V}_{\Delta t/2})} s(0, G_j) \right)$$

where $s(0, G_j) \subset \Omega_i$ is defined in Section 2.2.

(c) Calculate the modified fluxes

$$\tilde{F}_{\Omega_i \Omega_j} = \int_{s(0,G_j) \cap \tilde{\Omega}_i} w(\boldsymbol{x}, t_0) d\boldsymbol{x}, \ j \in \tilde{V}_{\Delta t}$$

with time step $2\Delta t$.

(d)
$$\text{Set}\quad F^1_{\Omega_i \Omega_j} = \begin{cases} \tilde{F}_{\Omega_i \Omega_j} - F^{1/2}_{\Omega_i \Omega_j} & , \quad \text{if } j \in (V_{\Delta t/2} - \tilde{V}_{\Delta t/2}), \\ \tilde{F}_{\Omega_i \Omega_j} & , \quad \text{else .} \end{cases}$$

2. If there exists a neighbour Ω_j, $j \in V_{\Delta t/2}$ of a smaller size than Ω_i :
   (a) Calculate the fluxes $F^{1/2}_{\Omega_i \Omega_j} = F_{\Omega_i \Omega_j}$, $j \in V_{\Delta t/2}$ with time step $\Delta t/2$ as in (6).

(b)
$$\text{Let} \quad \tilde{\Omega}_i := \Omega_i - \left(\bigcup_{j \in (V_{\Delta t/2} - \tilde{V}_{\Delta t/2})} s(0, G_j) \right)$$

where $s(0, G_j) \subset \Omega_i$ is defined in Section 2.2, see Figure 1 where Ω_8 is assumed to be a cell with side length $h/2$.

(c) Calculate the modified fluxes

$$\tilde{F}_{\Omega_i \Omega_j} = \int_{s(0,G_j) \cap \tilde{\Omega}_i} w(\boldsymbol{x}, t_0) d\boldsymbol{x}, \; j \in V_{\Delta t}$$

with time step Δt, see Figure 4, where $\tilde{G}_j = \boldsymbol{z}(t_0 + \Delta t, \tilde{\Omega}_i) \cap \Omega_j$.

(d)
$$\text{Set} \quad F^1_{\Omega_i \Omega_j} = \begin{cases} \tilde{F}_{\Omega_i \Omega_j} - F^{1/2}_{\Omega_i \Omega_j}, & \text{if } j \in V_{\Delta t/2}, \\ \tilde{F}_{\Omega_i \Omega_j}, & \text{else .} \end{cases}$$

3. Update $w_i^{n+1/2}$

4. If all Ω_j, $j \in V_{\Delta t/2}$ have equal size or are larger than Ω_i :
   (a) Calculate the fluxes $F_{\Omega_i \Omega_j}$, $j \in V_{\Delta t/2}$ with time step Δt as in (6).
   (b) Set $F^1_{\Omega_i \Omega_j} := F^1_{\Omega_i \Omega_j} + F_{\Omega_i \Omega_j}$.

5. Update w_i^{n+1}

In step 1 of the algorithm, basically a flux computation with CFL number two for cells with indices in $V_{\Delta t} - V_{\Delta t/2}$ is required. In step 2 of the algorithm the flux $F_{\Omega_i \Omega_j}$ at time Δt is divided into $F_{\Omega_i \Omega_j} = F^{1/2}_{\Omega_i \Omega_j} + F^1_{\Omega_i \Omega_j}$ where $F^{1/2}_{\Omega_i \Omega_j}$ is the flux at the half time step $\Delta t/2$. If the CFL number is less than one it is guaranteed by step 2(b) of the algorithm that $V_{\Delta t}$ contains the indices of all cells Ω_j which can get a contribution $F_{\Omega_i \Omega_j}$ from Ω_i in the second half of the time step Δt.

To validate the adaptation of MoT in one and two space dimensions we give an order analysis in the following sections.

3.2. Order analysis in 1-D

The one-dimensional advection equation with constant coefficients writes

$$w_t + w_x = 0.$$

It is solved with the second order MoT

$$w_{\Omega_i}^{n+1} = w_{\Omega_i}^n - \frac{1}{|\Omega|} (F_{\Omega_i \Omega_{i+1}} - F_{\Omega_{i-1} \Omega_i}),$$

where Ω_i is the interval $[x_{i-1/2}, x_{i+1/2}]$ and $|\Omega_i|$ its length. The flux $F_{\Omega_i\Omega_{i+1}}$ from cell Ω_i into Ω_{i+1} is defined by

$$F_{\Omega_i\Omega_{i+1}} = \int\limits_{x_{i+1/2}}^{x_{i+1/2}+\Delta t} w(x - \Delta t)dx = \int\limits_{x_{i+1/2}-\Delta t}^{x_{i+1/2}} w(x)dx.$$

Using equidistant grid points, it is sufficient for second order accuracy that u is reconstructed linearly taking centred differences for the approximation of the derivative w_x. This is because the second order terms always cancel, cf. [3]. On a non-equidistant grid u has to be reconstructed quadratically by

$$w(x) = \alpha_0 + \alpha_1 x + \alpha_2 x^2, \quad \frac{1}{|\Omega_j|}\int\limits_{\Omega_j} w(x)dx = w^n_{\Omega_j}, \ j = i-1, i, i+1,$$

to get second order accuracy.

At the transition between coarse and fine grid the flux $F_{\Omega_i\Omega_{i+1}}$ at time Δt is divided into

$$F_{\Omega_i\Omega_{i+1}} = F^{1/2}_{\Omega_i\Omega_{i+1}} + F^1_{\Omega_i\Omega_{i+1}},$$

where $F^{1/2}_{\Omega_i\Omega_{i+1}}$ is the flux at the half time step $\Delta t/2$.

Therefore, we approximate $F^1_{\Omega_i\Omega_{i+1}}$ by the difference $F_{\Omega_i\Omega_{i+1}} - F^{1/2}_{\Omega_i\Omega_{i+1}}$.

To describe the local truncation error for w^{n+1}_i, the Taylor expansion of

$$w(x, \Delta t) = w(x, 0) - \Delta x \lambda w_x(x, 0) + \Delta x^2 \frac{\lambda^2}{2} w_{xx}(x, 0) + \mathcal{O}(\Delta x^3)$$

where $\frac{\Delta x}{2} = |\Omega_{i+1}|$ has to be integrated

$$\frac{1}{|\Omega_{i+1}|}\int\limits_{\Omega_{i+1}} w(x, \Delta t)dx = w(x_{i+1}, 0) - \Delta x \lambda w_x(x_{i+1}, 0)$$

$$+\Delta x^2 \left(\frac{\lambda^2}{2} + \frac{1}{96}\right)w_{xx}(x_{i+1}, 0) + \mathcal{O}(\Delta x^3).$$

The exact cell average is compared with the solution of the numerical scheme

$$w^{n+1}_{i+1} = w^{n+1/2}_{i+1} - \frac{1}{|\Omega_i|}(F_{\Omega_{i+1}\Omega_{i+2}} - F^1_{\Omega_i\Omega_{i+1}}).$$

Here $F_{\Omega_{i+1}\Omega_{i+2}}$ is the second order flux calculated on the values $w^{n+1/2}_i$, $w^{n+1/2}_{i+1}$, $w^{n+1/2}_{i+2}$.

MATHEMATICA yields for this expansion

$$w^{n+1}_{i+1} = w(x_{i+1}, 0) - \Delta x \lambda w_x(x_{i+1}, 0) + \Delta x^2 \left(\frac{\lambda^2}{2} + \frac{1}{96}\right)w_{xx}(x_{i+1}, 0)$$

$$+\mathcal{O}(\Delta x^3).$$

This means second order accuracy for w^{n+1}_{i+1}.

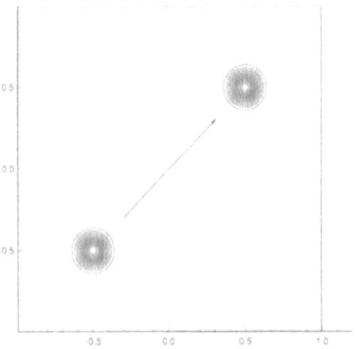

FIGURE 6.   Movement of the cone and grid in the example (SC)

3.3. Order analysis in 2-D

To illustrate the numerical order of the adapted MoT in two space dimensions the linear advection equation

$$\partial_t w + \partial_x(a(x,y)w) + \partial_y(b(x,y)w) = 0 \tag{7}$$

is investigated. We consider the Cauchy problem with initial data given by a smooth cone $w_0 \in C^2(\mathbb{R}^2)$ with center (α, β) and height 2

$$w_0(x,y) = \begin{cases} 1 + (1 - 4r)^6(\frac{560}{3}r^2 + 24r + 1), & r < \frac{1}{4} \\ 1, & \text{else} \end{cases},$$

where $r^2 = (x - \alpha)^2 + (y - \beta)^2$. For the local reconstruction of the solution w we used linear polynomials. This means we get an error of second order even on the finest grid at cells where the velocity changes sign. Moreover, an error of second order occurs at the transition between different grid levels. Due to [3] at these cells linear reconstruction is not sufficient to reach second order accuracy, because the second order terms do not cancel in these cases.

In the first example (SC) we choose constant coefficients $a = b = 1$ in (7). The cone is moved in a diagonal direction over a fixed grid, cf. Figure 6.

In the second example (RC) we look at variable coefficients $a = -y$, $b = x$. The advection equation transports the initial values defined by $(\alpha, \beta) = (.5, 0)$ counterclockwise around the centre of the coordinate system. At time $T = \pi$ the values have moved 180°. The cone is transported over a fixed grid that consists of fine cells in the right half of the coordinate system and of cells with double side length in the left half.

We get for (SC) convergence results of second order in the L_1 and L_∞ norm shown in Table 1. In (RC) a first order error occurs at the cells lying at the grid transition which diminishes in a global order of 1.5, cf. Table 1.

# points	cone (SC) error L_1	cone (SC) error L_∞	cone (SC) order L_1	cone (SC) order L_∞	cone (RC) error L_1	cone (RC) error L_∞	cone (RC) order L_1	cone (RC) order L_∞
64×64	$1.1\,10^{-2}$	0.41			$1.8\,10^{-2}$	0.60		
128×128	$2.8\,10^{-3}$	0.16	2.0	1.4	$6.9\,10^{-3}$	0.28	1.4	1.1
256×256	$4.7\,10^{-4}$	0.032	2.6	2.3	$2.3\,10^{-3}$	0.09	1.6	1.6
512×512	$9.1\,10^{-5}$	$6.0\,10^{-3}$	2.3	2.4	$9.6\,10^{-4}$	0.03	1.3	1.4

TABLE 1.   Error and order for (RC) and (SC)

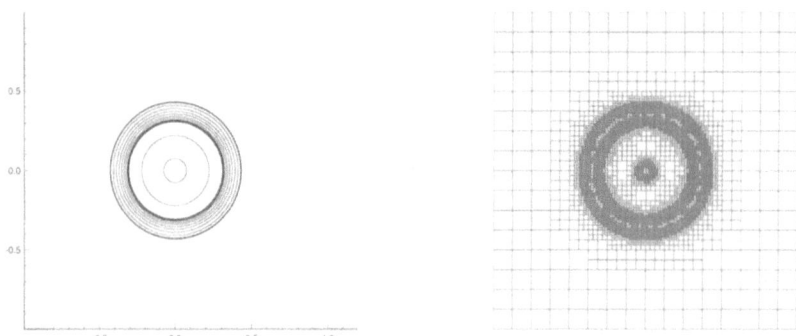

FIGURE 7.   Density and grid at time T=0.12

4. Numerical example

To demonstrate the performance of AMCIT for two-dimensional Euler equations we calculate the radial-symmetric explosion problem referenced in [10] with data

$$U(x, y, 0) = \begin{cases} U_l & \text{for } \sqrt{x^2 + y^2} < .13, \\ U_r & \text{for } \sqrt{x^2 + y^2} > .13. \end{cases}$$

The data starts with zero initial velocity everywhere and initial density and pressure $\rho_l = 2$, $p_l = 15$, $\rho_r = 1$, $p_r = 1$. We take 5 nested grids. The computational results at T=0.12 are displayed in Figure 7 for 1024×1024 gridpoints on the finest grid. The time step used was $\Delta t = 0.0003$ giving a maximum CFL number near of about 0.8. For the threshold ϵ in the error indicator we chose $\epsilon = 0.00025$.

The solution consists of a shock running outwards followed by a rarefaction wave and a contact discontinuity. At the centre a second shock runs inwards. Oscillations occur between the outward-moving shock and the contact discontinuity. This is caused by the minmod limiter which is performed only on the finest grid. Furthermore, the limiter is not independent of the direction, cf. [12]. Both the horizontal and the diagonal strip correspond well with the one-dimensional solution. Nevertheless, the two-dimensional solution is not as symmetric as the

| # points | elapsed time | | | # cells |
	AMCIT	finest grid	ratio	ratio
128×128	0.08h	0.14h	1.9	4.3
256×256	0.37h	1.20h	3.2	6.9
512×512	2.05h	9.92h	4.8	10.1
1024×1024	11.6h	83.1h	7.1	14.0

TABLE 2. Explosion Problem calculated on a Sparc 20 at 55MHz

one-dimensional result. Compared with the radial-symmetric results for shallow-water equations in [12], this is founded in the adaptation which happens to be not even symmetric in the generation of the grid, cf. Figure 7.

In Table 2 the performance of the sequential AMCIT for a varying number of grid points is compared with the code running on the finest grid. The threshold ϵ and the time step Δt have to be multiplied with 2 from each row to the next of Table 2 starting from the bottom to get the values corresponding to the coarser grids. Column 4 shows the ratio of the time consumption of the finest grid over AMCIT. Column 5 compares the part of the grid that is calculated by AMCIT with the finest grid; the maximum number of possible cells is divided by the number of cells actually calculated by AMCIT which is a measure for the saving of gridpoints by AMCIT. Both ratios show that the overhead for the adaptation is roughly 2.

References

[1] M. Fey, *Ein echt mehrdimensionales Verfahren zur Lösung der Eulergleichungen,* PhD thesis, ETH Zürich, (1993).

[2] M. Fey, *Decomposition of the multidimensional Euler equations into advection equations,* Seminar for Applied Mathematics, ETH Zürich, Research Report No. 95-14.

[3] M. Fey, *Das Transportverfahren von hoher Ordnung für skalare Erhaltungsgleichungen,* Manuscript, (1995).

[4] M. Fey, R. Jeltsch, A. -T. Morel, *Multidimensional Schemes for Non-Linear Systems of Hyperbolic Conservation Laws,* Numerical Analysis 1995, Edited by D. Griffiths and G.A. Watson, Longman, **1996**.

[5] M. Fey, R. Jeltsch, J. Maurer and A. -T. Morel, *The method of transport for nonlinear systems of hyperbolic conservation laws in several space dimensions,* Seminar for Applied Mathematics, ETH Zürich, Research Report No. 97-12.

[6] M. Fey and H.J. Schroll, *Monotone Split- and Unsplit Methods for a Single Conservation Law in Two Space Dimensions,* Institut für Geometrie und praktische Mathematik, RWTH Aachen, Research Report, No. 57, (1989).

[7] T. Gutzmer, *AMCIT – A New Method for Mesh Adaptation when Solving Time-Dependent Conservation Laws,* PhD thesis, ETH Zürich, (1998).

[8] T. Gutzmer, *Dynamic Load Balancing for Adaptive Mesh Coarsening in CFD*, Parallel Computational Fluid Dynamics '97, Proceedings of the Parallel CFD '97 Conference, Edited by A. Ecer, D. Emerson, J. Periaux and N. Satofuka, Elsevier Science B.V., to appear.

[9] A. Harten, *Adaptive Multiresolution Schemes for Shock Computations*, J. Comp. Phys., **115** (1994), 319–338.

[10] R. LeVeque, *Simplified Multi-Dimensional Flux Limiter Methods*, Proceedings of the ICFD Conference on Numerical Methods for Fluid Dynamics, Reading, **1992**.

[11] J. Maurer, *The Method of Transport for mixed hyperbolic-parabolic equations*, Seminar for Applied Mathematics, ETH Zürich, Research Report No. 97-13.

[12] A.-T. Morel, *A Genuinely Multidimensional High-Resolution Scheme for the Shallow-Water Equations*, PhD thesis, ETH Zürich, (1997).

Seminar for Applied Mathematics
ETH Zürich
Rämistrasse 101
Ch-8092 Zürich, Switzerland
E-mail address: gutzmer@sam.math.ethz.ch

International Series of Numerical Mathematics
Vol. 129, © 1999 Birkhäuser Verlag Basel/Switzerland

Adaptive Grid Methods for Reactive Flows

D. Hänel, P. Roth, M. Rose, C. Thill, U. Uphoff, and R. Vilsmeier

Abstract. The paper deals with a discussion of different adaptation concepts presently applied to simulations of reactive flows. The governing equations of inviscid, reactive flows are solved by Finite-Volume methods in combination with upwind schemes. Adaptive grid redistribution and hierarchical grid refinement methods are employed on structured grids. Hierarchical grid refinement concepts, like adaptive mesh refinement (AMR) and directional mesh refinement (DMR) have shown to be more flexible and efficient for multidimensional flows than grid redistribution. Unstructured, triangulated meshes offer good properties for grid adaptation due the fact that grid cells can be added, removed or deformed. Adaptation of unsteady features on unstructured grids are performed with a combinations of static and dynamic meshes. The different methods are demonstrated by results for detonations and other wave problems.

1. Introduction

Reactive flow phenomena range from stationary flames to complex, unsteady wave pattern and detonations, governed by more or less complex homogeneous or heterogeneous chemical kinetics. The simulation of reactive flows requires the solution of large systems of partial differential equations consisting of the usual conservation equations for a fluid phase and in addition of a number of species equations with source terms arising from chemical reactions. For diverging temporal scales of chemistry and flow, the system of equations becomes mathematically stiff, which requires special, implicit integration schemes. Similar scaling effects can arise in space, where combustion effects can take place in distinct zones of small size but with steep gradients of variables. If chemistry and flow are strongly interacting then complex unsteady combustion wave and detonation can appear. The exponential behaviour of variables due the sources terms (e.g. by Arrhenius laws) induces a high sensibility against numerical errors in these reaction zones.

In general, the simulation of reactive flows requires very accurate, efficient and robust methods of solution, more than required for common aerodynamic flows.

The use of adaptive grid methods is an effective way to satisfy essential requirements of accuracy for reactive flows. Adaptive grid methods are solution methods, where the grid is controlled during and by the solution by means of error estimates or other criterion. They are appropriate methods to deal with flow problems containing different characteristic spatial scales, where typically high resolution is required in rather small zones of the integration domain while in other parts moderate resolution is sufficient. Consequently, adaptive methods offer a global, high accuracy and sparse use of computer capacity.

2. Methods of solution

2.1. Governing equations

Chemical flows are studied presently under the assumption of compressible, inviscid flow. The governing equations (Euler equations) are derived from the conservation laws for mass, momentum, and energy of a fluid mixture.

The conservation equations for a control volume τ read in the integral form:

$$\int_V \vec{Q}_t \, d\tau + \oint_A \vec{H} \cdot \vec{n} \, dA = \int_\tau \vec{S} \, dV \tag{1}$$

The vector of conservation quantities $\vec{Q}$ has the components of the density ρ, the momentum $\rho\vec{v}$, the energy ρE and the species density ρ_s of a multi-component mixture. The terms $\vec{H}$ describe the corresponding fluxes acting normally on the surface A. The source term S contains chemical production rates in the case of chemically reacting flows.

$$\vec{Q} = (\rho, \rho\vec{v}, \rho E, \rho_s)^T \quad \vec{H}_{inv} = (\rho\vec{v}, \rho\vec{v}\,\vec{v} + p, \rho\vec{v} H_t, \rho_s\vec{v})^T \quad \vec{S} = (0, 0, \sum \dot{Q}_s, \dot{j}_s)^T$$

The solution of the conservation laws requires additional closure relations to express the thermal, the calorical, and the transport quantities arising in the flux $\vec{H}$ as function of the conservative variables $\vec{Q}$.

2.2. Numerical discretization

The conservation equations are discretized by Finite-Volume methods on structured or unstructured grids, which explore the integral form Eq. (1) on a small, finite control volume V_k. The conservative variables $\vec{Q}$ are volume-averaged values. The surface integral is approximated by the sum of numerical fluxes $\tilde{H}_k(\vec{Q})$ over all faces k of the control volume with the normal vector $\vec{n}_k$.

$$\frac{\Delta\vec{Q}}{\Delta t}V_k + Res_k(\vec{Q}) = 0 \quad \text{with} \quad Res_k(\vec{Q}) = \sum_{k=1}^{nk}\left(\tilde{H}_k(\vec{Q})\, A_k \cdot \vec{n}_k\right) - S_k(\vec{Q})\, V_k \tag{2}$$

In essential two different upwind formulations for the numerical fluxes $\tilde{H}_k(\vec{Q})$ have been used.

"The Advective Upwind Splitting Method" (AUSM) by Liou [1] is a splitting scheme, based on previous studies, [2]. This so-called hybrid splitting results in rather simple flux formulations well suited for structured and unstructured grids, as well. The AUSM scheme splits the inviscid flux F in a convective part, upwinded in flow direction and in a pressure part, upwinded corresponding to sonic spreading.

Test calculations in [1] have shown very accurate results, comparable with Roe's flux-difference splitting. A variant of this splitting is used by the authors in [3] for 2-D and [4] in 3-D in algorithms for unstructured grids. Extension to higher order accuracy is performed with MUSCL extrapolation and van Albada limiters, [5]. The AUSM scheme has shown a favourable low numerical dissipation, [6], well suited in Navier-Stokes solvers for subsonic and supersonic flows.

The flux-difference splitting method by Roe [7] is employed in a variant by Yee [8]. to inviscid, chemically reacting flow, which results in an efficient formulation for multiple species equations. The modified flux approach by Harten [9] and his entropy correction is applied for higher order computations on structured grids. On unstructured grids the Roe scheme is used in combination with the MUSCL approach.

2.3. Integration in time

Explicit Runge-Kutta time-stepping schemes are used for solving the flow part (Euler equations) and non-stiff chemistry. A Runge-Kutta scheme is employed which performs second order accurate in time and requires a minimum amount of computer storage. An intermediate Runge-Kutta step l for the discrete equations Eq. (2) reads

$$Q^{(l)} = Q^n - \alpha_l \, \Delta t \, Res(\hat{Q}^{(l-1)}) \quad \text{with} \quad l = 1, \cdots, N \tag{3}$$

In calculations $N = 4$ or 5 steps Runge-Kutta schemes are employed, with a theoretical maximum Courant number of $N-1$. The upwind scheme results in a slightly lower Courant number for reasons of stability. To accelerate the convergence to the steady-state solution local time steps are used which are dictated by the local stability limit and constant Courant number.

The *multi-sequence Runge-Kutta scheme* is a variant of the explicit Runge-Kutta version and is especially developed to reduce the severe stability restrictions in time-accurate flow computations on unstructured grids with cells of very different sizes as they usually appear in adapted grids. It is described in detail in [3]. The grid cells are sorted in groups of similar size. Within one global time step one Runge-Kutta sequence is performed on coarsest cells, on groups with smaller cells two or more sequences are applied to them. The parameter of the Runge-Kutta scheme and the organisation of the algorithm are synchronised to preserve time accuracy and are modified to perform multiple Runge-Kutta sequences with a minimum of additional work. Compared with the basic Runge-Kutta scheme, the multi-sequence scheme saves CPU time by a factor of 5 to 10.

Operator-splitting of Strang-type and combined implicit-explicit solutions are applied to problems of stiff equations, which appear in particular in reactive flow with detailed chemistry. The discrete equations Eq. (2) are split with respect to characteristic length and time scales into a flow part L_F and in a chemical part L_C and are solved in a sequence

$$L_F^{\Delta t/2} \, L_C^{\Delta t} \, L_F^{\Delta t/2} \, (Q) = 0 \tag{4}$$

3. Adaptive methods on structured grids

Structured meshes are well suited for most of discretization schemes and enable a high degree of vectorization and parallelization. They are characterized by a fixed number of grid points in a fixed order. But this order does not permit addition or depletion of grid cells without destroying the order. The only way of adaptation on a single mesh is therefore the method of adaptive grid redistribution, which means shift of nodes relative to their neighbours by preserving the node connections.

If additional, locally refined subgrids are superposed, then local grid adaption can be realized in a flexible manner. The most known of such hierarchical refinement concepts is the adaptive mesh refinement (AMR) method by Berger, Colella [10] and Quirk [11].

The two essential adaptive techniques on structured grids, the adaptive grid redistribution and the adaptive mesh refinement (AMR) technique and a variation, the directional mesh refinement (DMR) [12], are outlined in this following section.

3.1. Grid redistribution concepts

Grid redistribution on structured meshes is based on a fixed number of grid points, which are rearranged during the solution by applying an adaptation criterion. Redistribution schemes are generally implemented using either a control function [13], variational or spring-analogy schemes, [14]. Severe restrictions of redistribution schemes can arise during adaption from cells becoming too distorted in multi-dimensional problems. On the other hand, grid redistribution retains the original grid structure and all flow solvers for structured grids can be used without modification.

While adaptive grid redistribution in multi-dimensional problems is generally restricted, the same technique in one dimension is a rather flexible way for studying unsteady problems with complex physics. An 1-D, unsteady algorithm with dynamic grid redistribution has been developed in our numerical group [15] for analyzing effects of detailed chemistry in detonation waves. The solution is based on the Euler equations in coordinates relative to a moving grid, additionally with several species conservation equations and source terms for detailed chemistry. The numerical flux is modelled by the second order accurate Harten-Yee TVD-scheme [8]. Computations with detailed chemistry ($n_s \gg 1$) are performed with operator splitting and an implicit solver for stiff ODE's to overcome the stiffness.

The grid movement $\dot{x}$ is described by an additional equation, which is solved simultaneously with the flow equations. The mesh nodes are moved so that the arc-length based on selected variables is of equal size. The control function chosen for adaption is defined as a linear combination of gradients of variables (temperature, mass fraction, etc.) which are crucial for accurate simulations. Details of the formulations are given in [16].

One demonstrative example is presented for the time-dependent development of a detonation wave. The chemical reactions in a H_2/O_2 mixture were described by detailed reaction mechanisms consisting of 21 elementary reactions and 8 species.

The movement of grid points by adaptation is given in the $x - t$ plane in Fig. 1 (left) and the right figure shows the calculated pressure profiles at different times, which give an impression about the large gradients behind the front.

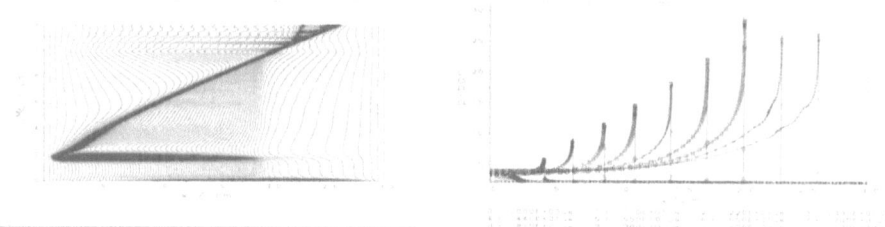

FIGURE 1. Computation of the development of an 1-D, unsteady detonation wave for detailed O_2/H_2 chemistry on a dynamic grid using grid redistribution. Space-time diagram of grid point movement (left) and pressure profiles at different times (right).

3.2. Adaptive mesh refinement

The adaptive mesh refinement (AMR) technique on structured grids is a widely used adaptation concept on structured grids. It was proposed by Berger, Colella 1989, [10], algorithmic improvements were made by Quirk [11] and others. The AMR concept has reached now a state where it can be applied more or less automatically to structured grid methods. An extension of the software package for conservation laws CLAWPACK to AMR is available now as a public domain software for solving 2-D flow problems, called AMR-CLAWPACK [17].

The adaptive mesh refinement concept employs hierarchical sets of successively finer subgrids in regions to be refined. The hierarchical structure tracks dynamically the flow features in an automatic manner, and adapts in both time and space. The finer subgrids cover and overlap coarser grid cells, where the coarsest grid is usually a global grid over the integration domain. Since the subgrids are orientated along the grid lines of the next coarser grid, rather simple connecting conditions can be defined. The AMR method refines simultaneously in all coordinate directions, thus cells of aspect ratio larger than in the basic grid do not appear at finer levels. The homogeneous refinement in AMR leads to a drawback if anisotropic features, like shear layers or even fronts are traced since in such cases cells of high aspect ratio are desirable. Thus the homogeneous refinement in AMR leads to a high complexity in particular in 3-D, because the memory requirement grows with the third power on higher refinement levels.

The AMR concept has been widely applied to inviscid or reactive flows adapting distinct wave and shock features, e.g. [18, 19, 20]. Applications in astrophysics are found e.g. in [21], to compressible Navier-Stokes solutions in [22], [23].

An application of the adaptive mesh refinement to reactive flow of a H_2/air mixture is presented in Fig. 2. The chemical reactions are described by 21 elemen-

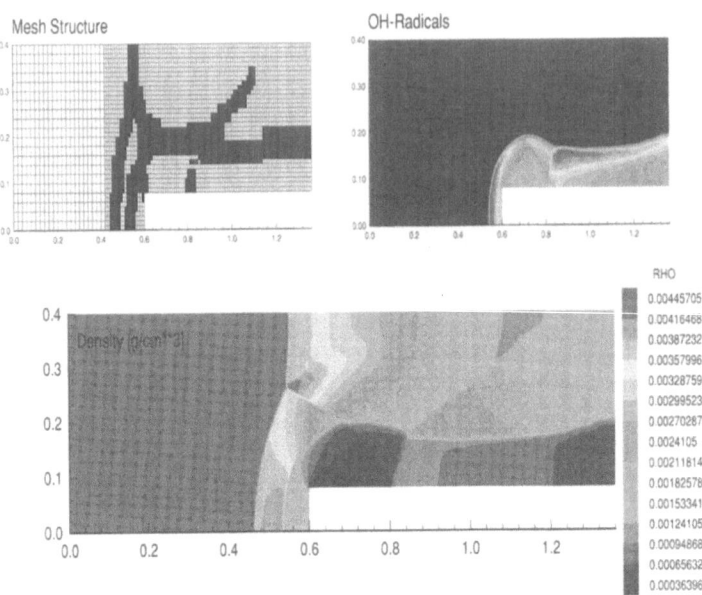

FIGURE 2. Reactive, supersonic flow at $Ma_\infty = 3.6$ over a forward facing step with detailed chemistry for a stoichiometric mixture of air/H_2. Shown is the mesh structure using AMR, the distribution of OH-radicals (burning range) and the distribution of global density (below).

tary reactions and 8 species. Supersonic flow forms a shock in front of a forward facing step, which ignites the gas mixture. The lower figure shows the "colours" of the global density with the formation of shocks and flame fronts. The zone of burning gas is represented by the "colours" of OH radicals in the upper, right figure. In addition, the mesh with refined zones is given in the upper, left figure. Details of formulations of the AMR method and applications to doppel-front detonations with heterogeneous gas-particle reactions are presented in [16].

3.3. Directional mesh refinement

An improvement of adaptive mesh refinement for anisotropic features offers the partially refinement in selected, individual directions, e.g. only in x- or in y-direction, so that cells of large aspect ratio are generated during adaption. This concept, called directional, adaptive mesh refinement (DMR), proposed in [12], has found to be better suited for viscous layers or front tracing, than the original AMR concept. The DMR concept enables the choice between three different types of refinement in 2-D and to 7 types in 3-D, so that higher cell aspect ratios on finer levels may be generated. Additionally to the directional refinement, the proposed algorithm avoids overlapping sub-cell structures, which results in essential memory savings compared to the AMR method. Difficulties with updating the fluxes, in particular due to hanging nodes, are overcome by using a least square method. A demonstrative example of this DMR concept is presented in Fig. 3 for supersonic, inviscid flow over a forward facing step in comparison with the same problem, using the AMR method, [12]. The memory requirement of DMR is about 1/3 of that of AMR for comparable accuracy.

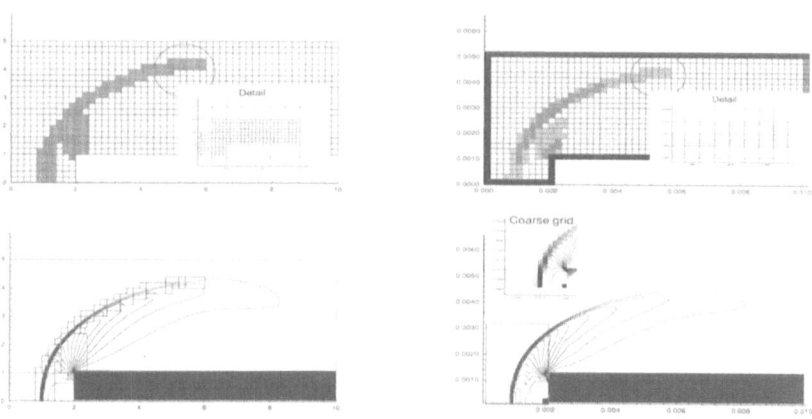

FIGURE 3. Supersonic, inviscid flow at $Ma_\infty = 3.$ over a forward facing step. Comparison of mesh structure and isobars using AMR (left) and DMR (right).

4. Adaptive methods on unstructured grids

Finite-Element as well Finite-Volume methods on unstructured, triangulated or tetrahedral grids have great advantages with respect to geometrical flexibility and adaptive meshing. Unstructured grids enable flexible handling of grid cell connections and thus, grid cells can be added, removed or deformed according to prescribed adaptation criterion. Adaptive solution concepts are very well developed for inviscid and viscous flows over complex 2-D and 3-D geometries. For general aspects of unstructured grids and more information about adaptation see also review papers, as e.g. in Deconinck, Barth [24], Venkatakrishnan [25].

Beside of the advantages of this mesh type, there exist a number of problems arising in this grid concept. Major problems are caused by the presence of different scale lengths, as they appear e.g. in viscous layers or near fronts. Reasonable resolution can only be achieved there by deformation of grid cells to high aspect ratios, i.e. by flat triangles or tetraeders, which impair the accuracy. Hybrid grids, a combination of triangular and quadrilateral elements, are a possible way to reduce this drawback. Corresponding developments, based on the algorithm described below, are in progress, [3]. An additional problem arises if such zones, like fronts, move in time and have to be adapted adequately.

An adaptive method on unstructured grids, is currently under development in our numerical group with the aim to overcome the problems mentioned above. This basic method solves the 2-D or 3-D Euler or Navier-Stokes equations for steady and unsteady flows, [6, 3, 26]. The equations are discretized by a Finite-Volume method on node-centred volumes with the AUSM or Roe splitting for the inviscid fluxes. Integration in time is performed with the standard Runge-Kutta scheme, Eq. (3) for steady state and with the multi-sequence Runge-Kutta scheme, [3], in unsteady cases. Mesh adaptation and flow computation are performed in a combined algorithm. The generation consists of several tools, as they are local mesh density, point insertion, edge reconnection and smoothing, which act all together on a closed triangulation. The process might be interrupted for a flow computation and continued for a following adaptation. Thus the steps of the mesh optimization problem are common for the generation as well for an adaptation of the mesh. Adaptation is performed by a method, called virtual stretching. Virtual stretching is also used to generate flat triangles in anisotropic flow regimes. Where, how strong and in which direction the mesh has to be changed is prescribed by one or more adaptation criterion in form of a vector, i.e. under consideration of absolute value and direction of the criterion. Criterion may be e.g. the gradient of density or the vorticity vector or both in combination. If the criterion is satisfied, then the mesh is locally transformed (stretched) to an isotropic mesh. Mesh generation proceeds now in the transformed plane. Transformation back to physical coordinates yields meshes with anisotropic resolution or refined zones. Special care is taken to combine different local resolution requirements, as several features are considered. More details about this method in 2-D are given in [3] and for 3-D in [26].

Extension to unsteady adaptation for moving features is possible in this concept, since the generation and adaptation mesh is a process that can be interrupted and continued at any time, and thus at any physical time levels. To avoid severe interpolation errors on dynamic grids, a combination of a static and a dynamic grid is used. The static mesh represents the essential steady-state features and is never changed. Adaptations are made using additional points, forming the dynamic mesh. These points are created and removed by a special procedure according to the moving flow features. The dynamic adaptation enables the computation of complex, unsteady flows, as they can appear in detonations. Results for an unstable detonation with an overdrive of $f = 1$ are presented in Fig. 4. A simplified one-step reaction is simulated in this example. The computation is performed in cycles, where each 10 time steps a new dynamic mesh is created. The total number of nodes of the static and dynamic grid is about 50.000 in average. Adaptation criterion is a combination of local pressure and density gradients. Fig. 4 shows details of the instantaneous density and of the adapted mesh.

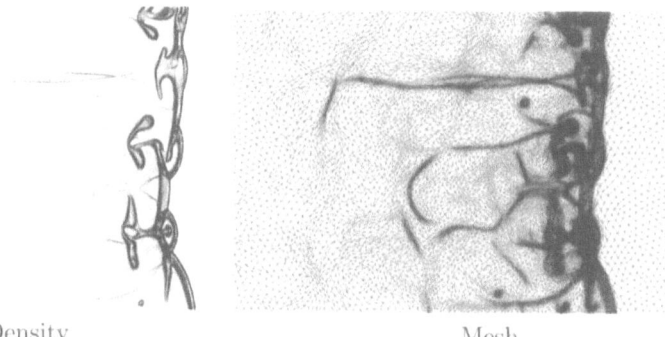

Density   Mesh

FIGURE 4. Computation of a 2-D, unstable detonation wave on adaptive, dynamic grids. Instantaneous lines of constant density (left) and adapted grid (right).

References

[1] M.S. Liou, *On a new Class of Flux Splitting Schemes. Lecture Notes in Physics*, Springer Verlag Berlin, **414** (1992), 115–119.

[2] R. Schwane and D. Hänel, *An Implicit Flux-Vector Splitting Scheme for the Computation of Viscous Hypersonic Flow*, AIAA-paper No. 89-0274, **1989**.

[3] R. Vilsmeier and D. Hänel, *Adaptive Solutions for Unsteady Laminar Flows on Unstructured Grids*, Int. J. for Numerical Methods in Fluids, **22** (1995), 85–101.

[4] R. Vilsmeier and D. Hänel, *A Field Method for 3-D Tetrahedral Mesh Generation and Adaption*, Proc. of 14th Int. Conf. on Num. Meth. in Fluid Dynamics, Bangalore (India), **1994**.

[5] B. van Leer, *Towards the Ultimate Conservative Difference Scheme*, A second-order sequel to Godunov's method. J. Comp. Phys., **32** (1979), 101–136.

[6] R. Vilsmeier and D. Hänel, *Solutions of the Conservation Equations and Adaptivity on 3-D Unstructured Meshes*, Proc. of 9th International Conference on Numerical Methods in Laminar and Turbulent Flow, **1995**.

[7] P.L. Roe, *Approximate Riemann Solvers, Parameter Vectors and Difference Schemes*, J. Comp. Phys., **22** (1981), 357.

[8] H.C. Yee, *A Class of High-Resolution Explicit and Implicit Shock-Capturing Methods*, in VKI Lecture Series 1989-04, Rhode-Saint-Genese, **1989**; also in H.C. Yee, *Upwind and Symmetric Shock-Capturing Schemes*, NASA TM-89464, **1987**.

[9] A. Harten, *High Resolution Schemes for Hyperbolic Conservation Laws*, J. Comp. Phys., **49** (1983), 357–393.

[10] M. Berger and P. Colella, *Local Adaptive Mesh Refinement for Shock Hydrodynamics*, J. Comp. Phys, **82** (1989), 67–84.

[11] J.J. Quirk, *An Adaptive Grid Algorithm for Computational Shock Hydrodynamics*, Ph.D. Thesis, Cranfield Inst. of. Technology, U.K., **1991**.

[12] C. Thill, U. Uphoff and D. Hänel, *Structured Mesh-Refinement Techniques for Reactive and Multi-Phase Flow*, Proc. of ECCOMACS 96, Paris, Sept., **1996**.

[13] P.R. Eiseman, *Adaptive Grid Generation,* Computer Meth. in Appl. Mech. and Eng., **64** (1987), 321–376.

[14] K. Nakahashi and G.S. Deiwert, *Self-Adaptive Grid Method with Application to Airfoil Flow,* AIAA-J.,**25** (1987), 513–520.

[15] U. Uphoff, D. Hänel and P. Roth, *Influence of Reactive Particles on the Formation of a One-Dimensional Detonation Wave,* Combustion Science and Technology, **110–111** (1995), 419–441.

[16] U. Uphoff, *Numerische Simulation von Verbrennungswellen in Gas-Partikel Gemischen,* Thesis at Univ. of Duisburg, Germany, **1997**.

[17] R. Le Veque and M. Berger, *CLAWPACK with Marsha Berger's Adaptive Mesh Refinement codes,* Web address: http://www.amath.washington.edu/ rjl/clawpack, **1997**.

[18] J. Bell, M.J. Berger, J. Saltzman and M. Welcome, *Three dimensional adaptive mesh refinement for hyperbolic conservation laws,* J. Sci. Comput., **15** (1994), 127–138.

[19] W. Speares and E.F. Toro, *A High Resolution Algorithm for Time-Dependent Shock Dominated Meshes with Adaptive Mesh Refinement,* Z. Flugwiss. Weltraumforschung, **19** (1995), 267–281.

[20] U. Uphoff, D. Hänel and P. Roth, *A Grid Refinement Study for Detonation Simulation with Detailed Chemistry,* Proc. of 6th Int. Conf. on Num. Combustion, New Orleans, March 4–6, **1996**.

[21] T. Plewa and M. Rozyczka, *Modern numerical hydrodynamics and the evolution of dense medium,* Rev. Mex. Astr. Astrofis., (1996). Proc. Starburst Activity in Galaxies, Eds. J. Franco, R. Terlevich, and G. Tenorio-Tagle, **1996**.

[22] J. Fischer and E.H. Hirschel, *Adaptive Navier-Stokes Calculations using a Combination of an Implicit Finite-Volume Method with a Hierarchically Ordered Grid Structure,* in Notes on Num. Fluid. Mech., Vieweg-Verlag Braunschweig, **38** (1993), 279–294.

[23] W.J. Coirier, *An Adaptively-refined Cartesian Cell-Based Scheme for the Eueler and Navier-Stokes Equations,* Ph.D. Thesis, The Univ. of Michigan, **1994**, and NASA TM 106754, **1994**.

[24] H. Deconinck, T.J. Barth (ed.), *Special course on unstructured grid methods for advection dominated flows,* AGARD Rep. 787, AGARD, Paris, **1992**.

[25] V. Venkatakrishnan: *A Perspective on Unstructured Grid Flow Solvers,* NASA CR-195025 ICASE Report No. 95-3 , Institute for Computer Applications in Science and Engineering, ICASE, **1995**.

[26] R. Vilsmeier and D. Hänel, *Computational Aspects of Flow Simulation on 3-D, Unstructured, Adaptive Grids,* in E.H. Hirschel (Ed.): Flow Simulation with High Performance Computers, Notes on Numerical Fluid Mechanics, Vieweg Verlag, Wiesbaden, **1996**.

Institut für Verbrennung und Gasdynamik,
Universität Duisburg,
D-47048 Duisburg, Germany
E-mail address: `haenel@ivg.uni-duisburg.de`

International Series of Numerical Mathematics
Vol. 129, © 1999 Birkhäuser Verlag Basel/Switzerland

The Riemann Problem of a System
for a Phase Transition Problem

Harumi Hattori

Abstract. We summarize the recent results concerning the Riemann problems of a hyperbolic-elliptic mixed type system related to a phase transition problem. Employing the entropy rate admissibility criterion and the entropy condition, we discuss two cases where the initial strains are given in the different phases and where they are given in the same phase.

1. Introduction

We summarize the recent results concerning the Riemann problem for a system describing a phase transition problem. We use both the entropy rate admissibility criterion and the entropy condition. The system we discuss is a hyperbolic-elliptic mixed type and given by

$$v_t - u_x = 0,$$
$$u_t - f(v)_x = 0, \tag{1}$$

where v, u, and f are strain, velocity, and stress, respectively. We assume that f is a smooth nonmonotone function of v as depicted in Figure 1. It is important to note that if f' is nonnegative, the system is hyperbolic and if f' is negative, the system is elliptic. In our case there are two intervals $(0,\alpha]$ and $[\beta,\infty)$ where the system is hyperbolic. They are called the α-phase and β-phase, respectively. We assume that f'' is negative in the α-phase and positive in the β-phase. The interval (α,β) is called the spinodal region and physically unobservable. The horizontal line for which the areas A and B are equal is called the Maxwell stress. The values of v in the α-phase and β-phase at which the Maxwell stress intersect f are denoted by v_α and v_β, respectively. The states $(0,v_\alpha]$ and $[v_\beta,\infty)$ are stable, $(v_\alpha,\alpha]$ and $[\beta,v_\beta)$ are metastable, and (α,β) is unstable. The values γ and δ in the α and β-phases are the values of v at which $f(\gamma) = f(\beta)$ and $f(\delta) = f(\alpha)$, respectively.

The author was supported by a U.S. Army DEPSCoR Grant DAAH04-94-G-0246 and an NSF Grant DMS-9704383.

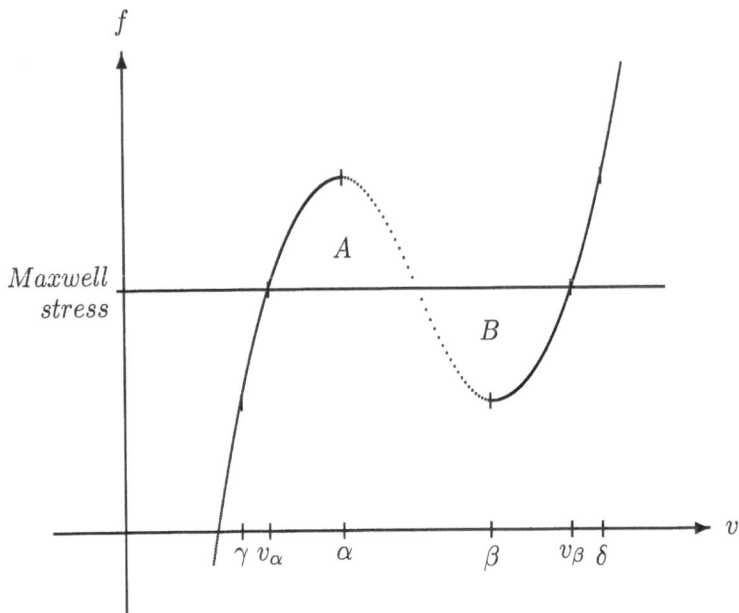

FIGURE 1.

The Riemann problem is a special initial value problem of (1) in which the initial data is given by

$$(v, u)(x, 0) = \begin{cases} (v_\ell, u_\ell) & x < 0, \\ (v_r, u_r) & x > 0, \end{cases} \tag{2}$$

where v_ℓ, u_ℓ, v_r, and u_r are constants. We study the Riemann problems employing the both criteria. We seek a self-similar solution in which constants states are separated by the elementary waves and the phase boundaries. We discuss the two cases. In Section 3 we discuss the case where v_ℓ and v_r are given in the different phases and in Section 4 we discuss the case where they are given in the same phase.

It is common to model phase transition phenomena in solid from continuum mechanics by using a nonmonotone constitutive relation. The Riemann problem of system (1) was discussed in various literature. James [9] initiated the Riemann problem for this type of problem. He proposed the one-parameter family of solutions for the Riemann problem if v_ℓ and v_r are given in the different phases. Abeyaratne and Knowles [1] proposed the kinetic relation for phase boundaries and discussed the Riemann problem using the kinetic relation and the initiation criterion. Hattori [5], [6] used the entropy rate admissibility criterion proposed by

Dafermos [2], [3] for hyperbolic systems. As far as the Cauchy problem is concerned, Le Floch [10] has shown the existence of global solutions in BV space if f is trilinear. Hattori [7] has shown the existence of weak solutions in the case where f is nonlinear. Slemrod [13], [14] and Fan and Slemrod [4] discussed the effects of viscosity and capillarity and proposed the viscosity-capillarity criterion. Shearer [12] considered the issue of nonuniqueness for the Riemann problem using this criterion. This paper is a contracted version of [8], where more details are discussed.

2. Preliminary

In this section we summarize the preliminary necessary for this paper.

1. *Elementary waves:* We call the rarefaction wave and the shock wave the elementary waves. The backward wave curve (B.W.C.) $B^r(v_o, u_o)$ is the set of (v, u) connected to (v_o, u_o) on the right by the backward rarefaction or shock wave. It satisfies the following relation:

Rarefaction curve: $\quad u = u_o + \int_{v_o}^{v} \lambda(w)dw, \qquad \begin{cases} v \leq v_o & \text{if } f \text{ is convex,} \\ v \geq v_o & \text{if } f \text{ is concave,} \end{cases}$

Shock curve: $\quad u = u_o - \sigma_b(v_o, v)(v - v_o), \qquad \begin{cases} v \geq v_o & \text{if } f \text{ is convex,} \\ v \leq v_o & \text{if } f \text{ is concave,} \end{cases}$

where $\lambda(w) = \sqrt{f'(w)}$ and $\sigma_b(v_o, v) = -\sqrt{\frac{f(v)-f(v_o)}{v-v_o}}$. The forward wave curve (F.W.C.) $F^r(v_o, u_o)$ is defined in a similar manner:

Rarefaction curve: $\quad u = u_o - \int_{v_o}^{v} \lambda(w)dw, \qquad \begin{cases} v \geq v_o & \text{if } f \text{ is convex,} \\ v \leq v_o & \text{if } f \text{ is concave,} \end{cases}$

Shock curve: $\quad u = u_o - \sigma_f(v_o, v)(v - v_o), \qquad \begin{cases} v \leq v_o & \text{if } f \text{ is convex,} \\ v \geq v_o & \text{if } f \text{ is concave,} \end{cases}$

where $\lambda(w) = \sqrt{f'(w)}$ and $\sigma_f(v_o, v) = \sqrt{\frac{f(v)-f(v_o)}{v-v_o}}$. We define $B^\ell(v_o, u_o)$ and $F^\ell(v_o, u_o)$ as the sets of (v, u) connected to (v_o, u_o) on the left by the corresponding waves. If the above inequalities are reversed, we obtain the corresponding relations.

2. *Phase boundary:* A phase boundary is the line of discontinuity in the xt-plane across which the phase changes. It satisfies the Rankine-Hugoniot condition. The phase boundary curve $P^r((v_o, u_o))$ (or $P^\ell((v_o, u_o))$) is the set of (v, u) connected to (v_o, u_o) on the right (or left) by the phase boundary and satisfies the following relations:

$$u = u_o - \sigma(v_o, v)(v - v_o),$$

where $\sigma(v_o, v) = \pm\sqrt{\frac{f(v)-f(v_o)}{v-v_o}}$ and v_o and v are in the different phases. If the cord joining $(v_o, f(v_o))$ and $(v, f(v))$ intersect f, the value of v is denoted by v_*.

3. *Admissibility criteria:* The weak solutions for (1) are not unique and to choose a physically relevant solution we employ admissibility criteria. There are two criteria that we use in this paper. The entropy rate admissibility criterion is the criterion that was proposed by Dafermos [2], [3]. This criterion roughly says that the rate of entropy (the energy) production is the smallest for the admissible solution. The entropy for (1) is given by

$$\eta = \frac{1}{2}u^2 + \int f(v)dv.$$

The rate of decay of the total energy is given by

$$E \equiv D_+\eta = \sum_{\text{jump discontinuities}} \sigma(v_-, v_+)A(v_-, v_+), \tag{3}$$

where $\sigma(v_-, v_+)$ is the speed of the jump discontinuity and

$$A(v_-, v_+) = [\frac{1}{2}(f(v_-) + f(v_+))(v_+ - v_-) - \int_{v_-}^{v_+} f(w)dw].$$

Here v_- and v_+ are the values of v on the left and right of a jump discontinuity. We denote

$$E(v_-, v_+) = \sigma(v_-, v_+)A(v_-, v_+).$$

The entropy rate admissibility criterion postulates that the solution is admissible if it solves (1) and minimizes (3).

The entropy condition is the criterion imposing that the entropy decreases across discontinuities. This is equivalent to requiring that $E(v_-, v_+) \leq 0$ holds across each discontinuity.

Whenever $A(v_-, v_+) = 0$, the cord joining $(v_-, f(v_-))$ and $(v_+, f(v_+))$ intersect f. We denote the value of v at which the cord intersect f by v_e. We assume that v_e is in the elliptic region for all values of v_- and v_+ in the hyperbolic regions. We discuss briefly the case where this condition is not satisfied at the end of Sections 3 and 4. We state the consequences of this assumption as lemmas without proof.

Lemma 2.1. *If the two phase boundaries move in the same direction, one of the phase boundaries violates the entropy condition.*

Lemma 2.2. *Suppose v_ℓ is specified in the α-phase and v_r is specified in the β-phase. If there are more than one phase boundary in the resolution of the Riemann problem, at least one of them violates the entropy condition.*

3. Riemann problem with one phase boundary

In this section using both the entropy rate admissibility criterion and the entropy condition, we discuss the Riemann problem (1) and (2) where v_ℓ and v_r are specified in the α-phase and β-phase, respectively. Lemma 2.2 implies that we need to consider only one phase boundary. We need to impose that $\sigma_p A(v_1, v_2) \leq 0$ be satisfied across the phase boundary. We also require that the speed of the phase boundary in absolute value is less than or equal to that of the backward and forward wave. We apply the entropy rate admissibility criterion among the solutions satisfying the above conditions. Therefore, we have the following optimization problem:

$$\min E \tag{4}$$

subject to

$$\sigma_p A(v_1, v_2) \leq 0 \tag{5}$$

and

$$\sigma_b \text{ or } -\lambda_1 \leq \sigma_p \leq \sigma_f \text{ or } \lambda_2. \tag{6}$$

The admissible solution is the solution to the Riemann problem (1) and (2) satisfying the above minimization problem. We denote this criterion the entropy-entropy rate admissibility criterion. This type of problem was discussed in [11] in the case where there are no shock waves.

We construct the admissible solution for given (v_ℓ, u_ℓ) and (v_r, u_r). For this purpose, we find the region of (v_1, u_1) where (5) and (6) are satisfied for a given forward wave curve $F^\ell(v_r, u_r)$. We call this region the feasible region. We define the curves called the stationary phase boundary curve, the equal area curve, and the equal speed curve. These curves corresponding to the equality signs in (5) and (6). We can also define these curves for a given backward wave curve.

Definition 3.1. *The stationary phase boundary curve (S.C.): This is the curve consisting of the points (v_1, u_1) satisfying*

$$f(v_1) = f(v_2), \quad u_1 = u_2 \tag{7}$$

as (v_2, u_2) moves along the forward wave curve. This v_1 satisfies $\gamma \leq v_1 \leq \alpha$ and exists if $\beta \leq v_2 \leq \delta$.

Definition 3.2. *The equal area curve (E.A.C.): This is the curve consisting of the points (v_1, u_1) satisfying*

$$A(v_1, v_2) = 0, \quad u_2 = u_1 - \sigma_p(v_2 - v_1) \tag{8}$$

as (v_2, u_2) moves along the forward wave curve.

Definition 3.3. *The equal speed curve-I (E.S.C.-I): This is the curve consisting of the points (v_1, u_1) satisfying*

$$u_2 = u_1 - \sigma_p(v_2 - v_1), \tag{9}$$

where

$$\sigma_p = \begin{cases} -\lambda_1, & v_1 \geq v_\ell \\ \sigma_b, & v_1 < v_\ell \end{cases}, \tag{10}$$

as (v_2, u_2) moves along the forward wave curve. If (δ, u_2) is on the forward wave curve, then this curve starts from (α, u_2) and if $v_1 < v_\ell$, the cord joining $(v_2, f(v_2))$ and $(v_1, f(v_1))$ passes through $(v_\ell, f(v_\ell))$.

Definition 3.4. *The equal speed curve-II (E.S.C.-II): Another type is the curve consisting of the points* (v_1, u_1) *satisfying*

$$u_2 = u_1 - \sigma_p(v_2 - v_1), \tag{11}$$

where

$$\sigma_p = \begin{cases} \lambda_2, & v_2 \geq v_r \\ \sigma_f, & v_2 > v_r \end{cases}, \tag{12}$$

as (v_2, u_2) *moves along the forward wave curve. If* (β, u_2) *is on the forward wave curve, this curve starts from* (γ, u_2).

Remark 3.5. We can define the above curves for a backward wave curve in a similar manner. In the case of E.S.C.-I, if (v_1, u_1) moves along the backward wave curve, σ_p is equal to the speed of the forward wave curve. In the case of E.S.C.-II, if (v_1, u_1) moves along the backward wave curve, σ_p is equal to the speed of the backward wave curve.

If $\sigma_p = \sigma_f$, the constant state (v_2, u_2) degenerates to a line in the xt-plane, and we interpret that there is no forward wave and that (v_r, u_r) is directly connected to (v_1, u_1). This phase boundary also satisfies the entropy condition.

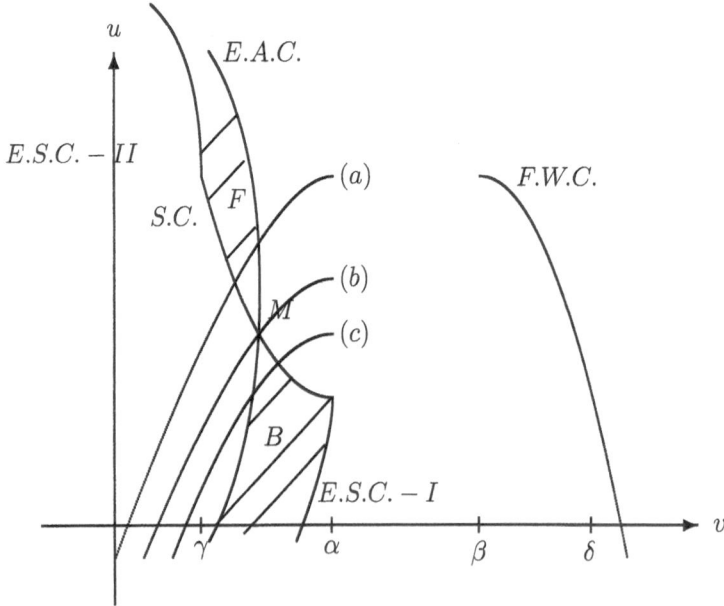

FIGURE 2.

Combining the above curves, we obtain the feasible region; see Figure 2. In this figure the shaded regions are the feasible regions. Depending on how the backward wave intersects with the shaded region, we obtain three cases.

(a) The backward wave intersects with the region F in Figure 2.
(b) The backward wave goes through the point M in Figure 2.
(c) The backward wave intersects with the region B in Figure 2.

In Figure 2, F (or B) stands for the fact that the phase moves forward (or backward) if the backward wave curve intersect this region and the v-coordinate of M is v_α. In cases (a) and (c) we move (v_1, u_1) along the backward wave curve in the feasible region and find the minimum of the entropy rate. If the backward wave curve cross the E.S.C.'s, there are cases where the cord joining $(v_1, f(v_1))$ and $(v_2, f(v_2))$ intersect with the graph of f at another v in the hyperbolic region. We denote this v by v_*. If v_1 and v_2 are close to v_α and v_β, respectively, v_* is in the spinodal region. As we move u_1, v_* moves out from the spinodal region. In this case we may have multiple solutions. One solution consists of v_ℓ, v_1, v_2, and v_r and another solution consists of v_ℓ, v_1, v_*, and v_r. It can be shown that the solution with v_* is not admissible except when $v_* = v_r$. In case (b), only the stationary phase boundary is allowed. If v_e is in the hyperbolic region, it is possible that the E.A.C. and E.S.C. intersect. If this happens we continue only the E.S.C. beyond the intersection point. In this case we may see as many as three phase boundaries.

4. Riemann problem with two phase boundaries

In this section we construct the solution to the Riemann problem where both v_ℓ and v_r are given in the same phase. We assume that they are given in the α-phase. There are two possibilities. One possibility is that there is no phase boundary and the middle constant state (v_m, u_m) is connected to (v_ℓ, u_ℓ) and (v_r, u_r) by the backward wave and the forward wave, respectively. This is what we observe in the hyperbolic case. Another possibility is that the solution has two phase boundaries separated by three middle constant states (v_1, u_1), (v_2, u_2), and (v_3, u_3) from left to right. In this case v_1 and v_3 are in the α-phase and v_2 is in the β-phase. There are four possible connections as in Section 3.

The optimization problem is given by

$$\min E \tag{13}$$

subject to

$$\sigma_{p_1} A(v_1, v_2) \leq 0, \quad \sigma_{p_2} A(v_2, v_3) \leq 0 \tag{14}$$

and

$$\sigma_b \text{ or } -\lambda_1 \leq \sigma_{p_1} \leq 0 \leq \sigma_{p_2} \leq \sigma_f \text{ or } \lambda_3, \tag{15}$$

where

$$E = E_b + \sigma_{p_1} A(v_1, v_2) + \sigma_{p_2} A(v_2, v_3) + E_f$$

and

$$E_b = \begin{cases} \sigma_b A(v_\ell, v_1), & v_\ell > v_1, \\ 0, & v_\ell \leq v_1, \end{cases} \quad E_f = \begin{cases} \sigma_f A(v_3, v_r), & v_r > v_3, \\ 0, & v_r \leq v_3. \end{cases}$$

Lemma 2.1 implies that it is enough to consider the case where two phase boundaries satisfy the condition in (15).

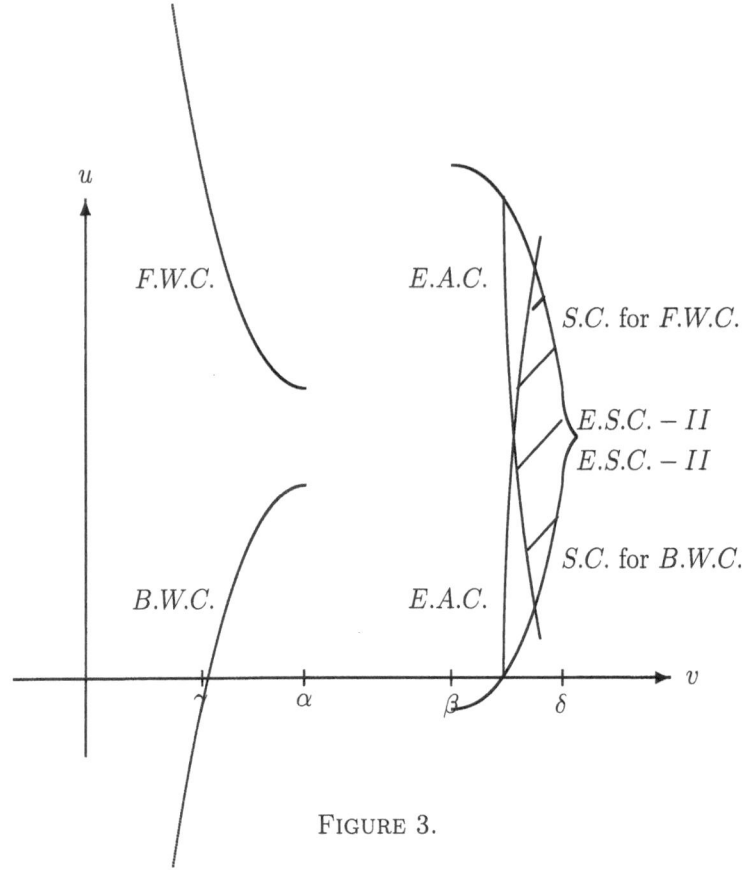

FIGURE 3.

As in the previous section we draw the backward wave curve and the forward wave curve from (v_ℓ, u_ℓ) and (v_r, u_r), respectively. Denote the intersection point by (v_m, u_m) if it exists. There are three cases depending on the relation between the two curves.

(d) Two curves do not intersect.
(e) Two curves intersect and $f(v_m) > f(v_\alpha)$. Therefore, v_m is in the metastable state.
(f) Two curves intersect and $f(v_m) \leq f(v_\alpha)$. Therefore, v_m is in the stable state.

In each case we identify the region of (v_2, u_2) where the solution satisfies (14) and (15). We call this region the feasible region. The results of Cases (d) and (e) are given in Figures 3 and 4, respectively. In Figure 4 the point S at which two stationary phase boundaries meet is denoted by (v_s, u_s). In Case (d) we need two phase boundaries to connect (v_ℓ, u_ℓ) and (v_r, u_r). In Case (e) it was shown in [7] that the solutions with two phase boundaries have lower entropy rates than the solution with no phase boundary provided that

$$\left| \frac{\partial E_b}{\partial v_2} \right| + \left| \frac{\partial E_f}{\partial v_2} \right| < \frac{\partial E_{p_1}}{\partial v_2} + \frac{\partial E_{p_2}}{\partial v_2} \qquad (16)$$

holds at $v_2 = v_s$ and furthermore, the right-hand side of (16) evaluated at (v_s, u_s) is positive, where the constant state (v_2, u_2) degenerates into a vertical line in xt-plane. This result resolves the nonuniqueness of the Riemann problem discussed in Shearer [12]. In Cases (d) and (e) we move (v_2, u_2) in the shaded regions and find the minimum of the entropy rate. In Case (f) there is no region where (14) and (15) are satisfied. We observe the admissible solution with no phase change.

It should be noted that Figures 2 through 4 are for the case where v_e is in the spinodal region. If the gap between the forward and backward waves is large and v_e is in the hyperbolic region, the equal speed curves and the equal area curves may intersect. If this happens We continue the E.S.C.'s. only. In this case we may observe as many as four phase boundaries.

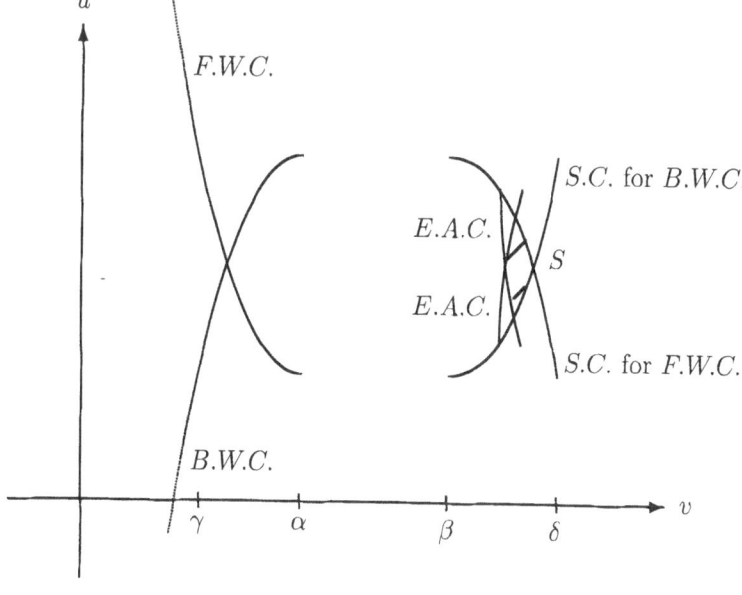

FIGURE 4.

References

[1] R. Abeyaratne and J.K. Knowles, *Kinetic relations and the propagation of phase boundaries in solids,* Arch. Rat. Mech. Anal. **114** (1991), 119–154.

[2] C.M. Dafermos, *The entropy rate admissibility criterion for solutions of hyperbolic conservation laws,* J. Diff. Eq., **14** (1973), 202–212.

[3] C.M. Dafermos, The entropy rate admissibility criterion in thermoelasticity, Atti Accad. Naz. Lincei Rend. cl. Sci. Fis. Mat. Natur., **8** (1974), 113–119.

[4] H. Fan and M. Slemrod, *The Riemann problem for systems of conservation laws of mixed type,* in IMA Vol. Math. Appl., (Springer, New York), **52** (1993), 61–91.

[5] H. Hattori, *The Riemann problem for a van der Waals fluid with entropy rate admissibility criterion: Isothermal case,* Arch. Rat. Mech. Anal., **92** (1986), 246–263.

[6] H. Hattori, *The Riemann problem for a van der Waals fluid with entropy rate admissibility criterion: Non-isothermal case,* J. Diff. Eq., **65** (1986), 158–174.

[7] H. Hattori, *The Riemann problem and the existence of weak solutions to a system of mixed-type in dynamic phase transition,* submitted for publication.

[8] H. Hattori, *The entropy rate admissibility criterion and the entropy condition for a phase transition problem,* submitted for publication.

[9] R.D. James, *The propagation of phase boundaries in elastic bars,* Arch. Rat. Mech. Anal., **73** (1980), 125–158.

[10] P. Le Floch, *Propagating phase boundaries: Formulation of the problem and existence via Glimm's scheme,* Arch. Rational Mech. Anal., **123** (1993), 153–197.

[11] T.J. Pence, *On the mechanical dissipation of solutions to the Riemann problem for impact involving a two-phase elastic material,* Arch. Rat. Mech. Anal., **117** (1992), 1–55.

[12] M. Shearer, *Nonuniqueness of admissible solutions of the Riemann initial value problem for a system of conservation laws of mixed type,* Arch. Rat. Mech. Anal., **93** (1986), 45–59.

[13] M. Slemrod, *Admissibility criteria for propagating phase boundaries in a van der Waals fluid,* Arch. Rat. Mech. Anal., **81** (1983), 301–315.

[14] M. Slemrod, *Dynamic phase transitions in a van del Waals fluid,* J. Diff. Eqns., **52** (1984), 1–23.

Department of Mathematics,
West Virginia University,
Morgantown, WV 26506-6310

International Series of Numerical Mathematics
Vol. 129, © 1999 Birkhäuser Verlag Basel/Switzerland

Computation of High-speed Flow Using Non-Oscillatory Scheme

S.B. Hazra

Abstract. Theory of non-oscillatory schemes has been used in conjunction with a finite-volume cell-vertex Navier-Stokes solver in this paper, in order to compute compressible viscous flow fields past airfoils which accurately capture shock without preshock oscillations. Flux difference splitting schemes of Roe(1981) has been implemented here as proposed by Jameson(1993). These schemes have the LED (local extremum diminishing) property which is equivalent to that of TVD (total variation diminishing) in one-dimension. Typical transonic and supersonic test cases have been studied. It is found that the scheme provides excellent shock resolution at high Mach numbers for steady solution of the viscous fluid flow equations expressed by Reynolds averaged Navier-Stokes equations in conservation law form augmented with Baldwin-Lomax (1978) turbulence model.

1. Introduction

Significant progress has been achieved in the past two decades in the area of numerical solution of Euler and Navier-Stokes equations. During eighties several second-order accurate shock-capturing schemes for solving Euler equations of gas dynamics were developed which could successfully and efficiently compute flow past complex configurations [1]–[5]. Among these, the finite-volume scheme of Jameson et. al. [1] became particularly popular due to its efficiency and robustness. It showed that steady-state flows containing moderately strong shock waves could be quite well predicted by a central difference scheme augmented by a carefully controlled blend of second and fourth order dissipative terms. Works on numerical solution of Navier-Stokes equations followed soon [6]–[14].

During this period, several authors attempted to construct numerical schemes that could eliminate spurious preshock oscillations that appeared in the computational solutions. Further, algorithm development was aimed at combining high accuracy with high resolution of shocks and contact discontinuities.

It has long been recognized that first order upwind differencing like that of Godunov scheme [16] can eliminate spurious oscillations in the neighborhood of shock waves at the expense of low accuracy in regions where the flow is smooth. Central difference schemes on the other hand, produce good solutions in smooth regions but are prone to oscillations in the neighborhood of shock waves, which can be suppressed only by the introduction of additional dissipative terms.

Stemming from the mathematical theory of hyperbolic conservation laws [17], Harten [22] proposed the concept of total variation diminishing (TVD) finite-difference schemes and constructed such a second order scheme by introducing antidiffusive terms and flux limiters to improve shock resolution. Similar idea may be found in the work of Boris and Book [18]. TVD schemes preserve the monotonicity of an initially monotone profile, because the total variation would increase if the profile ceased to be monotone. Consequently they prevent spurious oscillations. Van Leer [19] used flux limiters to produce a second order accurate scheme which would preserve the monotonicity of an initially monotone profile. Important contributions on TVD schemes have been made by Roe [20], [21], Osher [23], Yee [25] and many of the ideas on flux limiters have been unified by Sweby [24].

In a series of papers, Jameson [26]–[29] developed the theory of non-oscillatory schemes with particular reference to finite-volume space discretization with multi-stage time-stepping schemes for Euler equations of gas dynamics. TVD Navier-Stokes solution appear in Yee et.al. [31]. Two-dimensional transonic viscous flow fields in cascades or propulsion systems have been studied by Teipel and Wiedermann [30] using the TVD model of Harten [22] and Yee et. al. [31].

This paper presents application and implementation of flux difference splitting, originally introduced by Jameson [28], [29], in order to obtain high resolution of shock waves without preshock oscillations for two-dimensional Navier-Stokes equations expressed in conservation law form. Finite-volume cell-vertex discretization originally put forward by Chakrabartty [13] with algebraic turbulence model of Baldwin and Lomax [32] has been used in the present study. **The details regarding the governing equations, boundary conditions (transonic case only), finite-volume space discretization, the five-stage Runge-Kutta time stepping and the acceleration techniques are available in Reference [13] and are not repeated here.** The boundary conditions for supersonic case have been modified accordingly. Steady-state viscous transonic and supersonic flow past airfoils have been computed using an algebraically generated C-type grid following Jain [33].

2. Artificial dissipation

2.1. JST's dissipation term

We first explain the artificial dissipation term used in the basic code of Chakrabartty [12], [13] which has been modified here following Jameson [28] and explained in the subsequent sections. The finite volume discretization described in the subsequent sections of [13] amounts to central differencing and thus requires the addition of explicit dissipation terms for stability. There the dissipative flux $\vec{D}_{ij}$ is given by

$$\vec{D}_{ij} = \vec{d}_{i+1/2,j} - \vec{d}_{i-1/2,j} + \vec{d}_{i,j+1/2} - \vec{q}_{i,j-1/2} \qquad (1)$$

and in the scheme of Jameson et. al. [1] $\vec{d}_{ij}$ are a special kind of blend of second and fourth order differences expressed by

$$\vec{d}_{i+1/2,j} = \alpha_{i+1/2,j} \left\{ \epsilon^{(2)}_{i+1/2,j} \Delta_{i+1/2,j} \vec{W}_{i,j} - \epsilon^{(4)}_{i+1/2,j} \Delta^3_{i+1/2,j} \vec{W}_{i,j} \right\}. \tag{2}$$

Here $\epsilon^{(2)}_{i+1/2,j}$ and $\epsilon^{(4)}_{i+1/2,j}$ are adaptive coefficients designed to switch on enough dissipation where it is needed, $\alpha_{i+1/2,j}$ is the scaling factor and are defined as

$$\epsilon^{(2)}_{i+1/2.j} = k^{(2)} \max(\nu_{i+1.j}, \nu_{i,j}), \quad \epsilon^{(4)}_{i+1/2,j} = \max\{0, (k^{(4)} - \epsilon^{(2)}_{i+1/2,j})\}$$

$$\nu_{i,j} = \frac{|p_{i+1,j} - 2p_{i,j} + p_{i-1,j}|}{p_{i+1,j} + 2p_{i,j} + p_{i-1,j}}, \quad \alpha_{i+1/2,j} = \frac{1}{2} \left(\frac{V_{i,j}}{\Delta t^*_{i,j}} + \frac{V_{i+1,j}}{\Delta t^*_{i+1,j}} \right)$$

where $k^{(2)}, k^{(4)}$ are user specified constants, which controls the artificial dissipation, $V_{i,j}$ is the cell volume and $\Delta t^*_{i,j}$ is an estimate of the time step limit for a nominal Courant number $(= c\frac{\Delta t}{\Delta x})$ of unity (Chakrabartty [15]). The coefficient $\epsilon^{(2)}$ is proportional to the second difference of pressure in smooth regions of the flow proportional to the square of the mesh size, while $\epsilon^{(4)}$ is of order one.

The above dissipation model has been proved to be very effective in practice in numerous calculations of complex steady flows. But it is LED only under certain conditions. In the present work, the dissipative fluxes implemented by Chakrabartty [12], [13], have been modified in this work using the concept of flux difference splitting scheme of Roe [20].

2.2. Non-oscillatory schemes

For quite some time efforts have been made to modify existing schemes for computing solution of Eulers equations expressed in conservation form in such a way that they might not give rise to spurious preshock oscillations in the computational solution. Such efforts culminated in the total-variation-diminishing (TVD), total-variation-bounded (TVB), essentially-non-oscillatory (ENO) schemes. If in a scheme the local maximum does not increase and the local minimum does not decrease then such a scheme has been termed [28] a local extremum diminishing (LED) scheme.

Consider the discretization of a time dependent conservation law such as

$$\frac{\partial v}{\partial t} + \frac{\partial f(v)}{\partial x} + \frac{\partial f(v)}{\partial y} = 0,$$

for a scalar dependent variable v on an arbitrary mesh. Let v_j be it's value at the mesh point j. Suppose approximation of this equation can be expressed in semi-discrete form in terms of differences between v_j and other mesh values v_k as

$$\frac{dv_j}{dt} = \sum_k c_{kj}(v_k - v_j).$$

For a compact stencil of points, c_{kj} will be zero for most values of k. Such a scheme is called LED if

$$c_{kj} \geq 0.$$

In one dimension such schemes form a superset of the total variation diminishing schemes (TVD). The LED criterion can be directly applied to multi-dimensional problems. One such approach is based on flux difference splitting, as explained below.

2.3. Flux-difference splitting by characteristic decomposition

Roe [20] formulated flux-difference splitting by distributing the corrections due to the flux difference in each interval upwind and downwind. The difference scheme yields the symmetric limited positive (SLIP) scheme, applied to the differences of the characteristic variables. Let us introduce the flux-difference Δf for system of conservation laws

$$\frac{\partial w}{\partial t} + \frac{\partial f(w)}{\partial x} = 0$$

by

$$\Delta f_{i+1/2,j} = A_{i+1/2,j} \Delta w_{i+1/2,j}$$

where $\Delta f_{i+1/2,j} = f_{i+1,j} - f_{i,j}$, $\Delta w_{i+1/2,j} = w_{i+1,j} - w_{i,j}$ and $A_{i+1/2,j}$ is the Jacobian matrix calculated with Roe averaging,

$$\bar{q} = \frac{\sqrt{\rho_{i+1,j}}q_{i+1,j} + \sqrt{\rho_{i,j}}q_{i,j}}{\sqrt{\rho_{i+1,j}} + \sqrt{\rho_{i,j}}}$$

for any quantity q representing a flow variable.

Then a splitting according to characteristic fields is obtained by decomposing

$$A_{i+1/2,j} = T\Lambda T^{-1},$$

where the columns of T are the eigenvectors of $A_{i+1/2,j}$ and Λ is a diagonal matrix of the eigenvalues. Then

$$\Delta f_{i+1/2,j}^{\pm} = A_{i+1/2,j} \Delta^{\pm} w_{i+1/2,j}.$$

Now the corresponding dissipative flux is

$$\tfrac{1}{2} |A_{i+1/2,j}| \Delta w_{i+1/2,j}, \quad \text{where} \quad |A_{i+1/2,j}| = T|\Lambda|T^{-1}.$$

Simple stable schemes can be produced by the splitting

$$\Delta f_{i+1/2,j}^{\pm} = \tfrac{1}{2} \Delta f_{i+1/2,j} \pm \beta_{i+1/2,j} \Delta w_{i+1/2,j},$$

which satisfies the positivity condition on the eigenvalues if

$$\beta_{i+\frac{1}{2},j} > \tfrac{1}{2} \ \max |\lambda(A_{i+\frac{1}{2},j})|.$$

The basic ideas of nonoscillatory schemes as summarized briefly in the above subsections have been used for computing the modified Euler fluxes, by replacing the artificial dissipation term in Eq.(2) by the corresponding nonoscillatory dissipation term, the flux-difference splitting being applied separately in each coordinate direction. The new dissipative term reads as

$$\vec{d}_{i+1/2,j} = \alpha_{i+1/2,j} \left\{ \epsilon^{(2)}_{i+1/2,j} |A_{i+\frac{1}{2},j}| \Delta_{i+1/2,j} \vec{W}_{i,j} - \epsilon^{(4)}_{i+1/2,j} \beta_{i+\frac{1}{2},j} \Delta^3_{i+1/2,j} \vec{W}_{i,j} \right\}.$$

The viscous fluxes have been computed as stated in details in [13]. The dissipation model gives smooth convergence for viscous flow.

3. Results and discussions

Several calculations using the above flux-difference splitting scheme have been presented in Figure-1 and Figure-2. All the calculations have been performed on algebraically generated (256x61) C-grid [33]. In order to remove grid error, and the effect of far-field boundary condition, several test runs were made on smaller size grids and finally the grids used were chosen. The computations for each airfoil shape were carried out till the average residuals reduced below the order of 10^{-4}, which was taken as the convergence criterion. Roughly, 2000-time steps are required to reach this level of accuracy requiring about 29 minutes of CPU time on a Digital DEC-Alpha 3000/600 Desk-top computer system.

For the sake of convenience, we have referred here the solution using the Chakrabartty's [13]-scheme as "chk" and using flux-difference splitting as "fsp".

For computations with chk-scheme the parameters $k^{(2)}$ and $k^{(4)}$ for the second and fourth order dissipation have been fixed as $\frac{1}{2}$ and $\frac{1}{256}$ respectively. It may be noted that this choice of $k^{(4)}$ gives the result which shows best agreement with the experimental one. In fact, the results depend strongly on the values of the parameters, which provided initial motivation for the present study.

The first example we have computed is that of steady-state flow for a RAE2822 airfoil for $M_\infty = 0.73$, $\alpha = 3.19$ and $Re = 6.5E + 6$. In figure-1, top row, left, shows c_p distribution of this case where excellent agreement is achieved with the "fsp" (LED) computation. In figure-1, middle row, left, shows the Mach contour distributions using "fsp" scheme for this case.

The second example considered here is that of flow past a RAE5225 airfoil with $M_\infty = 0.737$, $\alpha = 2.33$ and $Re = 6.0E + 6$. In figure-1, top row, right, shows the c_p distribution of this case. Here also the result of LED(fsp) scheme shows excellent agreement with experimental result. In figure-1, middle row, right, shows Mach contour distributions of the computation using "fsp" scheme for this case. In the same figure, bottom row, a typical convergence history of this case has been shown where $log(\frac{\partial \rho}{\partial t})$ has been plotted against the number of time steps.

Convergence to a steady state has been achieved. Table-1 shows the comparison of different aerodynamic forces obtained by the present schemes under study for the two airfoils with experimental results [34], [35], [36]. Here also, the results obtained by "fsp" scheme shows significant improvement over "chk" scheme with respect to experimental results.

The results of supersonic cases using the present scheme are presented in figure-2. In this case, first example considered is the Rae2822 airfoil with $M_\infty = 1.3$, angle of attack $\alpha = 3.19$ and the Reynolds number $Re = 6.5E + 6$. In the top row of figure-2 the pressure (left) and Mach (right) contours of this case are shown. The second example considered is the NACA0012 airfoil with Mach number $M_\infty = 1.3$, angle of attack $\alpha = 2.33$ and the Reynolds number $Re = 6.5E + 6$. In the bottom row of figure-2 the pressure (left) and the Mach (right) contours of this case are shown.

Table-1

airfoil	scheme	M_∞	α	Re	CL	CD	CM
RAE2822	Expt	0.73	3.19	6.5E+06	0.803	0.0168	-0.099
	LED(fsp)	0.73	3.19	6.5E+06	0.8827	0.0148	-0.0892
	chk	0.73	3.19	6.5E+06	0.8449	0.0168	-0.0876
RAE5225	Expt	0.737	2.33	6.0E+06	0.6590	0.01292	-0.0832
	LED(fsp)	0.737	2.33	6.0E+06	0.6911	0.0099	-0.0939
	chk	0.737	2.33	6.0E+06	0.6701	0.0118	-0.0921

Comparison of lift, drag and moment coefficients of different airfoils using different schemes with experiments

4. Conclusion

The choice of artificial dissipation is very important in resolving the viscous flow phenomena properly. This plays a great role in the convergence and stability of the flow in supersonic region. Like inviscid transonic flow, for viscous transonic flow also Non-Oscillatory schemes can resolve properties better at a smaller computational cost. Experience with various examples computed and comparison with other theoretical and experimental results indicate that the present method is quite dependable.

Acknowledgement. I like to thank Prof. P. Niyogi of Indian Institute of Technology, Kharagpur, India, and Dr. S. K. Chakrabartty of National Aerospace Laboratories, Bangalore, India, for their help and cooperations during my stay in India. I thank the DAAD (German Academic Exchange Service) for their financial support during my stay in Germany. I also like to thank the organising committee for their partial financial support to attend this conference.

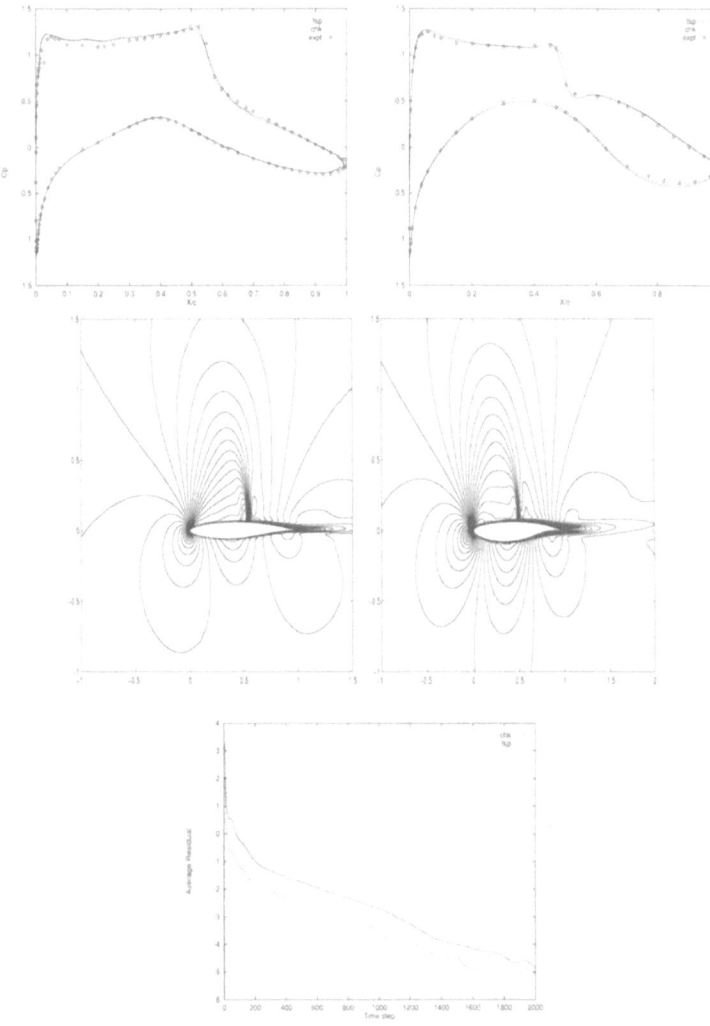

FIGURE 1. Surface pressure distributions (top row) of Rae2822 airfoil for $M_\infty = 0.73$, $\alpha = 3.19$ and $Re = 6.5E + 06$ (left), of Rae5225 airfoil for $M_\infty = 0.737$, $\alpha = 2.33$ and $Re = 6.0E + 06$ (right); corresponding Mach contours(middle row) using "fsp" scheme around Rae2822 airfoil(left) and around Rae5225 airfoil(right); and convergence history (bottom row) of Rae5225 airfoil

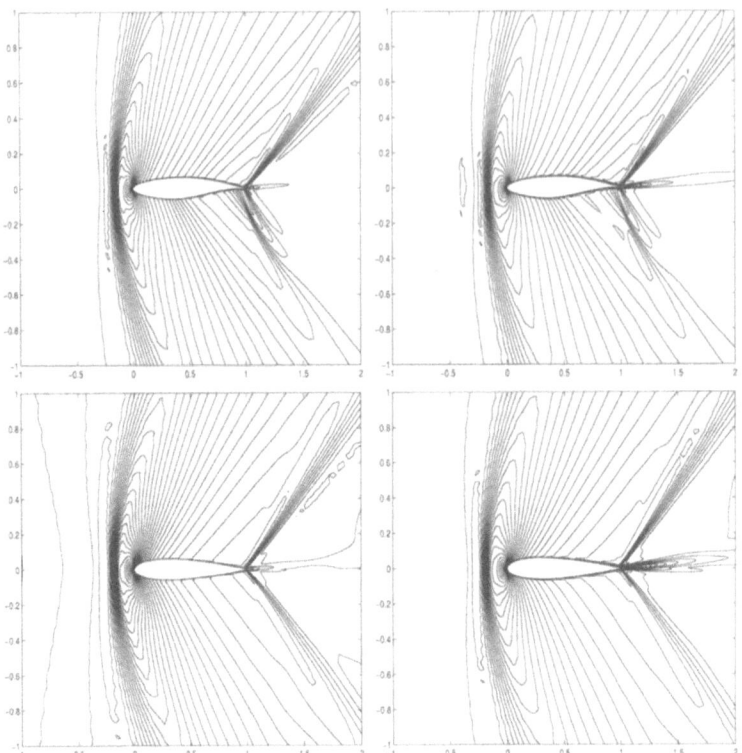

FIGURE 2. Pressure(left) and Mach(right) contours around Rae2822 airfoil for $M_\infty = 1.3$ $\alpha = 3.19$ and $Re = 6.5E + 6$(top row), NACA0012 airfoil for $M_\infty = 1.3$ $\alpha = 2.33$ and $Re = 6.5E + 6$(bottom row)

References

[1] A. Jameson, W. Schmidt and E. Turkel,*Numerical solutions of the Euler equations by finite volume methods using Runge-Kutta time stepping schemes*, AIAA Paper 81-1259, 1981.

[2] R.H. Ni, *A multiple grid scheme for solving the Euler equations*, AIAA J., **20** (1982), 1565–1571.

[3] M.G. Hall,*Cell vertex multigrid scheme for solution of the Euler equations*, RAE-TM-Aero 2029.*Proc. of conf. on num. methods for fluid dynamics* Univ. of reading, UK. April 1–5 (1985). IMA conf. series, Ox. Univ. press.

[4] A. Jameson and T. Baker, *Solution of the Euler equations for complex configurations*, Proc. AIAA 6th Computational Fluid Dynamics Conference, Danvers, (1983), 293–302.

[5] T.H. Puliam and J. L. Steger, *Recent improvements in efficiency, accuracy and convergence for implicit approximate factorization algorithms*, AIAA Paper 85-0360, AIAA 23rd Aerospace Sciences Meeting, Reno, January 1985.

[6] R.M. Beam and R. F. Warming, *An implicit factored scheme for the compressible Navier-Stokes equations*, AIAA J., **16** (1978), 393–402.

[7] L. Martenelli and A. Jameson, *Validation of a multigrid method for the Reynolds averaged equations*, AIAA Paper 88-0414, January 1988.

[8] R.W. MacCormack, *Current status of numerical solutions of the Navier-Stokes equations*, AIAA paper 85-0032, AIAA 23rd Aerospace Sciences Meeting, Reno, January 1985.

[9] W. Kordulla, *The Computation of Three Dimensional Viscous Transonic Flows with Separation*, Lecture Notes in Physics, (Springer-Verlag) **218**, (1984), 320–326.

[10] R.C. Swanson and E. Turkel, *A Multistage Time-Stepping Scheme for the Navier-Stokes Equations*, AIAA-85-0035.*AIAA 23rd Aerospace Science meeting*, June 14–17, Reno, Neveda, 1985.

[11] B. Muller, *Navier-Stokes Solution of Laminar Transonic Flow Over a NACA0012 Airfoil*, FFA report-140, 1986.

[12] S.K. Chakrabartty, *Numerical solution of Navier-Stokes equations for two-dimensional viscous compressible flows*, AIAA Journal, **27** (1989), 843–844.

[13] S.K. Chakrabartty, *A finite-volume nodal point scheme for solving two-dimensional Navier-Stokes equations*, Acta Mechanica, **84** (1990), 139–153.

[14] S.K. Chakrabartty, *Vertex-based finite volume solution of the two-dimensional Navier-Stokes equations*, AIAA Journal, **28** (1990), 1829–1831.

[15] S.K. Chakrabartty and K. Dhanalakshmi, *Computation of transonic flows with shock induced separation using algebraic turbulence models*, AIAA Journal, **33** (1995), 1979–1981.

[16] S.K. Godunov, *A difference method for the calculation of discontinuous solution of hydrodynamic equations*, Mat. Sbornik, **47** (1959), 271–306, Translated as JPRS 7225 by US Dept. of Commerce, 1960.

[17] P.D. Lax, *Hyperbolic Systems of Conservation Laws and the Mathematical Theory of Shock Waves*, SIAM Region. Ser. Appl. Math., SIAM Philadelphia, Pa, **11** (1973).

[18] J.P. Boris and D. L. Book, *Flux corrected transport I: SHASTA – A Fluid Transport Algorithm that Works*, J. Comp. Physics, **31** (1973), 38–69.

[19] B. Van Leer, *Towards the ultimate conservative difference scheme V: a second order sequel to Godunov's method*, J. Comp. Phys., **32** (1979), 101–136.

[20] P.L. Roe, *Approximate Riemann solvers, parameter vectors and difference schemes*, J. Comp. Phys., **43** (1981), 357–372.

[21] P.L. Roe, *Some contributions to the modelling of discontinuous flows*, Lectures in Applied Mathematics, **22**, Amer. Math. Soc., Providence, R.I., (1985).

[22] A. Harten, *High resolution schemes for hyperbolic conservation laws*, J. Comp. Phys., **49** (1983), 357–393.

[23] S. Osher, *Riemann solvers, the entropy condition and difference approximations*, SIAM J. Num. Anal., **21** (1984), 217–235.

[24] P.K. Sweby,*High resolution schemes using flux limiters for hyperbolic conservation laws*, SIAM J. Num. Anal., **21** (1984), 995–1011.

[25] H.C. Yee, *A Class of High-Resolution Explicit and Implicit Shock-Capturing Methods*, NASA TM 101088, Feb.1989. Also, Von Karman Institute for Fluid Dynamics Lecture Series 1989-04, Computational Fluid Dynamics, March 6–10, 1989, Rhode-St-Genése, Belgium.

[26] A. Jameson, *A non-oscillatory shock capturing scheme using flux limited dissipation*, Lectures in Applied Mathematics, Ed.: B.E. Engquist, S.Osher, and R.C.J. Sommerville, Amer. Math. Soc.,Part-1, **22** (1985), 354–370.

[27] A. Jameson, *Success and Challenges in Computational Aerodynamics*, AIAA paper 87-1184.

[28] A. Jameson, *Artificial diffusion, upwind biasing, limiters and their effect on accuracy and multigrid convergence in transonic and hypersonic flows*, AIAA-93-3359, 11-th AIAA Computational Fluid Dynamics Conference, Orlando, FL, July 6–9, 1993.

[29] S. Tatsumi, L. Martinelli and A. Jameson, *Design, implementation and validation of Flux Limited schemes for the solution of the compressible Navier-Stokes equations*, 32nd Aerospace Sciences Meeting of Exhibit, January 10–13, 1994, Reno, NV.

[30] I. Teipel and A. Wiedermann, *A TVD-Scheme for Computing Transonic Internal Flows*, CFD J., **(1)** (1992), 79–90.

[31] H.C. Yee, G. H. Klopfer and J. L. Montagne, *High-Resolution shock-capturing schemes for inviscid and viscous hypersonic flow*, J. Comp. Phys., **88** (1990), 31–61.

[32] B.S. Baldwin and H. Lomax, *Thin Layer Approximation and Algebraic Model for Separated Turbulent Flows*, AIAA Paper 78-257, 1978.

[33] R.K. Jain, *Grid generation about an aerofoil by the algebraic equation method*, DFVLR-IB-129-83/90.

[34] P.H. Cook, M.A. McDonald and M.C.P. Firmin, *Aerofoil RAE 2822 – Pressure Distributions and Boundary Layer and Wake Measurements*, AGARD-AR-138, 1979.

[35] P.R. Ashill, R. F. Wood and D. J. Weeks, *An Improved Semi – inverse Version of the Viscous Garabedian and Korn Method (VGK)*, RAE Tech. Rep. 87002, Jan. 1987.

[36] P.R. Ashill, D. J. Weeks and J. L. Fulker, *Wind Tunnel Experiments on Aerofoil Models for the Assessment of Computational Flow Methods*, AGARD CP 437, Dec. 1988.

AG Technomathematik,
University of Kaiserslautern,
67663 Kaiserslautern, Germany
E-mail address: hazra@mathematik.uni-kl.de

International Series of Numerical Mathematics
Vol. 129, © 1999 Birkhäuser Verlag Basel/Switzerland

Nonlinear MHD Processes in the Sun's Atmosphere

Alan W. Hood

Abstract. The outer atmosphere of the Sun, the corona, is an ionised high temperature plasma that is dominated by the coronal magnetic field. There are many important coronal phenomena that may be modelled as a magnetised fluid and described by the MHD equations. In their ideal form, the MHD equations are a system of nonlinear hyperbolic equations but it is the small dissipation terms that are important in many cases. Three examples are chosen to illustrate how numerical simulations can be used to investigate how dissipation can tap the energy in the coronal magnetic field. All involve the build-up of large gradients in the magnetic field through (i) the lack of smooth equilibria, (ii) waves propagating in an inhomogeneous plasma and (iii) the nonlinear evolution of an ideal MHD instability.

A major outstanding problem in solar physics is to explain why the solar corona is over hundred times hotter than the underlying plasma layers. This means there must be a heating mechanism to explain this high temperature and all realistic heating mechanisms depend on the local coronal magnetic field and require the dissipation of magnetic energy but, before any dissipation can occur, there must be a build up of large gradients in the magnetic field.

In the closed field structures of active regions, it is thought that the magnetic field, stressed by random motions of their photospheric footpoints, does not possess a smooth equilibrium but instead it relaxes to a configuration that possesses singularities in an ideal MHD plasma. These singularities are called "current sheets" and are the regions where large field gradients occur and where dissipation can occur through magnetic reconnection.

In the open field regions of coronal holes, on the other hand, magnetic footpoint motions simply propagate away as magnetic waves. However, Alfvén waves are anisotropic and different fieldlines have different wavelengths. If neighbouring waves propagate with different wavelengths, then they may phase-mix, build up large transverse gradients and then dissipate their magnetic energy.

Large sudden releases of magnetic energy are known as solar flares. The energy released in a large solar flare is the order of 10^{25} Joules making flares the most energetic events in the solar system. The main energy release occurs in a matter of an hour or so but shorter bursts of energy can be released in seconds. This is hard to explain unless there is driven magnetic reconnection. There is clear evidence that the coronal magnetic field after a flare is reorganised into a simpler configuration suggesting that reconnection has indeed occurred. The only source of energy available to drive these flares is the coronal magnetic field and flare theory needs to explain how the energy can be stored in the magnetic field and how it can be released on a timescale that is the order of the Alfvén time. It is believed that flares are the nonlinear development of an MHD instability. The linear MHD stability analysis of solar coronal fields shows that the dense photosphere provides a strong stabilising effect that allows the coronal field to store sufficient energy and its nonlinear development of an MHD instability that generates current sheets and drives magnetic reconnection at the local Alfvén speed. Some detailed large scale numerical simulations are presented.

1. Introduction and the MHD equations

The Solar Corona is an ionised plasma at a temperature of over $1 \times 10^6 \mathrm{K}$ and there are many fundamental problems in solar physics that can be modelled using the equations of Magnetohydrodynamics (MHD). In this paper, a brief introduction to the MHD equations is presented. Three problems involving numerical simulations are described. In a sense all three, although describing different phenomena, have the extraction of magnetic energy as their common theme. In Section 2, the braiding of closed magnetic field lines generates regions of strong currents that are dissipated and heat the plasma. Section 3 illustrates how the open field regions of coronal holes can be heated by Alfvén waves and the next section describes the non-linear evolution of an MHD instability as a model for a small solar flare. In an article of this length it is difficult to give a comprehensive reference list and, instead, the more important references are listed.

The equations of MHD (see Priest [8]) are

$$\text{the momentum equation} \qquad \rho \frac{D\mathbf{v}}{Dt} = -\nabla p + \mathbf{j} \times \mathbf{B} + \rho \mathbf{g}, \qquad (1)$$

$$\text{the continuity equation} \qquad \frac{\partial \rho}{\partial t} + \nabla \cdot (\rho \mathbf{v}) = 0, \qquad (2)$$

$$\text{the energy equation} \qquad \frac{\rho^\gamma}{\gamma - 1} \frac{D}{Dt} \left(\frac{p}{\rho^\gamma} \right) = 0, \qquad (3)$$

$$\text{the ideal gas law} \qquad p = \frac{1}{\tilde{\mu}} \rho R T, \qquad (4)$$

$$\text{the induction equation} \qquad \frac{\partial \mathbf{B}}{\partial t} = \nabla \times (\mathbf{v} \times \mathbf{B}) + \eta \nabla^2 \mathbf{B}, \qquad (5)$$

$$\text{Ohm's law} \qquad \mathbf{j} = \sigma \left(\mathbf{E} + \mathbf{v} \times \mathbf{B} \right), \qquad (6)$$

$$\text{Ampere's law} \qquad \mathbf{j} = \frac{\nabla \times \mathbf{B}}{\mu}, \qquad (7)$$

$$\text{and} \qquad \nabla \cdot \mathbf{B} = 0. \qquad (8)$$

In the above equations ρ is the plasma density, $\mathbf{v}$ the velocity, p the pressure, $\mathbf{j}$ the current density, $\mathbf{B}$ the magnetic induction, but more commonly called the magnetic field, $\mathbf{g}$ the gravitational acceleration, γ the ratio of specific heats, In the gas law $\tilde{\mu}$ is the mean molecular weight, R is the gas constant and T is the temperature. In the solar atmosphere, unlike the typical laboratory plasmas, the primary electromagnetic variable is the magnetic field and not the current.

If dissipation is neglected by setting $\eta = 0$ in (5), then the MHD equations are a nonlinear set of *hyperbolic partial differential equations*. Ideal MHD can be used to describe small amplitude oscillations in the solar interior and coronal magnetic fields. Nonlinear development of MHD instabilities can occur on a fast timescale but it is the inclusion of the small dissipation term that is extremely important in describing the release of magnetic energy in small-scale heating events, large solar flares and the acceleration of fast particles through magnetic reconnection.

The importance of the resistive term in (5) is governed by the size of the magnetic Reynold's number,

$$R_m = \frac{V_0 L}{\eta},$$

where V_0 and L are a typical velocity and lengthscale respectively.

If $R_m \gg 1$, dissipation and magnetic reconnection are only important where there are very strong currents, that is where there are large gradients in the magnetic field. In the solar corona R_m is the order of 10^{12} and so very short lengthscales are needed for magnetic reconnection. Short lengthscales are generated if, for example, singularities form in the equilibrium magnetic field.

1.1. MHD equilibria

Before discussing some of the interesting dynamical solar phenomena, it is important to realise that a simple description of the basic equilibria is not particularly straightforward. Setting $\mathbf{v}$ and time derivatives to zero and assuming derivatives in the y direction may be ignored, two dimensional equilibria can be described by the Grad-Shafranov equation. The magnetic field is expressed in terms of a flux function $A(x, z)$ such that

$$\mathbf{B} = \nabla A \times \hat{\mathbf{e}}_y + B_y(A)\hat{\mathbf{e}}_y, \tag{9}$$

and the pressure is given by

$$p(x, z) = P(A)e^{-z/\Lambda}, \tag{10}$$

where $B_y(A)$ and $P(A)$ are functions, possibly non-linear functions, of A and $\Lambda = RT/\tilde{\mu}g$ is the isothermal pressure scaleheight. The equation for the flux function A is the Grad-Shafranov equation, namely

$$\nabla^2 A + \frac{\partial}{\partial A}\left(\mu P(A)e^{-z/\Lambda} + \frac{1}{2}B_y^2(A)\right) = 0. \tag{11}$$

This is a non-linear elliptic partial differential equation that describes the equilibrium and there exists the possibility of catastrophe (also called either non-equilibrium or bifurcation) points, where there is no nearby solution for the given boundary conditions.

1.2. Ideal linearised MHD equations

Consider a uniform equilibrium, p_0, ρ_0 with the magnetic field $B_0 = B_0 \hat{\mathbf{e}}_z$. Then linearising the MHD equations about this equilibrium and taking Fourier components of the form $\mathbf{v} = \mathbf{v} e^{i(\mathbf{k}\cdot\mathbf{r}-\omega t)}$, gives the linearised equation of motion as

$$\omega^2 \mathbf{v} = c_s^2 (\mathbf{k}\cdot\mathbf{v})\mathbf{k} + v_A^2 \left[\mathbf{k}\times\{\mathbf{k}\times(\mathbf{v}\times\hat{\mathbf{e}}_z)\}\right]\times\hat{\mathbf{e}}_z, \tag{12}$$

where $c_s^2 = \gamma p_0/\rho_0$ and $v_a^2 = B_0^2/\mu\rho_0$, are the squares of the sound and Alfvén speeds respectively. Eq. (12) describes the three basic wave modes.

Taking the component of the vorticity along the equilibrium magnetic field, $\hat{\mathbf{e}}_z\cdot(\mathbf{k}\times\mathbf{v})$ gives the dispersion relation for Alfvén waves, namely

$$\omega^2 = (\mathbf{k}\cdot\hat{\mathbf{e}}_z)^2 v_A^2 = K^2 v_A^2 \cos^2\theta, \tag{13}$$

where K is the magnitude of the wavevector and θ is the angle between the equilibrium magnetic field and the wavevector. The important result is that Alfvén waves cannot propagate across field lines, when $\theta = \pi/2$. Thus, neighbouring field lines can have different frequencies and wavelengths because there is no way for Alfvén waves to communicate across the field.

The other wave modes, for the velocity components $(\mathbf{v}\cdot\mathbf{k})$ and $\mathbf{v}\cdot\mathbf{B}_0$, satisfy the dispersion relations

$$\frac{\omega^2}{K^2} = \frac{c_s^2 + v_A^2}{2} \pm \frac{1}{2}\left\{\left(c_s^2 + v_A^2\right)^2 - 4c_s^2 v_A^2 \cos^2\theta\right\}^{1/2}, \tag{14}$$

where the '+' sign corresponds to the fast magnetoacoustic wave and the '−' sign to the slow magnetoacoustic wave. The fast wave is almost isotropic in that it only exhibits a relatively weak dependence on the angle of propagation to the equilibrium magnetic field. However, the slow wave is similar to the Alfvén wave in that it does not propagate across the magnetic field. This inability to propagate information across the magnetic field is important in wave heating models of the solar corona.

2. Heating of closed magnetic field regions

Random photospheric motions, due to the ever-changing convection pattern, tend to 'braid' the coronal magnetic field. The stressed field tries to relax towards an equilibrium but Parker [7] suggested that there are no general 3D equilibrium without *current sheets*, that is discontinuities in the direction of the magnetic field. The existence of current sheets is important. Since the magnetic Reynold's number is very large, resistive effects are only important if the lengthscales are extremely small. But this is exactly what happens in a current sheet, since the perpendicular distance collapses to zero.

Numerical simulations of a random braiding of the photospheric footpoints have been performed in 3D by Galsgaard and Nordlund [2]. They find that strong currents rapidly build up in the coronal plasma and these are dissipated by magnetic reconnection. In addition, they show that magnetic energy that is fed into

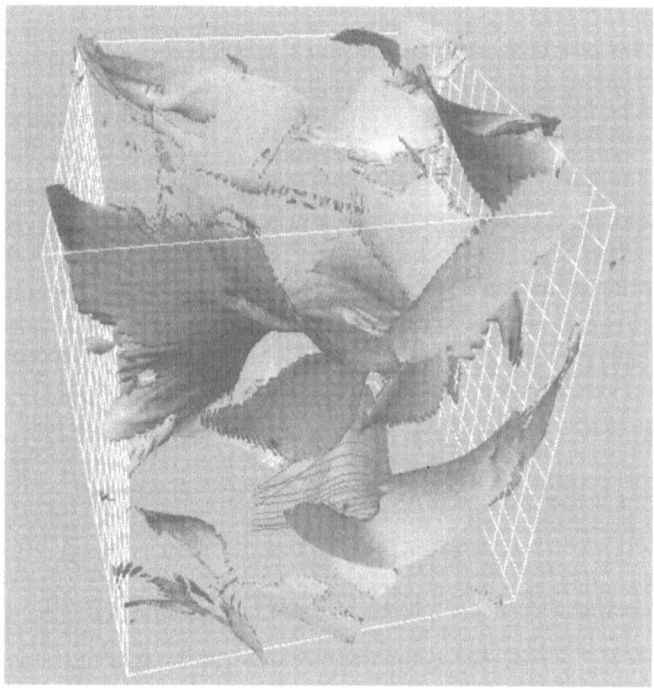

FIGURE 1. Isosurfaces of the current density showing the formation of strong currents, indicative of current sheets.

the system through the boundary motion (namely the Pointing flux) is all converted into ohmic heating. However, the limited resolution of the computational grid means that there is an artificial numerical resistivity that may be too large and, while there are regions of strong current, it is hard to say if current sheets have actually formed. To answer the question of current sheet formation, Longbottom *et al* [6] simplified the problem and investigated what happened when two simple footpoint shears were applied. They used a Lagrangian code that concentrated the gridpoints into the region where the large currents were building up. Plotting the maximum of the current as a function of the number of grids indicates whether the current is levelling off as the resolution is increased or whether the current continues to increase with increased resolution, indicating the formation of a current sheet. After the first shear the current formed a local concentration and the magnetic field was essentially 1D, with small boundary layers near the photospheric ends. However, if the second shear was sufficiently large, the current increased rapidly as the resolution was improved, suggesting that a current sheet had indeed formed. Obviously current sheet formation can never be proved by numerical simulation and an analytical proof is needed.

3. Heating of coronal holes

In coronal holes the magnetic field is open into interplanetary space. Thus, photospheric motions cannot build up magnetic energy in the coronal field as any stresses will simply propagate along the field as waves. If the background plasma density is no longer uniform but instead varies with the horizontal distance x, then the Alfvén speed will also be a function of x. Now, if photospheric motions oscillate the magnetic field with the same frequency, then different fieldlines will have different wavelengths. Thus, as the waves propagate along the open field, neighbouring field lines will become out of phase with each other and generate large horizontal gradients. These small lengthscales will eventually be smoothed out by resistivity and the magnetic energy will be extracted from the waves, thereby heating the plasma (see Heyvaerts and Priest, [3] and Hood, Ireland and Priest, [4]).

Assume there is a uniform, vertical background magnetic field

$$\mathbf{B}_0 = (0,0,B_0), \text{ but } \rho_0 = \rho_0(x) \quad \Rightarrow \quad v_A^2(x) = \frac{B_0^2}{\mu \rho_0(x)}.$$

To understand the process of Alfvén wave phase mixing, consider

$$\mathbf{v} = \left(0, v(x,z)e^{-i\omega t}, 0\right), \qquad \mathbf{B}_1 = \left(0, B(x,z)e^{-i\omega t}, 0\right).$$

The linear MHD equations, in the absence of viscosity and in the low beta limit, reduce to

$$-\omega^2 B = v_A^2(x)\frac{\partial^2 B}{\partial z^2} - i\eta\omega\frac{\partial^2 B}{\partial x^2}. \tag{15}$$

Note that in the ideal MHD limit, with $\eta = 0$, the solution for the perturbed magnetic field is

$$B = B_0(x)e^{ik(x)z}, \text{ and } k^2(x) = \frac{\omega^2}{v_A^2(x)}. \tag{16}$$

Now the x−derivative of B is $\partial B/\partial x = ik'(x)zB$, and it is clear that the gradient increases with height. A simple multiple scales analysis shows that

$$B(x,z) \approx B_0(x)e^{ik(x)z}e^{-z^3/z_d^3}, \tag{17}$$

(Heyvaerts and Priest, [3]) where the damping length z_d is given by

$$z_d = \left\{6\omega\left(\frac{k(x)}{k'(x)}\right)^2\frac{1}{\eta k^3(x)}\right\}^{1/3}. \tag{18}$$

This means that the damping length is proportional to the magnetic Reynold's number to the power one-third and to the horizontal inhomogeneity lengthscale to the two-thirds. Figure 2 shows that the oscillations in the perturbed magnetic field are damped with the damping length given by Eq. (18). Typical numbers for coronal holes show that phase-mixing can damp the Alfvén waves within a few wavelengths (Hood, Ireland and Priest, [5]).

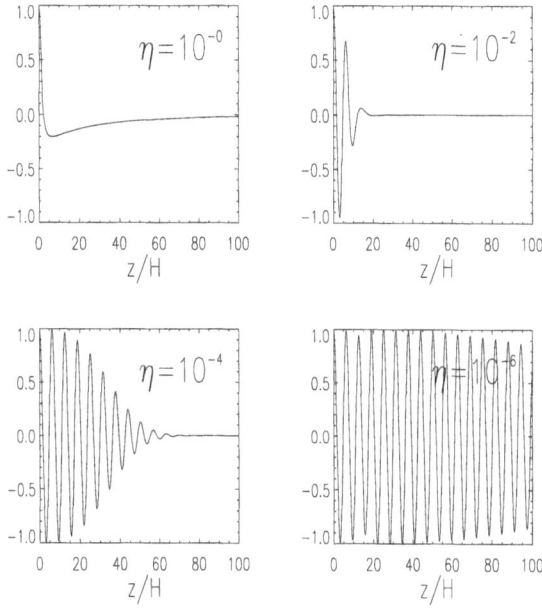

FIGURE 2. The perturbed magnetic field as a function of height for various values of the dimensionless resistivity. The damping of the Alfvén waves is clearly seen.

4. Ideal MHD instabilities in line-tied coronal loops

Coronal loops are observed to survive for many days, during which they show little change to their overall shape. Since the typical Alfvén timescale is on the order of a minute or so, it is clear that these magnetic loops must be stable to ideal MHD instabilities. However, it is also observed that these loops can become unstable and release a large amount of magnetic energy as a solar flare. Stability theory aims to explain how the magnetic field can remain in a stable state, during which time magnetic energy is slowly building up, and how an instability may be triggered once some critical conditions are exceeded. The important stabilising feature that must be included in all stability calculations and numerical simulations is the extremely dense photosphere that effectively anchors the magnetic footpoints through a process known as inertial line-tying. This may be modelled simply by assuming that all velocity components vanish at the photospheric ends of the loop.

A coronal loop may be modelled in cylindrical geometry by

$$\mathbf{B} = (0, B_\theta(r), B_z(r)).\tag{19}$$

The twist, $\Phi = LB_\theta/rB_z$ is the angle a field line passes through in going from one end of the loop to the other, and L, the length of the loop, are important parameters. Short loops tend to be stable whereas longer loops will become unstable to

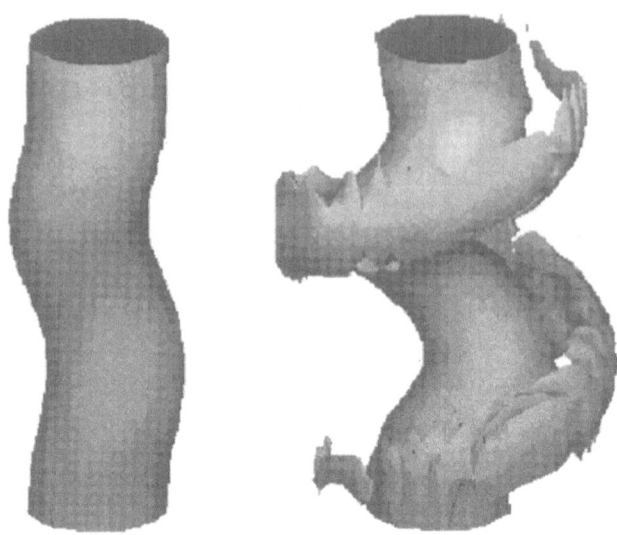

FIGURE 3. Isosurfaces of the current density at two times during
the nonlinear development of the kink instability.

ideal MHD instabilities. In the numerical simulation shown, the loop is unstable to
a linear displacement and the nonlinear evolution is investigated. The loop kinks
and in the process generates current sheet-like features (see Arber, Longbottom
and Van der Linden [1]). Figure 3 shows an isosurface of the current density. After
5 Alfvén times, the equilibrium current is still clearly seen but the loop is starting
to kink. However, the kinking of the loop starts to press field lines together in the
external region and, after 10 Alfvén times, strong currents are observed outside the
main body of the loop. With magnetic resistivity present, magnetic reconnection
occurs and magnetic energy is converted into plasma motions and heat.

It can be shown that the energy released is about two-thirds of the available
free energy, in excess of the potential lower limit, and is of the order needed to
drive small compact loop flares.

5. Conclusions

Three examples of MHD processes in the solar corona have been presented. In
each case, numerical simulations illustrate how the small dissipation terms in the
MHD equations produce important effects on the magnetic fields. In the absence
of dissipation, singularities in the magnetic field are produced through (i) a lack
of a smooth equilibrium, (ii) waves on neighbouring field lines being out of phase

with each other and (iii) an instability. In all cases, the dissipation removes the singularities and releases magnetic energy into the plasma resulting in heat and plasma motion.

The MHD equations provide a rich source of nonlinear phenomena and the recent satellite missions, such as Yohkoh and SOHO, have provided a wealth of new observations that require detailed modelling through numerical simulations and analytical methods. The next few years will be an exciting time in Solar MHD.

Acknowledgement. My sincere thanks to Drs Tony Arber and Klaus Galsgaard for providing Figures 3 and 1 respectively. My thanks to the organising committee for the invitation to speak at the conference. A most enjoyable experience.

References

[1] T. Arber, A. W. Longbottom and R. A. M. Van der Linden, *Current sheet formation and the effect of anomalous resistivity in unstable coronal loops,* Preprint (1998)

[2] K. Galsgaard and A. Nordlund, *Heating and Activity of the Solar Corona: 1 Boundary Shearing of an Initially Homogeneous Magnetic Field,* J. Geophys. Res. **101**, (1996) 13445–13460.

[3] J. Heyvaerts and E. R. Priest, *Coronal Heating by Phase Mixed Shear Alfvén Waves,* Astron. Astrophys. **117**, (1983) 220–234.

[4] A. W. Hood, *MHD Instabilities,* Solar System Magnetic Fields, Edited by E. R. Priest (Kluwer) **1986**, 80–120.

[5] A. W. Hood, J. Ireland and E. R. Priest, *Heating of Coronal Holes by Phase Mixing,* Astron. Astrophys. **318** (1996) 957–961.

[6] A. W. Longbottom, G. Rickard, I. Craig and A. Sneyd, *Magnetic Flux Braiding: Force-free Equilibria and Current Sheets* Astrophys. J. (1998) in press. Preprint (1997)

[7] E. Parker, *Spontaneous Current Sheets in Magnetic Fields,* Oxford University Press **1994**.

[8] E. R. Priest, *Solar Magnetohydrodynamics* D. Reidel, **1982**.

School of Mathematical and Computational Sciences,
University of St Andrews,
St Andrews
KY16 9SS, UK
E-mail address: alan@mcs.st-and.ac.uk

International Series of Numerical Mathematics
Vol. 129, © 1999 Birkhäuser Verlag Basel/Switzerland

The Relaxation Approximation to Hyperbolic System of Conservation Laws

Ling Hsiao and Ronghua Pan

In many physical situations one is required to consider a relaxation effect. For instance in gas dynamics the relaxation mechanism occurs when the gas is in the thermo-non-equilibrium.

Consider the hyperbolic system of conservation laws with relaxation

$$u_t + f(u,v)_x = 0$$
$$v_t + g(u,v)_x = \frac{V(u)-v}{\tau(u)}, \tag{1}$$

where the second equation contains a relaxation mechanism with $V(u)$ as the equilibrium value for v and $\tau(u)$ the relaxation time. We are interested in the zero relaxation limit $\tau(u) \to 0$. When certain stability criteria hold, the solution would converge to that of the equilibrium hyperbolic conservation law

$$u_t + f(u, V(u))_x = 0. \tag{2}$$

This is the problem of zero relaxation limit. A tightly related problem is the nonlinear stability of waves for this relaxation approximation, namely, the stability of elementary waves for the equilibrium hyperbolic conservation laws, related to the corresponding hyperbolic system with relaxation.

There are a lot of results on these two questions for the case when the equilibrium equation is a scalar hyperbolic conservation law (2). We refer to [1], [3], [16], [17] and [20] for L^∞, L^2 and L^1 stability, and [2], [3], [13], [21], and [23] for the zero relaxation limit in L^∞ and BV frameworks, respectively.

However, when the corresponding equilibrium is a system instead of a scalar one, the problem is more difficult and more challenges occur. This is the reason for us to be interested in some 3×3 model systems with relaxation for which the corresponding equilibrium system is the well-known p-system.

In the present lecture, we give the summary of results on the two questions (zero relaxation limit and stability analysis of elementary waves) with the following two model systems.

1. The reacting flow model proposed by R. J. LeVeque and others in [18]:

$$\begin{cases} (\rho r)_t + (\rho r u)_x = S \\ (\rho s)_t + (\rho s u)_x = -S \\ (\rho u)_t + (p + \rho u^2)_x = 0. \end{cases}$$

This model describes reacting gas in which there exist two modes. Where, ρr is the density of the major mode and ρs is of the minor mode, $r + s = 1$. u is the velocity and $p = \rho c^2(r + \beta s)$ is pressure which can be derived by Avogadro's law. Here c is the sound speed of the major mode. The parameter β may provide some tunous link with real physics. S is the source term satisfying

$$S = \frac{(r_E(\rho) - r)}{\varepsilon} = -\frac{(s_E(\rho) - s)}{\varepsilon},$$

where ε is the reaction time, $r_E(\rho)$ and $s_E(\rho)$ are equilibrium distributions.

The Lagrangian form of the above system is

$$\begin{cases} v_t - u_x = 0 \\ u_t + p_x = 0 \\ p_t + \frac{p}{v}v_t = \frac{p_E(v) - p}{\varepsilon}, \end{cases} \tag{3}$$

where, v is the specific volume, $p_E(v) = \frac{r_{1E}(v)}{v}$ and $r_{1E} = c^2(r_E + \beta(1 - r_E))$.

2. The rate-type viscoelastic model proposed by Suliciu in [24]:

$$\begin{cases} v_t - u_x = 0 \\ u_t + p_x = 0 \\ (p + Ev)_t = \frac{p_E(v) - p}{\varepsilon}, \end{cases} \tag{4}$$

where v and $(-p)$ denote strain and stress, u is related to the particle velocity, E is a positive constant called the dynamic Young's modulus, ε is a relaxation time.

It should be pointed out that the model system (3) is quasilinear while the system (4) is semilinear, therefore more difficulties occur when we deal with the system (3).

For both (3) and (4), the corresponding equilibrium system takes the form

$$\begin{cases} v_t - u_x = 0 \\ u_t + p_E(v)_x = 0, \end{cases} \tag{5}$$

with different forms of $p_E(v)$.

We will introduce the results on zero relaxation limit first and give the results on the stability of elementary waves afterwards.

1. Zero relaxation limit

Let us consider the initial problem for (3) or (4) with the following initial data

$$(v, u, p)(x, 0) = (v_0, u_0, p_0)(x), \tag{6}$$

such that

$$\lim_{x \to \pm\infty} (v_0, u_0, p_0)(x) = (v_\pm, u_\pm, p_\pm), \tag{7}$$

with $p_\pm = p_E(v_\pm)$. We assume that there exists an open set N, for $v \in N$ under considerations, it holds

(i) $p'_E(v) < 0,$
(ii) $p''_E(v) > 0,$
(iii) $|p'_E(v)| < \min\{\frac{p_+}{v_+}, \frac{p_-}{v_-}\},\ for\ (3),$ $or\ |p'_E(v)| < E,\ for\ (4).$

We have answered the question of zero relaxation limit for two cases. Either the solution of (5) is piecewise smooth with finite non-interacting shocks up to a time $T > 0$, or the state (v_-, u_-) and (v_+, u_+) are connected by centered rarefaction waves (1-mode or 2-mode). More precisely, we have obtained

Theorem 1.1. *[7] Under (i)–(iii), let $(v^0, u^0)(x, t)$ be a piecewise smooth solution of (5) with finitely many non-interacting shocks up to time $T > 0$, with the jumps suitably weak. Let $p^0(x, t) = p_E(v^0)(x, t)$. Then for each $\varepsilon \in (0, \varepsilon_0]$, with a positive constant ε_0, there exists a unique smooth solution $(v^\varepsilon, u^\varepsilon, p^\varepsilon)(x, t)$ of (3) with*

$$(v^\varepsilon, u^\varepsilon, p^\varepsilon) - (v^0, u^0, p^0) \in L^\infty([0, T], H^2).$$

Moreover, for any $\alpha \in (0, 1)$,

$$\sup_{0 \le t \le T} \int_{-\infty}^{+\infty} |(v^\varepsilon - v^0, u^\varepsilon - u^0, p^\varepsilon - p^0)|^2\ dx \le C_1 \varepsilon^\alpha,$$

and

$$\sup_{0 \le t \le T,\ d \ge h}\ |(v^\varepsilon - v^0, u^\varepsilon - u^0, p^\varepsilon - p^0)| \le C_h \varepsilon,\quad for\ any\ h > 0.$$

Here, C_1, C_h are positive constants independent of ε and $d \equiv dist(x,\ shocks)$. This result is proved by the matched asymptotic analysis method introduced in [4] and [26]. The key idea is that if we can construct a formal solution of (3) by matching the truncated Hilbert expansion (outer expansion) and shock layer expansion (inner expansion), then the existence of a solution of (3) and the convergence to the solution of (5) can be reduced to the stability analysis for the approximate solution.

Remark 1.2. The same result for the model (4) is given in [14]. Comparing to [14], in the proof of Theorem 1.1, the higher order expansion and estimates are needed in order to weaken the nonlinearity.

Theorem 1.3. *[11] Under (i)–(iii), let (v_-, u_-) and (v_+, u_+) be connected by centered rarefaction waves (1-mode or 2-mode) of (5), $(v^r, u^r)(x/t)$, with weak strength. Then for each $\varepsilon \in (0, \varepsilon_0]$, with the following initial data*

$$(v_0, u_0, p_0)(x) = \begin{cases} (v_-, u_-, p_E(v_-)), & x < 0, \\ (v_+, u_+, p_E(v_+)), & x > 0, \end{cases}$$

(4) has a unique global piecewise smooth solution $(v^\varepsilon, u^\varepsilon, p^\varepsilon)(x, t)$ such that

$$\sup_{x \in R^1,\ t \ge h}\ |(v^\varepsilon - v^r, u^\varepsilon - u^r, p^\varepsilon - p_E(v^r))| \to 0,\quad as\ \varepsilon \to 0,$$

for any positive number h.

Furthermore, the strength of the jumps in the solution $(v^\varepsilon, u^\varepsilon, p^\varepsilon)(x, t)$ decays exponentially fast as time tends to infinite. It should be pointed out that this work is strongly motivated by Xin's work in [27], where the zero viscosity limit to rarefaction waves for the one-dimensional Navier-Stokes equations of the compressible isentropic gases is proved. However, the approach is quite different here, due to the different dissipative effects between viscosity and relaxation and the special nonlinearity of (4). Since the dissipation of relaxation is much weaker than viscosity, the limit here is much singular than those in [27]. In [27], a key role is the observation of Liu and Hoff in [5] that the initial discontinuities of specific volume will propagate along the particle path and decay exponentially. In our case, the discontinuities will propagate along the three characteristics issued from the origin respectively and the decay estimates will be obtained by a straight characteristic analysis. Recently, Wang and Xin have succeeded in proving the small mean free path limit to centred rarefaction waves for Broadwell model in [25]. Since the form of the Broadwell model is quite similar to (4), the framework is similar. But the approach here is different from [25]. In [25], the second order correction of the Chapmann-Enskog expansion is used in constructing the smooth approximation for rarefaction waves, while we use the first order of the expansion directly. The pointwise estimate for first order derivatives of the solution is achieved by an iteration argument in [25], while in the present case we use the characteristic analysis to obtain a better estimate. [25] follows the procedure in [19] for making energy estimate where the Boltzmann entropy function is used. We employ the method of [8] without the need of entropy. The advantage of our approach is that it can be generalized to the general relaxation systems of conservation laws proposed by Jin and Xin.

2. Stability analysis for elementary waves

The stability of shocks and rarefaction waves has been proved for the model (3) or (4) under the assumption (i)–(iii) given above. We now assume $\varepsilon = 1$ in (3) and (4).

2.1. Stability for rarefaction waves

Nonlinear stability for 1-mode (or 2-mode) rarefaction waves under generic perturbations for (4) is given in the following theorem.

Theorem 2.1. *[8] Under (i)–(iii), let (v_-, u_-) and (v_+, u_+) be connected by 1-mode (or 2-mode) rarefaction waves of (5), $(v^r, u^r, p^r)(x/t)$, with $p^r = p_E(v^r)$. Let (V, U, P) be the smooth rarefaction waves of (v^r, u^r, p^r) introduced by Burgers equation, where $P = p_E(V)$. Then there exist positive constants δ_0 and ε_0, such that if*

$$|v_+ - v_-| + |u_+ - u_-| \le \delta_0$$

and

$$\|(v_0 - V(x,0), u_0 - U(x,0), p_0 - P(x,0))\|_{H^2} \le \varepsilon_0,$$

then (4) and (6) have a unique global smooth solution (v, u, p) tending to (v^r, u^r, p^r) uniformly in x as $t \to +\infty$.

The same result is expected to be hold for the model (3).

2.2. Stability for shock fronts

The nonlinear stability for shock fronts under "zero excessive mass" perturbation is proved in [22] and [28] for the model (3), in [6] for the model (4.

For generic perturbation, the linear stability for shock fronts is given in [10] for the model (3).

Theorem 2.2. *[10] Under (i)–(iii), let (v_-, u_-) and (v_+, u_+) be connected by a shock of (5), and the corresponding shock front of (3) is (V, U, P). Suppose that $p_E(v) = p_E(V) + p'_E(V)(v - V)$, and*

$$(v_0 - V(x, 0), u_0 - U(x, 0), p_0 - P(x, 0)) \in H^2.$$

Then there exist positive constants δ_0 and ε_0 such that if

$$|v_+ - v_-| + |u_+ - u_-| \le \delta_0,$$

$$\|(\sqrt{1 + x^2}(v_0 - V(x, 0), \sqrt{1 + x^2}(u_0 - U(x, 0))\|_{L^2} \le \delta_0$$

and

$$\|(v_0 - V(x, 0), u_0 - U(x, 0), p_0 - P(x, 0))\|_{H^2} \le \varepsilon_0,$$

Then there exists a unique global smooth solution to (3) and (6) satisfying

$$(v - V, u - U, p - P) \in C([0, +\infty), H^2)$$

and

$$\lim_{t \to +\infty} \ \|(v - V, u - U, p - P)\|_{H^1} = 0.$$

The same result for model (4) was obtained by Luo and Serre in [15]. In [15], in order to derive the energy estimates, one only needs to deal with a linear system of the error equations, while we have to work on the nonlinear one.

The nonlinear stability for 2-mode shock fronts under generic perturbations are given in the following theorem.

Theorem 2.3. *[9] Under (i)–(iii), let (v_-, u_-) and (v_+, u_+) be connected by a 1-shock and a 2-shock of (5), and the corresponding fronts of (4) are (v_1, u_1, p_1) and (v_2, u_2, p_2) respectively. Let the linear superposition of these two profiles with suitable translations be (V, U, P). Then there exist positive constants δ_0 and ε_0, such that if*

$$|v_+ - v_-| + |u_+ - u_-| \le \delta_0,$$

$$\|(\sqrt{1 + x^2}(v_0 - V(x, 0), \sqrt{1 + x^2}(u_0 - U(x, 0))\|_{L^2} \le \delta_0$$

and

$$\|(v_0 - V(x, 0), u_0 - U(x, 0), p_0 - P(x, 0))\|_{H^2} \le \varepsilon_0.$$

Then there exists a unique global smooth solution to (4) and (6) satisfying

$$(v - V, u - U, p - P) \in C([0, +\infty), H^2)$$

and

$$\lim_{t \to +\infty} \ \|(v - V, u - U, p - P)\|_{H^1} = 0.$$

The same result can be obtained for the model (3).

References

[1] I.L. Chern, *Long-time effect of relaxation for hyperbolic conservation laws*, Comm. Math. Phys., **172** (1995), 39–55.

[2] G.Q. Chen and T.P. Liu, *Zero relaxation and dissipation limits for hyperbolic conservation laws*, Comm. Pure Appl. Math., **46** (1993), 755–781.

[3] G.Q. Cheng, C.D. Levermore and T.P. Liu, *Hyperbolic conservation laws with stiff relaxation terms and entropy*, Comm. Pure Appl. Math., **47** (1994), 787–830.

[4] J. Goodman and Z.P. Xin, *Viscous limits for piecewise smooth solutions to systems of conservation laws*, Arch. Rational Mech. Anal., **121** (1992) 235–265.

[5] D. Hoff and T.P. Liu, *The inviscid limit for the Navier-Stokes equations of compressible, isentropic flow with shock data*, Indiana Univ. Math. J., **38** (1989) 861–915.

[6] L. Hsiao and T. Luo, *The stability of travelling wave solutions for a rate-type viscoelastic system*, to appear.

[7] L. Hsiao, H.L. Li and R H. Pan, *The zero relaxation limit of piecewise smooth solutions to the reacting flow model in the presence of shocks*, to appear.

[8] L. Hsiao and R.H. Pan, *Nonlinear stability of rarefaction waves for a rate-type viscoelastic system*, to appear.

[9] L. Hsiao and R.H. Pan, *Nonlinear stability of two-mode shock profiles for a rate-type viscoelastic system with relaxation*, preprint.

[10] L. Hsiao and R.H. Pan, *The linear stability of travelling wave solutions for a reacting flow model with source term*, to appear.

[11] L. Hsiao and R.H. Pan, *Zero relaxation limit to centered rarefaction waves for a rate-type viscoelastic system*, to appear.

[12] T.P. Liu, *Hyperbolic conservation laws with relaxation*, Comm. Math. Phys., **108** (1987), 153–175.

[13] T. Luo and R. Natalini, *BV solution and relaxation limit for a viscoelastic model with relaxation*, to appear.

[14] H. L. Li and R.H. Pan, *Zero relaxation limit to piecewise smooth solutions for a rate-type viscoelastic system in the presence of shocks*, preprint.

[15] T. Luo and D. Serre, *Linear stability of shock profiles for a rate-type viscoelastic system with relaxation*, to appear.

[16] T. Luo, *Asymptotic stability of planar rarefaction waves for the relaxation approximations of conservation laws in several dimensions*, J. Diff. Eqns., **133 (2)** (1997) 255–279.

[17] T. Luo and Z.P. Xin, *Asymptotic stability of planar shock profiles for the relaxation approximations of conservation laws in several dimensions*, preprint.

[18] R.J. LeVeque, H.C. Yee, P. Roe and B. v. Leer, *Model systems for reacting flow*, Final Report, NASA-Ames University Consortium NCA2-185, (1988).

[19] A. Matsumura, *Asymptotics toward rarefaction wave of solutions of the Broadwell model of a discrete velocity gas*, Japan J. Appl. Math., **4** (1987), 489–502.

[20] C. Mascia and R. Natalini, L^1 *nonlinear stability of travelling waves for a hyperbolic system with relaxation*, to appear.

[21] R. Natalini, *Convergence to equilibrium for the relaxation approximations of conservation laws,* Comm. Pure Appl. Math., (1997).

[22] R.H. Pan, *The nonlinear stability of traveling wave solutions for a reacting flow model with source term,* to appear.

[23] W. Shen, A. Tveito and R. Winther, *On the zero relaxation limit for a system modeling the motions of a viscoelastic solid,* preprint (1997).

[24] I. Suliciu, *On modelling phase transitions by means of rate-type constitutive equations, shock wave structure,* Int. J. Engng. Sci.,2 **8** (1990), 827–841.

[25] W.C. Wang and Z.P. Xin, *On small mean free path limit of Broadwell model with discontinuous initial data, the centered rarefaction wave case,* preprint.

[26] Z.P. Xin, *The fluid-dynamic limit of the broadwell model of the nonlinear Boltzmann equation in the presence of shocks,* Comm. Pure Appl. Math., **44 (6)** (1991), 679–713.

[27] Z.P. Xin, *Zero dissipation limit to rarefaction waves for the one-dimensional Navier-Stokes equations of compressible isentropic gases,* Comm. Pure Appl. Math., XLVI (1993), 621–665.

[28] W.-A. Yong, *Existence and asymptotic stability of traveling wave solutions of a model system for reacting flow Nonlinear Analysis,* TMA, **20 (12)** (1996).

Institute of Mathematics, Academia Sinica,
Beijing, 100080, P.R. China

International Series of Numerical Mathematics
Vol. 129, © 1999 Birkhäuser Verlag Basel/Switzerland

A Front Tracking Method for Conservation Laws with Boundary Conditions

K. Hvistendahl Karlsen, K.-A. Lie, and N.H. Risebro

Abstract. A front tracking method is used to construct weak solutions to scalar conservation laws with two kinds of boundary conditions — Dirichlet conditions and a novel zero flux (or no-flow) condition. The construction leads to an efficient numerical method. The main feature of the scheme is that there is no stability condition correlating the spatial and temporal discretization parameters. The analysis uses the traditional method of proving compactness via Helly's theorem as well as the more modern concept of measure valued solutions. Three numerical examples are presented.

1. Introduction

Let $\Omega \subset \mathbb{R}^d$ be a bounded open set with piecewise regular boundary $\partial\Omega$ and outward unit normal n. Let $x = (x_1, \ldots, x_d)$ and $f = (f_1, \ldots, f_d)$. We study scalar conservation laws

$$u_t + \nabla \cdot f(u) \equiv u_t + \sum_{i=1}^{d} f_i(u)_{x_i} = 0, \qquad u(x,0) = u_0(x), \tag{1}$$

for $(x,t) \in \Omega \times \langle 0, T]$ with either prescribed Dirichlet boundary data

$$u(x,t) = r(x,t), \qquad x \in \partial\Omega, t > 0 \tag{2}$$

or a zero flux (no-flow) boundary condition

$$f(u(x,t)) \cdot n = 0, \qquad x \in \partial\Omega, t > 0. \tag{3}$$

Approximate solutions are constructed by dimensional splitting, using front tracking for the one-dimensional equations. The resulting method is an extension of an unconditionally stable method proposed by Holden and Risebro [5], see also [6].

2. Dirichlet boundary condition

Consider (1) with prescribed boundary condition (2). For this problem, a singular solution may develop in finite time even for smooth initial data. Moreover, the hyperbolic boundary value problem (1) and (2) (even when (1) is linear) is usually not well-posed when the boundary condition is required to hold in the strong sense. For these reasons, the problem has been to find a physically reasonable framework which incorporates discontinuous solutions as well as a correct mathematical formulation of the boundary condition.

A common approach is the *vanishing viscosity* method, in which a small viscosity term $\varepsilon \Delta u$ is added to regularize the problem. Then the correct solution is chosen as the L^1 limit of solutions to the corresponding parabolic problem as ε tends to zero. This approach leads to the following Kruzkov type *entropy condition* [1]: We call a function $u(x,t) \in L^\infty \cap BV$ an entropy weak solution to (1) and (2) provided that for all $k \in \mathbb{R}$ and suitable test functions $\phi \geq 0$ with $\phi|_{t=T} = 0$, the following (entropy) inequality holds

$$\mathcal{L}_\phi(u) := \int_\Omega \int_0^T \Big(|u - k|\phi_t + \mathrm{sgn}(u - k)\big(f(u) - f(k)\big) \cdot \nabla\phi \Big)\, dt\, dx$$

$$- \int_{\partial\Omega} \int_0^T \mathrm{sgn}(r(s,t) - k)\big(f(\gamma u(s,t)) - f(k)\big) \cdot n(s)\phi(s,t)\, dt\, ds \quad (4)$$

$$+ \int_\Omega |u_0(x) - k|\phi(x,0)\, dx \geq 0.$$

Here γu denotes the L^1 trace. Thus, the boundary condition is satisfied in the following sense

$$\mathrm{sgn}(\gamma u(x,t) - k)(f(\gamma u(x,t)) - f(k)) \cdot n(x) \geq 0, \qquad \forall k \in I(r(x,t), \gamma u(x,t)),$$

for a.e. $(x,t) \in \partial\Omega \times \langle 0,T]$, where $I(\alpha, \beta)$ denotes the closed interval with bounds α and β. Furthermore, the entropy solution is unique and depends L^1 continuously on the initial and boundary data [1, 7].

2.1. One spatial dimension

Consider first the one-dimensional equation

$$v_t + f(v)_x = 0, \qquad (x,t) \in \langle a,b \rangle \times \langle 0,T] \tag{5}$$

with initial and boundary data

$$\begin{aligned} v(x,0) &= v_0(x), & x &\in \langle a,b \rangle, \\ v(a,t) &= v_a, \quad v(b,t) = v_b, & t &\in \langle 0,T]. \end{aligned} \tag{6}$$

We construct approximate solutions by front tracking [3]. Assume that the initial data is a step function, i.e., a series of initial Riemann problems. If the flux function f is piecewise linear then all Riemann problems have solutions within the class of step functions. At the left boundary $(x = a)$ we have a Riemann problem with initial and boundary conditions

$$\begin{aligned} v(x,0) &= v_0(a^+), & x &> a \\ v(a,t) &= v_a, & t &\in \langle 0,T]. \end{aligned}$$

The solution of this problem is defined by restricting the solution of the Riemann problem with left state v_a and right state $v_0(a^+)$ to $x > a$. A similar construction can be used at the right boundary $(x = b)$. All wave interactions lead to new Riemann problems, either in the interior of the domain or at the boundary. Therefore a global solution to the approximate problem can be obtained by tracking discontinuities and solving Riemann problems. This construction is well-defined in the sense that the number of wave interactions and interactions with the boundaries is finite even in infinite time; the proof is similar as for the initial value problem [3]. By construction, the front tracking solution satisfies the entropy condition for the approximate equation. Furthermore, the front tracking method is well-defined also when the boundary data are step functions, provided the step functions have a finite number of discontinuities. In fact, we have

Lemma 2.1. *Suppose that $v_0(x)$, $v_a(t)$, and $v_b(t)$ are piecewise constant functions with a finite number of discontinuities, have compact support and are bounded within $[m, M]$. Let f be a Lipschitz continuous, piecewise linear function with a finite number of breakpoints in $[m, M]$. Then the Dirichlet problem (5) and (6) has an entropy weak solution $v(x, t)$ that is piecewise constant in x for each fixed $t > 0$ and takes values in the set $\{v_0(x)\} \cup \{v_a(t), v_b(t)\} \cup \{\text{the breakpoints of } f\}$. Furthermore,*

$$\|v(x, t)\|_\infty \leq \max(\|v_0\|_\infty, \|v_a\|_\infty, \|v_b\|_\infty),$$
$$\mathrm{TV}(v(\cdot, t)) \leq \mathrm{TV}(v_0) + |v_a - v_0(a^+)| + |v_b - v(b^-)|, \tag{7}$$
$$\|v(\cdot, t_2) - v(\cdot, t_1)\|_1 \leq C|t_2 - t_1|.$$

The solution can be constructed by front tracking in a finite number of steps for any $t > 0$ and is stable with respect to initial and boundary data,

$$\|v^1(\cdot, t) - v^2(\cdot, t)\|_1 \leq \|v_0^1 - v_0^2\|_1 + t\|f\|_{Lip}(\|v_a^1 - v_a^2\|_\infty + \|v_b^1 - v_b^2\|_\infty). \tag{8}$$

We can use front tracking to construct approximate solutions to a general equation, by approximating the initial and boundary data by step functions and the flux function by a piecewise linear function. Then it follows by a standard compactness argument using (7) that the approximate solutions converge to a solution of (5) and (6). Moreover, the limit is an entropy weak solution satisfying the Kruzkov inequality (4).

Theorem 2.2. *Suppose that $v_0(x)$, $v_a(t)$, and $v_b(t)$ are $L^\infty \cap BV$ and that f is Lipschitz continuous. Then the Dirichlet problem (5) and (6) has an entropy weak solution $v(x, t)$ which can be constructed as a limit of front tracking solutions.*

2.2. Arbitrary spatial dimension

We construct approximate solutions of the Dirichlet problem (1) and (2) only in two dimensions as generalizations to multidimensions are straightforward. Assume that the domain Ω is covered by a Cartesian grid $\{i\Delta x, j\Delta x\}$, and that $\partial\Omega$ is represented by a piecewise constant curve on this grid. Let $S^{f_i}(t)$ denote the solution operator associated with (5) and (6) in the ith direction. Then the solution $u(x, t)$ of (1) and (2) is approximated by

$$u(x, n\Delta t) \approx u^n(x) = [\pi S^{f_2^\delta}(\Delta t)\pi S^{f_1^\delta}(\Delta t)]^n \pi u_0(x). \tag{9}$$

Here π denotes a grid cell average operator and f_i^δ is a piecewise linear approximation to f_i. The boundary function r is approximated by a function that is constant in x and piecewise constant in t along each cell at the boundary.

To investigate convergence of a sequence of approximations, we choose an appropriate time interpolation [6]

$$u_\eta(x,t) = \begin{cases} S^{f_1^\delta}(2(t-t_n))u^n, & t \in \langle t_n, t_{n+1/2}], \\ S^{f_2^\delta}(2(t-t_{n+1/2}))u^{n+1/2}, & t \in \langle t_{n+1/2}, t_{n+1}], \end{cases} \tag{10}$$

where $t_n = n\Delta t$, $\eta = (\Delta t, \Delta x, \delta)$, and $u^{n+1/2} = \pi S^{f_1^\delta}(\Delta t)u^n$. Then we can prove the following result:

Lemma 2.3. *Suppose that $\Omega \subset \mathbb{R}^d$ is rectangular. Then the front tracking approximations defined by (9) and (10) satisfy the following estimates*

$$\begin{aligned} \|u_\eta(\cdot, t)\|_\infty &\leq C_1, \\ \mathrm{TV}(u_\eta(\cdot, t)) &\leq C_2(T), \\ \|u_\eta(\cdot, t_2) - u_\eta(\cdot, t_1)\|_1 &\leq C_3(T)|t_2 - t_1|, \end{aligned} \tag{11}$$

where C_1, $C_2(T)$ and $C_3(T)$ are positive constants not depending on η.

Proof. The first inequality follows from (7) and $\|\pi v\|_\infty \leq \|v\|_\infty$. To prove the second, we enlarge the grid by a set of ghost cells outside the domain on which the solution is given by the boundary condition. This way, contributions from the boundary to the total variation are easily included and the second inequality can be proved for the auxiliary sequence on the enlarged grid; the proof is similar to the proof in [5] using the estimates in (7) and the stability inequality (8). The third result follows from the second and the finite speed of propagation of all waves. $\quad\square$

A standard compactness argument (Helly's theorem) gives that the sequence $\{u_\eta\}$ is convergent, and as for the one-dimensional case we can show that the limit is an entropy solution in the Kruzkov sense (4).

Theorem 2.4. *Suppose that u_0 and r are $L^\infty \cap BV$ and that the flux functions $f = (f_1, \ldots, f_d)$ are locally Lipschitz continuous. Let $\Omega \subset \mathbb{R}^d$ be rectangular. Then the Dirichlet problem (1) and (2) has an entropy weak solution $v(x, t)$ which is the limit of $\{u_\eta\}$ as $\eta \to 0$.*

The assumption that the boundary function r should be of bounded variation can be relaxed. Let us for the moment assume $d = 2$ and that $r : [a, b] \subset \mathbb{R} \to \mathbb{R}$ does not depend on t. Recall that $r \in BV$ if and only if for each $\Delta s > 0$,

$$\int_a^b |r(s + \Delta s) - r(s)|\, ds \leq C\Delta s,$$

for some constant $C > 0$. We introduce the spaces BV_α consisting of integrable functions on $[a, b]$ satisfying

$$\int_a^b |r(s + \Delta s) - r(s)|\, ds \leq C(\Delta s)^\alpha, \qquad \alpha \in \langle 0, 1].$$

Clearly, $BV \equiv BV_1 \subset BV_\alpha$. Assuming that $r \in BV_\alpha$ for some $\alpha \in (0, 1]$ will in an essential way affect the variation bound in (11). Recall that this bound relies on the stability result (8), so that in the present situation we can only show that

$$\mathrm{TV}(u_\eta(\cdot, t)) \leq C(T)(\Delta x)^{\alpha - 1}. \tag{12}$$

This estimate is in general not sufficient to ensure compactness of $\{u_\eta\}$. Instead we need to use the concept of measure valued (mv) solutions. Since $\{u_\eta\}$ is uniformly bounded, by Young's theorem we can infer the existence of a subsequence, still denoted by $\{u_\eta\}$, and a family of compactly supported probability (Young) measures $\nu_{x,t}$ such that the L^∞ weak-star limit $g(u_\eta) \to \bar{g}$ exists for any continuous function g, where the limit $\bar{g}$ is given by

$$\bar{g}(x, t) = \int_\mathbb{R} g(\lambda)\, d\nu_{x,t}(\lambda) =: \langle \nu_{x,t}, g \rangle, \qquad \text{for almost all } x, t.$$

Furthermore, the sequence $\{u_\eta\}$ converges strongly to u in $L^1(x, t)$ if and only if $\nu_{x,t}$ reduces to a Dirac measure located at $u(x, t)$. Following Benharbit, Chalabi, and Vila [2], we call $\nu_{x,t}$ an entropy mv solution of the Dirichlet problem (1) and (2) if for all $k \in \mathbb{R}$ and suitable test functions $\phi \geq 0$ with $\phi|_{t=T} = 0$,

$$\int_\Omega \int_0^T \left(\langle \nu_{x,t}, |\lambda - k| \rangle \phi_t + \langle \nu_{x,t}, \big(\mathrm{sgn}(\lambda - k)(f(\lambda) - f(k))\big) \rangle \cdot \nabla\phi \right) dt\, dx$$

$$- \int_{\partial\Omega} \int_0^T \mathrm{sgn}(r(s, t) - k) \langle \gamma\nu_{s,t}, f(\lambda) - f(k) \rangle \cdot n(s)\phi(s, t)\, dt\, ds \tag{13}$$

$$+ \int_\Omega |u_0(x) - k| \phi(x, 0)\, dx \geq 0,$$

where $\gamma\nu$ is the trace associated with ν. In general, the trace $\gamma\nu$ is associated with the Young measure ν in a non-unique way. However, it is shown in [8] that the expectation $\langle \gamma\nu_{s,t}, f(\lambda) \rangle$ appearing in (13) is uniquely defined. The definition (13) used here and in [2] is slightly different from the one used in [8].

Now, equipped with the (weaker) time estimate

$$\int_\Omega |u_\eta(x, t + \Delta t) - u_\eta(x, t)|\, dx = \mathcal{O}(1)\Delta t(\Delta x)^{\alpha - 1} = \mathcal{O}(1)(\Delta t)^\alpha, \tag{14}$$

which follows from the space estimate (12), we can show that the Young measure $\nu_{x,t}$ associated with the sequence $\{u_\eta\}$ is in fact an entropy mv solution of (1) and (2) in the sense of (13). Because of the uniqueness result for entropy mv solutions of Szepessy [8] (see also [2]), we can conclude that $\nu_{x,t}$ reduces to a Dirac measure located at $u(x, t)$, where $u(x, t)$ denotes the unique BV entropy weak solution of (1) and (2). Consequently, we have the following theorem (which, of course, holds for any number of space dimensions $d \geq 1$):

Theorem 2.5. *Suppose that u_0 and r are in $L^\infty \cap BV_\alpha$ for some $\alpha \in (0, 1]$, and that the flux functions $f = (f_1, \ldots, f_d)$ are locally Lipschitz continuous. Let $\Omega \subset \mathbb{R}^d$ be rectangular. Then, as $\eta \to 0$, the sequence $\{u_\eta\}$ converges in $L^1(x, t)$ to the unique entropy weak solution of the Dirichlet problem (1) and (2).*

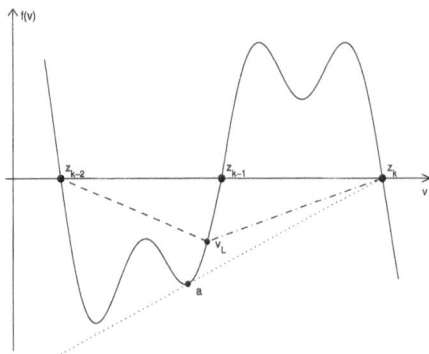

FIGURE 1. Illustration of a Riemann problem at the boundary.

3. Zero flux boundary condition

Consider (1) with prescribed initial and boundary data of the form (3). The flux function f must have compact support. Without loss of generality we assume that u_0 takes values in $[0, 1]$ and that $f(0) = f(1) = 0$. We seek weak solutions of (1) and (3) that satisfy

$$\int_\Omega \int_0^T (u\phi_t + f(u) \cdot \nabla\phi)\, dt\, dx + \int_\Omega u_0(x)\phi(x, 0)\, dx = 0.$$

Note that there are no boundary terms due to the no-flow boundary condition.

3.1. One spatial dimension

Consider now the one-dimensional problem (5) with initial and boundary data

$$\begin{aligned} v(x, 0) &= v_0(x), & x &\in \langle a, b \rangle, \\ f(v(a, t)) &= f(v(b, t)) = 0, & t &\in \langle 0, T]. \end{aligned} \qquad (15)$$

As above, we use front tracking as a means of analysis. To this end we must approximate f by a piecewise linear function f^δ and the initial data by a step function v_0^Δ. Assume that f has a finite number of zeros $z_1 = 0 < z_2 \cdots < z_N = 1$ and that f^δ is piecewise linear on each interval $I_k = [z_k, z_{k+1}]$ with $f^\delta(z_k) = f^\delta(z_{k+1}) = 0$. (In the following we drop the superscripts on f and v_0.) Without loss of generality, we can assume that all the intervals I_k have length ℓ. This is not a severe restriction and can easily be achieved by reparametrising the v-line.

We first discuss how to solve Riemann problems at the boundary. Consider a discontinuity hitting the right boundary at $x = b$. Assume that z_k is the boundary value at $x = b^-$ and that $f'(z_k) < 0$. The state v_L colliding with z_k must have $f(v_L) < 0$, see Fig. 1. Accordingly, we have that $a \le v_L < z_{k-1}$, where a is the (unique) largest state such that $f'(a) = -f(a)/(a - z_k)$. After the collision all waves must move left, and the new state adjacent to $x = b^-$ is z_{k-2}. This is indicated in Fig. 1: the right moving discontinuity before the collision (dash-dotted line) is replaced by a left moving discontinuity (dashed line). The picture is similar if the rightmost value is z_{k-2} and $v_L > z_{k-1}$. At the left boundary $x = a$ and at both boundaries initially we use a similar construction. Note that the boundary values initially are *constructed* from the initial data $v_0(a^+)$ and $v_0(b^-)$. We can prove that the front tracking algorithm has a finite number of steps by proving that a certain interaction functional is decreasing; see [3] for a similar proof. In fact, we can prove the result:

Lemma 3.1. *The front tracking solutions satisfy*

$$\underline{v} \leq v(x, t) \leq \bar{v},$$
$$\mathrm{TV}(v(\cdot, t)) \leq \mathrm{TV}(v_0) + 2\ell, \tag{16}$$
$$\|v(\cdot, t_2) - v(\cdot, t_1)\|_1 \leq C|t_2 - t_1|,$$

where

$$\underline{v} = \max_k \{z_k | z_k \leq v_0(x), x \in \langle a, b \rangle\}, \quad \bar{v} = \min_k \{z_k | z_k \geq v_0(x), x \in \langle a, b \rangle\}.$$

Moreover, the solution is stable with respect to the initial data.

Proof. It follows immediately from the above construction that the solution is bounded in L^∞. If $v_0(a)$ and $v_0(b)$ are zeros in the flux function, a careful case analysis of wave interactions shows that $\mathrm{TV}(v(\cdot, t)) \leq \mathrm{TV}(v_0)$. The result depends on the fact that all intervals I_k have equal length ℓ. Assume next that $z_{k-1} < v_0(a) < z_k$ such that $f(v_0(a)) \neq 0$. Then the boundary value $v(a, 0)$ must be either z_{k-1} or z_k. Similarly at $x = b$. Once the boundary values are determined, the zero flux problem can be reformulated as a Dirichlet problem and the second estimate follows from (7). The third estimate is a result of bounded total variation and finite speed of propagation.

To prove stability wrt. initial data we proceed in three steps. Let $w(x, t)$ and $v(x, t)$ denote the front tracking solution with initial data $w_0(x)$ and $v_0(x)$, respectively. Before the first interaction with the boundary, the zero flux problem can be reformulated as a Dirichlet problem and we have from (8) that

$$\|w(\cdot, t) - v(\cdot, t)\|_1 \leq \|w_0 - v_0\|_1 + tA\big(w_0(a), v_0(a), w_0(b), v_0(b); f\big).$$

Here A is positive and bounded above by $2\|f\|_{\mathrm{Lip}}$, but is not continuous with respect to its arguments, due to the construction at the boundary. However, a refined analysis shows that $A(\cdot) \to 0$ as $\|w_0(a) - v_0(a)\|_\infty$ and $\|w_0(b) - v_0(b)\|_\infty$ tend to zero. Then by introducing the concept of pseudopolygonals (see e.g., the monograph [4]), we prove that $\|w(\cdot, t) - v(\cdot, t)\|$ is non-increasing for every interaction with the boundaries and the stability result follows. $\qquad\square$

Thus we have constructed a sequence $\{v_\eta\}$ that can be shown to be compact by standard arguments. Each v_η is an exact solution of the approximate problem, and hence the limit is a weak solution of (5) and (15).

Theorem 3.2. *The zero flux problem* (5) *and* (15) *has a unique weak solution that can be constructed as the $L^1(x, t)$ limit of front tracking solutions.*

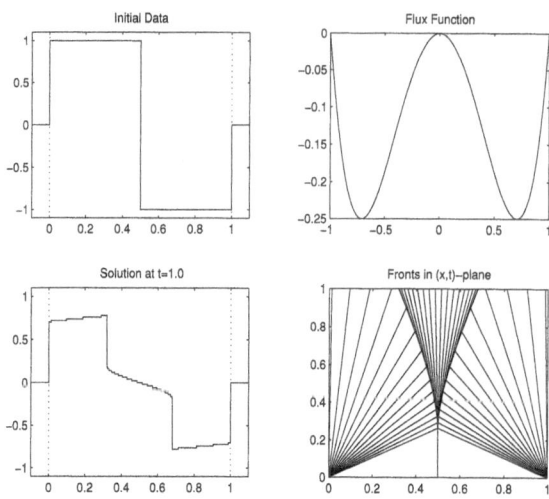

FIGURE 2. Example 1.

3.2. Arbitrary spatial dimension

As for the Dirichlet problem we can use dimensional splitting to construct solutions in two or higher space dimensions for (1) and (3). We cover the domain Ω by a Cartesian grid and make a piecewise constant approximation of the boundary $\partial\Omega$. For piecewise linear flux functions $f = (f_1, f_2)$ the approximate solution u_η can be defined as in (9) and (10).

The constructed solution is bounded; this follow from (15) and the properties of the projection operator. However, establishing a bound on the total variation is more difficult, since we have no control over the boundary trace *constructed* initially in each step. This is opposed to the Dirichlet case, where one could assume a certain regularity of the *prescribed* boundary condition.

4. Numerical examples

EXAMPLE 1   The first example is a one-dimensional Dirichlet problem with pre-scribed boundary values equal zero at $x = 0$ and $x = 1$, as given in Fig. 2. Initially there are three Riemann problems; from each boundary we get rarefaction waves that propagate into the domain and collide simultaneously with the stationary shock from the Riemann problem at $x = 0.5$. This produces a symmetric rarefaction wave propagating out towards the boundaries.

EXAMPLE 2   The next example is a one-dimensional zero flux problem, see Fig. 3. Each Riemann problem gives a rarefaction wave propagating to the left and two nearly aligned shocks that propagates right. A complex wave pattern develops as the waves reflect from the boundaries and interact with each other. Note that

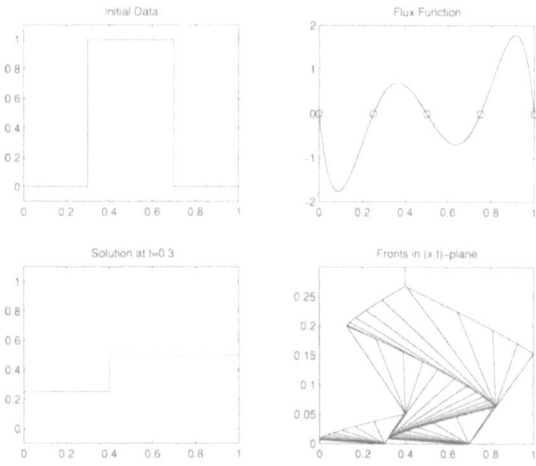

FIGURE 3. Example 2.

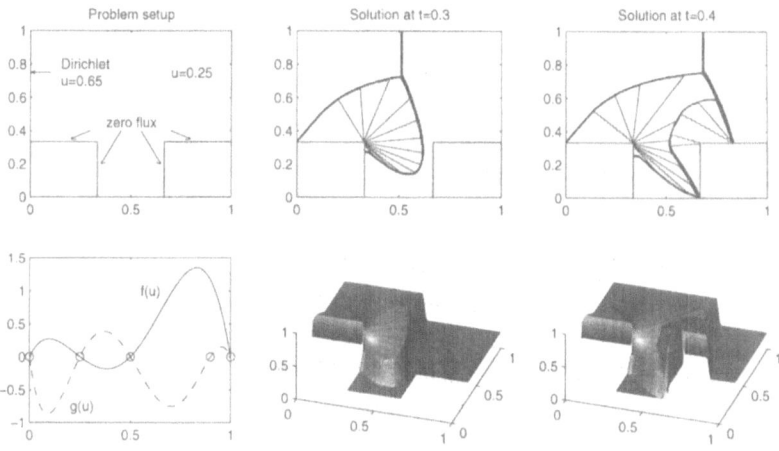

FIGURE 4. Example 3.

the boundary values change as waves are reflected at the boundaries. After time $t = 0.27$ the solution is stationary and consists of two zeros in the flux function.

EXAMPLE 3   Fig. 4 (left) describes the setup for the next problem. At the unspecified boundaries we impose absorbing boundary conditions; i.e., waves are allowed to pass out of the boundary with no effects on them. The Dirichlet boundary condition at the left boundary gives a shock wave that propagates into the domain.

The post-shock value corresponds to a nonzero flux in the y-direction and a complex wave pattern forms, consisting of rarefaction waves and three shocks meeting in a triple point. The rightmost shock propagates faster than the incident shock wave and is diffracted around the corner (Fig. 4, middle). Then the leading shock is reflected at the right vertical wall (Fig. 4, right). The reflected wave propagates backward and upward to meet with the triple point.

References

[1] C. Bardos, A. Y. LeRoux, and J. Nedelec. First order quasilinear equations with boundary conditions. *Comm. in Partial Differential Equations*, **4 (9)** (1979), 1017–1034.

[2] S. Benharbit, A. Chalabi, and J. P. Vila. Numerical viscosity and convergence of finite volume methods for conservation laws with boundary conditions *SIAM J. Numer. Anal.*, **32** (1995), 775–796.

[3] H. Holden, L. Holden, and R. Høegh-Krohn. A numerical method for first order non-linear scalar conservation laws in one-dimension. *Comput. Math. Applic.*, **15 (6–8)** (1988) 595–602.

[4] H. Holden and N. H. Risebro. Front tracking for conservation laws. Department of Mathematics, Norwegian University of Science and Technology. Lecture Notes.

[5] H. Holden and N. H. Risebro. A method of fractional steps for scalar conservation laws without the CFL condition. *Math. Comp.*, **60 (201)** (1993), 221–232.

[6] K.-A. Lie, V. Haugse, and K. H. Karlsen. Dimensional splitting with front tracking and adaptive grid refinement. *Numer. Methods Partial Differential Equations*. To appear.

[7] J. Málek, J. Nečas, M. Rokyta, and M. Růžička. *Weak and measure-valued solutions to evolutionary PDEs*, volume 13 of *Applied Mathematics and Mathematical Computation*. Chapman & Hall, London, 1996.

[8] A. Szepessy Measure valued solutions to scalar conservation laws with boundary conditions. *Arch. Rational Mech. Anal*, **107** (1989), 181–193.

Department of Mathematics, University of Bergen,
Johs. Brunsgt. 12, N-5008 Bergen, Norway
E-mail address: kenneth.karlsen@mi.uib.no

Department of Mathematical Sciences,
Norwegian University of Science and Technology,
N-7034 Trondheim, Norway
E-mail address: andreas@math.ntnu.no

Department of Mathematics, University of Oslo,
P.O. Box 1053, Blindern, N-0316 Oslo, Norway
E-mail address: nilshr@math.uio.no

International Series of Numerical Mathematics

Edited by
K.–H. Hoffmann, Technische Universität München, Germany
H.D. Mittelmann, Arizona State University, Tempe, USA

International Series of Numerical Mathematics is open to all aspects of numerical mathematics. Some of the topics of particular interest include free boundary value problems for differential equations, phase transitions, problems of optimal control and optimization, other nonlinear phenomena in analysis, nonlinear partial differential equations, efficient solution methods, bifurcation problems and approximation theory. When possible, the topic of each volume is discussed from three different angles, namely those of mathematical modeling, mathematical analysis, and numerical case studies.

Mathematics with Birkhäuser